History of Information Security

Series Description

The Springer book series History of Information Security publishes monographs about all aspects of the history of cryptology, signals intelligence, computer security, and information assurance, focusing on the interplay between technological change and organizational requirements. There are no limits in time period – monographs dealing with information assurance and warfare in antiquity are as welcome as those dealing with cybercrime – but subjects should be clearly defined, limited, and treated comprehensively. Prospective authors are invited to submit proposals or manuscripts, aimed at academic and/or non-expert audiences.

Series Editor

Karl de Leeuw [Founding Editor, recently deceased] (Amsterdam)

Advisory Board

John F. Dooley (Knox College)
Joseph Fitsanakis (Coastal Carolina University)
Ioanna Iordanoou (Oxford Brookes University)
Benedek Lang (Budapest University of Technology)
Jean-Jacques Quisquater (Université catholique de Louvain)
Betsy Rohaly Smoot (Independent Scholar / Retired NSA Historian)
John Tucker (Swansea University)
Joachim von zur Gathen (Bonn-Aachen International Center for Information Technology and Universität Bonn)

Marek Grajek

Enigma Myth Deciphered

Codebreakers, Commanders and Politicians

 Springer

Marek Grajek
Santa Maria del Cedro, Italy

ISSN 2662-7558   ISSN 2662-7566   (electronic)
History of Information Security
ISBN 978-3-031-65474-9   ISBN 978-3-031-65475-6   (eBook)
https://doi.org/10.1007/978-3-031-65475-6

This Springer imprint is published by the registered company Springer Nature Switzerland AG
The registered company address is: Gewerbestrasse 11, 6330 Cham, Switzerland

If disposing of this product, please recycle the paper.

Preface

The bibliography of texts devoted to the history of the German Enigma includes hundreds of titles and probably deserves to be published as an independent book. The popularity of the subject has led to the emergence of enigmology; a field of knowledge combining elements of historical and military sciences as well as mathematics and cryptology. This discipline has naturally developed its avant-garde—a group of researchers who focus their professional or hobby activities on issues related to Enigma. There are several traditional or electronic periodicals placing enigmology at the centre of attention. Real events have become the basis of the plot of several films and many novels, some of them documentaries, others of a decidedly sensationalist nature. After a few dozen years of secret, in the last decade of the twentieth century the archives flooded historians with documents from the era, but still leaving a knowledge gap stimulating the imagination. Hundreds of websites present a substantial set of information about the history of the machine, its ciphers and the fate of people connected with their history, mixing facts and fiction in almost equal proportions.

In the situation described, it will be honest both with the reader and with my own intentions if I start by setting out the motives for which I decided to write one more account of the events which have already been covered in such a wealth of studies. I certainly follow in the footsteps of my predecessors, being attracted by the magical aura of secrecy that still surrounds the Enigma story. The very name of the machine announces the participation in the Riddle. In fact, the story of Enigma contains even more secrets than it promises to the seekers beginning their journey. Just as important as the number is their nature: they concern surprisingly many spheres of human activity. Starting with purely personal fates, depicting the motivation of the protagonists of this story and drawing their silhouettes, through the dramas of teams working in isolation and under the pressure of secrecy, cabinet secrets of politicians whose names are firmly inscribed in the history of the past century, lecture halls of universities, where young people prepare for a mission very distant from the nominal field of their studies, the surprising outcome of battles and entire campaigns, the brilliant orders and outrageous mistakes of the generals, right down to the fates of entire nations... This variety of problems and images assures that the story of Enigma does not come soon, if at all, to a final conclusion.

Its secret nature influenced also the shape of most of the reports published so far. Accounts of the direct participants represent usually the most valuable historical sources. The decades of silence separating the earliest accounts from the events they cover had a strong influence on the shape of memories. Many of the key participants have left without leaving a record. Others published their memories when the outline of the events was known, but access to details and source documents remained limited. Relying, perforce, mainly on their own memory, they reported that part of the story which they directly participated in. The rule of knowledge separation, common in the secret services, made the available accounts present narrow fragments of the whole, and the inability to confront the memory with the sources stained them with shades of subjectivity.

The history of the Enigma has not been revealed to the general public as a result of the natural process, but rather of the Caesarean delivery. Who knows how much longer it would have taken to reveal the truth if it hadn't been for the publication in 1973 of the book of the French general, Gustave Bertrand, who thus had to play the role of a midwife at the birth of this sensational thread in World War II history. When generals take over the duties of a midwife, you can expect astonishing results. Even if pinching the British was not the main motive for Bertrand's action, judging by the reactions of those concerned, this is how they received the publication of his memoirs and responded with a series of accounts showing their own point of view. And probably there wouldn't be anything wrong with it if, in the face of revealing the fact that the German ciphers were broken, the British custodians of the Enigma's mystery had come to terms with the necessity, presenting a balanced and fully factual version of events. However, in the books representing their earliest response to Bertrand's revelations, facts were intricately intertwined with fiction or conscious disinformation. The victim of the banters between the former British and French Allies was the third party in this story—Polish codebreakers. More than a quarter of a century has passed since that early dispute over Enigma; in the light of the accounts and documents until now, no reasonable historian questions the priority of Rejewski, Różycki and Zygalski in making a cryptological breakthrough. It is enough, however, to go beyond the sphere of professional discussion to see the ghosts of the mythical engineer Lewiński or the anonymous Polish mechanic, employed in Berlin to assemble the Enigma, who—recruited by British intelligence and smuggled to Paris—laboriously reconstructed the device carving its model in wood. These and other fantasies, ad hoc created by British secret service veterans for the purpose of disinformation, enjoy good health, a long life and still affect the general perception of one of the greatest secrets of World War II.

So far, Polish attempts to restore the true picture of events have been not so much insufficient, but not very well directed. Politicians and historians sought official testimonies of facts, expressed by official representatives of the countries concerned and, in many cases, obtained them. However, today's opinion is shaped more by mass media, image and word rather than by the official celebration. As could be expected, the official acknowledgement passed without a broad and lasting echo. At the same time, Poland almost forfeited a duel in the mass media field. The most important books reporting the Enigma written by Polish historians and presenting the Polish point of view were published over a quarter of a century ago, on the basis of scarce sources available at that time. The role played by Polish mathematicians in breaking the Enigma has since then been more extensively and thoroughly analysed in publications by foreign authors. However, their books understandably focus on the influence of breaking the Enigma's ciphers on the fate of war, rather than on the theoretical and practical breakthrough that made their reading possible. In this respect, the significant contribution of Poles to the Allies' victory comes to an end in July 1939, when the results of their earlier work were communicated to the Allies. From that moment on, the accents in the works of enigmologists are usually distributed according to the criterion of the author's country of origin. The British authors expose the genius of their mathematicians, who transformed theoretical ideas provided by Poles into practical devices and methods, and created an organization that carried out decryption on an industrial scale. The American notes that the superbly designed British organization reached the end of its growth potential in the middle of the war, and that it was only the American engineering talent and experience in mass production that drew it from the deadlock. Both pictures contain a considerable dose of historical truth, but at the same time, they are separated by a long distance from the whole truth about Enigma.

Against this background, the intention to add another annex to the existing *corpus Aenigmae* is slightly better justified. Over the past 20 years, British and American archives have declassified hundreds of documents related to the history of Enigma. Many of them have been noticed and covered by historians, cryptologists or veterans associated with the history of the Enigma, but the results of most of these studies are only available in periodicals of limited circulation or on the Internet; during the same period, despite the abundance of new sources, there was no broad attempt at a new synthesis of the topic. This book did not originally aspire to be such a synthesis. However, as the materials for the planned, more modest publication were collected, their quantity at some point turned into quality. Analysing the previously known facts about the history of the Enigma and combining them with new sources, I noticed that under the weight of new testimonies and hypotheses, some opinions, considered as dogmas in the field of enigmology, are beginning to crack, while provoking questions about issues that predecessors overlooked or considered too obvious to try to unravel. This observation was followed by another reflection: why shouldn't a new attempt at synthesis come out of the place where the story of Enigma codebreaking began? Perhaps this would be a good opportunity to even out the disproportions in the details of relations concerning the participation of representatives of various countries of the anti-Axis coalition. In Anglo-Saxon historiography, the reader would find the memories of a carpenter building barracks in which Enigma's ciphers were broken and a woman who, as a young girl, distributed tea among the cryptologists working there. Meanwhile, the fates and profiles of three mathematicians, the main Polish heroes of the story, are presented in the available literature in a schematic way, not to mention the remaining Poles entangled in the mystery of the Enigma, whose participation has so far been marked by a footnote mention at best. We have already lost the opportunity to listen to the accounts of those who survived the war, but it is worth saving even what they wanted to tell us through people who knew them directly.

Thus, a new concept of the book has emerged, covering a wider scope than originally planned. I present a picture of the history of breaking the Enigma's ciphers, in which the role of Poles has been exposed more strongly than in any previously published book. This was possible mainly thanks to the invaluable help of the families of the Polish participants of the events, who made the family archives available and shared their memories. At the same time, I describe the Allied contribution to the operation of breaking the Enigma in more detail than in the positions available to the Polish reader so far, in many aspects and episodes going beyond the accounts contained in Anglo-Saxon historical literature. The detail is largely due to my predecessors—*enigmologists*, who did a lot of work on the source materials, providing many threads that I could include in the synthesis. Thanks to them, above all, the book refers to many facts known so far only in a narrow circle of specialists. These novelties have become a starting point for me to raise some fresh problems and to question some of the views accepted so far without any attempts at polemics or verification. In this respect, the book does not follow the trajectory of political correctness.

I must end my introduction with a confession which explains the presence of a separate layer of narration in the book, dedicated to cryptology. My professional experience is definitely closer to cryptology than history. Therefore, I took the liberty of resigning from some of the attributes of the historian's workshop (in an attempt to do so, the volume of footnotes quickly exceeded the size of the main text). Perhaps the shortcomings of the historical workshop will not override the value of a book for professionals—the report is based on a systematic analysis of sources, and my own hypotheses and conjectures have been singled out. I also hope that the attempts to explain for the use of interested readers the cryptologic nature of the Enigma's challenge and the answers given by the protagonists of the story will not cause the book to be rejected by readers who are not interested in the mathematical and technical side. The wealth of issues that I found interesting, although briefly presented, made the construction of the book resemble a bit the *silva rerum* scheme. Aware of this fact, the reader retains full freedom to choose the threads he or she finds attractive, and omitting the more specialized paragraphs will not make it difficult to read the whole.

Santa Maria del Cedro, Italy   Marek Grajek

Foreword for the English Edition

The years that have passed since the publication of the first edition of the book have confirmed that the effort to write it has not been in vain. The book managed to stimulate a discussion on the role of Polish codebreakers in the Allied victory in World War II and the related questions about the real hierarchy of Polish merits in this conflict and the nature of Polish patriotism. Judging by the echoes received, Polish readers accepted, with some surprise, the thesis that the crucial Polish contribution to the victory was not won on the battlefield, but resulted from the abstract work of cryptologists analysing mathematical complexities somewhere far from the front line. There has been a spontaneous movement of people of goodwill who have committed their time and effort to restore the memory of all the participants in this story. We owe them the touching moments of the triumphant return to the Homeland of the ashes of Karol Gwido Langer and Maksymilian Ciężki. Exhibitions and lectures on the achievements of Polish cryptologists are presented in many places around the world. Young people from all over the world take part in online cryptology competitions, getting to know the story of Marian Rejewski and his colleagues and finding out that mathematics in cryptologic incarnation can be different from school routine and fascinating.

But that's not the only thing that made me decide to return to the book. Its publication encouraged the families of its protagonists to share their memories and previously unknown documents, allowing for a more complete draft of their silhouettes. Documents have been declassified, the existence of which historians were aware of, but remained unavailable. The information contained therein did not render the statements or hypotheses presented in the first edition obsolete. However, they drew new, interesting aspects of the cryptologic adventure. Previously available documents sketched in detail the struggle against the ciphers of the German Kriegsmarine; the history of the war activities of *Hut 6* illustrates the scale of allied rule over German communication networks also with regard to Wehrmacht and Luftwaffe. The gradually declassified protocols of the post-war TICOM operation allow to look behind the scenes of the other side of the conflict, drawing on the one hand the successes of German codebreakers in their struggle against allied ciphers and on the other hand the reasons for their failure in the cryptologic race. This edition contains the conclusions of these analyses. However, they also contain references to sources that remain inaccessible; the book only brings us still closer to the truth about the Enigma story, without offering the whole truth.

Contents

Chapter 1
The Dawn of Machine Cryptography

The outbreak of the First World War heralded the end of the old world in many ways, but the scale and drama of the changes was reaching those concerned gradually. The soldiers were the first to learn the lesson of modernity. The massive use of machine guns and artillery changed the combat tactics, forcing armies to dig into the ground. The use of gases has familiarised mankind with the concept of weapons of mass destruction. The airplane carried the horror of war far to the back of the front, making the civilians, safe until now, an involuntary participant in the struggle. For the first time in the history of mankind, radio has allowed to coordinate actions on a global scale, contributing to turning the conflict into a world war. However, the use of the radio had as many advantages as disadvantages for the fighting parties. Anyone equipped with a radio receiver, an enemy or a friend, could capture messages from hundreds of kilometres away. Therefore, the premiere of the radio in its military function was accompanied by the practical application of cryptology on a scale unknown in previous history.

The methods of secrecy have been known to priests, alchemists and scholars, diplomats and military commanders for thousands of years. Until now they were one of the secrets of the power, and their knowledge and application was limited mostly to diplomatic cabinets and army headquarters. The first breach in the wall surrounding the secrets of cryptology was a result of the American Civil War. The telegraph, predecessor of the radio, later referred to as the Victorian Internet, was used on a massive scale. Telegrams sent over a distance of many thousands of kilometres had to be protected from unauthorized eyes, and one of the key factors of the North's victory was the security of its communications. While the use of the telegraph was a revolution in the command of the troops, the codes and ciphers used during the Civil War referred to traditional manual methods; mainly for this reason, the cryptographic revolution of the Civil War went almost unnoticed.

When radio replaced the telegraph at the advent of the First World War, the need for secrecy increased dramatically. Telegraphic messages were only intercepted by the enemy in exceptional circumstances; reason dictated that any radiogram could

© The Editor(s) (if applicable) and The Author(s), under exclusive license to
Springer Nature Switzerland AG 2025
M. Grajek, *Enigma Myth Deciphered*, History of Information Security,
https://doi.org/10.1007/978-3-031-65475-6_1

be intercepted by the enemy. Soon it turned out that the number of cipher clerks trained before the war did not meet the rapidly growing demand. The clerks trained on an *ad hoc* basis during the war were not as proficient as the specialists from the pre-war recruitment; they worked slower; acting in a hurry or under enemy fire they were making mistakes requiring the dispatches to be recoded and retransmitted. During the war, virtually all of the warring parties faced the decision to replace a cipher or code considered safe but complicated to use, with a solution more vulnerable to decryption, but simple enough to be used even by cursorily trained staff.

Simplyfying the design of ciphers was just one of the solutions. The Industrial Revolution of the nineteenth century, combined with the horrifying number of war victims, gave rise to a tendency to replace man with a machine. It was no surprise therefore, that attempts were made quite early on to replace the human cipher clerk with a ciphering device working as efficiently as flawlessly. It seems that the first challenge to automate the encryption and decryption process was taken up by the Dutch. The neutral Netherlands ruled the archipelago of hundreds of East Indian islands scattered over thousands of miles, equipped by nature with hidden bays and passages. The interests of the hosts in this region were guarded by a navy of only ten ships. The concern to maintain a neutral status resulted in their commanders being ordered not to take any action in case of encounters with foreign units without consulting the command in Batavia. The execution of the order required efficient communication between the ships and the base. In order to secure it, in early 1915, the Dutch navy lieutenants, Theo van Hengelo and Rudolf P.C. Spengler, were entrusted with the task of developing a ciphering machine. Although the experience of both officers so far has been limited to designing torpedoes, they have effectively tackled the challenge and designed a prototype machine that has become a protoplast of rotor encryption devices.[1]

The Dutch have simplified the encryption process by combining the functions of a ciphering device with an ordinary typewriter. Hand ciphers required skill and discipline: even one mistake could make part or all of the ciphertext unintelligible. The new device simplified the ciphering process; the open text was tapped on a keyboard, encryption process was carried inside the device, which printed the ciphertext like an ordinary typewriter. Shortly after the end of the war, both inventors returned to Europe on board of the cruiser *De Zeven Provinciën* with two machine prototypes in their luggage. The inventors tried to clarify the status of their design by asking the command to declare its military use or to agree to patent the machine in the civilian sphere. The further fate of inventors and their idea took the form of a court-bureaucratic thriller. Navy Command neither declared the use of the machine nor allowed for its patenting. After several months of silence on the part of their superiors, desperate inventors handed over the documentation of their invention to a law firm to prepare a patent application. By an astonishing coincidence, the previously silent minister of the Navy, Hendrik Bijleveld, reprimanded the officers,

---

[1] Karl de Leeuw (2003): *The Dutch Invention of the Rotor Machine, 1915–1923*, Cryptologia, 27:1, p. 73–94.

reminding them of their military status and presenting the opinion that the copyright on an invention developed in connection with the service belongs to the Navy. If the inventors were surprised by the coincidence in time between the contact with the lawyers and the minister's action, they kept their doubts to themselves. The deadlock has seemingly been overcome at the end of October 1919, when the admiralty decided to withdraw initial objections due to 'new circumstances'; van Hengelo and Spengler returned with energy to their patent application. However, when signing a contract with inventors the legal firm did not inform its clients about the conflict of interest resulting from representing the author of another project of similar purpose—Hugo Alexander Koch. Moreover, one of the main clients of the firm was the Dutch admiralty, which added another layer of the potential conflict of interest. As a result, an official patent application on behalf of van Hengelo and Spengler went to the patent office on 29 November 1919, while an application on behalf of Koch on 9 October: one month after both officers filed the invention documentation with the office and three weeks before the admiralty withdrew their claims.

Before the Dutch patent office announced (though not yet approved) the Koch patent on June 15, 1922, the law firm handling the two applications was shaken by an internal conflict, which resulted in the division of the firm, and van Hengelo and Spengler finding out that the lawyers had abused their trust and put their interests at risk. Through another lawyer they filed protests against granting a patent to Koch, but these were rejected by subsequent instances. The meeting of the Court of Appeal, which gave the final and unfavourable for them verdict, was chaired by …Hendrik Bijleveld, the same who had previously successfully torpedoed their application as a Navy Minister. The results of Karl de Leeuw's investigation indicate that Hugo Alexander Koch only provided a facade for the interests of the German company Chiffriermaschinen Aktiengesellschaft, whose history will soon be described.

In order to follow the chronology of events, we need to move from Holland to the United States. There, the next link in the chain was Edward Hugo Hebern; originally a carpenter, then a construction entrepreneur. We are unable to recreate the path that led him to cryptography, but already between 1912 and 1915 he developed and submitted for patenting several encryption devices. Perhaps he became interested in ciphers when he was serving his sentence in San Quentin prison between 1907 and 1909.[2] In 1915 he came up with the idea of connecting two typewriters with 26 wires, whose configuration could be freely modified. Pressing the key of the first machine printed the code letter in the second. The ability to change the configuration of the connections led Hebern to a logical consequence—the introduction of a moving rotor, whose internal connections performed the function of wires between the keyboard and the printer. From here a short path led to the concept of several moving rotors between the keyboard and a printer. In 1917 Hebern developed a design for an encryption machine, which he later gave the commercial name "Sphinx of the Wireless". It was not only the name of the device that proved the inventor's

---

[2] Zorpette Glenn, *The Edison Of Secret Codes*, Invention and Technology, 1994, Volume 9, Issue 3.

vivid imagination—he wrote in honour of his invention a nightmarish poetic ode ('sphinx of the wireless, guardian of treasure, brain of a nation, safety beyond measure, heart of a battleship, preserver of lives when brutal force against intellect strives'), which was probably best read... in an enciphered version. He obtained a patent for an electric encryption machine in 1921, the same year he founded a new company—Hebern Electric Code.

The company issued shares with a nominal value of one million dollars and started building a factory to employ 1.500 people. However, early 1920s was not a good time for cryptography in the United States; business needs were met by popular and cheap code books, so Hebern managed to sell only 12 copies of the machine and his company got into financial trouble. Its shareholders initiated legal action, which in mid-1926 ended with Hebern Electric Code declaring bankruptcy.

The inventor did not lay down his arms, but was not assisted even by the circumstances that would otherwise guarantee success. In the early 1930s, when the growing international tensions contributed to an increased interest in cryptology, US state agencies designed, built and used several types of devices clearly and obviously infringing Hebern's patent rights. This information reached somehow the inventor, who already after the end of the Second World War filed a lawsuit for compensation, but died in 1952, not having seen the resolution of the dispute. It was not until six years after his death that the government reached a settlement with the inventor's heirs, paying them a ridiculous sum of $30.000. In retrospect, it can be said that Hebern would probably have been more successful in exploiting another of his patents—the car turn signal.[3]

Let's take a moment to consider the evolution of the rotor encryption machine in the subsequent Hebern patents. The original idea of connecting two typewriters by wires made it easier for the cipher clerk to work and avoid mistakes in the encryption process, but generated a simple monoalphabetic substitution, which cryptanalysts have been dealing with for at least a thousand years. Patent 1.510.441[4] (cf. Fig. 1.1) concerned a device with one rotor, generating a polyalphabetic substitution with a period equal to 26 characters. The design of the rotor allowed for easy exchange of its connections (allowing to offer different customers a machine generating a different cipher), and switching from encryption to decryption mode was done by removing the rotor and inserting it into the device the other way round. This did not change the fact that the methods of breaking the cipher generated by the machine were known at least from the mid-nineteenth century. Patent 1,683,072[5] concerned a machine with five rotors, portable thanks to the abandonment of the typewriter and its replacement by light bulbs displaying successive cipher characters. The use of five rotors theoretically lengthened the cipher period to 11,881,376 positions, but did not make it secure enough; Hebern did not know that the direct

[3] US Patent 1.283.756, 05.11.1918.

[4] Date of notification 31.03.1921, award date 30.09.1924.

[5] Date of notification 20.11.1923, award date 04.09.1928.

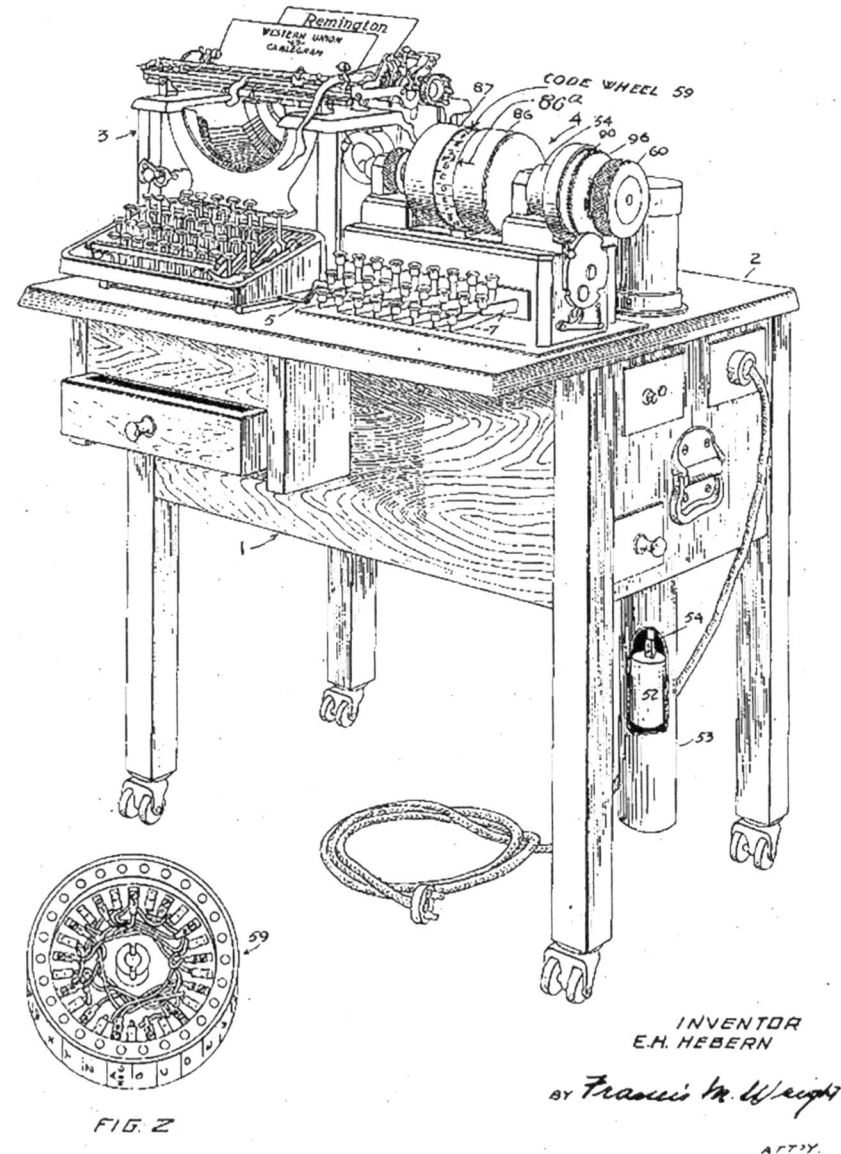

Fig. 1.1 Hebern's US patent 1.510.441. (Public domain)

reason for his failure to sell the device to the Army was the father of modern American cryptology, William Friedman, who broke its cipher[6] (Fig. 1.2).

There is a theory stating that the emergence of a pioneering idea increases the likelihood of its independent formulation elsewhere in the world; the history of rotor ciphering machines can act as one of its confirmations. Shortly after the inventions of van Hengelo/Spengler and Hebern (Koch's invention cannot be regarded as independent), the idea of a machine operating on a similar principle was developed by Arvid Gerhard Damm in Sweden and Artur Scherbius in Germany. Damm became interested in cryptography probably under the influence of his brother Ivar, a math teacher. Before the outbreak of World War One Damm filed three applications for ciphering devices with the German Patent Office. In the first months of the Great War he made an offer also to the British Foreign Office.[7] The British were unlikely to accept the proposal coming from Berlin, the capital of the country with which they were at war. Moreover, Damm's offer was a rather fancy design in which the movement of individual machine elements depended on the configuration of the chain loops driving them (see Fig. 1.3).

On 10 October 1919, Damm filed a patent application for a rotor encryption machine.[8] The machine had an impractical design and was not accepted by the

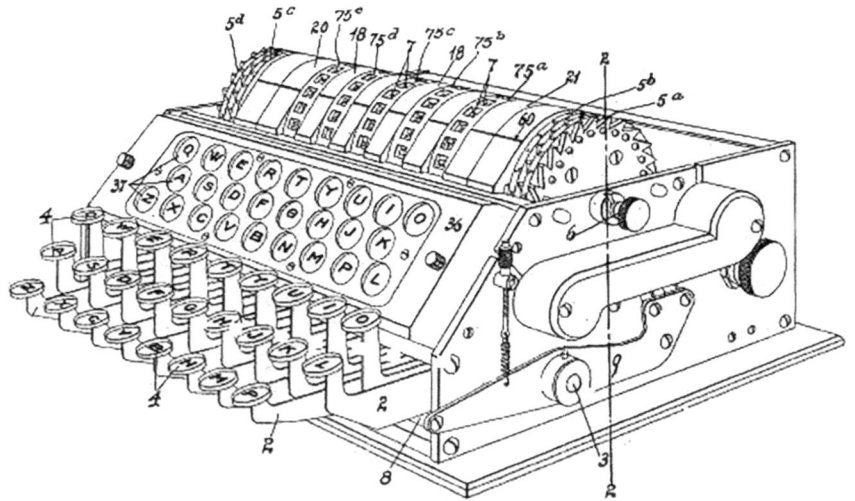

Fig. 1.2 Hebern's US patent 1.683.072. (Public domain)

---

[6] Ellison Carl M., *A Solution of the Hebern Messages*, Cryptology, 12: 3, 144–158. The problem posed to Friedman was significantly simplified by giving the cryptologist important elements of the solution.

[7] McKay C.G., *From the Archives. Arvid Damm Makes an Offer*, Cryptologia, Vol. XVIII Nr 3, s. 243–249.

[8] US Patent 1,502,376, application date 02.04.1920, granting date 22.07.1924, corresponding to Swedish patent SE 52,279, application date 10.10.1919, granting date 01.01.1922.

Fig. 1.3 Damm's US
patent 1.233.035. (Public
domain)

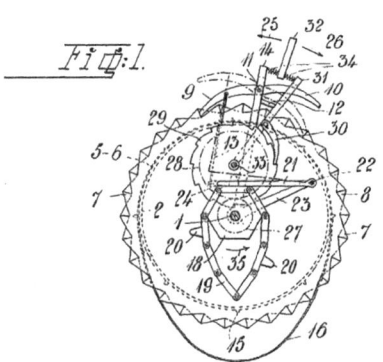

A. G. DAMM.
APPARATUS FOR PRODUCING SERIES OF SIGNS,
APPLICATION FILED JULY 21, 1915.

1,233,035.

Patented July 10, 1917.
5 SHEETS—SHEET 1.

market, but three years before the application Damm was successful in another sphere—he gained a group of interesting partners for cooperation in the Aktiebolaget Cryptograph[9] company he had founded in 1915. These included Emanuel Nobel—Alfred's nephew, K.H. Hagelin, director of the Nobel brothers' oil company in Russia, and Olof Gyldén, commander of the Royal Swedish Navy School. Damm's creative mind did not go hand in hand, it seems, with his engineering skills. The machines he designed suffered from technical faults that did not allow them to demonstrate their full potential. As a result, the sales, and with it the company, were lame (Fig. 1.4).

Right at that moment a providential man appeared in Aktiebolaget Cryptograph, in the person of Boris Hagelin, son of one of the main shareholders. During Damm's absence in Sweden he learned that the Swedish army was considering buying a German ciphering machine. His engineering experience allowed him to react immediately and effectively. He has reworked one of Damm's ideas, simplifying its design and bringing it closer to a competitive machine. These steps were not appreciated by the head of the company after his return to the country, but allowed the firm to win the tender and get a large order. Damm tactfully passed away at the verge of the company's success, leaving the project to his inventive subordinate. The Hagelin family decided to buy out the company, which has become one of the few commercially successful cryptography projects so far. The company's legal successor, Crypto AG, was operating in Zug, Switzerland, until 2018. Boris Hagelin managed it almost to the last days of his life in 1983.

Meanwhile, at the end of World War I in Germany, two partners were looking for an idea that would allow them to spread their wings. The more creative of the two was Arthur Scherbius, born in 1878 in Frankfurt am Main as the son of a merchant.

---

[9] Hagelin Boris C. W., *The Story of the Hagelin Cryptos*, Cryptologia, 18(3), July 1994, p. 204–242.

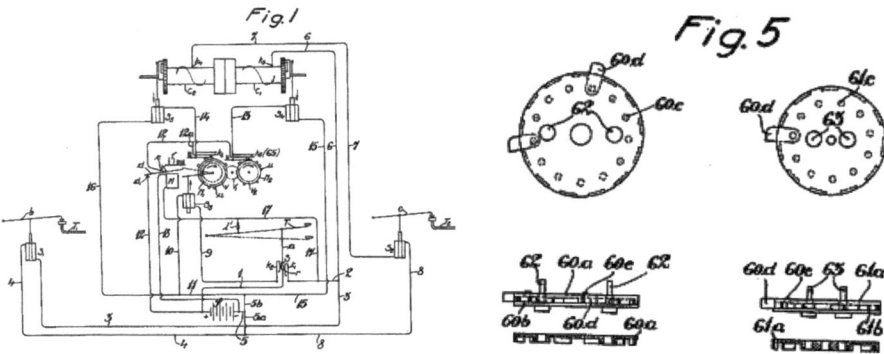

Fig. 1.4 Damm's US patent 1.502.376. (Public domain)

After studying at the Technical Universities of Munich and Hannover, he obtained his doctorate presenting a dissertation on the control of water turbines. He worked in several companies in Germany and Switzerland, obtained several patents in the field of technical ceramics and finally joined forces with Richard Ritter in 1918. At Scherbius & Ritter, the first partner was responsible for ideas and engineering, the second for the commercial aspects of the business. The area of Scherbius' activities stretched quite widely—from turbine control to heated pillows. World War One caused Scherbius to extend his interests to cryptography. While the war was still going on, he developed the concept of a device that transformed code numbers into pronounceable syllables. The solution was addressed mainly to business and diplomacy; the telegraphic regulations of that time allowed for cheaper sending of messages consisting of pronounceable texts. On 23 February 1918 Scherbius filed a patent application for a rotor encryption machine (see below), constructed on the same principle as the van Hengelo and Spengler, Damm and Hebern machines, with the difference that in the first version of the machine the author only provided for encryption of numbers, so the rotors were equipped with 10, not 26 contacts.[10] The prototype was quite unwieldy and could only be used in stationary offices. The company offered the device to the German Ministry of Foreign Affairs and Navy HQ, but the tension of the last days of the war and the revolutionary chaos that followed were not conducive to new ideas. Nevertheless, the partners did not give up hope of success and transferred the copyright to the newly founded company Gewerkschaft Securitas (by an amazing coincidence, a company registered in the Netherlands by Hugo Koch had the same name), replaced on 9 February 1923 by a new company, Chiffriermaschinen Aktiengesellschaft.

The new company has shown initiative in promoting the machine, which was given the commercial name Enigma (from Greek αίνιγμα—riddle). The bulky Enigma A model weighing over 50 kilograms was exhibited at the congresses of the

---

[10] Quirantes Arturo, Model Z: A Numbers-only Enigma Version, Cryptology, 28: 2, pp. 153–156. The design of the machine indicates that its primary purpose was to encrypt the code.

International Postal Union in 1923 in Bern and 1924 in Stockholm. The ancestor of the whole family of ciphering machines was equipped with four rotors placed on a common axis in a constant order. The difficulty in attacking the cipher generated by the machine resulted from the fact that each of the rotors rotated after a keypress a different number of positions, and additionally the fourth rotor rotated in the opposite direction to the others. Presented in 1924, model B was a moloch, whose keyboard contained 57 characters and, taking into account the "small/big letters" mode switch, was able to encrypt 83 characters. However, the machine's encryption rotors had only 26 contacts, enabling encryption of 26 characters. Company engineer, Paul Bernstein, patented a way to switch the device between different modes of operation using the J and Q[11] letters, which are rare in German. A major novelty in model B was the fact that the rotors could be removed from the machine and their order on the axis could be changed. Many historians kept on repeating that the device was withdrawn from the open market and kept secret after the mentioned UPU congresses. In fact, the story of secrecy concerns another encryption device of German design, the T-52 cipher teletype cipher manufactured by Siemens & Halske, which was withdrawn from the market in 1934. Civilian versions of the Enigma were still widely available even during World War II (Fig. 1.5).

In 1926, the company launched the C model, which has been a milestone in history. The unwieldy and error prone typewriter mechanism was abandoned. Instead, the machine was equipped with a set of lamps; pressing a key illuminated one of the lamps. This solution increased the reliability of the machine and reduced its dimensions and weight, and over time allowed it to be used in the field. An equally important innovation introduced in model C was the application of the idea by Willi Korn, the deputy and Scherbius' successor as the main engineer of the company. He proposed to equip the Enigma with a so-called "reflector"—a stationary drum whose function was to reflect electrical signals having reached it through the rotor battery, forcing them to go through the same battery again before reaching the lamps. The main reason for the introduction of the reflector was the problem of switching from encryption to decryption mode and vice versa in earlier machine versions. In the B version, the rotors had to be removed from the machine and placed on the axis in the reverse order, and then the complementary key to that used during the encryption had to be set in the machine. Complicated rules defining the selection of the key were the source of mistakes made by cipher clerks. Later on, we will take a closer look at how the use of reflector solved this problem and at the same time created a different, more important one. Finally, in 1927, the Enigma D model appeared, in which reflector did not rotate during operation, but could be mounted in any of 26 possible positions. In the same year it was deemed unnecessary to maintain the fiction of the Koch patent—the Chiffriermaschinen Aktiengesellschaft acquired also the Dutch copyright (Fig. 1.6).

---

[11] Patent DE425566. If necessary, the characters J and Q were replaced by I and K respectively in the cipher box.

Fig. 1.5 Enigma K, final civilian version of the machine. (cryptomuseum.com)

The improvement of the machine's construction did not translate into the commercial success. The trends that caused Hebern's bankruptcy in the USA were also evident in Weimar Germany. It was not until 1925 that the German navy purchased several copies of the specially adapted machine, introducing it in 1926 for use in the Kriegsmarine under the name of Funkschlüssel C. Early versions of Enigma could meet the requirements of the Navy, but did not correspond to the realities of the army. The large and heavy machine could be mounted on board a ship or in a stationary radio station on land, but it did not meet the requirements of a mobile warfare.

The fact that Kriegsmarine was the first to apply the machine ciphers was probably due to the revelations published a little earlier. Among the statesmen of the twentieth century, Winston Churchill stood out for his interest and understanding of intelligence work. However, he also had a second passion, which in time was to bring him the Nobel prize in literature—the vein of the historian and writer. When publishing his memoirs of the Great War in 1923, he revealed that during the war the

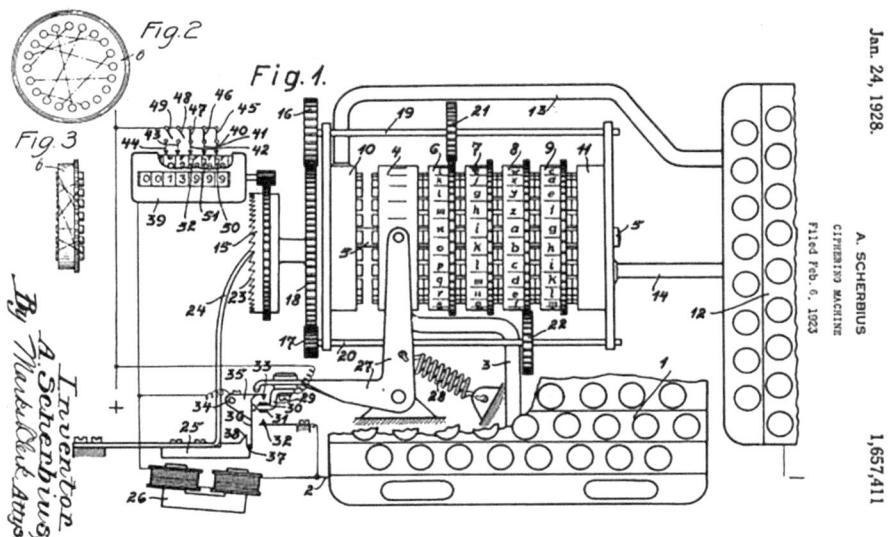

Fig. 1.6 Scherbius's US patent 1.657.411. (Public domain)

British had read on the regular basis the dispatches of the German navy.[12] He justified this surprising indiscretion with reference to an earlier official history of Kaiserliche Marine, which confirmed the capture by the Russians of German ciphers aboard the cruiser "Magdeburg" and their transfer to the Allies.[13] Churchill's indiscretion was complemented by Admiral "Jacky" Fisher, commander of the Royal Navy during the First World War, who in his memoirs published a year later also referred to the reading of the German messages as the basis for British naval victories.[14] Even without a complete picture of the allied codebreakers' successes, the message contained in both books was clear enough for Kriegsmarine; the future war required more effective and reliable methods of securing its communications.

Although Churchill's revelations did not concern her directly, the German Army could not hope that she was not affected by communication security problem. The basic Reichswehr cipher at that time was the system called "Doppelwürfelverfahren", representing a double transposition based on two different keywords.[15] Breaking it required considerable effort and time, but the methods of attack were widely known and used with good results. We do not know which side of the initiative came from,

---

[12] Churchill Winston, The World Crisis, Mc Millan, Toronto 1923, p. 503.

[13] The Germans knew the further fate of the seized code books from a Russian lieutenant, Galipin, who was taken prisoner in 1915 and during interrogation revealed his role in recovering the books and handing them over to the Allies.

[14] Fisher John, Memories, Hodder and Stoughton, London 1919, p. 108.

[15] Rejewski Marian, *Memories of my work in the Cipher Bureau of the Second Division of the Main Staff in 1930–1945*, Wydawnictwo Naukowe UAM, Poznań 2011, p. 17–24.

but the inventors of the Enigma, together with Chiffrierstelle,[16] the Reichswehr unit responsible for the design of codes and ciphers, started to consider changes in the machine's construction required by the army. They consisted in reducing the number of rotors to three, replacing the printing mechanism with a set of 26 lamps and adding a battery to allow the device to be used in field conditions. However, the most important change suggested by military cryptographers was the introduction of a new element: a plugboard resembling an old-fashioned telephone exchange. It was a well-thought-out decision; without a plugboard, the machine corresponded roughly to Hebern's design, whose code was being broken at the same time by William Friedman. However, the attack method he used was ineffective against a machine equipped with a plugboard. The main partner of Scherbius and his engineering staff in these discussions was the then head of ChiStelle, Major Rudolph Schmidt, whose name we shall meet again in quite unexpected circumstances. The reduction of the size and weight of the machine together with its mobility allowed for its acceptance—on July 15, 1928 the Enigma G version was approved for use in the Army; Reichswehr thus found a solution to the problem posed by Churchill's indiscretions.

In 1926, the German officers probably did not feel that with Enigma they were introducing a key element for the army's future strategy and the character of the next war. The victorious generals tend to sanctify freshly acquired military knowledge, elevate it to the altars of the staff academies and reject any deviations and doubts about the accepted doctrine. After the First World War, the French did it, starting to build the Maginot line. Officers from other countries were more creative in drawing up the rules for the next war. Before the German front collapsed in the summer of 1918, Marshal Foch's victory schedule planned the final offensive only for 1919. Its key element—"Plan 1919", developed by the Colonel of the British Tank Corps, J.F.C. Fuller, provided for the breaking of German lines by several thousand tanks.[17] Fuller's plan was a departure from the strategy of the trenches, generally considered to be a degeneration of the rules of war, and a return to manoeuvring warfare. The truce of November 1918 did not allow for a practical test of the new idea, but the first full-scale conflict in Europe after the end of the Great War proved that the idea of the manoeuvring war did not come to an end. During the Polish-Soviet War of 1919–1920 the armies alternately covered several hundred kilometres in just a few weeks—torn off boots and barefoot infantry became one of the symbols of this war. During this conflict the Poles were the first in the world to successfully test the concept of armoured-motor raids to the rear of the enemy. However, the post-war discussion on the future of the tank was dominated by its inventors—the British. The supporters of the new weapons, J.F.C. Fuller, Liddell Hart and General Giffard Martel—argued that the potential of a tank can only be used by adapting the tactics of other types of troops to the capabilities it offers. Their adversaries, cavalrymen of origin and experience, adopted the position that

---

[16] Chiffrierstelle was often abbreviated as Chistelle or simply Chi.

[17] Messenger Charles, Art of Blitzkrieg, Allan Ltd., 1994, p. 30.

they are ready to accept only such a tank, that would feed on oats. Their attitude towards tactical value of the tank can be summarized by the statement of the then head of the British Imperial Staff: "I don't read Fuller's books! And I don't think I will ever read. It would only annoy me".[18]

Deliberations similar to the controversy surrounding the role of tanks also took place among the officers of the air forces. The apostle of the doctrine of war assuming an independent and decisive role of aviation was the Italian general Giulio Douhet. He assumed that the strategic air force would lead to such a significant disorganization of the opponent, that other services could occupy his territory in police rather than military action. Douhet's ideas were developed by the British Chief of Staff of Aviation, Hugh Trenchard. The experience of World War I convinced him that "the psychological effect of the bombing in relation to its material effects is like twenty to one".[19] Endowed with a practical mind, instead of constructing theories he focused on building strategic bombers' force. On his path he encountered an argument convincing British politician, who considered the bomber an attack weapon, hardly fitting into the arsenal of peaceful democracy. Arguing that it was easier to destroy the enemy's planes by bombing their bases rather than by shooting them over Britain's own territory, Trenchard managed to build the foundations of the RAF's Bomber Command. At the same time, in the USA, William Mitchell was demonstrating the capabilities of the maritime aviation by bombing and sinking several decommissioned ships. Both his ideas of using aviation and the way they were presented proved to be indigestible for the military establishment, whose representatives forced Mitchell to leave the ranks of the army. However, the result of his attempts was the subsequent construction of the American carrier fleet and its later victories in naval battles, in which both fleets separated by hundreds of miles did not ever see each other.

The debut of new weapons on the battlefields of World War One signalled a change in the nature of wars. The victorious countries have developed ideas of the application of the new tools of war, but have not managed to integrate the results into a coherent military doctrine, leaving the task to the defeated ones. Heinz Guderian did not invent the Blitzkrieg concept from scratch. In fact, his concept of armoured and mobile combat refers directly to Fuller's "Plan 1919". Guderian's merit was the combination of concepts developed so far independently; the use of armoured weapons, aviation, motorized infantry and artillery. In the realities of 1918, the unit that broke through the enemy's defences started to operate not only in an operational but also informational vacuum. Lack of communication with the HQ did not allow to call for fire support or reinforcements. Transforming tactical success into strategic breakthrough was impossible under these conditions. Tanks wandering around the rear of the enemy, without information about his actions, without the supplies and fire support would turn from hunter into the game. They had to be provided with reconnaissance, delivery of supplies and the ability to

---

[18] Messenger Charles, op. cit., p. 71.
[19] Messenger Charles, op. cit., p. 34.

identify targets for the cooperating aviation, which required fast, reliable and secure communication, not available in the realities of the Great War. Guderian, however, had unique experience, which allowed him to solve the problem of coordinating the activities of different kinds of troops: for a large part of the past war he commanded a field wireless station. He intended to equip every staff, tank, and airplane of his own army with a radio. Securing their communication required a portable, easy to use and fast method of ciphering despatches. In this way the radio, and with it the Enigma machine, became the logical keystone of Blitzkrieg's doctrine. Critics of Guderian's concept pointed out that equipping armoured and motorized units, building aviation capable of providing them with aerial support, training the thousands of specialists necessary for their operation and supplying them with huge amounts of ammunition and fuel made demand on the economy unsupportable by the German industry. However, Guderian was able to find a supporter, whose vote was settling any discussions. In the beginning of 1934 Hitler was observing the exercise of several tank prototypes, summarizing his impressions shouting: "This is what I need! This is what I want!".[20] The very first meeting of Hitler and Guderian sealed the fate of the new military doctrine.

Hitler's interest in the tank and Blitzkrieg doctrine came just in time to guarantee Enigma's career, but too late to ensure the success of its inventor and his company. At the turn of the 1920s and 1930s, orders reaching Chiffriermaschinen Aktiengesellschaft at a rate that did not allow it to survive. Its co-founder and driving force was lacking. Arthur Scherbius died in 1929 in somewhat old-fashioned circumstances; as a result of injuries suffered when he was unable to control the horses. On the very verge of success, on 5 July 1934, the company assets were transferred to Chiffriermaschinen Gesellschaft Heimsoeth und Rinke, a new company founded by the directors of its predecessor, Rudolph Heimsoeth and Elsbeth Rinke. The demand for ciphering machines was soon to reach such a scale that the production of Enigma under licence had to be shared between Olympia Büromaschinen, Ertel-Werk für Feinmechanik, Konski & Krüger and Atlas-Werke AG.

---

[20] O'Neill, The German Army and the Nazi Party 1933–1939, Cassel 1966, p. 94.

Chapter 2
The Mechanical Challenge

Before the era of machine ciphers, a codebreaker could count on success at the price of reaching the limit of his mental strength. In many cases, cryptologists unknowingly crossed this border, overpaying triumph over the code with a nervous breakdown. The most famous were the adventures of Georges Painvin after breaking the German **ADFGVX** cipher in April 1918 (which allowed to fight off the German offensive of last chance) and William Friedman after breaking the cipher of the Japanese Purple machine in 1940 (which did not prevent the tragedy of Pearl Harbor). In both cases, a few months' rest allowed the codebreakers to recover; the story is silent about the incidents in which this proved impossible.

The ciphering machine confronts the codebreakers with a task surpassing in complexity everything they had encountered before. The most complex codes consisted of tens of thousands of codewords, breaking them required finding the first slot in the code cover, after which the structure of the language itself and some patience allowed to do the rest of the work. The need for manual work limited the complexity of traditional, hand ciphers. When trying to break them, the codebreaker was faced with a huge number of combinations, but still possible to be grasped by a trained mind. Therefore, the codebreaking agencies in the period of World War One were recruiting people familiar with the huge numbers; accountants and stock brokers. The machine's infinite patience allows it to repeat the same operation many times without fatigue and without error. As a result, the number of combinations generated by a cipher machine is expressed by numbers unimaginable to an most humans, even the London City's brokers. In order to demonstrate the scale of the problem facing anyone attacking the Enigma cipher, let us look at how the design of the machine resulted from the methods of traditional cryptology, to what extent it improved them and what role the individual elements of the machine played in securing the cipher. The term Enigma refers to the whole family of ciphering machines, of which we will only familiarize a few members. The basis for reviewing the design and operation of the device will be its most common model Enigma

I (known also as the model M3 or *services Enigma*), used during World War II by the German military.

The oldest known known cipher is nowadays referred to as monoalphabetic substitution. Its functioning comes down to replacing each character of the open text alphabet with the corresponding and always the same character of the cipher text. In the Bible a cipher was used in which the first character of the Aramaic alphabet, *aleph*, was replaced by the last, *taw*, the second, *beth*—by the penultimate, *shin*, and so on; as a result, the cipher is known as *atbash*. Julius Caesar replaced in his confidential correspondence each sign of the Latin alphabet with a character three positions further down in the alphabet; the cipher went down in history under his name. There are many variants of the described cipher, however, in all of them only one cipher alphabet is defined (hence the term "monoalphabetic cipher"), and each character of the open text always corresponds to the same character of the cipher text. Let us note that Edward Hebern's original concept of connecting two typewriters with variable maze of wires was equivalent to using a monoalphabetic cipher. The methods of breaking this cipher have been known for at least a thousand years, so his initial proposal wasn't up to the challenge of the twentieth century.

In the second half of the fifteenth century an Italian, Leon Battista Alberti, proposed the use of a different method of encryption, and at the same time laid the foundations for the later development of encryption machines. The essence of his proposal was to use two mutually movable rings, on which the open and cipher alphabet were written respectively. After encrypting one or more characters of open text, the rings were moved by one or more positions, thus creating a new cipher alphabet and using it to encrypt the next character(s). As many pairs of the open/cipher alphabets were used in the ciphering process, this method is nowadays referred to as a polyalphabetic cipher. Let us note again that the original version of the Hebern single rotor machine was the equivalent of a polyalphabetic cipher. In his machine the pairs of clear/cipher alphabets started to repeat themselves during the ciphering after rotor's full turn.

This feature represented a serious weakness of Hebern's construction, as in the middle of the nineteenth century Charles Babbage demonstrated that a polyalphabetic cipher could be broken by decomposing the cipher text into parts corresponding to the repeating monoalphabetic substitutions. Fortunately, the concept of a moving rotor offered a practical solution also to this problem; it was enough to ensure that the substitutions (almost) never repeat themselves. For this purpose, several moving rotors had to be used; the number of their possible relative positions could be much larger than the length of any cipher text. This is why in his next machine Hebern used as many as five rotors, generating a total of nearly 12 million different monoalphabetic substitutions.

The principle of the rotor ciphering machine's operation will be easier to present using the example of a simplified, two rotor machine, whose alphabets contain only the characters **A, B, C, D, E** and **F** (see Fig. 2.1).

There are 6 contacts on each side of the rotor. Inside the rotor, the wires connect pairs of contacts on both sides. The left rotor turns the letter **A** into **B**, **B** into **A**, etc., realizing permutation described as (**AB, BA, CE, DC, EF, FD**); the right rotor

Fig. 2.1 6-letter Enigma
rotors in their starting
position. (Author)

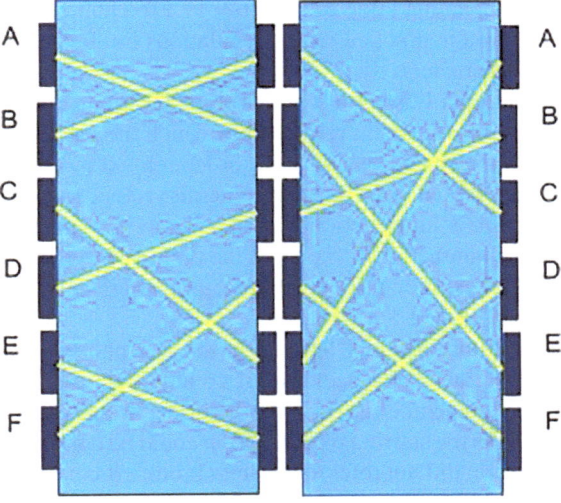

Fig. 2.2 Left rotor
advanced by one position.
(Author)

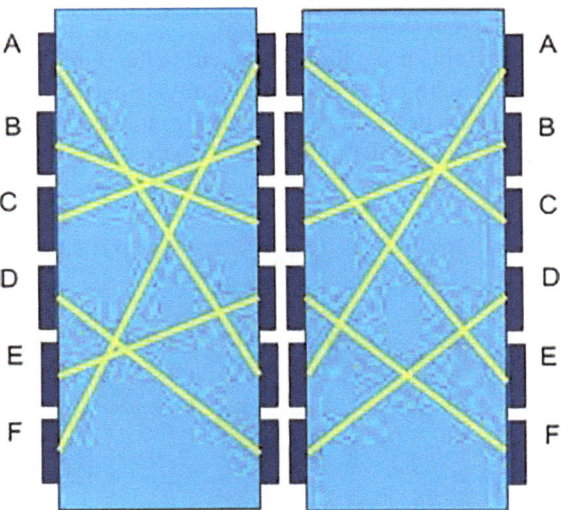

implements the permutation (**AC**, **BE**, **CB**, **DF**, **EA**, **FD**). In the reciprocal position
illustrated in the figure above, the rotors jointly realize the premutation (**AE**, **BC**, **CA**,
DB, **ED**, **FF**). After enciphering the first character left rotor advances by one position
(see Fig. 2.2) changing the effective permutation realized by both rotors to (**AA**, **BB**,
CE, **DD**, **EF**, **FC**).

After each full turn of the left rotor, the right rotor moves by one position; the
effective permutation is changed after each step of each rotor. In the example of a
simplified machine, it starts repeating substitutions after 6*6 = 36 characters. When
using, as in the real Enigma, three rotors with 26 contacts, the substitution will only

start repeating after 26*26*26 = 17.576 characters. However, such a situation never occurred in practice, as the regulations limited the length of the Enigma message to 250 characters.

Enigma I used only three rotors during its operation, but it was equipped with a set of five different rotors, from which the cipher clerk was choosing three, putting them on a common axis in an order defined by the daily key to the cipher. The rotors were marked with Roman numerals; rotors I, II and III were being used since the machine's introduction into use, rotors IV and V were added at the end of 1938. Rotor numbered VI shown in the picture below comes from the Enigma model used by Kriegsmarine (in which the cipher clerk selected three out of the eight available rotors) (Fig. 2.3).

Another element of the key to the cipher was the starting position of the three rotors. Each rotor had a set of 26 letters (in the Kriegsmarine Enigma) or numbers (in the machine used by the Wehrmacht and Luftwaffe) on the perimeter of each rotor, so the starting rotor position could be defined as **MAG** or **13-01-07**. However, this term did not refer to the actual rotor core position. Paul Bernstein suggested that the rotor should be equipped with a movable ring that could be rotated in relation to the rotor core (translating this description into a reduced, 6-character Enigma we get the diagram shown in Fig. 2.4).

Each ring had a turnover notch, so the ring position determined also rotor's turnover position (see Fig. 2.5). Initially, in each ring, the notch was placed opposite a different letter (number). Machine constructors hoped that this solution would ensure the irregular movement of successive rotors, complicating the Enigma analysis. As we will see, in fact, this structure has had the opposite effect.

Fig. 2.3 Kriegsmarine Enigma rotor. (enigmamuseum.com)

Fig. 2.4 Rotor core and ring. (Author)

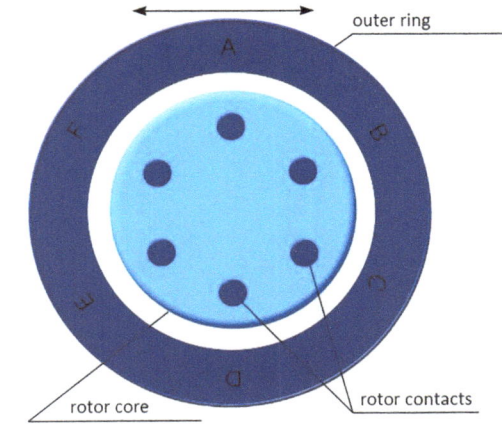

Fig. 2.5 Rotor turnover notch visible in the upper right part of the ring. (enigmamuseum.com)

In addition to the three moving rotors, the Enigma had two drums which remained stationary during encryption; the entry drum and the reflector (Fig. 2.6).

The function of the entry drum was to perform preliminary permutation of the wires leading from the keyboard to the rotors. In the commercial Enigma model, the entry drum led signals from the keyboard to the rotors in the letter order of a typical German typewriter: QWERTZUIO... etc. Cryptanalysts struggling with the secrets of the military Enigma rightly assumed that in this model the wiring of the entry drum is different from the commercial one.

The reflector (German: *Umkehrwalze*) played an interesting role in the device. Absent in the earliest patents, it represented a late addition to the design suggested in 1926 by Willi Korn (who after Scherbius' tragic death of took over the role of chief engineer). Reflector was equipped with a set of 26 contacts placed only on one of its side surfaces, inside the contacts were connected in 13 pairs. The reflector caused each electrical signal to pass through a set of rotors twice; from the input

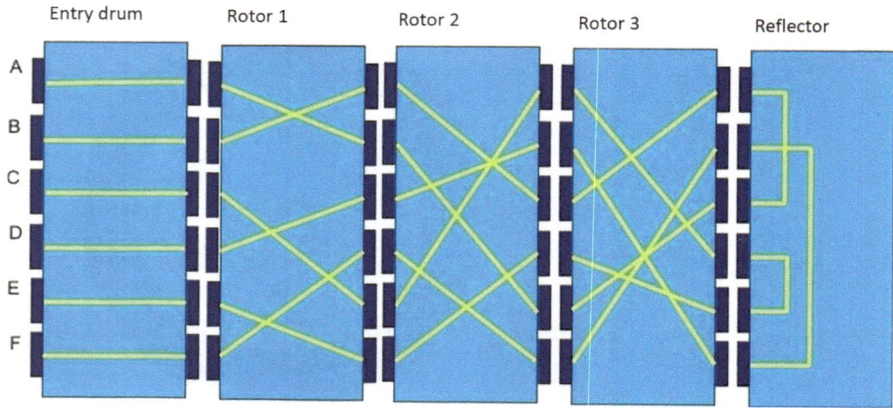

Fig. 2.6 6-letter Enigma. From left: entry drum, rotor battery, and reflector. (Author)

drum to the reflector and then back, in the opposite direction. The new element very cleverly solved the problem of switching the machine from encryption to decryption and vice versa. Enciphering the letter **A** at the rotor position from Fig. 2.5 (without reflector) yields **E**, but enciphering the letter **E** yields the letter **D**. The permutation implemented by the rotors is not reciprocal; rotors in any given position cannot be used alternately for enciphering and deciphering. The inventors of rotor machines dealt with this problem in two ways. The first solution was to remove the rotors from the machine and insert them back in reverse order. The second one was the installation of a mode switch, which during the enciphering fed the signals into the rotor battery from one side, while during the deciphering—from the other. Let's take a look at the enciphering process in the machine with a reflector (Fig. 2.6). The letter **A** is encrypted as **B**, the letter **B** as **A**. Similarly, the letter **C** is encrypted as **E** and **E** as **C**. The permutation performed by the machine with a reflector has a recip-rocal feature that allows texts to be encrypted and decrypted without having to change the machine mode! The cipher clerk sets the machine according to the key, presses the keys corresponding to the letters of plain text and reads the successive letters of the cipher text. The recipient places the machine in the same starting posi-tion and enters the letters of the cryptogram in turn, reading out the plain text. Adding a reflector was a practical solution that speeded up the work and reduced the number of errors. However, the use of reflector had an important side effect: no character could be transformed into itself. The letter **A** could not be encrypted as **A**, **B**—as **B**, etc. The price of simplifying the cipher clerk's work turned out to be high.

Although the wiring of rotors in the military Enigma was different from the com-mercial model, the structural elements presented so far were basically the same for both variants. However, before commissioning Enigma for military use army cryp-tographers demanded several modifications to make the military model more secure than its commercial equivalent. The basic difference between both types of machine was a plugboard resembling an old-fashioned, manually operated telephone exchange. The wires corresponding to the 26 letters coming out of the rotor set were connected to the sockets on the front of the machine (Fig. 2.7). The operator connected the sockets in pairs, using the cables supplied.

As a result of the plugboard connections, the characters corresponding to the connected sockets were swapped in places. Thus, the connection of the socket marked **A** with socket **C** resulted in a lamp corresponding to letter **A** illuminating if a signal corresponding to letter **C** appeared at the output of the rotor set. The principle of reciprocity, described on the occasion of discussing the reflector, required that also in the plugboard the letters should be exchanged in pairs: the exchange of the letter **A** for **C** entailed the necessity of a symmetrical exchange of the letter **C** for **A** (Fig. 2.8).[1]

The plugboard can be treated as a stationary drum with variable internal connections. The key applicable on a given day determined how many and which letters should be connected. In the first years of using Enigma, a typical key required 5 pairs of letters to be plugged. Later on, the number of swapped characters increased, at the end of the war the keys used a variable number of connections, from 5 to 11 (Fig. 2.9).

After getting acquainted with the basic elements of the machine's construction, it is time to present it in its entirety and learn how to use it. One of the basic rules of cryptology requires that the cipher should remain safe even if the adversary knows the cipher construction (i.e. ciphering device itself falls into the enemy hands). Security of the cipher should rely entirely on the secret and frequently changed

Fig. 2.7 Enigma plugboard. (CCA 2.0, School of Mathematics—University of Manchester)

---

[1] Each of the plugs had two wires. Each plugboard socket contained a spring, which in the absence of a plug, connected wires corresponding to the same letter (this is why in English Enigma descriptions the letters not connected with the plugs were referred to, in charming Anglo-German jargon, as *self-steckered*, i.e. connected to self).

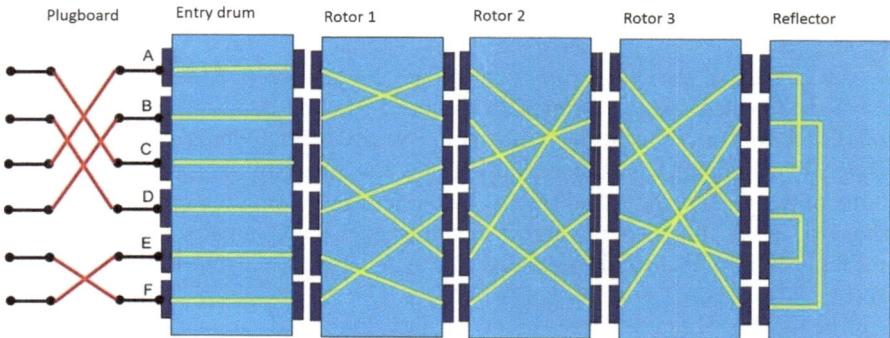

Fig. 2.8 Six-character Enigma with plugboard. (Author)

Fig. 2.9 Enigma I.
(cryptomuseum.com)

cipher key. The number of possible keys determines the number of possibilities that a cryptanalyst attacking the system must analyze. In case of Enigma this number is determined by the following elements:

- selection of three of the five available rotors and their order on the machine axis,
- settings the rotor rings,
- starting position of the rotors,
- number and selection of letter pairs swapped by the plugboard.

Those interested in mathematics or simply large numbers will find detailed calculations of the number of theoretically possible keys to the Enigma cipher in the box below. For the rest it may be sufficient to say that this figure equals approximately 619,953,965,349,522,804,000,000. This number of possible keys was determined assuming knowledge of the machine's construction. If the design of the machine remains unknown, the number of possible combinations rises to about $3*10^{114}$. By comparison—the number of atoms in the Universe available to our observations is estimated at only 10^{80}. German cryptologists analyzing the security of the Enigma cipher obviously believed in the power of large numbers. They assumed that the only practical attack on the cipher is the use of contemporary equivalents of computers—electromechanical sorters and tabulators.[2] Knowing their speed, they have calculated that even the highest available hardware concentration would not allow to break the key in a reasonable time.

Enigma Combinatorics: State Space and Key Space

Let us analyze the functioning of Enigma I model in two cases. In the first one we will assume that the cryptanalyst knows the very principle of operation of the machine, however, he does not have access to information concerning its internal construction, in particular he does not know the wiring of the machine rotors. In this way, we will determine the possible number of states a machine may be in. In the second scenario, we will assume that the person attacking the cipher has a copy of the machine, but does not know the cipher key. We will determine the maximum number of theoretically possible keys.

Scenario I

Each Enigma rotor has 26 pins on its both surfaces. If we choose the first pin on one surface, we can connect it to any of the 26 pins on the other surface. If we choose the next pin, we can connect it to any of the 25 pins on the other surface only. Therefore the number of possible wirings of any rotor is equal to 26 • (26 − 1) • (26 - 2) • ... • (26 - 24) • (26 - 25) = 26!, which translates into 403,291,461,126,605,635,584,000,000.

The first rotor is selected from among 26! theoretically possible. By selecting the first rotor we eliminate one of the possibilities, so we select the second one from among (26! - 1) possible. Similarly, the third rotor brings (26! - 2).

(continued)

---

[2] *Some analysts in their imagination allowed such an idea* - the post-war testimony of Wilhelm Fenner, head of one of the sections of the German cryptology agency Chiffrierstelle, TICOM, DF-187A. *Enigma used according to the instructions is impossible to break. (...) It can be broken with the use of a huge complex of Hollerith machines, but the probability of breaking is marginal* - post-war testimony of Erich Hüttenhain, OKW/Chi cryptologist, TICOM I-2.

The same mechanism allows to take into account the entry drum, getting (26! - 3) possibilities. The total number of possible rotor combinations represents the product of all four values and is equal to:

26,453,071,587,484,435,966,565,383,154,187,825,269,596,587,475,679,144,
168,629,439,112,098,429,555,222,911,717,025,686,471,982,186,496,000,000.

Irrespective of the selection of rotors, the three selected rotors can be arranged in 6 ways.

There are 13 connections inside the reflector. When the wire is connected to the first pin, the other end of the wire can be connected to any of the other 25 pins. After selecting the first connection pair, the next one is selected from 23 pins, the third one from 21 etc. The total number of possible reflector combinations is therefore 25 * 23 * 21 - ... * 1, i.e.:

$$7\ 905\ 853\ 580\ 625.$$

The ring in each of the rotors can be set in any of 26 positions. The position of the ring in the first and second rotor affects the number of available states, as it determines the rotor turnover position. The third rotor is followed by the fixed reflector, so the position of its ring does not affect the number of possible states. As a result, the ring settings bring 26 * 26 possible states, that is:

$$676.$$

Before starting work, each of the three rotors is set in one of 26 possible positions, which translates into $26 \cdot 26 \cdot 26 = 26^3$ combinations, i.e:

$$17\ 576.$$

The last element of the machine contributing to the number of possible states is a plugboard. The number of connection (let's mark it as **p**) can vary from 0 to 13. With 26 plugboard sockets, the selection of **p** pairs can be done in (26/2**p**) ways. Determining the number of possible connections is similar to determining the number of possible pairs of connections in reflector; after selecting the first socket, the place of connection of the second end of the cable is selected from among (2p - 1) possibilities. The selection of subsequent sockets gives (2p - 3), (2p - 5), ..., 1 possible states. The total number of possible states is equal to (26/2p) - (2p - 1) - (2p - 3) - ...- 1. The table below shows the number of possible combinations depending on the value of **p**.

(continued)

The value of p	Number of combinations	The value of p	Number of combinations
0	1	7	1 305 093 289 500
1	325	8	10 767 019 638 375
2	44 850	9	53 835 098 191 875
3	3 453 450	10	150 738 274 937 250
4	164 038 875	11	205 552 193 096 250
5	5 019 589 575	12	102 776 096 548 125
6	100 391 791 500	13	7 905 853 580 625

It is worth noting that the number of possible combinations reaches a maximum for 11 connections, and not, as one might expect, with 13 pairs. German cryptologists were apparently aware of this relationship, recommending the use of 10 plugboard connections.

Combining the partial results (and assuming $p = 11$), we obtain the number of possible Enigma states equal to

3 064 537 774 667 781 245 162 783 206 916 868 949 205 691 845 820 125 435 032 225 270 994 269 522 075 900 473 712 575 645 083 889 591 356 267 102 713 648 819 347 298 713 600 000 000 000 000

approximately equal to $3 \cdot 10^{114}$.

Scenario II

We assume that the attacker knows the machine's internals. We also assume that a plugboard contains 10 connections. The number of possible cipher keys is determined as follows.

The knowledge of the rotor and reflector wiring means that they do not contribute to the number of key states. The number of combinations generated by the plugboard connections has been determined in the previous step being equal to 150 738 274 937 250.

The selection of 3 out of 5 possible rotors causes the first rotor to be selected from among 5, the second one from the remaining 4, and the third one from 3. The selection can therefore be made in $5 * 4 * 3 = 60$ ways. The three selected rotors can be placed on the machine axis in 6 different orders, and the rings can be set in $26 * 26 = 676$ different ways.

In the previous scenario, without knowing the internals of the machine, we assumed that the number of possible starting rotor positions is equal to $26^3 = 17,576$. The knowledge of the design forces us to correct this value due to the

(continued)

mechanism of the so-called "doublestepping". We assumed so far that the movement of the rotors is regular. However, the Enigma design caused that when the middle rotor moved by one position, the left rotor also advanced by one position. The change of position of the rotors was therefore as follows:

ADS		
ADT		
ADU		
ADV		
AEW		
BFX	←	double step
BFY		
BFZ		

The doublestepping reduced the number of possible rotor combinations by 26 * 26 positions. Effectively, the number of their possible positions was therefore $26^3 - 26^2 = 16,900$.

Total number of possible cipher keys is equal to the product of the above values, i.e.:

$$619\ 953\ 965\ 349\ 522\ 804\ 000\ 000.$$

This number does not shock the modern computer user, but in the reality of the Second World War it was almost as abstract as the number of Enigma states in the previous scenario. The devices used by the Allies to break the Enigma key were able to check about 20 possibilities every second. It would take about a trillion years to find the key using the brute force approach.

Knowledge of mathematical subtleties was not necessary to operate the device. Enigma had two basic tasks: to make the cipher secure and to make the operator's work simple enough to reduce errors to a minimum. But that doesn't mean that the operator's job was trivial. The problems started at the stage of preparing the clear text for encyphering. The Enigma alphabet had 26 Latin letters, not including digits and special characters; spacing, period, comma, etc. It was necessary to develop a procedure allowing the missing characters to be represented using available letters and to avoid ambiguity in their interpretation. This task was solved in slightly different ways in the different services.

Numbers were originally represented as adjacent letters on the machine's keyboard, distinguishing the numerical notation by adding before and after the number the Y character. On the machine keyboard the number 2 is adjacent to the letter W, 3 to E and 0 to P, so that e.g. the number 2330 was represented as YWEEPY. This system caused errors and misunderstandings, so during the war it was changed to a more reliable method. In the new system each digit was transcribed according to its

German pronunciation. As a result, the number from the previous example would be **ZWEIDREIDREINULL**. As the sequences of zeros appearing in many numbers would facilitate the attack, they were being replaced with words **CENTA** (for hundred), **MILLE** (for thousand) and **MYRIA** (for ten thousand). For instance, the number 4000 was represented by **VIERMILLE**, number 40011 by **VIERCENTAEINSEINS**). Due to the frequent occurrence of digits in the messages, it was recommended to use their different spelling and abbreviations: e.g. **VIER** or **VIR**, **ZWEI** or **ZWO**, **FUENF** or **FUNF**.

The Enigma's keyboard lacked the space symbol; it was either omitted or replaced with rarely used letter **X**. Due to the rare occurrence of the letter **Q** in German and the frequent use of the bigram **CH**, its occurrences in clear text were replaced with **Q** (e.g. the word *Richtung* was written as **RIQTUNG**). Wehrmacht and Luftwaffe on one side, and Kriegsmarine on the other, applied different conventions for the representation of special and punctuation marks. The **X**, **Y** and **Z** characters, which are rare in German, were mainly used for this purpose. In Wehrmacht and Luftwaffe, the dot at the end of the sentence was replaced by an **X**, the dot in a different context (e.g. numerical) by the **YY** sequence; the **ZZ** sequence represented the comma, the **KLAM** sequence (from the German *Klammern*) the bracket, and the **FRAGZ** sequence (from *Fragezeichen*) the question mark. In Kriegsmarine, the dot was replaced by an **X**, a comma by **Y**, a colon by **XX**, a question mark by **UD**, a dash by **YY**, and a parenthesis by **KK**. Urgent messages were distinguished in Kriegsmarine by inserting in the heading of the message one of the tags **BIENE**, **WESPE** or **MUECKE**, and in Wehrmacht and Luftwaffe the phrase **KRKR** (from *Kriegsmeldung*). Routine message headlines identifying sender and recipient were represented by the sequences **VONVON** or **VVV** (from German *von*—from) and **ANAN** or **AAA** (from German *an*—to). Proper names and foreign words were distinguished by repeating and enclosing them with **X** characters (e.g. the word Paris was written as **XPARISXPARISX**). As the repeatable elements usually occur at the beginning of the message and its end, the procedure tried to eliminate their negative effect by recommending that, before and after the actual text of the message, the operators prepends and appends a few words unrelated to the message content, separating the padding from the text with three- or four times repeated pair of letters. Let's assume that the last word of the relevant message was **BDU** (abbreviation for the submarine fleet commander—*Befehlshaber der Unterseeflotte*). The end of the message could take the form: **BDUKBKBKBWASSEREIMEREICHBAUM** (where **BDU** is a part of the actual message, **KBKBKB**—separator, **WASSEREI MEREICHBAUM**—padding). This well-designed precaution occasionally turned against its authors—the operators too often and too literally treated the recommendation to pad the message with "an accidental word of neutral meaning, e.g. a bucket, a telephone, an oak, a wardrobe", repeating in the ciphertexts words taken literally from the manual. Characters crucial for the proper understanding of the message were distinguished by triple repetition. For example, the letter **U** in the submarine designation was represented by **UUU** (the U-550 submarine designation was represented by **UUUFUNFFUNFNULL**). At the end of the whole procedure, the Kriegsmarine cryptographer divided the prepared open text into groups of four and the Wehrmacht and Luftwaffe operators into groups of five characters.

Let's sum up the brief course of the Enigma operator with an example. Let's assume we have intercepted and decrypted the message getting its (almost) clear text:

```
    BINE VVVS QOLT ITZA AAFL OTTE XMEI NSTA NDOR TDRE ISMN ORDJ
CAPG RISC APGR ISJY GEHE MITT TTEI NSFU NFFU NFNA QEIN SVIR FUNF
XQSQ SQSF ERNS PREC HERE ICHB AUM
```

The initial group **BINE** is a shorthand form of the urgency marker **BIENE**. The **VVVSQOLTITZ** section means that the message was sent by (**VVV**) the ship named **SCHOLTITZ**, similarly **AAAFLOTTE** means that it was sent to (**AAA**) the entire fleet (**FLOTTE**). The sender reports that it occupies a position (**MEINSTANDORT**) three nautical miles (**DREISM**, **SM**—shorthand for *Seemeilen*) north (**NORD**) of Cap Gris (the name Cap Gris is placed in quotation marks—**J**—and repeated). After the dot (**Y**) the sender reports that he is being accompanied (**GEHEMIT**) by the T155 torpedo boat (**TTTEINSFUNFFUNF**) with a course (**NAQ** from *nach*) 145 (**EINSVIRFUNF**) degrees. Triplet **QSQSQS** separates the actual text from the padding, including words **FERNSPRECHER** and **EICHBAUM**.

After rewriting the open text of the message, it was time to encipher it. To this end, Enigma should be prepared for work. Each cipher clerk received a sheet containing unique machine settings within a given communication network for each day of the key validity period. The illustration below shows the original cipher key sheet valid in one of the Wehrmacht communication networks in October 1944.

In most communication networks using Enigma, the keys were changed at midnight (in the late phase of World War II the practice of changing the key several times a day was implemented). We can imagine how a signals officer approaches the Enigma at midnight from October 30 to October 31. After opening the cover of the device, he takes out the axle and removes the rotors **VI, II** and **III** (cf. the keys on 30 and 31 October in Fig. 2.10). He then selects the rotors **IV, V** and **I**, setting their rings in positions **21, 15** and **16** respectively. He assembles the rotors in the correct order on the axis, inserts the axis into the machine and closes the machine cover.

From this moment on, the operation of the device is taken over by a cipher clerk. At first, he connects the letters **KL, IT, FQ, HY, XC, NP, VZ, JB, SE** and **OG** in the plugboard. He then sets the starting position of the rotors according to the key. The table of keys in the illustration above dates back to 1944, when the encryption procedure was slightly modified, so the column marked "Kenngruppen" provides four possible starting positions for the rotors. However, during the early period the key only provided for one variant of the rotor start position—for example, let's assume that the first of the four potential rotors start positions, **JKM**, is valid. The cipher clerk therefore places the Enigma rotors in the starting position **JKM**, common for all the machines working in the same network and referred to as the *machine key*.

Encrypting all the messages starting from the same rotors position would create so called *depth*, greatly facilitating the job of the potential adversaries. Therefore, every message was enciphered starting from the unique rotors starting position, referred to as the *message key*, individually chosen by the cipher clerk. The cipher

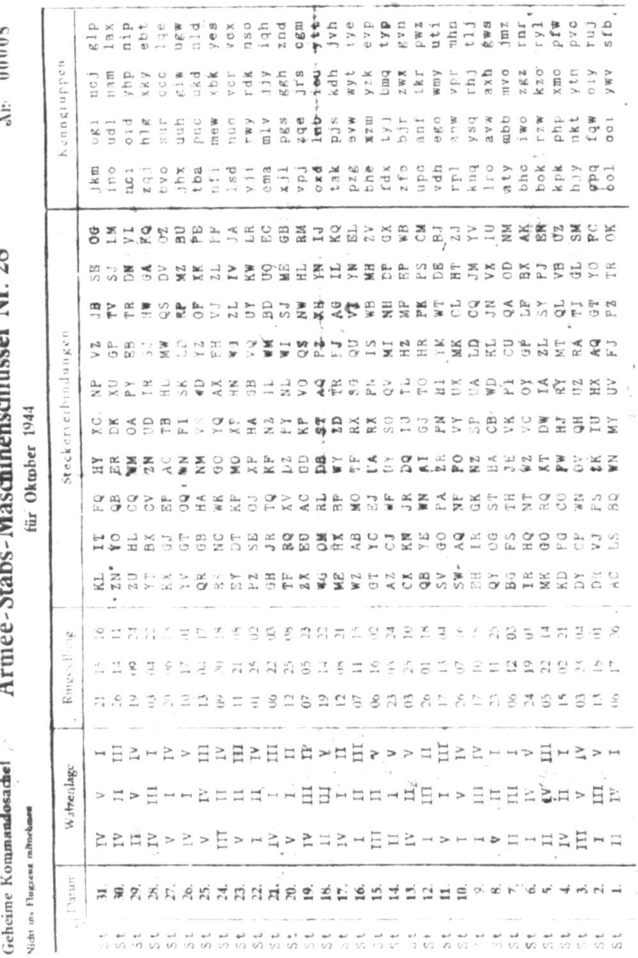

Fig. 2.10 Werhmacht Enigma key for October 1944. (cryptomuseum.com)

clerk at the sending end needed to securely communicate his chosen message key to the receiving station. All he needed to do was to encipher the message key starting from the rotors position defined by the machine key. When the Enigma was being commissioned in 1926 the radio technology was still rather unreliable; any distortion of the transmitted message key would make the entire message unreadable. German cryptologists proposed a simple and effective precaution preventing this from happening; the message key was being enciphered twice, not once. When the cipher clerk on the receiving end deciphered the message header including two identical message key copies, he could safely proceed to deciphering the message contents; otherwise, he would require the message retransmission. In the cryptology any repetition represents a potential weakness. We shall soon learn that it was precisely the described procedure thar cost the Germans the Enigma secret.

Let's come back to the example assuming that the cipher clerk has chosen an individual message key in the form of characters **PGJ**. He encrypted it twice, starting from the machine key **JKM** and receiving

PGJPGJ
XKYEAP.

He then moved the rotors to a position **PGJ** key and started enciphering the clear text—in our example the text of the message redacted in the example above:

KFKW DTKX LVFD SDIE HYEV AULZ YGQT SYDT DXHO KWIK EDNA RNEI IHXW
JKPM PNFP ROIX JLRR XXSC SKLP EFCQ TWIH DJQO JKPK FGDP KXRR EYDT
XWIO YXEQ QMQR KAFY UBICLAJ.

Both on the sending and receiving end, Enigma was operated by two soldiers. The first one was tapping the keys corresponding to the letters of clear text (or on the receiving side—the ciphertext), the second one was taking down the characters highlighted on the machine panel. After the text was encrypted, the operator prepended it with the enciphered repeated message key (**XKYEAP**), and passed the complete ciphertext to the radio operator, who transmitted the text in Morse code. On the receiving end the radio operator forwarded the received text to the cipher clerk, who started by deciphering the message header receiving (assuming correct reception) six characters.

PGJPGJ.

Having received identical copies of the message key, the cipher clerk was assured that the message header had been received correctly and he could set the Enigma rotors to the message key positions (**PGJ**) and proceed decrypting the message body:

BINE VVVS QOLT ITZA AAFL OTTE XMEI NSTA NDOR TDRE ISMN ORDJ CAPG
RISC APGR ISJY GEHE MITT TTEI NSFU NFFU NFNA QEIN SVIR FUNF XQSQ
SQSF ERNS PREC HERE ICHB AUM.

When Edward Hebern offered his machine to the US Army and Navy, both services commissioned their own codebreakers to analyze the security of the cipher. In
both cases the verdict was negative: Agnes Driscoll on behalf of the Navy and
William Friedman, working for the Army have managed to break the cipher. We
don't know much about the research on the Enigma security by the German military
codebreakers. German cryptology agencies were among the first to employ mathematicians. In the case of Enigma, the results of their work were somewhat ambiguous. On one hand, they accurately identified the major weaknesses of the commercial
model[3] recommending supplementing it with a plugboard. On the other hand,
German mathematicians laboriously calculated the number of possible combinations of machine elements and the key to the cipher, and after they had obtained the
astronomical values quoted above, they decided that the military Enigma cipher was
secure. They haven't noticed that the same math that reassured them about the security of the cipher can offer a shortcut reducing the complexity of the challenge to a
level that allows an effective attack on the cipher.

---

[3] Inspectorate 7/VI OKH oraz Pers Z, crypto section of the German Foreign Office), NARA RG
457, HCC Box 1405 and Paschke Adolf, *Das Chiffrier- und Fernmeldewesen im Auswärtigen Amt:
Seine Entwicklung und Organisation*, Bonn 1957.

Chapter 3
Flying Start

Mystery has not been a Polish specialty in history. When in the majority of European courts the Black Cabinets were busy breaking codes and reading diplomatic correspondence of other countries, Polish democracy publicly debated state affairs. It was only during the period of Poland's partition that independent Polish thought was transferred to the underground—generations of conspirators were becoming familiar with codes and ciphers. However, when in November 1918 the reborn Polish state started to build its own structures, among the beginnings of various services inherited after the partitioning countries there were no structures dealing with ciphers, and it wouldn't be any wonder if the Polish Army would consider the organization of the signals intelligence as a whim. The situation of a nascent country made it necessary to measure its needs with the number of bayonets and sabers required at numerous flash points on uncertain borders rather than with the ciphers broken.

The feeling of instability was not, after all, exclusively a Polish specialty in postwar Central Europe. The Great War itself, the destruction and marches of troops were enough to disorganize the communities and economies of the region's countries. The collapse and disintegration of tsarist Russia and Austro-Hungarian Empire has destabilized much of the continent. The countries created on their ruins fought for shares in the inheritance or simply for survival. Armies of the Soviet revolution and of the tsar's epigones have been wandering through the lands of Russia in a deadlier battle than the recently ended war, and the young Soviet empire has been pulsating like an amoeba for several years. Among the rapid changes in the situation, the timely and reliable information about the intentions of the main actors of the drama and the events on its vast stage was worth several divisions. The Polish command had significant advantages in the intelligence game. Years of independence conspiracy made marshal Piłsudski and his colleagues understand well underground work, its needs and techniques. During the First World War, they managed to integrate the hitherto scattered currents of independence activity within the Polish Military Organisation (Polska Oranizacja Wojskowa), whose scope covered the

© The Editor(s) (if applicable) and The Author(s), under exclusive license to
Springer Nature Switzerland AG 2025
M. Grajek, *Enigma Myth Deciphered*, History of Information Security,
https://doi.org/10.1007/978-3-031-65475-6_3

most important theatres of the ongoing conflict, from Paris to Vladivostok and from Finland to Istanbul. This effective and extensive intelligence organization gave the country invaluable services in the early days of its independence, though at the cost of many lives. For the Polish command, however, it was obvious that its effectiveness would drop should the confrontation between Poland and Soviet Russia escalate into the regular war. Intelligence reports coming from Russia confirmed that Soviet peace initiatives represent only a smokescreen for the concentration of troops on Polish borders. Reason dictated that preparations should be made for action in the new situation. When the front line stands between the command and the network of agents, the effectiveness of human intelligence decreases. On the other hand, regular warfare forces the fighting parties to use the radio, creating the conditions for intercepting and decrypting its messages. But Poland, destroyed by war, provided its soldiers with weapons and necessary material with the utmost difficulty. Trying to organize a signals intelligence in these conditions was a task for dreamers.

However, there has never been a shortage of these in Poland; officers, soldiers and civilians, coming from all parts of the country, trained in various armies and educated by universities all over Europe. On November 13, 1918, Private Pradellok from the Wielkopolska region drew the attention of the Polish staff to the importance German radio station in Warsaw he was serving in. This allowed three Polish officers to take over the station at night of 18/19 November. The same night Lieutenant Bronislaw Sroka used this facility to transmit a signal notifying the world of the creation of an independent Polish state. The radiogram was intended primarily for foreign governments, but the first answer received was "Polish Poznań greets Warsaw". A slightly ahead of the events telegram came from Stanisław Jóźwiak, who, on behalf of the Soldier's Council in Poznań, had been commanding the local radio station since November 10, and on December 27 offered its services to the command of the Wielkopolska Uprising. Earlier than Warsaw, on 4 November, the station in Kraków had fallen into Polish hands, followed by Lublin, Przemyśl, Lwów, Grudziądz and Toruń. In the first months of independence, these stations were staffed with a somewhat improvised teams of wireless operators trained in the partitioning armies and student volunteers. Not only did they manage to ensure communication between the main centers of the young state, but also, starting from March 1919, they were able to intercept foreign messages, initiating thus the functioning of Polish signals intelligence.

These actions may be linked to Major Karol Bołdeskuł, who entered the Polish army on 12 January 1919. Born in 1877 in Kołomyja district, Bołdeskuł represented a personality typical of the Polish borderlands; his father's name suggests Romanian roots, his mother's—Hungarian. A graduate of the Polish secondary school in Lviv, he was trained at the Theresian Academy in Wiener Neustadt. A key period of his military service was 1916–1917, when he was serving as a head of Austrian signals intelligence on the entire Eastern Front. Having taken up the duties of Head of the sixth Branch of the General Staff on April 1, 1919, he vigorously started the organization of the signals intelligence service. The most important Bołdeskuł's contribution of was making the supreme command aware of the possibilities offered by signals intelligence. He could quote solid arguments: during the Great War the

Austrians and Germans owed much of their success to breaking Russian ciphers. The cipher systems of the white Russian armies came from the same tradition, the Bolsheviks' systems could not break with them completely—an attempt to break them both was certainly worth time and effort. Bołdeskuł persuaded the command to assign a large part of the precious radio equipment to the interception service: out of 26 available sets 6 were used exclusively for the interception. The first condition for the later success was met.

The second condition was finding the right people for the job. Young section did not record significant successes until the summer of 1919. She owed reading of a few Soviet cipher to the fact that Polish troops captured the keys to the "Mayak" and "Mars" ciphers. This situation could have continued if it had not been for the wedding of one of the section's officers', Lieutenant Stanisław Sroka, sister. Willing to dance at her wedding, he asked his friend, Lieutenant Jan Kowalewski, to substitute him at the night shift. Kowalewski, a chemist by education, had had some intelligence experience during his service in General Żeligowski's division in Kuban, but his knowledge of cryptology was then limited, as he himself admitted, to having read Alan Edgar Poe's "Golden Bug". Probably the boredom of the night shift caused that instead of sorting the incoming messages and passing them to the proper sections, Kowalewski started to analyze them and cracked Denikin's army cipher before dawn. The very next day he was assigned to the Cipher Bureau and entrusted the organization of its codebreaking section. The effects came almost immediately. Kowalewski started his new job on 1 August 1919, and already in the middle of the month he broke the first Soviet ciphers. Soviet, Ukrainian and Denikin's army telegrams were encrypted using traditional, manual ciphers. A characteristic feature of Bolshevism—its paranoid care for the secret—was manifested mainly in frequent changes of cipher keys. In most networks, the keys were changed every two weeks, constantly providing Kowalewski with new challenges. He was throwing himself on every intercept with neophyte's zeal, rarely giving his colleagues the pleasure and honor of breaking the code. He also achieved considerable skill in his new craft; it took him less than four hours to break one of the keys, another one was broken on the basis of the only message available. For most keys, the time needed to break ranged from a few hours to three days. As a result, 75–85% of the Soviet and Denikin's army's messages were read on the day of sending or the next one, and up to 25% of the messages were read in parallel with their intended recipient.

Proficiency came just in time. In January 1920 Polish stations almost simultaneously received the Soviet offer to start peace talks and the messages informing about the transfer to the Polish front of units that had just defeated the white Volunteer Army. The Polish staff has long assumed that after the resolving the campaign against Denikin the Bolsheviks will focus their attention on the western border. Decrypted messages confirmed these assessments and revealed the falsehood of the Soviet peace proposals. In a few months Poles managed to organize the last link of the signals intelligence—the distribution of information from the decrypts to the staffs of the field armies. It seems that this element was planned and implemented by Bołdeskuł's deputy, Major Ignacy Matuszewski. He had several key assets at his disposal. As one of the founders of the Polish Military Organisation he appreciated the importance of the information and the secret of reading the enemy's ciphers. In the inner circle of conspirators surrounding marshal Piłsudski, probably only Walery Sławek enjoyed comparable authority in matters of intelligence. Thanks to direct access to the C-in-C and good cooperation with his chief of staff, Matuszewski effectively translated intelligence into orders and directives for the forces in the field. Starting in February 1920 Polish troops undertook a number of local actions aimed at disorganizing the concentration of the enemy forces. In this period Polish signals intelligence became a weapon so effective and reliable, that orders to destroy the telegraphic network and force the enemy to use the radio were sent to partisan units operating at the Soviets rear. Signals intelligence played an important role in the preparation of the Kiev operation, recognizing in detail the enemy's O-d-B, but after the start of the fighting it quickly lost its importance. Many Russian troops abandoned their radio stations. Messages of the others represented evidence of the effect of the Polish attack rather than useful information: in an often-quoted message the commander of the 58th Division reported "I am here and here, but where my troops are – I don't know".

The maneuverable nature of the war taking place in the following months further increased the importance of signals intelligence. The multitude of messages exchanged between the Soviet headquarters attacking or retreating provided material for cryptanalysis. It seemed that Kowalewski managed to train enough specialists to be able to assign them directly to front and army headquarters, speeding up the flow of information. However, in the summer of 1920, when the Soviet divisions stood at Warsaw's gateway and the decryption of their messages became particularly important, the staff of the cipher section turned out to be too slim to cope with the challenge. Fortunately, the critical situation of the army and the country caused the general mobilization of the country's resources; numerous volunteers joined the ranks including professors and students of Polish universities. Among them were future mathematical celebrities, professors Stanisław Leśniewski, Stefan Mazurkiewicz and Wacław Sierpiński. They were predestined to work on Soviet ciphers not so much by their mathematical skills as by their excellent knowledge of Russian (Leśniewski and Mazurkiewicz graduated from Russian schools, Sierpiński was interned in Wiatka during the Great War). We do not know the circumstances that led to the assignment of the three to the cipher section, but we do know that this decision was an important component of the later triumph over the Soviet army.

The role of the codebreakers was only revealed after many decades. It can be illustrated by just two episodes. The first one concerned an exceptionally long message intercepted on 12 August 1920 and enciphered with a new, so far unused key. Any reasonable cryptographic service changes the key before a start of the new operation. This, combined with the considerable length of the message, led to the conclusion that the cryptogram might contain an operational order for the next stage of the Soviet attack. Breaking the cipher usually requires a lot of material to be gathered, but in the described situation Kowalewski and Mazurkiewicz started the attack immediately, on the basis of the single intercepted message. To Kowalewski's surprise "in less than an hour the cipher started to yield to attack. This message was not yet fully read when we knew from its fragments that it represented an order for a decisive assault on Warsaw".[1] Codebreakers realized that contrary to the expectations of the Polish command, the Soviet armies did not intend to attack Warsaw concentrically, bypassing it instead from the north, repeating General Paskiewicz's maneuver from the period of the November Uprising in 1831 and at the same time exposing themselves to a Polish counterattack on their left wing. Without waiting for the final of the work, Kowalewski passed the general theses of the order directly to marshal Piłsudski, who turned the knowledge of the enemy's intentions into a spectacular victory.

The second episode was linked to the operations of Siemion Budionny's Mounted Army. She was ordered to stop the Battle of Lviv and march hastily towards Lublin. By carrying out this order in time Budionny's army could threaten the rear of Polish troops engaged in the decisive operations against the Soviet Western Front. Instead, Polish codebreakers were following the violent exchange of messages related to Budionny's and Stalin's failure to carry out the orders, permitting Polish armies to continue their offensive without worrying about their rear. The last accent of the codebreakers' contribution to the victory was the appointment of Lieutenant Jerzy Suryn to the Polish delegation for peace negotiations in Riga. Suryn's nominal task was to encipher and decipher the messages exchanged between the delegation and Warsaw. However, he was also involved in breaking the ciphers used by the Soviet delegation in communication with Moscow, offering the Polish delegates a precious advantage in the negotiations.

In the critical phase of the Battle of Warsaw a coincidence of Polish and British signals intelligence activities could have fatal consequences for the Polish offensive. The British were focusing on the diplomatic messages rather than the military ones. On 17th August 1920 the London "Times" published an article entitled "Red Trickery", unambiguously referring to the content of the broken Soviet cryptogram. The obvious testimony of the weakness of the Soviet codes is reported to have caused a violent dispute between Lenin and Chicherin. The foreign affairs commissioner demanded that Lenin takes decisions allowing to raise the professionalism of the Soviet cryptologic service to a new, higher level, contributing to better security of the Soviet communications. Lenin took the lead in the dispute, arguing that the cryptologists' revolutionary spirit is more important than their professionalism. Thanks to his attitude the Poles were able to keep their advantage in the field of cryptography and on the battlefield, and the British were able to commit further indiscretions.

Signals intelligence represents a complex structure, based on specialized equipment, perfect organization and above all, people with unique skills and experience. The improvisation of effective signals intelligence service by the young Polish state was one of the many expressions of energy and enthusiasm with which the Poles started to build and extend their regained independence. However, the energy of nations is reflected in the actions of individuals. Without the experience of Major Bołdeskuł and his organizational talents, signals intelligence would have probably

---

[1] Jan Kowalewski, *Ciphers the key to victory in 1920.*, On the trail, 1969, XXII, No. 7–8.

reached maturity too late to play an important role in war. Without a chance to reveal Jan Kowalewski's codebreaking talent, the fate of the war could have been different. When a year after the victory General Sikorski decorated Lieutenant Kowalewski with the Virtuti Militari cross, winking at him he said "*For the victorious war*". Since Kowalewski's and his team's success the Poles have associated the exotic codebreaker's specialty with victory. The success achieved in the cradle had a great influence on the further fate of Polish cryptology.

Chapter 4
Adolescence

Victory has compelling power. It grants the victor a monopoly on the truth, telling the defeated to look for those guilty of failure. The winners receive the glory of yesterday without obligation to reflect on tomorrow. Poland, triumphant in the war with the Bolsheviks, could not enjoy such luxury. Polish victory contained a bitter note of menace. Piłsudski undertook the Kiev campaign believing that there was no free Poland without an independent Ukraine. The forces of the young state proved sufficient to defend and strengthen its own independence, but they were not enough to achieve the strategic goal of the campaign - the consolidation of an independent Ukrainian state. When the Polish soldier put an end to the Soviet plans to wholesale export of the revolution on a European scale, Stalin decided that its smuggling across the border was not the worst substitute. The long Polish-Russian border has become a place of confrontation of different concepts of man, state and actually - everything. When Henry Stimson, taking over as U.S. Secretary of State, was closing in 1929 the "Black Cabinet", which had been providing significant services to the country's politics to date, he stated that "Gentlemen do not read other gentlemen's mail". Poland could not respect this principle, largely due to a deficit of gentlemen in the region.

The actual founder of Polish codebreaking service, lieutenant Jan Kowalewski, after the end of the war served in the ranks of Silesian insurgents. His success in breaking Soviet ciphers must have been a bit of a public secret, as in 1923 he was invited to Japan, where he became a lecturer in cryptology at the Imperial Military Academy in Tokyo. The Cipher Bureau (Biuro Szyfrów, BS[1]) that he abandoned was transforming from war improvisation to normality. The first manifestation of the change was its new head—lieutenant and classical philologist Jakub Plezia. Given that his experience in the codebreaking exceeded that of entire BS staff, it is surprising that Plezia only appeared there in 1920. He served in the Austrian army

---

[1] For the sake of clarity, from now on I will use this generally accepted name of this service, which it adopted in 1930, omitting details of subsequent organisational changes.

throughout the Great War. Injured during the Gorlice offensive in 1917, in July of this year he was assigned to signals intelligence, first on the eastern front, then in Italy. When he joined the Polish army in the autumn of 1918, he emphasized his experience as a codebreaker. However, his declaration was well ahead of time and Plezia was sent to the Polish Liquidation Commission as an interpreter. Another of his attempts was also in vain: his report entitled "Cipher and radiotelegraph" initially met only a polite response by the military authorities. His experience, however, was registered, and after Kowalewski's departure Plezia proved to be the best candidate as his successor. Probably unconsciously, the heads of the Polish intelligence service have thus incorporated the Cipher Bureau into the mainstream of codebreaking services of this age. Jakub Plezia was the founder of a scientific dynasty in classical philology. In addition to his command of classical languages, he was fluent in English, French, German, Russian, Hungarian and Italian, and was also able to speak Czech and Ukrainian. Immediately after World War One, linguistical skills continued to be the main asset in dealing with ciphers. However, this state of affairs was soon to change, under the influence of Jakub Plezia's successors in the Polish Cipher Bureau (Fig. 4.1).

Major Gwido Langer and second lieutenant Maksymilian Ciężki were representatives of a fascinating human mosaic that shaped the atmosphere of the armed forces of the resurrected Poland. Raised in Cieszyn, Langer at the age of seventeen chose the career of a professional soldier in the Austrian army. World War I found him at the Theresian Military Academy in Wiener Neustadt, from where he marched

Fig. 4.1 Founder of Polish Cipher Bureau, Maj. Jan Kowalewski (sitting second from right), future head of its German section, Ltn. Maksymilian Ciężki (standing first from right) ans unknown officer of Japanese Imperial Navy (sitting at the center). (Barbara Ciężka)

to the front as a freshly commissioned second lieutenant of the 16th Landwehr Infantry Regiment. He fought for nearly 2 years and was wounded several times, before having been taken PoW by the Russians in June 1916 near Sapanów. Released from the PoW camp in Siberia only in December 1918, he and his compatriots found themselves immediately among the revolutionary chaos and joined the ranks of Polish Siberian Division. Originally, he became a lecturer at the officer school, but when the division joined the combat, he took a position in line. In the vicinity of Tomsk, by directing traffic at one of the railway stations, he contributed to saving the division from the dire straits, but soon afterwards, in January 1920, near Krasnoyarsk, he was taken prisoner again, this time by Soviets. Interned at the PoW camp in Tuła, he joined two other officers in an escape which led them through the country suffering the paroxysms of revolution and civil war to Poland, where he joined the army again in early November 1920. He commanded a company, then a battalion in the fourth Infantry Regiment, was a staff officer in the Mountain Division, graduated from the High Military School, and then in 1929 he went on to work in intelligence, taking up the position of the Head of the Cipher Bureau in early 1930 (Fig. 4.2).

The military career of his future deputy, Maksymilian Ciężki, took place in a smaller space, but also included interesting traits. Born in 1898 in Szamotuły, he was one of nine children of a farmer who, after settling in the town, tried his hand as a construction entrepreneur. He must have preserved some links with the farming, as in due time he enrolled young Maks to the local agricultural school. The outbreak of the Great War made Ciężki's fate follow a different path than planned by his father. Young Maximilian crossed the line between the private and public parts of his life in 1914, creating in his hometown a boy scouts' troop and becoming its leader. He was called up in 1917 to the German army, he went through the Kaczmarek-Regiments[2] road, fighting on the front in France until February 1918. Transferred to the signals school in Ohrdruf, he used his leaves to turn his former scouts' team into a conspiracy group, buying weapons from demoralized German soldiers. Their armory proved helpful shortly after Ciężki's demobilization. Ciężki returned to Szamotuły, joined the freshly established Workers' and Soldiers' Council and threw himself into work, transforming the scouts' teams in the neighboring villages into troops of the Polish Security Guards. When the Wielkopolska Uprising broke out, he and another member of the Council headed a unit that took over nearby Wronki at night from 30 to 31 December. Severe lung disease caused him to leave the line duty shortly afterwards. After his recovery, in April 1919, he recalled his signals training and joined the staff of the radio station in the Poznań Citadel. At the turn of 1919/1920, having completed the training at the Officer School of Signals in Zegrze, Ciężki was commissioned as a lieutenant and assigned to the Central

---

[2] Kaczmarek-Regiments—common name given to the regiments of the German army recruited in the regions inhabited mostly by the Poles, among whom Kaczmarek was the most popular family name.

Fig. 4.2 Gwido Karol
Langer, 1921. (Hanna
Kublicka)

Lithuania Military Command,[3] then to the central Warsaw wireless station, and finally in 1923 to the second Department of the General Staff (i.e. intelligence service). At that time Jan Kowalewski had no longer formal links with the Cipher Bureau, however, several photographs suggest that he maintained close contacts with the successors of his achievements, and in particular that relation linking a mentor and a student developed between Kowalewski and the young Maks Ciężki. Consequently, it was Ciężki, who transmitted the traditions and experiences of the section, which was soon to bring spectacular results (Fig. 4.3).

Langer and Ciężki had, as the career officers, little to do with classical philology. Antoni Palluth, postman's son born in Pobiedziska, probably took a bit of Greek and Latin at St. Mary Magdalene's college in Poznań. He took his school duties seriously, deserving the exemption from tuition fees not only for himself but also for his brother. At the end of his college in the summer of 1918, he was awarded a special

---

[3] Central Lithuania Military Authority—temporary military authority of the area around Vilnius, occupied by the mutinous Polish troops in 1920 (Fig. 4.3).

Fig. 4.3 Maksymilian Ciężki at work, ca 1922. (Barbara Ciężka)

prize—a book with a personal dedication from Emperor Wilhelm. He did seem to feel grateful—in December 1918 he joined the ranks of Wielkopolska uprising troops. That's how he got to the staff of the central radio station at Poznań Citadel (where he met Maks Ciężki), but he didn't stay there for long, joining first the field station on the northern uprising front, and then, successively, the stations of the first and second Cavalry Divisions during the Polish-Soviet war. After demobilization he started the studies in engineering at the Warsaw Technology University, extending at the same time his signals experience as one of the earliest Polish radio hams (his station was using call sign TPVA). When the lack of resources forced him to interrupt his studies in 1925, Ciężki helped him to find a new way of life—Palluth was employed at the German section of the Cipher Bureau's (Fig. 4.4).

The triumvirate which was to play the role of architects in the future victory over Enigma ciphers, differed from its counterparts in other countries. At the British Government Code & Cipher School (GC&CS) prevailed common for the British establishment system of "old boys". This system assured that new team members would be charming companions in the club conversations and good partners in the golf or bridge party. The same system was based on the assumption that someone who managed to master the subtleties of Greek and Latin conjugations and declinations, would be able to master also other, minor secrets of power. Knowledge of the language structures acquired during the classical studies was indeed an asset in breaking the previous generation of codes and ciphers, contributing to the success of "Room 40" during the First World War. The same model of education and recruitment meant that no one in the circle of British codebreakers would see the need to change their methods early enough. The codebreakers were not isolated in their assessment. Most mathematicians, who were to play a major role in the upcoming events, saw their discipline as an abstract field, with no links to practice or even

Fig. 4.4 Antoni Palluth,
mid-1930s. (Jerzy Palluth)

more so to the military. As early as 1940, Godfrey H. Hardy, the tutor of many generations of British mathematicians wrote in his "Mathematician's Apology": "*Real mathematics has no effects on war. No one has yet discovered any warlike purpose to be served by the theory of numbers or relativity, and it seems very unlikely that anyone will do so for many years*".[4]

Against this background Polish army, created in the fire of battle as an amalgam of different traditions and experiences, appeared as a democratic institution, allowing for different career paths and directions. Ciężki and Palluth's technical passions would probably be considered at GC&CS as craftsman's job, not suitable for a gentleman. However, the future has shown that it was precisely their passion for engineering that has become one of the keys to breakthrough in codebreaking. A logarithmic slider is more often an engineer's tool than a Latin declination or a Greek poem meter. Faced with a new problem, he describes it in the language of mathematics and expects its help in solving it. The subtleties of the social structure of both codebreaking services were certainly not the only reason for the success of the Poles and the failure of the British. Years later, Langer summarized the sources

---

[4]G. H. Hardy, A Mathematician's Apology, Cambridge University Press 1940, p. 44.

of his Bureau's success with a concise statement: "We were better motivated." However, although motivation encourages the search for a solution, it does not suggest the paths leading to it.

In the early 1920s Polish Cipher Bureau was routinely watching secret communications of Soviet Russia and Germany—the countries that posed the most serious threat to the young Polish state. Organized during the war of 1920, the network of intercept stations in the east was complemented by stations intercepting the messages of the western neighbor, located in Starogard, Poznan and Krzesławice near Kraków. At the end of the twenties a new cipher appeared in German networks, resistant to the attack methods used so far. The staffing constraints and the focus on the adversary's land forces meant that Polish intelligence did not monitor Kriegsmarine's communications, so the emergence of Enigma in 1926 in her service went unnoticed. Only the use of the machine by Reichswehr alerted Polish codebreakers. Polish intelligence has managed to establish that on 15 July 1928 the Germans introduced a cipher called *Maschinenschlüsselverfahen Enigma G,* that the device uses a feature called *Stöpselstellung*, and that each unit using it regularly receives keys to the cipher, which consists of three numbers not exceeding 26. As early as 1928, the cipher analysis was entrusted to a team consisting of Maksymilian Ciężki, Lieutenant Wiktor Michałowski and a civilian employee of the Cipher Bureau, certain Czajsner. Statistical analysis of the intercepted cryptograms confirmed the machine character of the cipher. Ciężki was aware that traditional methods of attack would probably turn out to be helpless against the machine code, but his first attempt to use an alternative approach miserably failed. He demonstrated a sample of the cipher to engineer Ossowiecki, specialist in paranormal phenomena, known for his ability to read letters closed in several envelopes placed one in another. When it turned out that the cipher envelope was an impenetrable barrier even for Ossowiecki, Ciężki realized that an attack on Enigma would require a long-term strategy. Looking for ideas, he had to remind himself of the key role of the three mathematicians in the decryption of Soviet ciphers in the hot days of summer 1920. Their success did not guarantee a lucky repeat, but the experience of the Cipher Bureau did not suggest any other direction of search; he decided to reach for the help of mathematics and mathematicians. It is difficult to say whether to respect more the courage of Ciężki's concept or his superiors who approved his plan. Ciężki intended to organize a cryptology course for a group of mathematics students, and then select from among its graduates most promising candidates for codebreakers, providing them with the opportunity of apprenticeship in a new profession before challenging the Enigma. The implementation of this plan required a significant and long-term organizational and financial effort; when presenting the idea to his superiors Ciężki had to remark that he could not guarantee its success. In spite of the required expenditure, despite the lack of a guarantee of success, at the end of 1928 the heads of Polish intelligence service agreed to the plan. We can fairly assume that without Kowalewski's earlier success and the role a group of mathematicians had played therein the decision would have been different.

When in 1920 the three mathematics professors volunteered to serve in the army, the major merit of the intelligence service was to release them from counting supplies at the depot; in 1929 Ciężki had to find the candidates for the codebreakers himself. He started by presenting a sample of the German cipher to one of the Bureau's veterans, Professor Stefan Mazurkiewicz from the Warsaw University. The cipher of double transposition was being broken by Poles for years, but the professor assessed it as unbreakable. His opinion was a signal that Ciężki had to look for a solution in other academic centers. Many hypotheses were put forward to explain the organization of the course in Poznań—none of them explains Ciężki's decision by itself. In fact, the Cipher Bureau organized two courses for future code-breakers at the same time, and the same logic was used in the choice of their location. Kowalewski had experience from the tsarist army, so he was naturally predestined to break Russian ciphers. Maksymilian Ciężki was trained in the German ranks, against which he later fought in the ranks of the Wielkopolska Uprising (just like his colleagues from the section—Wiktor Michałowski and Antoni Palluth). Polish intelligence service sought to recruit for the Cipher Bureau people who knew the language, the organization of communications and, above all, the mentality of a potential adversary. The logical consequence was to organize a course for candidates for the Soviet section based in Warsaw, and for the German section in one of the intellectual centers of the former Prussian partition. The University of Poznan was at that time one of the youngest universities in the country. In 1928, a significant part of the students represented Poles from Greater Poland, Silesia and Pomerania, who had received primary and secondary education in German schools. As a result, the most frequently repeated hypothesis concerning the location of the course is the assumption that the choice of the location was deter-mined by the knowledge of the German language. Even if accurate, the argument is not entirely convincing. If language was the decisive criterion, the course would probably be organized at the Jagiellonian University in Krakow, which had a longer tradition and also guaranteed practical bilingualism of its students.

Another hypothesis indicated the scientific level of mathematics at the Poznań University. Most Polish universities at that time suffered from certain organizational sclerosis, making it difficult for young scientists to break through to the top—a phenomenon that is probably felt until today. The creation of a new academic insti-tution offering independence and stability acted as a magnet: the young university attracted an interesting group of scientists. In the field of mathematics, these were primarily representatives of the Lviv School of Mathematics, which had already developed a position of global recognition. Professor Zdzisław Krygowski was the driving force in everything that concerned mathematics in Poznań at that time. Born in Lviv, a graduate of the Jagiellonian University continued his studies in Berlin and Paris. After returning to his home country, he joined the Lviv Polytechnic, acting as its rector in the years 1917–1918. The proposal from Poznań was apparently not attractive: it came from the university in the course of its organization and con-cerned only the assuming responsibility for the one of the two chairs of mathematics at the Faculty of Philosophy. However, Krygowski was an individual who could find recognition in any environment. He quickly managed to attract and bring up a circle of fellow academics known later as the Poznań school of mathematics. In 1926, just before the events described, he took up the position of president of the Polish Mathematical Society. Knowing the recognition that Polish mathematicians enjoyed in the world at that time, chairing their association is a good illustration of Krygowski's and his circle's position.

However, the reputation of the young university and its mathematical faculty does not explain itself why it was entrusted with the organization of the course, either. The University of Lviv enjoyed far greater recognition in the mathematical world, students remembering Austrian schools were also mostly bilingual, and the location of the university away from Germany made it easier to ensure secret. The decision about the organization and location of the course was most probably made by Maksymilian Ciężki and his friend Antoni Palluth. Both born in Greater Poland, met for the first time and became friends in Poznań, during the Greater Poland Uprising. The choice of the place for the course had to be determined in equal measure by their technical interests, which required looking for a solution to the problem in the sciences, the position of Professor Krygowski's chair of mathematics and language issues. However, the factor that united the others and thus decisive was the relationship of both officers with Poznań. Ciężki's choice of as the head of the German section of the Cipher Bureau was not a coincidence; the heads of intelligence service knew well that the story of his own life would provide an appropriate level of motivation. Ciężki and Palluth, on the other hand, had the right to assume that the memory of German rule and of the struggle for freedom in 1918 and 1919 would also provide the right motivation for the students of Poznań University.

In December 1928, Professor Krygowski asked his assistants to choose a group of students fluent in German and achieving good results. The beginnings of the course were accompanied by a bit of internal conspiracy, as all activities had to be kept secret from Krygowski's all-knowing secretary of German origin. At the inaugural meeting with the students two officers of the Cipher Bureau, Major Franciszek Pokorny (the then head of the Cipher Bureau) and Lieutenant Ciężki, appeared in civilian clothes. The faculty of mathematics occupied part of the premises of the Poznań Castle, erected in the early twentieth century as one of the residences of the German Emperor. The location of the course there would have contained a dose of charming irony, but in reality, the classes were organized in more discreet military premises. Twice a week the participants of the course went by tram in the direction of the Poznań Citadel, where a signals battalion was accommodated and an intercept station was operating. The lecturers were Palluth, Pokorny and Ciężki. Palluth quickly made his way from a radio amateur to an efficient cryptanalyst in the Cipher Bureau. According to the memories of his family members, he trained Japanese codebreakers still before the Poznań course. Palluth and Pokorny focused on the theoretical aspects of cryptography and cryptanalysis. Ciężki was giving the training a practical dimension, using the intercepts from the local station. Years later, Marian Rejewski noticed that the scope and material of the course were based strictly on Marcel Givierge's book *Cours de Cryptographie*, published in 1925: the Cipher Bureau took care to organize the course on the basis of the most up-to-date materials. After presenting the students with the theoretical basics of construction and breaking various types of ciphers, it was time for practical exercises. Among them, the greatest emphasis was placed on the methods of breaking the Enigma's predecessor in the German army—the double transposition cipher (Fig. 4.5).

Of the more than twenty participants, three were given the opportunity to play a key role in further Enigma-related events. The oldest member of this group was Marian Rejewski, born in 1905 in Bydgoszcz. G.B. Shaw used to say that common sense is instinct, but enough of it is genius. If this is true, Rejewski has simply inherited the essential component of his future successes. Both on the side of the father, a landowner from Gniezno and then a tobacco merchant from Bydgoszcz, and in the family of his mother, the owners of a brewery in Pogórze near Toruń, common sense and stepping hard on the ground were valued qualities. His youth spent in the Prussian partition allowed him not only to master the German language perfectly, but also, as later events showed, to make a number of insightful observations concerning the mentality of his German neighbors. The Rejewski family socialized mainly among the Polish families in Bydgoszcz. The ban on speaking Polish in public places, announced by the Prussians as part of the Kulturkampf, allowed Poles to feel comfortable only during private meetings and joint Sunday trips out of town. Thanks to them, a circle of friendship and acquaintance was formed, in which Marian Rejewski met his future wife when he was only eight years old. The choice of mathematics as a field of study was a logical consequence of Rejewski's character and the system of values he adhered to in his environment. An additional reason for this decision could have been the fact that one of Rejewski's relatives was a

Fig. 4.5 Building on the right—German Kaiser's residence in Poznań, on the left—local military command. When the German emperor received the parade of the local garrison depicted on the postcard, no one could have imagined that in a few years' time Polish students would be learning mathematics in his former residence and breaking of the German ciphers in the neighboring building. (*Poznań na starych pocztówkach, Jan S. Zaus, Księży Młyn, Łódź 2008*)

co-founder of the Vesta insurance company in Poznań, the first life insurance company in the Prussian partition of Poland. As a practical and future-planning man, Rejewski intended to specialize in insurance mathematics and assume the duties of an actuary.

Henryk Zygalski came from a family whose main branch had been living for years in south-western Greater Poland; Wolsztyn, Kębłowo and Odra; another branch settled near Gniezno, where they entered into affinity with the Rejewski family among others. Henryk's parents turned from agriculture into a stable existence of the middle-class bourgeoisie: Michał and Stanisława Zygalski ran a sewing workshop in Poznań. Their son passed his matriculation exam in 1926 at the same Maria Magdalena secondary school, whose graduate was Antoni Palluth, and immediately after his matriculation he started studying mathematics at the University of Poznań.

Jerzy Różycki of the Rola coat of arms was the only one of the three who did not come from the lands of the former Prussian partition, bringing the spirit of the Polish Borderlands to the team. He came from a family with great traditions; Karol Różycki was the commander of a cavalry regiment in Józef Dwernicki's corps in the November Uprising in Volhynia in 1831; his son Edmund was the organizer of Polish units in Ukraine during the January Uprising in 1863. Jerzy's father, Zygmunt Różycki, did not manage to meet the family legend and after graduating from medical school in St. Petersburg he became a medical officer in the Tsar's army. A compulsive gambler, he had to rebuild the family's existence from scratch several times after he lost all his assets in hazard. Jerzy was born in 1909 in Olszana, in the then Kiev province. Already during the World War One he started his education at the Polish Gymnasium in Kiev, but it was interrupted by the revolutionary turmoil from which his parents took refuge in Poland in 1918. They settled originally in Wyszków, where Jerzy graduated from the secondary school in 1926. His scientific career was a natural choice in the family of Masters in Pharmacy, so in the following year Różycki began studying mathematics at the University of Poznań. However, his versatile interests made him willing to betray mathematics in favor of geography, astronomy and foreign languages (Fig. 4.6).

Cryptology is like a ship sailing through the vast and monotonous ocean. It presents itself attractively in the harbor, where the novice is engaged on duty in the hope of male adventure, fight against the elements and the breath of exoticism in distant ports. Then there are days filled with boredom watch and monotony of activities deprived of romance. Only very few are allowed to achieve their dream, exotic port, or experience the excitement of breaking the code. No wonder a group of cryptology adepts was shrinking during the course. Rejewski was one of the first to quit. A bit surprisingly, as during the course he became known as a natural cryptology talent. Endowed with an excellent memory and associative imagination, he immediately combined facts in which others found relationships as a result of extensive work. After completing his studies and defending his thesis in March 1929, Rejewski decided to pursue his idea for life and continue his studies in Göttingen, improving his knowledge of statistical methods and insurance mathematics. The rest of the

Fig. 4.6 Marian Rejewski, Jerzy Różycki (upper right), and Henryk Zygalski. Imagination and a sense of humor were components of the workshop of future codebreakers. Only Rejewski, two years older, invariably remained serious. (Marian Rejewski—Janina Sylwestrzak, Jerzy Różycki— Jerzy Różycki, *Jeden z pogromców Enigmy, Elżbieta Szczuka, Oficyna Wydawnicza RYTM, 2023, Henryk Zygalski—Anna Zygalska-Cannon*)

participants continued their activities until April of the same year. After completing the course, the eight most promising graduates received a proposal from the Cipher Bureau to continue their training in practical terms. The offer concerned people who still had a year or two of studies ahead of them, but Ciężki had found a formula allowing them to combine studies with the secret work. The Cipher Bureau established its branch in Poznań, where cryptology adepts were required to work for twelve hours per week. In order to make it easier to combine studies with work for the army, the branch was located in the basement of the office building adjacent to the Castle, across the street. The messages were delivered from the BS headquarters in Warsaw or from the Poznań Citadel intercept station. The main task of the group was reconstructing the keys to the double transposition cipher and pass them on to the head office, where the cleric staff was engaged in deciphering and translating the messages. It was a monotonous work, however, allowing them to learn the subtlety of the cipher and the habits of German cryptographers.

When Rejewski visited Poznań after a year in Göttingen during the summer holidays of 1930, he was offered an assistantship with Professor Krygowski. Rejewski decided to interrupt his studies in Göttingen and take advantage of Krygowski's proposal. It seems that the reason for the change of plans was a discrepancy between the hopes he had for his studies and reality. Rejewski left for Germany from the country which tried with great effort to heal the wounds inflicted by the war and develop the regained independence. He was leaving Poznań during the final period of preparations for the General National Exhibition, which was to document the achievements of the first decade of this work. The university of Göttingen appeared to him as a treasury of mathematical knowledge and a place sanctified by the memory of Carl Friedrich Gauss, on whose grave he laid flowers upon arrival. On the spot he found a cheerful bunch of wealthy and light-hearted people from all over the world, treating studies as a convenient excuse to prolong their carefree youth. Even the level of lectures was a disappointment after Poznan. Rejewski entered the new environment in the recent months of Hilbert's activity, so in his memoirs he emphasized not the lectures, but the moments spent in the university's rich library. Taking up a job as an assistant was probably more a token of Rejewski's realism than his ambitions. This is indicated by the fact that soon after assuming his new duties he started asking the participants of the codebreaking course about its results and later fates. He addressed among others Różycki and Zygalski, who were still working on ciphers. Their military superiors must have remembered Rejewski's talents, as they immediately accepted his offer of cooperation. Thus, at the end of 1930 in a branch of the Cipher Bureau in Poznań, a team was formed, whose work was soon to push cryptology onto new, previously unknown tracks. The apprenticeship time in the profession was coming to an end, and soon the trial time was to come.

Chapter 5
Breakthrough

The violence, global outreach and the scale of the casualties suffered by countries participating in World War One have created a belief that the nightmare cannot be repeated. The League of Nations, inspired by President Wilson, was to become the assurance of peace. And although the United States finally chose isolation, the war-tired countries of Europe wanted to believe in a peaceful tomorrow. But peace was not destined to the twentieth century. The economic crisis of the early 1930s was a herald of the looming storm and was largely its cause. For the USA, late and somewhat forced participation in the war seemed to be a great business. The period of post-war economic recovery allowed President Hoover to announce the eradication of poverty. A little prematurely; on 24 October 1929 a joint action of the greatest bankers of the era managed to prevent the collapse of stock exchanges, but only five days later it turned out that they were not gods of money either. The period of prosperity preceding the collapse democratized the financial markets, so the turmoil of the "Black Tuesday" strained the fortunes of the great and annihilated the savings of the little ones. Beginning in the USA, the economic crisis started a march that soon encompassed most countries of the world. Everywhere it was accompanied by a wave of bankruptcies, rising unemployment and the dramatic aggravation of social problems, scrupulously exploited by political extremists.

In Germany, tired of lost war, revolution, hyperinflation and several coup attempts, the possibility of rebuilding the country has just emerged, when the impact of the crisis has dashed hope of stabilization. A wave of bankruptcies has raised unemployment from 400,000 to 3 million. When Chancellor Brüning dissolved the Reichstag on July 16, 1930, the NSDAP belonged rather to the category of political folklore—it gained 800,000 votes in the last elections. Hitler himself was full of fears about the future—in the ranks of his party a burnout syndrome, lethal to any action movement, was observable. The campaign plan for the September elections provided for the organization of 34,000 rallies, but if not the crisis after the election Hitler would still fit the term given to him by satirist Kurt Tucholsky: "This man does not exist. He's just the noise he makes". On election day, 14 September 1930,

M. Grajek, *Enigma Myth Deciphered*, History of Information Security,
https://doi.org/10.1007/978-3-031-65475-6_5

the NSDAP won over 6.5 million votes, becoming Germany's second political force after the SPD.

In parallel with the electoral fever in Germany, Rejewski was getting into work in a branch of the Cipher Bureau in Poznań. The development of events in Germany could not remain without influence on the activities of Polish intelligence—work on the Enigma was discretely resumed. The Poles already knew that the new cipher was utilizing the machine originating from a commercial Enigma; the case allowed them to learn more about the new challenge as early as 1929. In January this year, a package addressed to the local branch of one of the German corporations reached the customs office in Warsaw. It would probably have gone unnoticed if it hadn't been for the attention-grabbing behavior of the recipient's representative. He showed up at the customs office before it could send a notice of the arrival of the parcel, and what is more, he arrived accompanied by a representative of the German embassy. The two guests began to eagerly convince the Polish officials that the parcel, sent by mistake to Warsaw, should be immediately sent back to the sender in Berlin. The attention paid to the seemingly innocent consignment aroused suspicion among the Polish customs officers, who assured the guests that in view of the end of the office's work the consignment would be returned to Berlin, but only on Monday morning. This gave the Polish intelligence service, duly notified by the customs officials, a whole weekend to open the package discreetly, examine its contents and restore it to its original state. Since the declared contents of the package were radio equipment, the intelligence officer called on Saturday morning Antoni Palluth and one of his associates to help them investigate the equipment. They both did not return home until Monday; in the meantime, the device was dismantled, its parts photographed and measured, then assembled again and packed. It was an incident of no importance to the knowledge of Polish intelligence about the enemy's ciphers—the package contained the commercial Enigma—but the first contact of the intelligence service with cipher devices was to find its further development. When the Poles returned to the struggle with German machine ciphers, they decided to start by discreetly purchasing a commercial model of the device. To do this, Palluth gained cooperation a friendly Swedish businessman, and made sure that the copy purchased at the Leipzig trade fair went to Warsaw. Some Enigma's historians repeat the account that the Poles were supposed to try to decrypt the intercepted German telegraphs using the just purchased machine, and they experienced disappointment. This would be more or less equivalent to buying the same cello as the one used by YoYo Ma and being surprised that Bach's suites do not flow spontaneously from under the bow—later events prove that Polish cryptologists were far from being naive.

The purchase was arranged by a company that has become the executive arm of Polish intelligence in many operations requiring technical advancement, civil status or simply discretion. As early as 1928, chiefs of the intelligence service remarked that buying some specialist equipment on the open market would reveal too much about its areas of interest. Fortunately, the Cipher Bureau had both employees with the required skills and funding to extend their capabilities. A decision was made to establish a specialized company in which Polish intelligence would be a silent or at least a preferred partner. The natural candidate for the leader of the project became an experienced signals intelligence officer, Antoni Palluth. In 1927 the superiors began to build for him a legend of an innocent, civilian entrepreneur. They transferred Palluth to the reserve promoting him to second lieutenant. Considering that at the turn of 1932/33 Rejewski saw Palluth's name directly above his own on the Cipher Bureau's payroll,[1] his civilian status represented only the cover.

The other associates were found in the radio hams community. In June 1926, during the First National Radio Exhibition, the Danilewicz brothers, Ludomir and Leonard Stanisław, were awarded the gold medal for the construction of a short-wave radio station. After this success Palluth offered them cooperation, however, the proposal did not bring results until two years later. The Danilewicz family came to Poland in 1919, escaping from the Kuban steppes from the revolutionary chaos. The head of the family, Jan Danilewicz, was a surgeon, originally in Niewinnomyskaja (where his children were born—the oldest Aldona, followed by Ludomir and Leonard Stanisław), then in the hospital in Rostov-on-Don. After escaping to Poland, by ship across the Black Sea to Bulgaria, Austria and the Czech Republic, the family tried without success to build a new life in Warsaw. Then the Danilewicz family moved to Zakopane, where skiers provided a steady flow of patients for their X-ray office, and finally settled in Sosnowiec, where Jan took up a stable post of a railway physician. The fates of the brothers had split for some time when Ludomir moved to Warsaw, catching up on a backlog of war in St. Casimir's Gymnasium, and Leonard Stanislaw remained in Sosnowiec attending Staszic Gymnasium. However, the parting did not last long—soon the brothers met again among the students of the Warsaw University of Technology and the newly established Polish radio ham club. The brothers' early business ventures, the "Dacho" company (Danilewicz-Chomicz) and Wytwórnia Aparatów Elektrycznych, did not stand the test of time. Only the company founded in 1928 on the initiative of Palluth, whose co-founders were the Danilewicz brothers and Edward Fokczyński (his workshop at 34 Nowy Świat Street became the first seat of the company) was successful. The name under which it went down in history, AVA, was a combination of call signals of radio stations of Danilewicz (TPAV) and Palluth (TPVA). AVA's main area of activity has become the production of specialist radio equipment. A series of 10 short-wave radio stations has become the beginning of the reception and broadcasting center of the Polish signals intelligence. AVA has produced radio stations for many Polish Navy ships, including the destroyers „Wicher", „Burza", „Grom" and „Błyskawica", and several submarines. Later company's engineers designed several types of airborne radios and suitcase transmitters for undercover agents. Chance has caused that AVA was to go beyond its original specialization and also contribute to the development of machine cryptanalysis. The Polish signals intelligence has undergone several personal and organizational changes during this time. In January 1930, Franciszek Pokorny was replaced by Gwido Karol Langer as its head. In the middle of 1931, the own cipher section was merged with the signals intelligence division—it was on this occasion that the Cipher Bureau gained its name, under which it went down in history. An important aspect of the changes was to focus all tasks related to signals intelligence and cryptology under one roof. With the

---

[1] Letter of M. Rejewski to Colonel Lisicki of 12.02.1978.

country's limited resources, this was a solution to maximize the effects. At the same time, conflicts of competence and competition typical for intelligence services in other countries were avoided.

The results of the 1930 election revealed a polarization of moods in Germany which did not allow a clear forecast of the country's future. Hitler declared that the election result alone was more important to him than the 107 seats in the Reichstag ("We are not, in principle, a parliamentary party"). However, judging from several actions testifying to the sudden conversion of the NSDAP to legalism, it did hurt him that the country's second most important political force was not taken into account in the maneuvers around the positions of president and chancellor. And when the invitation finally came in the autumn of 1931, it ended in disappointment and disgrace; President Hindenburg's entourage discreetly communicated his judgement to the public: "this Czech corporal is a strange guy, who could be a minister of the post office, but certainly not a chancellor." In fact, Adolf Hitler was not even a citizen of the country to whose leadership he aspired at the time. This absurd state of affairs had to be changed quickly in spring 1932. In May, the term of office of President Hindenburg ended. The old Marshal was tired of playing the role of a symbol of German unity, especially if this were to involve electoral bullshit, but he was convinced by his entourage that the situation of the country requires another sacrifice on his part. Also, on the opposite side of the barricade there were some hesitations—Hitler hamletized whether his party, of clearly anti-parliamentary attitude, should once again stand up to the electoral race. He made the decision to participate in the elections three weeks before the fact and only then did he rush for German citizenship. Despite an election campaign that the world had not seen before, the use of radio, cinema and airplane, and the increase in the number of supporters to over thirteen million, Hitler did not become an equal opponent to the old Marshal in either the first or second round of the election. The July Reichstag elections in this special year, when the Germans went to the polls five times, did not bring a solution. The NSDAP won nearly 14 million votes and has become the strongest party in the country. However, the joy lasted for a short time—the parliament was dissolved on the first day of its term. The next election was an act of desperation for all participants. For democrats, in the face of growing political polarization and the obvious inefficiency of the democratic machinery. For the voters, due to the fatigue that has reduced turnout at ballot boxes. For the Nazis, in the face of lack of money and weariness of the party ranks. The only effects of the next elections were a drop in support for the fascists and the promotion of the communists to the position of the third force. However, the fate of Germany, and with it a large part of Europe, was no longer to be decided at the polling station.

When the one-day Reichstag was gathering in Berlin, a team of Polish codebreakers was set up in the new premises. Polish intelligence service could not remain indifferent to developments in Germany. Experts in the subject maintained that knowledge about German state could be reduced to knowledge of its army, and the role of Defense Minister Kurt von Schleicher in behind-the-scenes political machinations confirmed this opinion—more access to information about the Reichswehr was urgently needed. At the beginning of September, the Cipher Bureau decided to close its Poznań branch, move its staff to the headquarters and make its work fully operational. The decision to move from Poznań to Warsaw came relatively easily to the two younger members of the team; after completing their studies Różycki and Zygalski had no obligations, and cryptology enchanted them during the course. Rejewski was torn between his academic career and work for the army, pure and applied mathematics, cryptology and loyalty to his mentor, Professor Krygowski. Perhaps if he had been faced with a choice two years ago, he would have chosen the academic option. However, these two years, during which he combined his activity in both spheres, allowed him to assess his preferences and make a choice. To the benefit of much of the civilized world, cryptology has won. At the beginning of September 1932, all three of them moved into the rooms of the Army HQ in the Warsaw Saxon Palace. Their first task represented still training; they were to break the four-letter code of the German Navy. During a month's work they demonstrated everything that later became the team's trademark—reliable intuition combined with knowledge of the opponent's mentality, harmonious cooperation and mutual inspiration, as well as distance to their own concepts, which allowed them to quickly recognize the wrong leads. The attack on the code was complicated by the scarcity of intercepted material and the suspicion that almost all the observed traffic is of a training nature. The key to success has been the observation that relatively many code groups standing at the beginning of the telegram begin with the letter **Y**. Poles aptly assumed that the code had an alphabetical structure,[2] and the mentioned groups of the code correspond in the clear text to the questioning pronouns (*wer, wo, wie, wann, warum*), usually standing at the beginning of the sentence.

This assumption allowed to choose a special pair from among the intercepted dispatches. The first message was one of the shortest, consisting of six code groups only. The second, even shorter, consisted of only four groups, with the call signs and time of sending indicating that it represented a response to a six-group telegram. The **YOPY** group at the beginning of the first telegram indicated that the telegram was a question, while the 4 groups in the return telegram could indicate that the answer was a four-digit number. The four-digit number suggested a date, so the **YOPY** group was probably asking "when". Trying to match the date question with six words, cryptologists showed excellent knowledge of German mentality. Knowing how much emphasis is placed on historical tradition in the German army, they assumed that the question must relate to a date, a fact or a person significant in the history of the country. With this assumption, the code groups almost themselves have arranged into a clear text *Wann wurde Friedrich der Grosse geboren?* (In which year was Frederick the Great born?), and the answer—1712. Having gained an anchor point codebreakers gradually reconstructed a large part of the code book, which allowed the BS for six months, until the change of code, to read the traffic of the German Navy. When the breakthrough in the code's shell had already been made and its deepening was only a matter of time and patience, Ciężki invited Rejewski to a face-to-face conversation and offered him a new challenge, initially in

---

[2] A code is alphabetically structured when both the words of plain text and the corresponding code groups are arranged alphabetically. This construction was used in simple codes, so that the same code book could be used during both encoding and decoding. For this reason, alphabetical codes are also called single book codes.

secret before his colleagues and working on it only after hours. From the beginning of October, after his usual work, Rejewski was moving to a separate room, where Ciężki gathered all the knowledge that Polish intelligence has managed to gather so far about the Enigma.

Meanwhile, in Germany immediately after the presidential elections of April 1932, General Kurt von Schleicher initiated a series of intrigues, the first stage of which was to force the resignation of his political mentor, General Wilhelm Groener, then minister of interior and defense. As a consequence of Groener's resignation, the Brüning's cabinet collapsed and Hindenburg appointed a government under the weak leadership of von Papen, in which von Schleicher himself took over as defense minister. After the October elections to the Reichstag, a pageant of politicians began, partly trying to break the deadlock, partly—to deepen the chaos and use it for their own purposes. As befits a defense minister, the first volley was fired by von Schleicher himself; when Hindenburg reappointed von Papen as chancellor on December 2, von Schleicher declared that the head of government had lost the army's confidence, leading to a cabinet crisis, from which he himself emerged as designated chancellor. He built his hopes for the formation of a parliamentary government with the participation of the NSDAP on the intimacy of the then number two in the Nazi hierarchy, Gregor Strasser. However, his proposal to join von Schleicher's government cost him the accusations of splitting the party, resignation from all positions within the NSDAP, and two years later his life. Thus, von Schleicher's masterful plan failed. Meanwhile, von Papen, as a former chancellor, showed an activity that he could not develop when holding this office. In the first days of 1933 he secretly met Hitler. Both politicians agreed that they see von Schleicher as their common enemy, which allows them to cooperate. However, their hypothetical cabinet needed the army's permission, so the condition for its creation was to find a candidate for the post of defense minister, who could allow the Reichswehr to swallow von Schleicher's fall. Von Papen nurtured hopes that in the course of the talks he would manage to soften Hitler's position and satisfy his ambitions with the position of vice chancellor, taking over the lead himself. His will for revenge did not allow him to hesitate even when it became clear that Hitler would not accept any other agreement than assuring him position of the chancellor. Hindenburg was a little concerned about the direction of the talks, but von Papen reassured the president that from the back seat he would be able to control the actions of this politically undeveloped simpleton, Hitler: "There is no danger. We hired him to work for us". When the business partners gained another von Schleicher's enemy, general von Blomberg, as a candidate for the minister of defense, nothing prevented the president from appointing and swearing in a new government under Adolf Hitler on 30 January 1933. The first stage of Hitler's fight was crowned with triumph. The next one was to start in less than a month, on the day of the Reichstag fire.

When votes were being counted in Berlin after the last elections in 1932, the lonely Rejewski in the room of the Cipher Bureau took stock of his knowledge of the new challenge. There wasn't much of it; a copy of the commercial Enigma showing the principle of its operation, a lot of intercepted ciphertexts and intelligence reports confirming that the cipher key contains three numbers no bigger than

26 and an element called *Stöpselstellung*. From the available information the following conclusions could be drawn; a military machine differs from a commercial model by construction element containing wires or jumpers responsible for the *Stöpselstellung*. The basic setting of the machine includes the order of the rotors on the axis (knowledge taken from the commercial machine) and the initial setting of the rotors (key consisting of three numbers). Such modest knowledge was not enough to launch an attack, an alternative approach had to be found. However, there remained a third component—the messages enciphered with Enigma intercepted by Polish stations.

Rejewski's predecessors have already noticed that each message starts with six letters representing an individual message key, repeated twice. Individual message key was selected freely by the cipher clerk on the sending side and had to be communicated to the receiving party. Basically, it was enough to attach only one copy to the message. However, the methodical Germans considered that any mistake in transmitting or receiving the key would make the message unreadable and decided to stay on the safe side adding one more copy of the key. This decision increased the number of characters encrypted in the same machine setting from three to only six, which did not seem to represent a significant weakness. Rejewski was of a different opinion. German cryptologists had to have great confidence in the power of the Enigma, since they introduced one of the sins of the ciphering systems—the repetition. Rejewski noticed that every first and fourth letter of the message represents an encrypted first letter of the key. The same relationship was valid between the second and the fifth and the third and sixth letters of the message respectively. All the individual keys were enciphered at the same machine setting, defined for the given day, which made them comparable with each other. Rejewski also noted that many of the intercepted messages had the same beginning. Should the cipher clerk had really chosen the individual and unique keys of every message, this would not have happened. The observed repetitions signaled the second mistake made by the German cipher clerks—the selection of repetitive and non-random message keys.

Regardless of the mistakes made by the enemy, Rejewski quickly met the expectations, which made the Cipher Bureau officers devote time to training mathematicians. One of Professor Krygowski's favorite subjects was group theory and permutation theory. He managed to pass his passion on to the students, because already in the first step of his work Rejewski proceeded to describe the examined cipher in terms of permutation theory, introducing thus cryptanalysis into a new stage of development. He started by analyzing the way Enigma transforms the first letter of the message key into the first and fourth letter of the cipher. He noticed that if enough messages were intercepted on a given day, an interesting pattern could be formed. If in one message the letter **A** was transformed into the letter **E** in the fourth position, in another message the first letter **E** was transformed into the fourth **S**, in the next **S** into **G** and finally **G** back into **A**: subsequent transformations formed a closed cycle: **A** → **E** → **S** → **G** → **A**. More such cycles could be found in other messages, and by combining them with the mathematical theorems formulated by Rejewski, it was possible to reconstruct the message key on the basis of its enciphered version only (the interested reader will find a more detailed description of the identification of the cycles and their significance for the analysis of the Enigma code in the box below).

Determining the Key Cycle of the Message

Let's assume that during the day we intercepted messages including the following six letter headers:

DBA VLB	EHQ IJN	RXX WTW	DBA VLB	EHQ IJN	RXX WTW
VLB PFV	IJN JPY	WTW RAP	VLB PFV	IJN JPY	RAP TWT
PFV FQI	JPY MSA	BCP CGS	PFV FQI	JPY MSA	BCP CGS
FQI KVK	MSD UWU	CGS BYM	FQI KVK	MSD UWU	CGS BYM
KVK XET	UWU NIZ	AYM ACO	KVK XET	NIZ UNION	AYM ACO
XET GOJ	NIZ QZR	SDO SDD	XET GOJ	NIZ QZRR	SDO SDD
GOJ ZUG	QZR LRE		GOJ ZUG	QZR LRE	
ZUG YMF	LRE HNH		ZUG YMF	LRE HNH	
YMF OBC	HNH THL		YMF OBC	HNH THL	
OKC DKQ	TAL EXX		OKC DKQ	TAL EXX	

Analyzing the first line of the first column we notice that the (otherwise unknown) first letter of the message key is encrypted as D in the first position and as V in the fourth. We choose the next line where the initial letter of the key is enciphered as V in the first position and find that in the fourth position it is enciphered as P. Following in the same way down the first column, we get the characters F, K, X, G, Z, Y and O. From the key in the last line of the first column, we find that the character O goes to D, from which we started the process. The cycle sought is closed, taking the form (DVPFKXGZYO).

This cycle does not include all the letters of the alphabet, so we choose a key that starts with a letter that does not belong to the cycle just found, e.g. EJP IPS, and by following the steps described above we obtain a cycle (EIJMUNQLHT). This cycle also does not include all characters of the alphabet, so when selecting keys starting with missing letters, we obtain cycles (RW), (BC) and one-character cycles (A) and (S).

All the cycles indicated above are based on the sequence of letters in the first and fourth key position. We record this fact as

$$\Pi_A\Pi_D = \text{(DVPFKXGZYO) (EIJMUNQLHT) (RW) (BC) (A) (S)}$$

Analyzing analogically the sequence of characters in the second and fifth and third and sixth key positions we can determine the cycles generated by them:

$$\Pi_B\Pi_E = \text{(BLFQVEOUM) (HJPSWIZRN) (AXT) (CGY) (D) (K)}$$
$$\Pi_C\Pi_F = \text{(ABVIKTJGFCQNY) (DUZREHLXWPSMO)}$$

Note that cycles of equal length always occur in pairs. This is a consequence of the operation of the enigma reflector and the resulting reciprocity of the cipher; if in a given position the letter A is encrypted as S, then in the same position the letter S will always be encrypted as A.

(continued)

The expressions $\Pi_A\Pi_D$, $\Pi_B\Pi_E$ and $\Pi_C\Pi_F$ represent the products of permutation. Determination of the message key on the basis of its enciphered version requires the identification of its components, i.e. permutations Π_A, Π_B, . . . , Π_F. This can be achieved by recording one of the cycles of the product of $\Pi_i\Pi_{i+3}$ under the other in reverse order and looking for the right phase of mutual ordering. For the product of $\Pi_A\Pi_D$, it is a practical method, requiring only 2–10 tests. The product of the $\Pi_C\Pi_F$ is slightly worse in this respect, requiring 12–13 trials. However, Rejewski was able to find a shortcut, taking advantage of the lack of discipline of German cipher clerks. Noticing the messages with repeated first six characters, he assumed that the cipher clerks have a habit of choosing keys consisting of the same letter repeated three times— **AAA**, **BBB**, **XXX**. In some cases, especially when there were cycles of length 1 in the permutation products, it was easy to verify whether the hypothetical key corresponds to a known cycle structure. If the letter of the ciphertext is present in a given cycle, the corresponding letter of plain text should be looked for in a second cycle of the same length. This criterion also allows some hypotheses to be rejected by negative verification; if the hypothetical key character and the corresponding cipher character belong to the same cycle or to two cycles of different lengths, the hypothesis should be rejected.

Let's assume we have reason to suspect that the cipher clerk has chosen an individual message key in form **AAA**. The first character of the key, the letter **A**, appears in the $\Pi_A\Pi_D$ as a single letter cycle. The corresponding one-character cycle is the cycle (**S**), so the letter **S** represents the first character of the key in its enciphered form.

The concept of the cyclic structure of Enigma ciphers was an unprecedented achievement of cryptanalysis. The method of key reconstruction developed by Rejewski, based on the knowledge of the corresponding cycles, has put the cryptologist's previous practice upside down. In most known techniques of attacking a cipher, the cryptanalyst first tried to buy, steal, guess or otherwise reconstruct the clear text of the message. By matching the clear with the ciphertext, he tried to reconstruct the very ciphering procedure, only having both elements at his disposal did he try to reconstruct the key to the cipher. Rejewski did an unbelievable thing—having only a fragmentary knowledge of the machine's construction and ciphering procedure, he managed to find the key to the cipher without reproducing the text of the open message. The success was rapid and fascinating, but its full meaning was to be revealed only in the future.

So far, Rejewski's method had two important limitations. First of all, its functioning was based on the mistakes made by the German cipher clerks, who kept on choosing repetitive keys. During Rejewski's earliest attack on the Enigma cipher, single letter repetition (**AAA**, **XXX**, etc.) was most popular. The Poles took advantage of the enemy's error in such a way that after determining the cyclic structure of the cipher for a given day, they encrypted a set of keys most frequently repeated and searched for them among the intercepted messages. The second, more significant limitation of the method was the fact that without the knowledge of the machine's construction, even the knowledge of the key did not allow to achieve the code-breaker's basic goal—the reconstruction of the clear text of the message. The Cipher Bureau had a key to the cipher, but still did not possess the lock that it was opening.

This time too, the Poles received help from the enemy, albeit by an indirect route, through Paris. In February 1921, France and Poland signed a military alliance, which due to the military impotence of Germany at that time was rather unspecific. General Faury became commander of the Polish Military Academy, Polish officers (among them Jan Kowalewski) studied at French military schools, but in many areas the cooperation was only symbolic. One of the exceptions to this rule was the coop-eration between the signals intelligence sections of both partners. The French code-breaking service, very effective during World War One, was scattered after the war among four or five independent organizations whose cooperation, to put it mildly, was not harmonious. The French had to see the weakness of their own codebreak-ing, so they decided to create, next to the section dealing with the classic attacks on potential adversaries' ciphers, a section whose primary task was to trade in other countries' codes and ciphers. The head of the so-called "D" section was a veteran of the First War, wounded at Gallipoli, Captain Gustave Bertrand. Given the tasks of the section, he was the right man in the right place. His first adventure with cryptol-ogy was working in 1919 in the cipher section of the Allied Staff in Constantinople. He then moved to the French embassy in Turkey, where partly as a result of innate predispositions, partly influenced by the atmosphere of the East, he developed the mentality of a Levantine carpet trader. Alexander the Great reportedly used to say that no fortress is secure if in its walls there was a gate wide enough to let through a donkey loaded with gold. Bertrand's approach to acquiring secrets represented a similar way of thinking. In the market for intelligence secrets any partner offering an interesting product or a good price was acceptable; Bertrand was not so much a codebreaker as a codebroker. Such an attitude allowed him in 1933 to catch up with-out reserve with Soviet intelligence. During his first meeting with the Soviets their representative, Natan Porecki (using pseudonyms Raymond or Ignacy Reiss), pre-tended to be a representative of American intelligence service. However, when at the next meetings he revealed his true principals, it did not spoil the good, commer-cial atmosphere accompanying the exchange of secrets concerning the Czechoslovakian, Hungarian and Italian ciphers.[3] A little earlier, during his career in Turkey, Bertrand had made a connection which in the coming years was to deter-mine the relationship between French and British intelligence as much or more than international agreements. When Bertrand served at the French embassy in Constantinople, Wilfred Albert Dunderdale, known among his friends as Biffy, was the head of the MI6 outpost in that city. Directing the Turkish outpost was for him almost equivalent to a return to his homeland, as he was born in 1899 in Odessa, where his father represented the interests of Vickers-Armstrong company. Even before the Soviet revolution expelled the Dunderdale family from Russia, Biffy enlisted in the Royal Navy in 1914. He must have gained the good reputation of his superiors, as he was offered an independent position as head of the MI6 post as early

---

[3] Christopher Andrew, Wasilij Mitrochin, '*Mitrochin Archive*', WWL Muza SA, Warsaw 2001, p. 106.

as 1922. We know nothing about the common adventures of Bertrand and Biffy from that period, but the nature of their later relations indicated that they were able to accumulate a capital of trust enabling them to cooperate regardless of the current political situation.

In June 1931, the French embassy in Berlin was approached by an anonymous German offering to sell secret documents. The embassy official, suspecting the provocation, asked the interlocutor to send a more detailed commercial offer to the address of a secret French intelligence service in Paris. The letter, which soon arrived, included an offer to sell instructions for a new ciphering machine. The offer was taken seriously, but out of caution, for the first meeting, on 8 November 1931 in Verviers, Belgium, no intelligence officer was sent, but a little mercenary, a little adventurer and a mystery dealer, known as "Rex". "Rex", born in Germany as Rudolph Stallman, avoided trenches during WWI by taking refuge in the UK, where he made a living from illegal gambling. Sentenced by a British court, he moved to France, where he married Miss Lemoine in 1918, taking not only his wife's name but also the nationality of her country. He has conducted extensive business life, ranging from gold mines in French Guyana to trading the secrets of many European countries. The French intelligence was one of the significant recipients of the goods he offered, and also a silent partner in some undertakings. His partner in the conversation proved to be certain Hans-Thilo Schmidt, a clerk employed in the cryptology unit of the German army—Chiffrierstelle. A veteran of World War I, former officer, who after the war did not find a place in Reichswehr. He has been busy with various activities, but has not been successful in providing his wife and two children with a standard of living that met the expectations of the German middle class. When the economic crisis contributed to the bankruptcy of his wife's parents' hat factory, he felt compelled to seek the protection of his brother, Rudolph Schmidt, whom we met a little earlier when, as a head of the Chiffrierstelle, he played an important role in the Reichswehr's commissioning of the Enigma. When Hans-Thilo asked for help, Rudolph was already a lecturer at the German military academy, but thanks to his former connections he was able to offer his brother a clerical post at his previous place of service.

The salary of a clerk did not allow for extravagance, so Hans-Thilo left his wife and children at their house in Ketschendorf near Berlin, where life was cheaper, while in Berlin he suffered not so much from parting with his family, as from the difficulties of using all the possibilities that the metropolis offered to a single man. In his case, trivial marital misconduct led directly to betrayal of the country. Given the practical, commercial attitude of both parties, "Rex" and Schmidt easily agreed on the details of the deal. In the next meeting on 19th and 20th December in the same hotel in Verviers on the French side participated both, Bertrand and "Rex". Schmidt handed over the promised package of documents and was registered in the French intelligence files as an agent codenamed HE. He was also known as "Asche" or "Asché"; it is not known whether his cryptonym HE was an acronym for the somewhat sinister pseudonym "Asché" or whether "Asché" was just a way of reading his cryptonym. The first delivery of documents included a technical description of the Enigma and instructions for use of the machine; "Asché" returned to Berlin richer with 10,000 marks, twenty times his monthly salary at Chiffrierstelle. If

Bertrand knew the full story of Enigma, he could have bitterly pondered over the irony of the story that made the secret of the machine revealed by the brother of the man who decided to put it into use in the German army. However, the Frenchman was not particularly interested in history, and it soon turned out that the path from the "Asché" documents to the mystery of the German ciphers was not as simple as he had initially imagined. The letter "D" in the name of the unit headed by Bertrand could indicate that its task was to decrypt, but neither its boss nor the rest of the staff were codebreakers and could not use the information just received. Bertrand tried to interest both, the other codebreaking units of the French army and the British one. The army codebreakers have excused themselves from working on the Enigma with lean staff and resources and different work priorities.

In an attempt to find a partner on the British side Bertrand presented Biffy with an offer to participate in the business on the basis of the joint use of the materials provided by 'Asché' and to co-financing this business. Dunderdale received Bertrand's materials with interest, but returned from London after only three days, discouraged by the opinion of the specialists he had consulted, who ruled that the documents provided were void of practical meaning. This opinion was not completely unfounded. The documents provided by Schmidt were the equivalent of the car user's manual describing starting and stopping the engine, replacing wheels, bulbs and fuses, and preparing the vehicle for towing. However, one cannot expect anyone to be able to drive a car solely on its basis, not to mention its repair or building a copy. In particular, none of the documents contained the key detail—the internal connections of the machine's rotors. Without this information, it was impossible to reconstruct Enigma, and without it, the manual was only a curiosity. Bertrand, looking for other opportunities to use his booty, remembered the Polish-French military alliance and took up the effort of a trip to Poland. He arrived in Warsaw on December 7, 1931, offering Langer and Ciężki a description of the machine and instructions for its use. As he noted in memoirs, the Poles were far more optimistic about the information received than his own countrymen and the British.

Langer's optimism had quite a weak foundation—so far, the Ciężki's section has not made any progress in attacking Enigma ciphers. Langer pointed out to Bertrand that the logical supplement to the materials received would be the keys to the cipher, the samples of which he received in subsequent shipments in May and September 1932. On this second occasion, Bertrand once again came to Warsaw in person, in order to deduce from the climate of the talks what progress the Poles possibly made on the basis of the materials provided. However, he returned to Paris disappointed; no gesture from Langer's side indicated the possibility of quick success. Meanwhile, developments in the international situation made the insight into Germany's real intentions increasingly urgent—on 11 December 1932, representatives of the five powers signed a protocol in Geneva granting the Weimar Republic equal rights in arms, equivalent in practice to the rejection of most of the military restrictions of the Versailles Treaty. Just two days earlier, with a great sense of timing, Ciężki passed on to Rejewski a technical description of the machine, instructions for its use and the keys valid for September and October of this year from the latest Bertrand's delivery.

Until now, cryptanalysts have focused on the linguistic or statistical properties of the cipher itself. Rejewski was the first to construct a mathematical model of ciphering machine's internals. The permutation theory worked so well in the case of the message key that it became the natural candidate for describing the machine itself. Rejewski characterized each element of the machine—rotors, plugboard, entry drum—as an unknown permutation and their combined action as a product of permutations. He noted some of the Enigma's properties, which could facilitate the process of reconstructing the machine through theoretical reasoning only. The first feature was the cyclic structure of permutation generated by Enigma. Each setting of the machine rotors resulted in a different cyclic permutation structure. Rejewski referred to such a unique structure, whose elements were the number of cycles and the number of characters in each of them, as the characteristics of the day. The second feature of the constructed model was the rotors' movement. Let us recall that the first rotor of the Enigma moved by one position after enciphering each letter, only after its full rotation the next rotor also advanced to the next position. This meant that in 21 out of 26 cases, only the first rotor moved during the enciphering of the six-character key; the remaining rotors remained stationary and their participation in the process was constant. Describing the unknown machine in terms of mathematical equations represented a complete novelty in the world of the cryptology and the codebreaking. However, further progress was not immediate and the unknown variables in those equations represented the first obstacle encountered by Rejewski. Most of us are used to equations where variables represent number types we know from our education; integers or real numbers. In Rejewski's equations variables represented the unknown permutations. Any theory permitting solutions of this equation type was missing and, to make further progress, Rejewski had to work it out himself. His mathematical education stood the test—Rejewski managed to formulate and prove a number of mathematical theorems which offered the basis of the solution of his set of equations, among them the crucial theorem on the product of two permutations.

The mathematical Enigma model led Rejewski almost to his goal—reconstruction of the machine's internals, and in particular internal connections of the rotors. As a result of mathematical transformations of equations describing machine Rejewski managed to eliminate the transformations realized by the reflector, and the second and third rotor. He still didn't know the connections of the entry drum and the plugboard. He was helped by the information provided by Bertrand: it contained the keys for September and October 1932, including the plugboard settings. After eliminating this variable from his equations, there was only one unknown permutation left: the wiring of the entry drum. In a commercial machine its internal connections were arranged in the order of the keys on the machine keyboard. Lacking a better hypothesis, Rejewski assumed the permutation from the commercial machine, but despite repeated attempts the solution of his equations invariably led to contradiction. This might happen due to a wrong hypothesis concerning the entry drum or the shift of the middle rotor during the encryption of the message key. Rejewski did not have a criterion that would permit him to identify which was the case. Therefore, he kept on testing the hypothesis basing on the messages from the following days hoping that he would end up on messages without the middle rotor movement—to no avail. Many years later he recalled being on the verge of abandoning his approach and acknowledging the failure of his attack. However, before he gave up the party, driven partly by desperation and partly by intuition, Rejewski risked checking another hypothesis. If in the commercial Enigma the Germans did not use completely random connections, maybe they also reached for some order in the military model? In his first attempt, Rejewski assumed the alphabetical structure and to his surprise the equations, as if by the touch of a magic wand, stood in an orderly line and revealed the solution! In this way he managed to determine the wiring of the first Enigma rotor. In 1932 the order of the rotors in the machine was changed once a quarter. Luckily for Rejewski, the keys provided by Bertrand covered two months belonging to different quarters. In the following quarter, another rotor took a fast position—it was enough to repeat the already tested procedure and determine the connections of the second rotor. Knowing the details of two rotors, the determination of the wiring of the third one and the reflector was trivial.

Reconstruction of Rotor Connections

Rejewski expressed the transformations introduced by each element of the Enigma and the machine as a whole, as permutations. In the case of Enigma, permutations operated on an Ω set containing 26 letters of the alphabet: {A, B, C, D, ..., X, Y, Z}. The Π permutation is the transformation that each element of a Ω set transforms unambiguously into an element of the same set. The permutation effect of Π on the element α is represented as $\alpha\Pi$.

For each Π, an inverse Π^{-1} is defined, such that for each α belonging to Ω the relationship $(\alpha\Pi)\Pi^{-1} = \alpha$ occurs. In other words, subjecting the element α successively to the permutation Π and the inverse permutation results in the same element α. The Π permutation action repeated twice will be referred to as Π^2, three times as Π^3, etc. the product of Π_1 and Π_2 permutations will be the permutation defined as $\alpha(\Pi_1\Pi_2) = (\alpha\Pi_1)\Pi_2$ and the identity will be the permutation $\mathbf{1} = \Pi\Pi^{-1}$.

In the Enigma machine, the signal corresponding to the letter of plain text passes first through a plugboard, which realizes the Π_S permutation, then through an entry drum (which realizes the Π_X permutation), three encryption rotors, which realize the Π_1, Π_2 and Π_3 permutations in turn, and reaches the reflector (Π_R permutation). From the reflector, the signal returns through the rotor battery in reverse order, which therefore perform the reverse permutations of Π_3^{-1}, Π_2^{-1}, Π_1^{-1}, Π_X^{-1} and Π_S^{-1} respectively. Moreover, when the machine key is pressed, the first rotor is moved to the next position. To take this into account, we define the Π_T permutation, which transforms each character of the alphabet into the next character (A $\rightarrow$ B, B $\rightarrow$ C etc.). Below, we will focus our attention on six unknown permutations, which are a transformation of the first six characters of the message (corresponding to a repeated key): We will mark these permutations as Π_A, Π_B, Π_C, Π_D, Π_E and Π_F. Assuming that only the first rotor advances during the encryption of the key, we can express the permutations of Π_A—Π_F as follows:

(continued)

$$\Pi_A = \Pi_S \Pi_E \Pi_T \Pi_1 \Pi_T^{-1} \Pi_2 \Pi_3 \Pi_R \Pi_3^{-1} \Pi_2^{-1} \Pi_T \Pi_1^{-1} \Pi_T^{-1} \Pi_E^{-1} \Pi_S^{-1}$$

$$\Pi_B = \Pi_S \Pi_E \Pi_T^{2} \Pi_1 \Pi_T^{-2} \Pi_2 \Pi_3 \Pi_R \Pi_3^{-1} \Pi_2^{-1} \Pi_T^{2} \Pi_1^{-1} \Pi_T^{-2} \Pi_E^{-1} \Pi_S^{-1}$$

$$\Pi_C = \Pi_S \Pi_E \Pi_T^{3} \Pi_1 \Pi_T^{-3} \Pi_2 \Pi_3 \Pi_R \Pi_3^{-1} \Pi_2^{-1} \Pi_T^{3} \Pi_1^{-1} \Pi_T^{-3} \Pi_E^{-1} \Pi_S^{-1}$$

$$\Pi_D = \Pi_S \Pi_E \Pi_T^{4} \Pi_1 \Pi_T^{-4} \Pi_2 \Pi_3 \Pi_R \Pi_3^{-1} \Pi_2^{-1} \Pi_T^{4} \Pi_1^{-1} \Pi_T^{-4} \Pi_E^{-1} \Pi_S^{-1}$$

$$\Pi_E = \Pi_S \Pi_E \Pi_T^{5} \Pi_1 \Pi_T^{-5} \Pi_2 \Pi_3 \Pi_R \Pi_3^{-1} \Pi_2^{-1} \Pi_T^{5} \Pi_1^{-1} \Pi_T^{-5} \Pi_E^{-1} \Pi_S^{-1}$$

$$\Pi_F = \Pi_S \Pi_E \Pi_T^{6} \Pi_1 \Pi_T^{-6} \Pi_2 \Pi_3 \Pi_R \Pi_3^{-1} \Pi_2^{-1} \Pi_T^{6} \Pi_1^{-1} \Pi_T^{-6} \Pi_E^{-1} \Pi_S^{-1}$$

While the permutations of Π_A, ..., Π_F remain unknown, their products $\Pi_A\Pi_D$, $\Pi_B\Pi_E$ and $\Pi_C\Pi_F$ are known after the determination of the cyclic key structure for the day. On the basis of the above-mentioned patterns, the permutation products can also be presented as:

$$\Pi_A \Pi_D = \Pi_S \Pi_E \Pi_T \Pi_1 \Pi_T^{-1} \Pi_2 \Pi_3 \Pi_R \Pi_3^{-1} \Pi_2^{-1} \Pi_T \Pi_1^{-1} \Pi_T^{-1} \Pi_T^{3} \Pi_1 \Pi_T^{-4} \Pi_2 \Pi_3 \Pi_R \Pi_3^{-1} \Pi_2^{-1} \Pi_T^{4} \Pi_1^{-1} \Pi_T^{-4} \Pi_E^{-1} \Pi_S^{-1}$$

$$\Pi_B \Pi_E = \Pi_S \Pi_E \Pi_T^{2} \Pi_1 \Pi_T^{-2} \Pi_2 \Pi_3 \Pi_R \Pi_3^{-1} \Pi_2^{-2} \Pi_T^{2} \Pi_1^{-1} \Pi_T^{3} \Pi_1 \Pi_T^{-5} \Pi_2 \Pi_3 \Pi_R \Pi_3^{-1} \Pi_2^{-1} \Pi_T^{5} \Pi_1^{-1} \Pi_T^{-5} \Pi_E^{-1} \Pi_S^{-1}$$

$$\Pi_C \Pi_F = \Pi_S \Pi_E \Pi_T^{3} \Pi_1 \Pi_T^{-3} \Pi_2 \Pi_3 \Pi_R \Pi_1 \Pi_T^{-1} \Pi_2 \Pi_2^{-1} \Pi_T^{3} \Pi_1^{-1} \Pi_T^{3} \Pi_1 \Pi_T^{-6} \Pi_2 \Pi_3 \Pi_R \Pi_3^{-1} \Pi_2^{-1} \Pi_T^{6} \Pi_1^{-1} \Pi_T^{-6} \Pi_E^{-1} \Pi_S^{-1}$$

In the constructed model, we assume that during encryption of the first six characters, the position of the second and third rotor remains constant. This allows for the replacement of the fixed expression $\Pi_2\Pi_3\Pi_R\Pi_3^{-1}\Pi_2^{-1}$ in the above equations by the permutation Π_Q, which reflects the virtual rotor of the machine, which consists of a second and third real rotor and a reflector. As a result of this simplification, the equations take form:

$$\Pi_A \Pi_D = \Pi_S \Pi_E \Pi_T \Pi_1 \Pi_T^{-1} \Pi_Q \Pi_T \Pi_1^{-1} \Pi_T^{3} \Pi_1 \Pi_T^{-4} \Pi_Q \Pi_T^{4} \Pi_1^{-1} \Pi_T^{-4} \Pi_E^{-1} \Pi_S^{-1}$$

$$\Pi_B \Pi_E = \Pi_S \Pi_E \Pi_T^{2} \Pi_1 \Pi_T^{-2} \Pi_Q \Pi_T^{2} \Pi_1^{-1} \Pi_T^{3} \Pi_1 \Pi_T^{-5} \Pi_Q \Pi_T^{5} \Pi_1^{-1} \Pi_T^{-5} \Pi_E^{-1} \Pi_S^{-1}$$

$$\Pi_C \Pi_F = \Pi_S \Pi_E \Pi_T^{3} \Pi_1 \Pi_T^{-3} \Pi_Q \Pi_T^{3} \Pi_1^{-1} \Pi_T^{3} \Pi_1 \Pi_T^{-6} \Pi_Q \Pi_T^{6} \Pi_1^{-1} \Pi_T^{-6} \Pi_E^{-1} \Pi_S^{-1}$$

In the previous box we described how Rejewski was able to factor the products of $\Pi_A\Pi_D$, $\Pi_B\Pi_E$ and $\Pi_C\Pi_F$ into their components Π_A, ..., Π_F, using German errors or the exhaustive search method. Assuming the factors are known, the original six equations can be presented as:

(continued)

$$\Pi_A = \Pi_S \, \Pi_E \, \Pi_T \, \Pi_I \, \Pi_T^{-1} \, \Pi_Q \, \Pi_T \, \Pi_I^{-1} \, \Pi_T^{-1} \, \Pi_E^{-1} \, \Pi_S^{-1}$$
$$\Pi_B = \Pi_S \, \Pi_E \, \Pi_T^{2} \, \Pi_I \, \Pi_T^{-2} \, \Pi_Q \, \Pi_T^{2} \, \Pi_I^{-1} \, \Pi_T^{-2} \, \Pi_E^{-1} \, \Pi_S^{-1}$$
$$\Pi_C = \Pi_S \, \Pi_E \, \Pi_T^{3} \, \Pi_I \, \Pi_T^{-3} \, \Pi_Q \, \Pi_T^{3} \, \Pi_I^{-1} \, \Pi_T^{-3} \, \Pi_E^{-1} \, \Pi_S^{-1}$$
$$\Pi_D = \Pi_S \, \Pi_E \, \Pi_T^{4} \, \Pi_I \, \Pi_T^{-4} \, \Pi_Q \, \Pi_T^{4} \, \Pi_I^{-1} \, \Pi_T^{-4} \, \Pi_E^{-1} \, \Pi_S^{-1}$$
$$\Pi_E = \Pi_S \, \Pi_E \, \Pi_T^{5} \, \Pi_I \, \Pi_T^{-5} \, \Pi_Q \, \Pi_T^{5} \, \Pi_I^{-1} \, \Pi_T^{-5} \, \Pi_E^{-1} \, \Pi_S^{-1}$$
$$\Pi_F = \Pi_S \, \Pi_E \, \Pi_T^{6} \, \Pi_I \, \Pi_T^{-6} \, \Pi_Q \, \Pi_T^{6} \, \Pi_I^{-1} \, \Pi_T^{-6} \, \Pi_E^{-1} \, \Pi_S^{-1}$$

From the documents provided by the French, Rejewski knew the Enigma's base setting for the day, which included the plugboard settings. This meant that the Π_S permutation is known, which allows to transfer it to the left side of the equations:

$$\Pi_S^{-1}\Pi_A\Pi_S = \Pi_E \, \Pi_T \, \Pi_I \, \Pi_T^{-1} \, \Pi_Q \, \Pi_T \, \Pi_I^{-1} \, \Pi_T^{-1} \, \Pi_E^{-1}$$
$$\Pi_S^{-1}\Pi_B\Pi_S = \Pi_E \, \Pi_T^{2} \, \Pi_I \, \Pi_T^{-2} \, \Pi_Q \, \Pi_T^{2} \, \Pi_I^{-1} \, \Pi_T^{-2} \, \Pi_E^{-1}$$
$$\Pi_S^{-1}\Pi_C\Pi_S = \Pi_E \, \Pi_T^{3} \, \Pi_I \, \Pi_T^{-3} \, \Pi_Q \, \Pi_T^{3} \, \Pi_I^{-1} \, \Pi_T^{-3} \, \Pi_E^{-1}$$
$$\Pi_S^{-1}\Pi_D\Pi_S = \Pi_E \, \Pi_T^{4} \, \Pi_I \, \Pi_T^{-4} \, \Pi_Q \, \Pi_T^{4} \, \Pi_I^{-1} \, \Pi_T^{-4} \, \Pi_E^{-1}$$
$$\Pi_S^{-1}\Pi_E\Pi_S = \Pi_E \, \Pi_T^{5} \, \Pi_I \, \Pi_T^{-5} \, \Pi_Q \, \Pi_T^{5} \, \Pi_I^{-1} \, \Pi_T^{-5} \, \Pi_E^{-1}$$
$$\Pi_S^{-1}\Pi_F\Pi_S = \Pi_E \, \Pi_T^{6} \, \Pi_I \, \Pi_T^{-6} \, \Pi_Q \, \Pi_T^{6} \, \Pi_I^{-1} \, \Pi_T^{-6} \, \Pi_E^{-1}$$

Rejewski's intuition and knowledge of his opponent's nature helped him to make his next step. On the right side of the equations, the permutation Π_I, Π_X, and Π_Q remained unknown. Let us recall that the Π_X represents the transformation realized by the entry drum, which proved to be an identity, that can be omitted from the equations. Two permutations Π_I and Π_Q remained unknown. Rejewski rewrote the equations in form:

$$\Pi_t = \Pi_T^{-1}\Pi_S^{-1}\Pi_A\Pi_S\Pi_T^{1} = \Pi_I\Pi_T^{-1}\Pi_Q\Pi_T^{1}\Pi_I^{-1}$$
$$\Pi_u = \Pi_T^{-2}\Pi_S^{-1}\Pi_B\Pi_S\Pi_T^{2} = \Pi_I\Pi_T^{-2}\Pi_Q\Pi_T^{2}\Pi_I^{-1}$$
$$\Pi_w = \Pi_T^{-3}\Pi_S^{-1}\Pi_C\Pi_S\Pi_T^{3} = \Pi_I\Pi_T^{-3}\Pi_Q\Pi_T^{3}\Pi_I^{-1}$$
$$\Pi_x = \Pi_T^{-4}\Pi_S^{-1}\Pi_D\Pi_S\Pi_T^{4} = \Pi_I\Pi_T^{-4}\Pi_Q\Pi_T^{4}\Pi_I^{-1}$$
$$\Pi_y = \Pi_T^{-5}\Pi_S^{-1}\Pi_E\Pi_S\Pi_T^{5} = \Pi_I\Pi_T^{-5}\Pi_Q\Pi_T^{5}\Pi_I^{-1}$$
$$\Pi_z = \Pi_T^{-6}\Pi_S^{-1}\Pi_F\Pi_S\Pi_T^{6} = \Pi_I\Pi_T^{-6}\Pi_Q\Pi_T^{6}\Pi_I^{-1}$$

By multiplying the pairs of the above equations, he obtained a set of five equations with two unknown Π_I and Π_Q:

(continued)

$$\Pi_1 \Pi_u = \Pi_1 \Pi_T^{-1} \left(\Pi_Q \Pi_T^{-1} \Pi_Q \Pi_T \right) \Pi_T^{1} \Pi_1^{-1}$$

$$\Pi_u \Pi_w = \Pi_1 \Pi_T^{-2} \left(\Pi_Q \Pi_T^{-1} \Pi_Q \Pi_T \right) \Pi_T^{2} \Pi_1^{-1}$$

$$\Pi_w \Pi_x = \Pi_1 \Pi_T^{-3} \left(\Pi_Q \Pi_T^{-1} \Pi_Q \Pi_T \right) \Pi_T^{3} \Pi_1^{-1}$$

$$\Pi_x \Pi_y = \Pi_1 \Pi_T^{-4} \left(\Pi_Q \Pi_T^{-1} \Pi_Q \Pi_T \right) \Pi_T^{4} \Pi_1^{-1}$$

$$\Pi_y \Pi_z = \Pi_1 \Pi_T^{-5} \left(\Pi_Q \Pi_T^{-1} \Pi_Q \Pi_T \right) \Pi_T^{5} \Pi_1^{-1}$$

From the above system of equations, Rejewski eliminated the common factor $\Pi_Q \Pi_T^{-1} \Pi_Q \Pi_T$, obtaining four equations with one unknown Π_1 (while replacing the common factor $\Pi_1 \Pi_T^{-1} \Pi_1^{-1}$ with a new permutation, denoted Π_V):

$$\Pi_u \Pi_w = \Pi_1 \Pi_T^{-1} \Pi_1^{-1} \left(\Pi_1 \Pi_u \right) \Pi_1^{1} \Pi_T^{1} \Pi_1^{-1} = \Pi_V \left(\Pi_1 \Pi_u \right) \Pi_1^{-1}$$

$$\Pi_w \Pi_x = \Pi_1 \Pi_T^{-1} \Pi_1^{-1} \left(\Pi_u \Pi_w \right) \Pi_1^{1} \Pi_T^{1} \Pi_1^{-1} = \Pi_V \left(\Pi_u \Pi_w \right) \Pi_1^{-1}$$

$$\Pi_x \Pi_y = \Pi_1 \Pi_T^{-1} \Pi_1^{-1} \left(\Pi_w \Pi_x \right) \Pi_1^{1} \Pi_T^{1} \Pi_1^{-1} = \Pi_V \left(\Pi_w \Pi_x \right) \Pi_1^{-1}$$

$$\Pi_y \Pi_z = \Pi_1 \Pi_T^{-1} \Pi_1^{-1} \left(\Pi_x \Pi_y \right) \Pi_1^{1} \Pi_T^{1} \Pi_1^{-1} = \Pi_V \left(\Pi_x \Pi_y \right) \Pi_1^{-1}$$

Here's a bit of a theory that we've skipped before. Let's start with the definition of conjugated permutations: Two Π_K and Π_L permutations defined on the set Ω are called conjugated if there exists another Π_P defined on the same set, such that $\Pi_K = \Pi_P \Pi_L \Pi_P^{-1}$. According to the next definition, the list of lengths of all cycles specified for a given permutation is called the cyclic structure of that permutation. Rejewski formulated the theorem stating that two permutations of Π_K and Π_L defined on the same set of Ω are conjugated if and only if they have the same cyclic structure. This theorem allowed him to reconstruct the factors knowing their product.

Returning to the main argument: Let us note that each of the four equations derived from the last transformation shows a structure $\Pi_K = \Pi_P \Pi_L \Pi_P^{-1}$, known to us from the definition of conjugated permutations. The unknown cyclic permutation Π_V is conjugated to the known Π_T^{-1}. By recording the $\Pi_w \Pi_x$ permutation under $\Pi_u \Pi_w$ in reverse order in all possible ways, Rejewski generated possible solutions for the expression $\Pi_1 \Pi_T^{-1} \Pi_1^{-1}$. Following the same procedure for $\Pi_x \Pi_y$ and $\Pi_w \Pi_x$ he identified one solution among the solutions received, which was also present among the ones identified in the previous step—this was the correct one. Finally, by recording the permutation Π_T underneath it, he received the sought-after permutation Π_1, cyclically shifted from the actual one. Establishing the correct mutual phase required the verification of at most 26 hypotheses and allowed to determine the actual permutation of the first rotor.

The theoretical foundation of Rejewski's breakthrough was the permutation theory and in particular Rejewski's theorem on the product of permutations. The most important difference between the commercial and military Enigma versions was the plugboard added in the latter model. It was a prudent decision of the military Enigma designers. We will soon learn that the ciphers of a commercial machine, without a plugboard, were being broken by the codebreakers of several countries even before the war. Meanwhile, the plugboard was an element of the military Enigma, which contributed the largest share to the number of possible keys. Translated into the language of science, the Roman principle *divide et impera* recommends that if you cannot solve a complex problem, it is worth trying to divide it into less complex partial problems, and solve each of them separately. Rejewski's earliest observation about the cyclical structure of permutations generated by Enigma became the key to such an approach. The permutations introduced by rotors are changed by their movement, however, the settings of the plugboard remain fixed. If the rotor permutations are arranged in a closed cycle, each character belonging to this cycle is transformed by a switchboard twice—once on the way from the keyboard to the reflector and again on the way back. Both transformations cancel each other out; the effective permutation is solely dependent on the rotor wiring. A change of letter pairs in a plugboard only affects which letters belong to a given cycle, but does not affect the number of cycles and the number of letters in each of them! Let us assume that the codebreaker has established the following cyclic structure for a certain machine setting: **(DVPFKXGZYO) (EIJMUNQLHT)(RW)(BC)(A)(S).** When finding a cyclic structure in the form of **(HVARZXGKYO)   (EIJMUNQLDT)(FW)(BC)(P)(S)** on any other day, there is good reason to believe that in both cases the rotors of the machine were arranged identically and the plugboard settings resulted in four pairs of letters being swapped. Later on, we shall encounter Rejewski's observations on the cyclic structure of the cipher as a theoretical basis for many attacks on the Enigma cipher. However, the effects already discussed justify the opinion describing Rejewski's theorem on the product of permutations as "the theorem that won World War II".

Christmas 1932 must have been a distress time for Rejewski. It is a Polish custom to spend Christmas days with your loved ones, in an atmosphere of peace and joy. Rejewski could not yet celebrate the solving the problem he was facing. Perceptible closeness to success and the awareness that he was going the right direction did not allow to enjoy peace either. During a few days of hectic work between Christmas and New Year's Eve he managed to fill in the missing details. He solved his system of equations, which was equivalent to reengineering of the Enigma internals through pure mathematics only. The knowledge of the construction in combination with the previously developed method of breaking the message keys allowed him to read the first ciphertext before the end of the year. For the second time in just a few weeks he has done something that machine constructors and the codebreakers of other countries considered impossible. The Enigma cipher was to remain secure even if a machine itself was captured by the enemy. Rejewski's success permitted the rotor data to be forwarded immediately to AVA factory, where they were first

Fig. 5.1 Polish Enigma clone. In fact, the photograph shows a copy of the machine produced in France between 1940 and 1941, based on documentation provided by the Poles. The differences from the original machine are obvious: the plugboard on the top rather than front panel of the machine, and the alphabetical layout of the keyboard and lamp panel. (Muzeum Historii Polski)

used to modify the commercial Enigma,[4] and then to start work on the design of Polish Enigma clones. The tool, whose destiny was to contribute decisively to Hitler's defeat, was forged less than a month before he came to power (Fig. 5.1).

On the way to his success Rejewski used two shortcuts, which much later were used to diminish the scale of his achievement. The first one was the unknown wiring of the entry drum. In order to overcome the obstacle, Rejewski put aside his mathematics for a moment and reached for his psychological intuition, but intellectual honesty did not allow him to accept in the long run a success based on intuition only. When the breaking of Enigma's cipher became a routine activity, he decided to return to the problem, and developed a purely mathematical method of reconstructing the entry drum wiring. Some of the later continuators of his work, and after them many historians, failed to match Rejewski's intellectual honesty stressing the crucial role of luck in his initial success. The second shortcut made by Rejewski during his early attack concerned the plugboard settings. During this attack he used the keys to the cipher provided by the French. In fact, the keys for September and

---

[4] In the commercial Enigma model, the rotor connections were changed according to the scheme established by Rejewski. Since the commercial Enigma did not have a plugboard, its operation was simulated using caps on the machine's keys, changing according to the plugboard settings.

October 1932 represented the only external information that Rejewski used during his attack. Again, the mathematician was not satisfied with the solution to the problem alone; he was looking for a way offering the solution without any external information, based on pure mathematics. And indeed, he managed to design a method promising success also without the knowledge of the plugboard settings. His method was complicated and involved a considerable amount of work, requiring an analysis of the messages intercepted over a period of about a year. The Cipher Bureau was never forced to try it out in practice. When, after many years, Rejewski's success came to light, he was asked by the journalists and historians whether the documents provided by the French represented and indispensable part of the solution. After over 40 years from the events Rejewski most probably could not recall the construction of his more advanced attack and politely confirmed. However, his wartime report, declassified by the French special service in 2015, presents the details of his purely theoretical method, which have been examined by the contemporary cryptographers, who confirmed its effectiveness.[5] Later on, Rejewski expressed also the view that the documents delivered by the French permitted him to achieve the breakthrough at least a year earlier than in alternative scenarios, letting to understand that they were not indispensable for success.

---

[5] It should be noted that the effectiveness test was conducted on a set of data corresponding to Rejewski's assumptions. Without the original 1930's intercepts, it is impossible to confirm whether the data meeting the assumptions of the method actually occurred.

Chapter 6
Solitude of a Long-Distance Runner

The intellectual effort connected with breaking ciphers resembles work on a new scientific theory. Moving among the apparent chaos of facts, the codebreaker tries to find elements of order in it, determine the causes of the phenomena and their mutual relations, translate them into the simplest possible theory, allowing others to follow his thoughts and verify the validity of his reasoning. However, the nature researched by most scientists is neutral. Its rights are waiting for the discoverer, neither announcing their existence nor hiding it too much. The essence of the cipher construction is mystery, attempting to hide the truth and order behind many curtains, deliberately designed to deceive, confuse, lead astray and intimidate their researcher with the enormity of work. The scientist usually works without significant pressure from the outside world. In his workplace time flows with its own rhythm, possible success or failure are primarily of personal importance. Codebreakers could rarely afford the luxury of distance to their work. The fate of battles, campaigns and entire wars depended on their work, the acquisition or loss of an ally, the tearing out of an enemy's secrets or the betrayal of one's own. Half a misery when a codebreaker was working on a system whose functioning was known to him at least in principle. His success depended on his infinite patience in repeating the same monotonous operations hundreds and thousands of times, his ability to capture subtle relationships between facts distant in time and space, and his intuition to see the possible shortcuts or the enemy's errors. Sometimes, however, he faced a challenge completely new, unknown to his predecessors, mysterious and seemingly impenetrable. What's worse, such challenges usually happened at the turning points in history. Enemy's change of the code or cipher often heralds new and violent events. It puts the codebreaker in a psychologically difficult situation: the apogee of his intellectual effort happens when everyone's eyes are focused on him, when his entire environment insistently reminds how much depends on his knowledge, talent and concentration. The enormous concentration of mental effort in December 1932, the nature of the attacked cipher completely different from what has been known so far and the significance of the breakthrough achieved could

M. Grajek, *Enigma Myth Deciphered*, History of Information Security,
https://doi.org/10.1007/978-3-031-65475-6_6

easily justify Rejewski's weariness. However, he himself seemed to treat his achievement as a solution to a more complex problem at the university exam on a group theory. In less than a month, working alone, he made one of the most spectacular breakthroughs in the history of cryptanalysis. However, the time of lonely effort came to an end, the relief was on its way. When Rejewski put the first broken Enigma ciphertext on Ciężki's desk, further isolation inside the team was unnecessary. In the face of the paramount importance of the Enigma's ciphers, the tasks that the two remaining codebreakers were busy with have lost their importance. In the first days of 1933, Rejewski, Różycki and Zygalski began working together to further exploit the breakthrough. They also quickly realized that there were enough challenges and sources of future glory for everyone.

The Enigma problem initially consisted of three layers. The first one was the machine itself and its unknown construction. Some of the information necessary to reconstruct its structure was obtained by codebreakers analyzing the commercial model and the information obtained by Polish intelligence service. The instruction provided by Bertrand confirmed the conjectures and completed the few missing details, including the rotor rings. However, the most secret element of the Enigma's construction was the rotor wiring. Bertrand in his post-war memoirs mentioned that despite repeated requests, it proved impossible to obtain this information from "Asché". The French therefore asked the spy to provide a list of those employed in the production of the machine in the hope that one of them might be blackmailed or bribed. This did not work either: the machine manufacturers only supplied the army with components, and the final assembly was carried out in a protected area where each employee was constantly being watched. When the classical intelligence workshop failed, a mathematician came to help—the rotors were reconstructed by Rejewski on a purely theoretical basis. At the opposite end of the scale to the machine's construction there were individual message keys. Someone who knew the design of the machine and its settings could read the messages without difficulty. However, from the intruder's point of view the cipher was protected by two curtains: an unknown design of the machine and its settings on a given day. Paradoxically, Rejewski found a method that allowed him to reconstruct the message key, even though he knew neither the construction of the machine nor its settings. His observation regarding the cyclic structure of the cipher let him ignore the machine's plugboard settings, at least temporarily. After this achievement the only element missing was the method of finding the right machine key among roughly 159 million million million of its possible settings. The history of the next few years was a chronicle of a race between Polish and German cryptologists, in which the prize was a machine key.

In the earliest period of work, the starting point in the attack on the machine's key was to determine the individual message key. It was usually found thanks to the predilection of German cipher clerks for trivial keys, such as **AAA**, **BBB** etc. However, the German radio security service issued a ban on keys containing repeated letters—messages with this structure disappeared from the ether. Enigma operators switched instead to the keys based on the keyboard layout. Keys of the type **AAA** were replaced with **QWE**, **ASD** and **FGH** (the reader will want to take a look at the computer keyboard to understand their nature). The Poles quickly noticed the innovation and adjusted their methods accordingly. When the security service forbade the use of consecutive characters in the keyboard rows, the operators switched to letters selected from the keyboard diagonals (**WSX**, **RDX**, etc.). Even when both were banned the recommendations, which were supposed to increase the security of the cipher, paradoxically made the task easier for the Poles. The prohibition of repetition has been poorly worded or understood, eliminating keys in which any letter was repeated (**ABA**, **CCD**, etc.). The Cipher Bureau quickly developed a method to use this information to find the key of the message. Another way was statistical analysis of keys. It turned out that the German operators had an unconscious predilection for certain characters and aversion to others. In keys, the letters **A** and **Q** were the most frequent, vowels were on the second position in the frequency table, **L** and **O** were on the third position, while **Y** and **J** rarely appeared. The limitation of the statistical method was the fact that in each network the operators demonstrated different preferences.

Finding the message key was only the first stage of work—rotor order, starting rotor positions and the plugboard settings remained unknown. Rotor order was usually determined using the "clock method" developed by Jerzy Różycki. He took advantage of another mistake of their adversary, this time made not by the operators, but by the Enigma constructors. The ring of each rotor had a turnover notch that caused the rotor to move to the next position. In each of the rotors used until September 1938, the notch was located at a different point on the perimeter of the ring.[1] The engineers probably thought that the movement of the rotors would become more irregular, making it more difficult to analyze the cipher. Różycki's method was based on an insight into the statistical properties of language. If we take two fragments of text in German (or in fact in any language), of about 100 characters, and write them one under the other, we would find about 8 identical letters in vertical columns.[2] However, if instead of the German text we take 100 random letters, we would find only about four such pairs. If both texts are enciphered with the same key, this feature is preserved in the ciphering process. However, if the texts are enciphered with different keys, we will get a result typical for a random character string.

Różycki's "Clock Method"

Let us take two fragments of German-language text, writing them one under the other and highlighting the positions where the same letters appear in both rows:

(continued)

---

[1] Rotor I caused the next rotor to jump in the position corresponding to the letter R, rotor II - letter F, rotor III - W. The catches in the later introduced rotors were located as follows: rotor IV - the letter K, rotor V - the letter A. The constructors, however, learnt a lesson from their mistake too late, because in the rotors VI, VII and VIII introduced to use in Kriegsmarine, the notches were placed at the same points, with the letters A and N.

[2] For languages other than German there is a similar relationship, although the number of repetitions in each case may be different and specific to the language. Phenomenon being described was originally invented around 1922 by American cryptographer, William Friedman.

ESDUERFENALSOKE INE CHIFFRIERVERFAHRENVERWENDE TWERDE
KLARTELEGRAMMES IND INNORMALERSPRACHEABGEFASSTCODETE

The situation where the same letter appears in the columns of the text is called coincidence. In the fragment of 50 characters, we recorded 5 coincidences; the 10% coincidence of characters slightly exceeds 8%, typical for the German language. Next, let's encrypt both fragments first with the same key (**AAA**) :

FXJRRTDYVTOAPJSKMZVQGCDGXZS IXMNGNBZPOQFSWTWKWHNTWV
WEZUXNUYYXKNXBEKMMDZBDKDFYXXFGCGVSZLLXNHXAYMOCKCOV

And various keys (e.g. **AAA** and **AAH**):

FXJRRTDYVTOAPJSKMZVQGCDGXZS IXMNGNBZPOQFSWTWKWHNTWV
ZATVELYSHPLLYY IVQVBZFJMLGOXLJHUHXKUHZLTXHPXBWBGYVC

The operation of encrypting both texts with the same key has retained the number and positions of character coincidences from plain text. In texts encrypted with different keys there was only one coincidence, which is a result below the theoretically expected 4%, and it is a good indication of the randomness of the cipher generated by Enigma.

The selection of keys in the above example was not accidental. For the application of the clock method, cryptologists searched among the intercepted messages for such ones whose message key differed only in the third position (e.g. **ZDA** and **ZDR**, **OJH** and **OJW** etc.). Messages meeting this condition could easily be reduced to a common position, where their fragments were encrypted starting from the same position of the rotors. If the keys used to encrypt the two texts of the example differ e.g. by 7 positions, in order to bring the two ciphertext to a common position, they should be shifted by the same number of characters. Depending on whether or not the central rotor of the machine has shifted during the encryption of the two texts, the ciphertext can be matched in one of the two ways shown below.

FXJRRTDYVTOAPJSKMZVQGCDGXZS IXMNGNBZPOQFSWTWKWHNTWV
          ZATVELYSHPLLYY IVQVBZFJMLGOXLJHUHXKUHZLTXHPXBWBGYVC

FXJRRTDYVTOAPJSKMZVQGCDGXZS IXMNGNBZPOQFSWTWKWHNTWV
ZATVELYSHPLLYY IVQVBZFJMLGOXLJHUHXKUHZLTXHPXBWBGYVC

(continued)

The number of coincidences occurring in the first pair (4) in comparison with the second pair (1) confirms the correctness of the shift in the first pair. In this way, we have excluded rotors II and V from the rightmost position. The example above shows that the middle rotor did not advance between positions **AAA** and **AAH**. If rotor II was in the fast position, middle rotor would advance in position **AAF**, with rotor V—in position **AAA**. By analyzing other pairs of ciphertext, two of the remaining three possible rotors could be eliminated and thus the identity of the machine's high-speed rotor could be determined.

The purpose of the clock method was to determine which rotor takes the fast position in the machine. The clock method was one of the few methods of attacking the Enigma cipher developed by Poles, which was not based on group theory. In fact, it was identical to the method of the so-called coincidence index, invented before 1922 by an American cryptologist, William Friedman. It would be interesting to learn whether Jerzy Różycki copied the Friedman's work or maybe he rediscovered the coincidence index independently? Friedman described his discovery in a brochure published in 1922 in France and the USA and classified immediately afterwards. In the available reports Różycki's attack was described using terminology different from this used by the original inventor. Rejewski mentioned in his memories that the program of the Poznań cryptology course was based strictly on Marcel Givierge's book "Cours de Cryptographie". In this book Friedman's method was only mentioned in a footnote, without going into the details. It might indicate that Jerzy Różycki could have developed his method independently of its original author.

Once the fast rotor has been identified, logic dictates the determination of the other two. However, in the early period of their work the order of the remaining rotors was of secondary importance for the Poles: it changed only once a quarter, so it was more important to reconstruct the daily settings of the plugboard. The so-called grate method was used for this purpose. Its precise description requires entering into mathematical details, which are presented for the interested readers in the box below. The method took advantage of the fact that six pairs of letters connected in the plugboard left a significant part of the enciphered text unchanged.

Grate Method
The starting point for the grate method is the assumption that within the first 6 characters of the message the middle rotor did not advance. We also take advantage of the fact that the plugboard does not change all the characters. Let's assume that the plugboard realizes the identity transformation, which allows to transform the system of Rejewski's equations into:

(continued)

$$\Pi_Q = \Pi_T^{\,x}\,\Pi_I^{\,-1}\,\Pi_T^{\,-1}\,\Pi_A\,\Pi_S\,\Pi_T^{\,x}\,\Pi_I\,\Pi_T^{\,-x}$$
$$\Pi_Q = \Pi_T^{\,x+1}\,\Pi_I^{\,-1}\,\Pi_T^{\,-2}\,\Pi_B\,\Pi_S\,\Pi_T^{\,x+1}\,\Pi_I\,\Pi_T^{\,-x-1}$$
$$\Pi_Q = \Pi_T^{\,x+2}\,\Pi_I^{\,-1}\,\Pi_T^{\,-3}\,\Pi_C\,\Pi_S\,\Pi_T^{\,x+2}\,\Pi_I\,\Pi_T^{\,-x-2}$$
$$\Pi_Q = \Pi_T^{\,x+3}\,\Pi_I^{\,-1}\,\Pi_T^{\,-4}\,\Pi_D\,\Pi_S\,\Pi_T^{\,x+3}\,\Pi_I\,\Pi_T^{\,-x-3}$$
$$\Pi_Q = \Pi_T^{\,x+4}\,\Pi_I^{\,-1}\,\Pi_T^{\,-5}\,\Pi_E\,\Pi_S\,\Pi_T^{\,x+4}\,\Pi_I\,\Pi_T^{\,-x-4}$$
$$\Pi_Q = \Pi_T^{\,x+5}\,\Pi_I^{\,-1}\,\Pi_T^{\,-6}\,\Pi_F\,\Pi_S\,\Pi_T^{\,x+5}\,\Pi_I\,\Pi_T^{\,-x-5}$$

where x is an unknown exponent. The Π_A—Π_F permutations are determined using previously described methods. Then, for all equations above, we determine the Π_Q values for subsequent values of exponent x. For all x-values except one, the determined Π_Q permutations are different. The value x, at which the calculated Π_Q values are equal, determines the starting position of the first rotor (remember the Π_Q value, which will prove useful in further steps).

In practice, in order to improve the procedure, successive powers of all rotor permutations were written on the sheets of paper in which the windows were cut out (the method took its name from the sheets resembling a grate). The Π_A—Π_F permutations were printed on a second sheet, which was shifted under the grate respectively—in one of the positions the Π_Q permutations visible in all six windows were similar. The action of the plugboard resulted in the fact that the Π_Q values visible in all windows were not equal, however, usually enough similarities were found to identify the proper solution. Those letters which did not match indicated the plugboard settings.

A by-product of the grate method was the determination of the Π_Q permutation, representing the combined action of the reflector, second and third rotor. This permutation proved to be useful in identifying both rotors and determining their relative position. Codebreakers have prepared a catalogue of all possible values of this permutation, covering 6 * 26 * 26 = 4056 positions. After determining the plugboard settings, they searched the Π_Q value obtained in the catalogue.

After determining the order of the rotors and their starting positions, only one unknown remained—the positions of the rotor rings. The method for its determination was the least sophisticated. Rejewski and his colleagues knew well the stereotyped language of the German military traffic. Many messages began with the phrase **ANX** (**AN**—German "To", **X**—the replacement for space character). The codebreaker was choosing any message presumably starting with **ANX**, and then, pressing its first character on the machine's keyboard, he manipulated the fast rotor ring until he received the letter **A** as a result. He was able to verify the correctness of his conclusion typing the second and third letter of the message—if the result was **N** and **X**, the setting was correct. In the worst-case scenario he had to check 26 ring settings in this way, which was considered a practical solution at this stage.

As we could see the reconstruction of machine key was quite an arduous process. The codebreaker started his work by identifying the cipher's cyclic structure on a given day. On its basis he was able to determine the individual message keys and the permutations corresponding to the first six characters of the ciphertext. In the next stage he used the clock method to identify the fast rotor and could proceed to determine the settings of the plugboard using the grate method, additionally using the determined Π_Q permutation value to identify the second and third rotor. At the last stage he determined the position of the rings using the **ANX** method. In fact, not all activities had to be carried out every day. Initially the order of rotors changed every quarter. Ring settings—once a month. The base rotor positions were reconstructed using the characteristic cycles. As a result, the only daily operation was finding the plugboard settings using the grate method.

When Rejewski was busy solving his equations, Bertrand continued his meetings with "Asché". During 1932 he met with him three times: on 8 May again in Verviers, from 2 to 3 August in Berlin and from 19 to 20 October in Liège. The materials provided by the spy belonged to two categories. The first were the keys to the Enigma for the following months of the year. "Asché" had an easy access to the keys: their distribution was his responsibility at ChiStelle. Between 1931 and 1939, the representatives of the French intelligence service held a total of 19 meetings with their agent, selecting for this purpose places offering tourist or entertainment attractions, located in Belgium, Denmark, Switzerland, Czechoslovakia and France. The French, with a great deal of psychological intuition, have decided that combining the liaison missions with some entertainment will balance the stress. "Rex" was a regular participant of these meetings, being alternately accompanied by French intelligence officers dealing with the issues to be discussed. Among them Bertrand emphasized the importance of breaking German ciphers, Perruche—information of a military-political nature concerning the remilitarization of Germany. "Asché" was an expensive source, so his handlers were expecting measurable benefits in return. Bertrand could not demonstrate the results—the broken German messages. In this situation, control over the activities of "Asché" gradually passed into Perruche's hands and focused on the political issues. Bertrand was to benefit from the cooperation with his agent, although he was supposed to wait several years for the dividend to be paid. For the time being, he has travelled Europe frantically, exploring the possibilities and intentions of potential partners, trying to arrange local coalitions and offering assistance in the form of intelligence materials. Bertrand's immediate goal was to surround Germany with a chain of intercept stations capable of taking over a significant part of its radio traffic. He focused his efforts primarily on the two countries in the region: Poland and Czechoslovakia. Until the outbreak of the war, Bertrand visited Poland thirteen times and Langer visited France as many times. For the purpose of mutual contacts, they took pseudonyms taken from the partner's languages: Bertrand became "Bolek" and Langer was known as "Luc". Bertrand was also successful in Czechoslovakia, where he made contact with the head of military intelligence, František Moravec, known in the intelligence triangle trio as

"Raoul".[3] However, Bertrand's primary objective, structuring the trilateral French-Czechoslovak-Polish cooperation, was unattainable. Both Langer and Moravec were former Austrian officers, which facilitated agreement, but mutual mistrust of both countries resulting from the conflict of interests in Langer's native Cieszyn Silesia did not allow for the development of open cooperation. Bertrand's only achievement in this area was to bring about the establishment in 1938 of the emergency wireless network known as B.L.R.—from the first letters of the pseudonyms of the trio.

Bertrand's visits to the Baltic states did not bring measurable results. There was an intercept station in Latvia, organized with the help of French intelligence. However, the warm welcome was not followed by real cooperation; the hosts seemed to see a potential adversary in Soviet Russia rather than in Germany. The visits to Estonia and Lithuania had no effect: no country had any noteworthy signals intercept activities. During the meeting initiating the cooperation between Bertrand and Langer, the latter promised sharing with the French any results that his service might achieve using the materials received. However, months passed, and no decrypted German messages were coming from Poland. Bertrand could understand why—codebreaking services of his own country could not cope with the Enigma cipher either. However, he was surprised by Langer's noticeable indifference to the material transmitted from ChiStelle. The possible explanation of this puzzle was both disturbing and comforting: do the Poles no longer need keys provided by "Asché"?

Bertrand would have been surprised if he had known that only a tiny part of his deliveries had contributed to Rejewski's success. When the codebreaker started his attack, Polish intelligence had already gathered considerable information about the machine on its own. "Asché" provided Bertrand with the documents concerning the model of the machine called Enigma I, introduced on 12 May 1930. In his memoirs, Rejewski uses terminology that explicitly refers to the Enigma G, temporarily used by Reichswehr before the introduction of Enigma I: this is clearly indicated by the use of the term *Stöpselstellung* instead of *Steckerbrett* for the plugboard. While the manuals delivered by the French represented only a confirmation of the knowledge acquired by the Polish intelligence service itself, the keys to the cipher delivered in the following years could have saved the Poles a lot of work. However, none of them has ever reached codebreakers' desks. Langer and Ciężki decided to play poker locking the keys in their own safe. One can only guess the motives of their decision. They must have feared that news about their Bureau's success could take too wide a circulation, when shared with Bertrand. Their fear, that the source of the cipher keys might dry up in case of war, when the information from the broken messages becomes particularly valuable, might also play some role. Perhaps Ciężki demonstrated somewhat Jesuit attitude: having promised Bertrand to share any results

---

[3] "Raoul"—František Moravec, 1895–1966, at the time of the events in question, Colonel, later— Brigadier General of the Czechoslovak Army, from 1938 to 1945 head of Czech Military Intelligence (Zapovodajskeho Oddeleni). Bertrand was able to find his soul mate in Moravec - the Czech was as eager to cooperate with the Soviet services as he himself.

obtained with the help of France, he was determined not to make commitments he was not inclined to respect. However, the decisive role was certainly played by the desire to motivate their codebreakers to develop their own methods of key recovery, independent of the delivery of external data. Rejewski, Różycki and Zygalski have exceeded his expectations with surplus.

For the Cipher Bureau 1933 was the time of sowing. The breakthrough from the end of the previous year did not allow to read ciphertexts regularly enough and in a sufficiently short time to have an operational value for the intelligence service. At the beginning of 1934 came the harvest time. Until then the codebreakers have exploited their discoveries, gaining experience in recovery of the cipher keys. When their methods proved effective and AVA provided the first Enigma clones, it was possible to separate the research from the current exploitation. BS4 hired a group of trusted employees who, equipped with several AVA-made Enigma clones, took over the task of deciphering the messages, allowing the mathematicians to focus on developing new methods of breaking the ciphers. The codebreakers' participation in the daily functioning of the Cipher Bureau was henceforth limited to reconstructing the key for a given day and handing it over to the clerical staff. It was a labor-intensive activity only in the first days of the new quarter, when all elements of the new key had to be reconstructed. Of course, everyone at the Cipher Bureau was aware that the current use of the Enigma does not take advantage of all the possibilities offered by machine. The simplest way to increase the security of a cipher is to change its key more frequently. The Poles were worried that the currently used methods would not allow to read the Enigma in a reasonable time, if all elements of the new key—the rotor order, ring and plugboard settings and the starting position of the rotors—are changed daily. It was necessary to develop a more efficient way of reconstructing the key to the cipher. Unique cyclic structure of Enigma's cipher provided a good starting point. If it was possible to determine the cyclic structure for all possible combinations of machine rotors and their starting positions, the reconstruction of the key would only require determining the number and length of cycles on a given day and finding thus obtained structure in the catalogue. The road to the realization of his brilliant idea was bumpy: the determination of the cyclic structure for each rotor combination was laborious, and an attempt to develop a catalogue using hand methods would require years. Codebreakers realized thus that the logical adversary for a ciphering machine is another machine facilitating attack. This proposal led to the construction of the device, which became known as a cyclometer, and in time it became the beginning of another revolution in the codebreaking (Fig. 6.1).

The cyclometer consisted of two interconnected replicas of Enigma, with the fast rotor of the second machine shifted 3 positions forward from the first one. Such a reciprocal positioning of the rotors made it possible to recreate a situation which, with the normal use of the Enigma, would occur after enciphering the first copy of the message key. The cyclometer operator selected the order of the rotors and set them in the position, for which he intended to determine the cyclic structure. Then, using the switch next to one of the letters of the alphabet he was feeding the electric current into the battery of rotors. Current flowed through both units causing the panel lamps belonging to the same cycle to illuminate. The operator was noting

Fig. 6.1 Cyclometer (reconstruction). (codebreakers.eu)

their number and turning on the switch at any letter not belonging to the first cycle, again writing down the number of lamps lit. He repeated this operation until the sum of the cycle lengths was 26. After determining the cycles linking the first and the fourth letter of the key, the operator moved the rotors of both machines by one position and repeated the procedure for the second and the fifth and finally for the third and sixth letter of the key. The result of the test consisted of three numbers, defining the cyclic structure (a more detailed description of the construction and functioning of the cyclometer is presented in the box below). It took the codebreakers almost a year to develop a catalogue for all 105,456 rotor positions. From that moment on, it took only a dozen or so minutes to determine the Enigma's daily key. After intercepting the number of messages sufficient to determine the cyclic structure of the key (usually 80–100 messages), a card with matching structure was found in the catalogue and the starting position of the rotors was determined. The cyclic structure was not always unambiguous: sometimes several rotor positions resulted in the same cyclic structure. However, it took only a few moments to try out the received variants on one of the messages.

After the work on the catalogue was finished, the cyclometer was put on the shelf and the main codebreakers' tool became six boxes, each containing 17.576 catalogue cards. The hand-operated, electromechanical cyclometer was not an advanced design, but it deserves to be remembered—it was the first device to counter the ciphering machine. It became the ancestor of a family of cryptanalysis support machinery that began to appear in the arena of struggle, giving the codebreaking gradually a semi-industrial character. Their application marked the end of linguistic cryptanalysis, replaced with mathematics and engineering delivering more and more sophisticated devices supporting the process breaking the ciphers.

Simplified Cyclometer

A very simplified model of the Enigma, which not only operates in the six-letter alphabet, but also consists of only one rotor and a reflector, will help us to present the functioning of the cyclometer. As the machine plugboard does not affect the number and length of cycles, so we can completely ignore it. By connecting two machines to each other and adding a lamp in each circuit, we obtain a simplified model of the cyclometer, presented below (Fig. 6.2).

Then imagine breaking the cable at the letter F and inserting the battery into the circuit. The current will flow through the rotors and reflectors of both machines as shown below, illuminating the two lamps located at the letters F and C (Fig. 6.3).

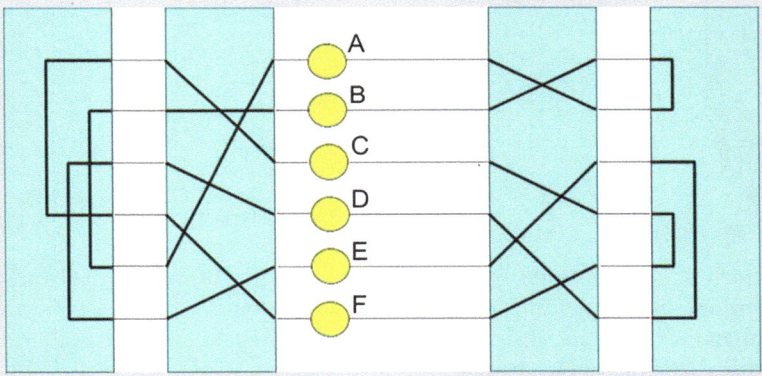

Fig. 6.2 Simplified, single rotor and 6-letter cyclometer

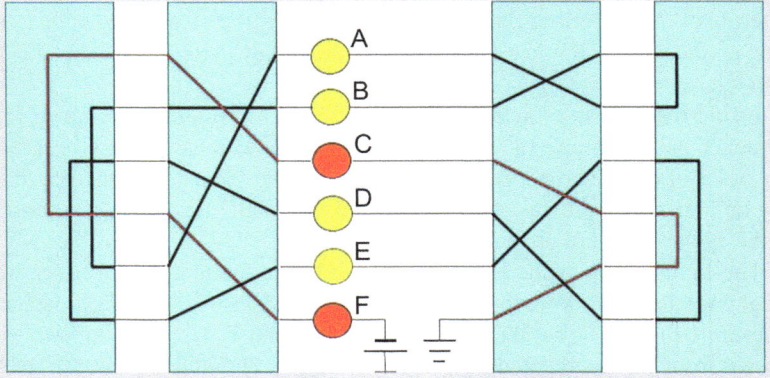

Fig. 6.3 Simplified cyclometer displaying 2-letter cycle

(continued)

This means that at the selected position of the machine's rotors, the letters F and C belong to cycle 2 letters long. By connecting the battery to the wires corresponding to other letters, which do not belong to the cycle identified above, we will obtain the remaining cycles for the given rotor position, referring to the first and fourth letters of the key. Then we advance the rotors of both machines by one position and repeat the process, determining the relationship between the second and the fifth, and finally between the third and sixth key letters in the same way.

In order to ensure that the catalogue can be arranged and the cyclic characteristics can be searched for efficiently, all possible characteristics have been marked with a unique number. For the 26 characters of the Enigma alphabet, 101 different cyclic characteristics are possible:

Cycle identifier	Cyclical structure
1	(13)(13)
2	(12)(12)(1)(1)
3	(11)(11)(2)(2)
4	(11)(11)(1)(1)(1)(1)
....	....
101	(1)(1)(1)(1)(1)(1)(1)(1)(1)(1)(1)(1)(1)

Each starting position of the Enigma rotors has been assigned a triplet describing the relationship between the first and fourth, second and fifth and third and sixth letters of the machine key. Each catalogue card took a form similar to the one shown below:

AAC-   (9, 4, 56),

where the **AAC** describes the starting position of rotors, 9 describes the structure of the cycles linking the first and fourth key letters, 4 – second and fifth, etc.

Inside the box, in which all 17,576 cards corresponding to a given rotor sequence per machine axis were placed, the cards were sorted e.g. in ascending order of characteristic numbers. After defining the cyclic structure of the key on a given day, it was sufficient to find the catalogue card corresponding to the cyclic structure just identified.

The catalogue of cyclic characteristics did not contain any data concerning plugboard settings, however, knowledge of rotor positions made it easier to determine them. The products of $\Pi_A\Pi_D$, $\Pi_B\Pi_E$ and $\Pi_C\Pi_F$ permutations were known from the intercepted cryptograms. After removing all wires from the plugboard, all letters of the alphabet should be pressed consecutively in the fixed position of the rotors, read from the characteristics catalogue. This allowed the permutation of the Π'_A to be determined. By moving the rotors

(continued)

three positions forward in a similar way, the Π'_D permutation could be recovered. The knowledge of both permutations allowed to calculate their product of $\Pi'_A\Pi'_D$. The $\Pi_A\Pi_D$ permutation is the transformation of the $\Pi'_A\Pi'_D$ by an unknown Π_S permutation corresponding to the connections of the plugboard. It is also known that the Π_S permutation has a special form, consisting of cycles 1 or 2 characters long. This allows to position the known products of $\Pi_A\Pi_D$ and $\Pi'_A\Pi'_D$ in all possible ways one under the other to determine the Π_S.

Both the cyclometer and the catalogue appeared at the right time. Until 1935 the structure of German communication networks and ciphering procedures were stable and predictable. Polish codebreakers noted that for some time during 1933 and 1934 German traffic included messages which could not be broken using usual methods. Broken messages from the same period referred to a more complex Enigma model (Enigma II) with eight rotors, designed for use in higher-level staffs. Before the codebreakers were able to address a new challenge, the experiment failed. The new machine proved to be unreliable; Polish listeners were amused when, instead of an enciphered message, the Germans were sending a clear text message *Chiffriermaschine kaputt*.[4]

In 1935, the first signs of German remilitarization appeared in the air. The Treaty of Versailles forbade the country to organize the air force. This restriction was being initially bypassed via civilian Lufthansa, the paramilitary Deutscher Luftverband and the organization of the construction, testing and training center in Russian Lipieck. During the first two years of his reign, Hitler stabilized his position within the country sufficiently to undertake the first gamble with the external world. On 26 February 1935 he gave Göring the order to organize the air force. Judging by the speed of new force's creation, its secretly laid out foundations had to be solid. Four years later the Luftwaffe was to become the most powerful air force in the world. For the time being, on August 1, 1935, Polish stations received the first messages broadcast within the new Luftwaffe communication network. Just a few months earlier, the emergence of a new network using the separate Enigma key, would be a problem for the Cipher Bureau. Now, with the catalogue of cyclic characteristics, effort of reconstructing the new key was reduced to a minimum. The new network promised a wealth of information. German land army traditionally preferred the wire communication—Luftwaffe depended on radio almost entirely. Even more so in its early organization stages, when its units were stationed in the improvised airfields. Polish codebreakers welcomed the advent of the new radio network also for another reason. A new service meant new people, learning new responsibilities and making mistakes in the course. Luftwaffe's operators did not disappoint the hopes of the Cipher Bureau.

However, not all the changes in the German cipher systems were received by the Poles indifferently. Starting from February 1936, the Cipher Bureau observed a

---

[4] The encryption machine is damaged.

number of changes in German procedures with real concern. Introduced modifica-
tions seemed to be addressing the codebreakers' achievements precisely, as if the
Germans knew about the Bureau's successes and tried to put an end to them.
Rejewski and his colleagues were drafting some conspiracy theories about the possible causes of this race. A story circulated, probably without any ground, that the
Polish delegate in the League of Nations used an argument during the discussion on
German armaments, that could only have come from the broken messages. Our
present knowledge allows to assess changes in Enigma's operational procedures as
a result of natural evolution of the cryptographic system. Maybe the Cipher Bureau's
commanding officers decided that the morale and productivity of the team would be
boosted if the conspiracy theories are not denied too vigorously.

Meanwhile, since 1 February, 1936, the Enigma operators started to change the
rotor order every month, instead every quarter as before. On 1 October of the same
year, further changes took place: the rotor order was changed daily, and in the plugboard, instead of the previous six pairs, from five to eight pairs of letters were connected. Modifications did not make an impression on the Poles. The catalogue of
characteristics provided comfort and effectiveness. However, in the second half of
1937, deciphered messages gave real cause for concern; they signaled that Germany
was planning to introduce a new reflector. The reflector itself was not a problem—
the method used in the past to reconstruct its wiring remained valid. The use of the
new reflector, however, meant that the catalogue of cyclic characteristics, which
required year's effort to complete, would become obsolete. On 2 November the
blow fell—the identified cycles were not found in the catalogue. The wiring of the
new reflector (marked by the Enigma manufacturer as Bruno, Umkehrwalze B or
UKB) was worked out within a few days, but the laborious creation of the catalogue
had to start again. The current decryption could not suffer, so the work on the catalogue was carried out in parallel to the key breaking with the old, laborious and
time-consuming methods.

Meanwhile, in September 1937, codebreakers were faced with a new challenge—the Sicherheitsdienst (SD) communication network. The SD operators were
better trained than their military colleagues. Luckily, the German cryptology suffered from the lack of coordination. Should the new network start its operation after
the new reflector has been introduced, its keys would represent certain problem for
the Cipher Bureau. However, the SD network appeared two months before the catalogue became useless. On the first day of its operation, the Poles identified the rotor
settings on the basis of the cyclic structure, but they stumbled in the very next step
when they tried apply the **ANX** method. None of the intercepted messages started in
the way standard for other networks. In the next step, codebreakers tried another
crib, common for the majority of the messages decrypted so far—the word
EINS. They succeeded after a few attempts; they found a ring setting that allowed
them to decipher a fragment of the cipher as **EINS**. However, the success was partial; apart from the four initial letters, the rest of the message was illegible.
Codebreakers have guessed the cause of the problem: the SD protected its messages
with a double veil. Each clear text was first encoded in a four-letter code and then
encrypted with Enigma. Having a hint, they managed to decipher a number of messages and identify a number of code groups. Breaking SD code and cipher was to
bring a fair dividend in the future.

After the end of 1937, the German section of the Cipher Bureau moved to its new location. At the time of Rejewski's first success the room known as the "Room under the Clock" at the army HQ building, whose door was covered with a curtain, provided sufficient security. Since then, the BS4 team has grown to a dozen or so people, and the operation took an almost industrial character; it was increasingly more difficult to ensure its security in the bustling Saxon Palace. A decision was made to move the staff and equipment to the premises of the newly built secret wireless center in Pyry near Warsaw. The complex in the Kabaty Forest was not designed for the codebreakers. Before 1939, Polish defense policy assumed that the Soviet Union represented the greatest potential threat to the country's security. Any information from the east was valuable for Polish intelligence service, but the drastic security measures taken by the Soviets forced Poles to carry out any intelligence work mostly from neighboring countries. The Poles have surrounded Russia with a ring of intelligence stations located in the Balkans, Turkey, Iran, Afghanistan, China and Manchuria. Pyry complex was designed as a wireless center connecting the HQ of the Polish intelligence service with its stations and agents abroad. Unintentionally, it also played the role of the show window for AVA's technology. Founded in 1928, the company grew rapidly, meeting a considerable part of the Polish armed forces' demand for specialist radio equipment. In the mid-1930s Ludomir Danilewicz undertook a series of experiments on quartz oscillators, which he used in measurement receivers, identifying the transmitter on the basis of precise measurement of its frequency. In 1934 the company moved to its new headquarters at 25 Stępińska Street, designed by the owner's sister, Aldona Danilewicz. At that time AVA employed around 100 people, four years later their number rose to 250 and was to exceed 300 just before the outbreak of war. At the same time Tadeusz Heftman, a talented radio amateur and Danilewicz's friend from the Sosnowiec gymnasium, was appointed its technical director.

Company management methods were ahead of their time. Appreciating the importance of good work organization, its bosses decided to employ Stanisław Guzicki, who at the Warsaw University of Technology cooperated with Prof. Karol Adamiecki, the forerunner of the scientific work organization in Poland. Guzicki did not hide his pro-communist sympathies, but both, AVA and its military supervisors, took a pragmatic approach. In the vetting process counterespionage service quickly established that his beliefs were not accompanied by activities against the state. However, the case handler informally instructed the candidate to subscribe to one of the right-wing newspapers, which would provide both of them with a bureaucratic alibi. Presumably as a result of Guzicki's recommendations AVA employees were offered a decent standard health care, and their families could take advantage of summer holidays at the company's expense. There was also a custom of celebrating holidays together. AVA managed to win the loyalty of its employees assuring that its secrets remained secure when some of the crew fell into German hands during the war.

An unusual profile of operation was sometimes a source of exceptional adventures. Antoni Palluth experienced one of them when during the tests of the wireless station constructed for one of the Polish submarines the ship was sucked to the seabed and could not break free. Palluth occasionally performed missions exceeding by far the scope of the business management. His son Jerzy remembered years later how his father used to pack a Browning into his suitcase, preparing for a long road through several intermediate countries to a secret meeting with a Polish agent somewhere in Europe. Fortunately, his son did not know that the weapon was accompanied by superiors' recommendation to make proper use of it in case of a compromise. Labels on toys brought as a gift for his sons represented a particular chronicle of these journeys. The lifestyle combining the standard responsibilities of a company's manager and secret intelligence operations, must have been very stressful. According to the family tradition about a year before the outbreak of the war Palluth planned to give up his secret duties and focus on running the company. However, his supervisors reacted quickly responding that there are no ex-secret agents. General design of the communication center in Pyry and a significant part of its radio equipment represented the result of work by the engineers and employees of this strange, military-civilian company. Radio receivers and transmitters were safely hidden in the underground shelters, telescopic antennas extending above the ground only for the duration of the communication session, usually at night, buildings hidden from air observation in the forest and protected by barbed wire fence. During the World War Two German signals officers were brought to Pyry to study a well-organized communication center. For the time being, however, codebreakers from the German section of the Cipher Bureau found a safe place, while the Soviet section moved to an inconspicuous building at Jaworzyńska Street in Warsaw. Soon after the relocation the codebreakers faced a new challenge, this time prepared by their own superiors. At the beginning of 1938, the head of military intelligence service, Colonel Stefan Mayer, decided to evaluate the effectiveness of the signals intelligence service and ordered an exercise lasting for two weeks. During the exercise, BS4 was able to break about 75% of the intercepted German messages in just few hours. In his post-war memoirs Rejewski was a bit grumpy that with a few extra posts it was possible to achieve even higher efficiency, he admitted, however, that three quarters of broken messages represented the level that no other codebreaking service in the world could probably get close to.

The first half of 1938 was undoubtedly the most productive and happy period in the work of the team. Their achievements gained the recognition of their superiors, expressed both in the form of promotions and decorations and in the material form. The chief of the General Staff appreciated the successes of the Cipher Bureau awarding the key codebreakers and officers cash prizes. The unexpected income helped to stabilize family life, build new houses or buy appartments in good quarters, look with optimism at the future of children planned or born. Langer, promoted to the rank of Lieutenant Colonel, built a magnificent villa at Miłobędzka Street in Warsaw, where his only daughter, Hanna, was born soon. One of the rooms was entirely dedicated to an impressive library of books covering cryptology and adjacent fields; engineering, linguistics and mathematics. Maksymilian Ciężki, who in the meantime had been commissioned as Major, bought from his older brother Władysław a large plot of land in Piątkowo near Poznań, built there a spacious villa, and started arranging the surrounding garden. His three sons, Zdzisław, Zbigniew and Henryk, were born respectively in 1923, 1926 and 1929, so new estate in Piątkowo could become the center of the family life. Initially, it played this role only during the summer holidays, as duties kept Ciężki mostly in Warsaw. Rejewski, who in 1934 married the daughter of the Bydgoszcz dentists, Irena Maria Lewandowska, whom he met for the first time as a child. Chief of Staff's award allowed him to move in 1935 to his own apartment at 8 Mickiewicza Street, in the then fashionable district of Warsaw—Żoliborz. Just in time, because in June 1936, Rejewski's first son, Andrzej Zygmunt, was born. In the autumn of 1938, there was a second child on the way (daughter, Janina Maria, born in February 1939).

Różycki continued his studies at the University of Poznań, obtaining his second degree in 1937, this time in geography. His professional and scientific achievements were followed by family stabilization; in 1938 he married Maria Barbara Mayka, whom he is said to have met for the first time a dozen or so months earlier playing basketball. He used the money from the prize to purchase premises at Wilson Square for the family pharmacy. Henryk Zygalski also decided to build a new house in his hometown of Poznań, at Matejki Street. So far, he did not really know when and in whose company, he would be able to live in it; until now he has not found a life partner, and working in BS did not allow for frequent trips to Poznań. The apparent wealth of the Cipher Bureau's employees was sometimes troublesome considering the need to keep its cause confidential. The Różycki family passed on the memory of Jerzy's conversation with his mother, who, in view of the mystery surrounding her son's work, asked him with motherly concern whether his money was not a fruit of the crime? Jerzy confirmed, remarking however that he steals the information only. He could not say more at the moment, but one day his mother shall be proud of him. Unfortunately, codebreakers' improving material situation was visible also for their environment causing envy, and the mystery surrounding Bureau's work and successes did not allow to disperse the dense atmosphere.

However, the team had no time to worry about the reception of their work. In May 1937 BS4 recorded changes in the system used by the German Navy. Until now, it was using the same ciphering procedures as the army. In May, instead of the well-known repeated key, characters appearing in the message headers stopped to form a meaningful cyclic structure. The new system was broken after just one week, partly due to a mistake by German operators, partly due to experience of Polish codebreakers. On May 8th, a torpedo boat using the callsign **AFA** sent a signal indicating that it had not received instructions for using the new cipher. It was ordered to temporarily encipher its messages using the old system, which the Poles easily broke. This gave BS4 information about the machine settings, but it still couldn't reconstruct the key from that day. Looking at traffic of the same day Rejewski noticed three messages, whose source and order indicated three parts of a multi-part message. This gave him a crib concerning the beginning of the second and third parts of the message. German cipher clerks were ordered to mark the continuation of the earlier message with a marker **FORT** (from the German *Fortsetzung*) and the repeated hour of the transmission of the previous part. The first part of the message was sent at 23.30, so the second message should start with letters **FORTYWEEPYWEEPY**.[5] It turned out that all three messages were enciphered with

---

[5] Enigma keyboard did not include the digits. They were substituted by letters: 1 – Q, 2 – W, 3 – E,…, 0 – P and enclosed with Y, so 2330 was equivalent to YWEEPY.

the same machine setting, recorded on April 30. Rejewski's experience allowed him to break about 100 messages intercepted between May 1 and 8. The crib that permitted to attack Enigma cipher using the multi-part messages, gave the method its name—the attack was to be known as "Fortyweepy". Having broken some 15 messages a day during a period of a week Polish team was able to reconstruct the outline of the new Kriegsmarine's ciphering procedure. Most important feature of the new system was enciphering the individual message key using special bigram tables instead of Enigma. This novelty represented a clever move by the Kriegsmarine cryptographers. With Wehrmacht or Luftwaffe Enigma, it was enough to reconstruct the daily machine settings to read all the messages during that day. New Kriegsmarine procedure was different: even the reconstruction of the daily machine settings did not permit to decipher the message itself, as its individual key was enciphered using the unknown bigram tables, completely unrelated to the Enigma cipher. Without the bigram tables every message key had to be attacked individually and laboriously, which made the attack impractical.

However, messages broken in May 1937 permitted the Polish team to distinguish three cipher variants: ordinary (*Marineschlüssel*), officer (*Offizierschlüssel*) and staff (*Admiralschlüssel*). For some time BS4 has read most of the regular Kriegsmarine traffic and some of the messages in Offizier variant. More or less complete reconstruction of the bigram tables required breaking a considerable number of individual keys. According to Rejewski's memories, the volume of Kriegsmarine messages being intercepted by Polish stations was insufficient, and Polish interest in the war at sea too limited to dedicate resources to the new challenge.

All the more so considering that the cryptographers of other German services were also breaking out of the trenches of the routine. At peak of the Sudetenland crisis, on September 15, 1938, both Wehrmacht and Luftwaffe have introduced a new ciphering procedure, that made almost all Polish practical methods of breaking the Enigma ciphers obsolete. Until now the cipher clerk was supposed to choose an individual message key for each message sent, but all the individual message keys were enciphered using the common Enigma settings. In his later account Rejewski noted that enciphering the message key using the common daily machine settings represented a fundamental German mistake, as it allowed not only to break the key but also helped to reengineer the machine's internals. According to Rejewski it would be more secure to transmit the key to the recipient as an open text. The Germans, a little too late to avoid damage, finally followed Rejewski's advice. The cyclic structure provided the basis for most of the attack methods used by Polish team. Its reconstruction required on the average interception of some 80–100 messages enciphered at the same machine settings. In the new system each message was enciphered using individual machine setting—collecting a sufficient number of messages and determining the cyclic structure of the cipher became impossible. The German innovation was so clearly aimed at the catalogue of characteristics that concerns among Polish codebreakers about the security of their secrets significantly increased. For BS4 team the next four months were the period of the most hectic activity in the entire history of the section.

New Procedure for Encrypting the Message Key
In the old system, the cipher clerk sets the enigma according to the network key, chooses an individual message key, e.g. **PGJ**, and then enciphers it twice, receiving (open text in the upper line, cryptogram in the lower line):

PGJPGJ

ARWYXF

In the new system, the cipher clerk selects two random trigrams, for example **RXL** and **PGJ**. He then sets up the Enigma according to the network key, which at present did not include the starting position of the rotors. The rotors are then set in positions described by the first three characters selected, i.e. **RXL**. Starting from this position, the operator enciphers the second trigram twice, receiving for instance a string **VFHTXZ**. Then he writes down the header of the message, which in the new system is nine characters long. It contains the first trigram (in the example **RXL**) in clear text, followed by six characters of an enciphered message key (**VFHTXZ**). The message header takes thus shape:

RXLVFHTXZ

In the old system, all the message headers were encrypted with the same key, which allowed them to be compared and collated. Assuming that the operator did not repeat selected three characters, in the new system each message header was encrypted with a different key, which did not allow for comparison.

So far, Polish attacks have exploited a weakness involving the repetition of a message key, encrypted with the same machine key. The German cryptologists demonstrated some inconsistency. They eliminated the encryption of the individual message key with the same machine key, but retained the repetition of the message key. Rejewski and his colleagues immediately noticed that even within the limited space of six characters some cycles can still be identified. Cycles of a very specific nature, only one character long, transforming a key letter into the same letter when enciphered for the first and second time. In the new system, each message started with three trigrams, like **RXH HDR SDA**, where the **RXH** represented the starting position of Enigma rotors given in clear text, and the **HDR** and **SDA** represented a repeated message key enciphered starting from this position. In both copies of the enciphered message key letter **D** appears twice within the distance of three letters. Situation, when certain, unknown letter of the message key gets transformed into the same letter in both copies of its enciphered version was described by codebreakers as a female (or as a 1-cycle, as female is nothing else than a cycle one character long).[6] Females were frequent in the Enigma cipher, some 40% of rotor positions generated one. One cannot hope to break the cipher finding just one or two females. However, the codebreakers suspected that they form unique patterns. If one could find a way to compare the patterns defined for several rotor starting positions, females promised a way leading to the break-in. Henryk Zygalski proposed to compile the catalogue of females in form of paper sheets, where every female was registered as the hole in the sheet. In the new Enigma procedure, the starting position of the rotors was transmitted in open text, so the codebreakers knew which sheets to select for comparison. By superimposing several or a dozen sheets selected in this way, suitably offset, it was possible to find a position where the patterns of the females in all the sheets corresponded to each other. Implementation of Zygalski's proposal required huge effort—identification and punching of over 40.000 females, and any change in Enigma's internals could obsolete the resulting catalogue. Memory of a year of work spent on the preparation of the catalogue of cyclic characteristics, rendered obsolete by the simple change of Enigma's reflector, was discouraging. There should be a better way to defeat Enigma.

In less than a month of hectic work the team worked out the idea of a device which went into history under the name of Rejewski's bombe. It took the AVA engineers another month to transform this idea into the engineering design, to manufacture and deliver six copies of a new device. To keep the secret, the individual components were manufactured by different AVA divisions and external companies and finally delivered to the center in Pyry. There, they were assembled in a room accessible only to BS commanders and three codebreakers, as well as selected AVA employees, including Palluth and AVA's trusted technician, Czesław Betlewski. In November 1938 the bombes proved their efficiency—Cipher Bureau was back in the business.

Each bombe represented an aggregate of six Enigmas connected in pairs, whose rotors were driven synchronously by an electric motor. In each pair, the fast rotor of the second machine was shifted three positions in relation to the first Enigma, which reflected the setup during message key encoding. Bombe was running through all possible rotor positions searching for so called "spectacles". Rejewski described as "spectacles" the situation, when an unknown letter of the message key was enciphered as the same letter in both copies of the key, respectively in the first and fourth, second and fifth and third and sixth positions. Once the codebreakers were able to identify among the intercepted messages at least three "spectacles", the operator could start setting the bomb. For that purpose, he was setting its rotors in the positions corresponding to the machine key given in clear text in the respective message headers (in the example below—**RTJ**, **DQX** and **HPL**). Then, at the front panel of the device, he selected the letter representing the "spectacles" (in the example—letter **W**) and started the device. Once started, the bomb was running through all the 17.576 possible rotor position within less than two hours. The signals from Enigma pairs were compared with each other. When the control circuits detected the selected letter appearing at the outputs of all three machine pairs simultaneously, it caused the bomb to stop. As a rule, the bomb stopped several times during the single run; its operator was recording the rotor positions at each stop and restarting the

---

[6] In Polish language "female" is pronounced as "samiczka", from transforming any letter into the <u>same</u> letter.

Fig. 6.4 Rejewski Bomb
(reconstruction). (Enigma
Cipher Center, photo
Łukasz Gdak)

device. After the bomb run was completed, all recorded stops had to be checked—
one of them could represent the actual solution. The bomb examined all possible
rotor positions in particular rotor order. The three rotors could be arranged in six
ways; changing the order of the rotors in the bombs and restarting the device would
extend the time of the key search to over twelve hours. The problem was solved by
ordering six bombs, each with rotors set in a different order (Fig. 6.4).

Start position of the rotors (open text)	Double-encrypted message key
RTJ	WAH WIK
DQX	TWO MWR
HPL	RAW KTW

Cyclometer was the harbinger of the new era in the codebreaking, era of the
automated cryptanalysis. The bomb crossed the t's and dotted the i's. It represented
a prototype for entire new family of the codebreaking equipment, allowing for a
quick and fully automatic search of the machine's key space. In order to identify a
possible solution, the bomb exploited a logical relationship between the elements of

the message header. It was a pioneering device to such an extent that the codebreakers experienced difficulties in choosing a meaningful name for their design. Rejewski recalls in his memoirs that it was called a bombe[7] "for lack of a better idea". Tradition has kept alternative versions of its name's origin. One of them goes that Czesław Betlewski coined the name from the ticking of a mechanism resembling the sound of a time bomb mechanism. However, we know that among the AVA employees the device was known rather as a "mangle", probably due to the similarity of the mechanism powering the axes of six Enigma replicas to the mangle rollers. According to the most popular version, Różycki became its godfather when the codebreakers came up with the idea during the discussion over a dessert—an ice bomb. Even if it is not possible to ascertain its true origin, the bombe exerted on the future of cryptanalysis an influence adequate to its name.

The very idea of the bombe was improvised rapidly, and that resulted in important limitations of its usage. Its most important limitation was the assumption that the letter being searched for was not changed by the plugboard. Despite the outward resemblance "spectacles" were very different from the "females" known from Zygalski sheets. "Females" represented cipher's purely cyclic property and were thus unaffected by the plugboard settings. "Spectacles" were not based on the permutation theory; they represented a purely statistical property of the Enigma cipher and as such they were vulnerable to the plugboard settings. At the time the bombe was designed, this was not a significant limitation: Enigma keys routinely used five or six out of thirteen possible letter pairs, which offered good chances of success. The second limitation was the assumption that during the ciphering of the message key the middle rotor remained stationary.

Principle of Operation of the Rejewski Bombe
Let's assume that among the messages intercepted during the day there were three cryptograms with the following headers:

Machine key	Double-encrypted wiring key
RTJ	WAH WIK
DXQ	TWO MWR
HPL	RAW KTW

By initial position we mean the position of the machine's rotors when encrypting the first letter of the first copy of the message key. Let us mark the initial position of the first, 'fast' machine rotor when encrypting the first telegram by P_{11}, the second and third P_{12} and P_{13} respectively (analogously P_{21}—the initial position of the middle and P_{31}—the last rotor of the machine when encrypting the first message). The position that the *n-th* rotor will take after encrypting the *x* letters of the *m-th* message is marked as P_{nm+x}.

(continued)

---

[7] In the original, the device was referred to by the Polish word 'bomba', i.e. 'bomb'. However, it found its way into cryptologic jargon via a document in German, adopting its German wording also in the English.

From the message headers quoted above we know that the rotors in positions P_{11} and $P_{11}+3$ give the same letter **W**. The same relationship exists for (P_{12+1}) and (P_{12+4}) and also for (P_{13+2}) and (P_{13+5}).

Without knowing the order of the rotors on the machine axis and the ring settings on the rotors, the three-letter machine key, sent in an open text message, is not sufficient to reconstruct the message key. However, by sending the machine key in open text, the opponent gives us additional information about the relationship between the position of the rotors when encrypting all three messages. We don't know which rotor takes the fast position, but we do know that all the message headers intercepted on the same day had been enciphered with the same rotor in the fast position. Knowledge of the machine key allows to reconstruct the relative, mutual position of the rotors for the first, second and third message. The positions of the first rotor in the first (**J**) and second (**Q**) message are 7 letters apart. Similarly, the position of the first rotor in the second (**Q**) and third (**L**) encryption is 21 characters apart. We can write that

$$P_{12} = P_{11+7}$$
$$P_{13} = P_{12+21}$$

By substituting P_{12} and P_{13} values we can state that the letter **W** appears in the encrypted message key twice at the following rotor positions:

$$P_{11} \quad \text{and} \quad P_{11+3}$$
$$P_{11+8} \quad \text{and} \quad P_{11+11}$$
$$P_{11+4} \quad \text{and} \quad P_{11+7}$$

Translating the observations formulated above into bombe settings, we set the rotors in three pairs of machines as follows:

- The fast rotor of the first Enigma in the first pair is set in any starting position,
- The fast rotor of the first Enigma in the second pair is set to a position shifted by 8 characters from the first Enigma in the first pair, while the fast rotor of the first Enigma in the third pair is set to a position offset by 4 characters from the first Enigma in the first pair
- All fast rotors of the second Enigmas are set in a position advanced by 3 characters in relation to the fast rotor in the first machine of that pair.
- The central and slow rotor in all pairs is set based on the distance between the respective key letters. For example - the difference between the settings of the central rotor for the first and second message is 4 characters (distance between **X** and **T**). The central rotors in both machines of the second

(continued)

pair are set in a position offset by 4 marks in relation to the central rotors of the first pair. Similarly, we determine the mutual offset of the remaining rotor pairs.

After all the rotors have been set as described above, we can start the machine and note the positions of the rotors for which the bombe has stopped (i.e. detected the letter **W** at the output of all three pairs of machines simultaneously). The stop positions (there could be more than one) mark the probable machine key. The correct position is determined by deciphering a fragment of the message at each stop.

At this point, the codebreaker knows the order and position of the rotors, but the ring and switch settings are missing to know the full machine key. Determining the position of the rings is trivial. The positions of all three rotors and the letter of the open text machine key are known. It is enough to keep the rotors in their fixed position, rotate the rings until **RTJ** (or **DXQ** or **HPL**) characters appear in the Enigma windows.

Interestingly, in his post-war memories Rejewski treated his idea surprisingly superficially. He was describing in the detail the mathematical background of the other attacks on Enigma cipher, pretending to have forgotten the details of the bombe. One could almost feel the author's distance to his own idea. That attitude could result from several sources. The bombe was a spectacular but costly and short-lived success. It saved the day for the BS4 when other methods of attack became obsolete in September 1938, but after just few weeks of enthusiasm it turned, at least apparently, into a costly failure, when another change in the way the Enigma was used reduced the effectiveness of the bombes. But this failure does not seem to be the only, or even the main reason for Rejewski's reserve. In spite of his several years long practice as a codebreaker in his heart Rejewski was still a mathematician. From this perspective the need to reach for help of the mechanical device, looking for the solution somewhat in blind, must have been a disappointment. Rejewski clearly perceived the bombe as the failure of his beloved mathematics and therefore was distancing himself from his own idea.

When the BS4 team was still celebrating a quick recovery of control over Enigma ciphers, the atmosphere was disturbed by the report of a Polish agent in Germany. He delivered the information that each Enigma is equipped with not three, but five different rotors and the Germans plan to start using the remaining two soon. The blow fell on December 15, 1938. The number of rotors used in the Enigma has not changed, but the cipher key was now created by selecting three out of the five available rotors. The Poles quickly solved the problem of the unknown wiring of the new rotors, with a little help of the adversary. When a new ciphering procedure was introduced by the Wehrmacht and Luftwaffe on 15th September, the SD network continued to use the old version. Three months later new rotors were applied also in the SD network, still using the old ciphering procedure. That mistake permitted the Cipher Bureau to recover the wiring of the new rotors using old and tested methods. Recovery of the new rotors was only a partial consolation in the new reality. The total cost of six Rejewski's bombes got close to the annual budget of entire Cipher Bureau. The introduction of two new rotors meant that the rapid reconstruction of the key required the use of not six, as before, but sixty bombes. Ordering sixty bombes necessary in the new situation was far beyond the capacity of Polish intelligence service. All the more so because another change in the key structure, introduced on 1 January, 1939, has practically defeated the bombes. From that day on, Enigma plugboard settings included 8 to 10 connections. One of the conditions of bombe's effectiveness, invariability of the letter representing "spectacles" by the plugboard—was increasingly unlikely to be met.

Fortunately, when Rejewski was busy working on the concept of the bombe, and AVA was manufacturing the machinery basing of his ideas, Zygalski simultaneously was developing his concept of the catalogue of "females". The mid-September change of the ciphering procedure made useless not only the old catalogue of cyclic characteristics, but in fact nearly all existing methods of breaking the Enigma based on the cipher's cyclic structure. In the new scenario every message header was enciphered at the individual rotor starting position; the codebreakers could not accumulate cipher material permitting recovery of its cyclic characteristic. The old cycles, with a length of up to 13 letters, simply did not fit into frames of six letters long message headers. Females appeared in the headers in large numbers, they even seemed too numerous to be useful in practice. The detection of one female did not provide information about the machine settings. The essence of Zygalski's idea was to develop a catalogue that would make it possible to compare females observed in many messages. The procedure developed by Zygalski required assuming certain order of rotors, and then comparing patterns corresponding to a number of females by overlaying sheets from the set representing selected rotor order. If the assumption was correct, after stacking several sheets (usually ten to twelve), there was only one hole showing through. Absence of the hole through marked the false assumption, the stacking process had to be repeated using the set corresponding to another rotor order.

The sheets named after their inventor had one crucial advantage and several disadvantages. Females represented cipher's cyclic property and were thus unaffected by the plugboard settings—feature crucial after the changes introduced on 1 January. Preparation of the Zygalski sheets required a lot of effort and time. For every distinct rotor order a set of 26 sheets needed to be prepared, each sheet including 51 * 51 fields. On the average females represented 40% of fields in every sheet, so a single sheet required determining the position and perforating about a thousand holes. This gigantic work was done by three codebreakers by hand, cutting holes in millimeter paper sheets with a razor blade. It is difficult to assess whether the waste of their time and effort was due to the will to keep the secret or the financial crisis of the Cipher Bureau, which was forced to spend about 100.000 zlotys on the construction of bombes.

Zygalski's Sheets

Zygalski's sheets are a form of a catalogue of cyclic characteristics, adapted to the way the Enigma was used after 15 September, 1938. In the previously used catalogue, the basic problem was the reconstruction of the rotor's starting position. In the new system, the initial position of the rotors was given in the open text in the message header—its knowledge was useless without information on the order of the rotors on the machine axis and the setting of the rings. The purpose of using Zygalski's sheets was to reconstruct the order of the rotors and the rings settings.

In the previous system, a significant number of messages encrypted at a common starting position of the rotors allowed to reconstruct the full cyclic characteristics of the code. After 15 September, codebreakers had only one message encrypted at a specific position of the rotors and could only analyze the cyclic structure of the cipher within 6 characters of the encrypted message key. The 6 characters did not contain cycles of considerable length, so they focused their attention on cycles one character long. Let's assume that during the day we intercepted the following message headers:

```
KAI     RWKRGV
SEL     GSNWSX
PBA     MSDRAD
```

In these headers, the repeated letters of the encrypted key, R, S and D respectively, were highlighted. Their occurrence means that the unknown letter x of the individual message key has been encrypted twice three characters apart as $R(S, D)$. In other words, the permutation performed at a specific starting position of the rotors contains a cycle of length 1, transforming an unknown key letter twice into the same character. Cryptologists described colloquially such a situation as a *female*. As we know, the length and number of cycles in the permutation performed by the machine depend only on the initial position of the rotors—the settings of the plugboard do not affect them. By analogy to the old catalogue of cyclic characteristics, you can imagine a catalogue of all rotor positions which in effect generate a *female*. However, there are relatively many such positions: about 40% of rotor positions generate a *female*. A single *female* does not identify the starting position of the rotors as effectively as the cyclic characteristics from the old catalogue. However, if a number of *females* can be compared with each other, assuming a certain order of the rotors, this assumption can be confirmed or rejected.

For this purpose, codebreakers have prepared sets of sheets, on which the positions of the rotors in which a *female* appears are marked by a hole. A separate set of sheets was prepared for each possible rotor order. One set included 26 sheets: a separate sheet had to be prepared for each of the 26 positions of the third (slow) rotor. Fortunately, there was no need to make separate

(continued)

permutations for the first and fourth, second and fifth, and third and sixth letters of the key: it was sufficient to shift the sheets developed for the first and
fourth letters of the key accordingly. For each possible rotor position the sheet
included 26 positions of the central rotor (marked on the horizontal axis) and
the fast rotor (vertical axis). Due to the need to shift one sheet relative to the
other in the process of sheet stacking, sheets were 51 × 51 characters wide
(with some rows and columns repeated). The following is an example sheet,
corresponding to the rotor sequence VI-I-III, the fast rotor position corresponding to letter K (the actual sheet is a fourfold repetition of the sheet in the
illustration) (Fig. 6.5).

Before starting the sheet stacking process, preparatory steps must be taken.
The first one is to select from among the intercepted messages as many as
possible containing *females*. Then the assumptions concerning the order of
the rotors on the machine axis and the position of the left rotor ring should be
made, which are subject to verification in the further procedure. For each of
the 6 possible orders, 26 hypotheses for the left-hand ring alignment of the
machine must be checked, which in the worst case means the verification of
156 hypotheses.

The assumption concerning the rotor order allows to eliminate from among
the selected messages those in which the middle rotor has shifted within the
message key. For this purpose, let us mark the starting position of the *i-th*

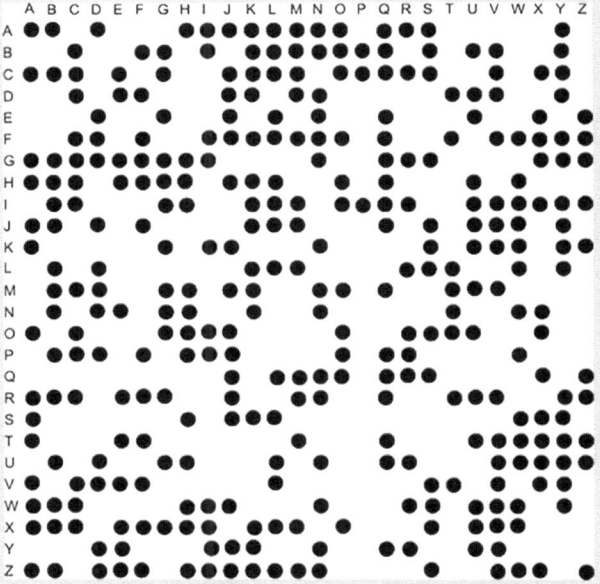

Fig. 6.5 Sample Zygalski sheet. Symbol ● represents a hole in the sheet, or the
'female'

(continued)

rotor by s_i, the position in which the *i-th* rotor advances by z_i, and the position of the *i-th* rotor ring by p_i. The assumption concerning the order of the rotors implies the value of z_1, and the key transmitted in open text—s_1. Let us calculate the value

$$m_1 = z_1 \quad s_1 \quad 1 \, (\text{mod } 26),$$

and then, from among the selected keys, remove those for which the condition occurs

$$ent \left[(j - m_1 + 26) / 26 \right] \neq 0 \, (\text{for } j = 0, \, \dots, \, 5)$$

The remaining keys are divided into three categories, corresponding to *females in the* first and fourth, second and fifth and third and sixth key positions. Let's assume that as a result of this procedure we received the following group of message headers:

```
RZV IWAILX
OSK ASBAGZ
PNY HTOHJM
UGN KGRKTB
```

From among the sheets corresponding to the assumed order of the rotors, select the sheet corresponding to the hypothesis concerning the position of the left rotor ring (mark the assumed position of the ring as H and assume in the first step that $H = \mathbf{A}$). We also choose the first of the keys containing *the female* in the first and third position. The selected sheet corresponds to the setting of the ring expressed by the equation

$$p_3 = \mathbf{R} - \mathbf{R} = \mathbf{A},$$

in which the $\mathbf{R}$ value comes from the rotor setting given by the clear text in the message header. At the same time, the numbers of rows w and columns k in the selected sheet correspond to the positions of the rings

$$p_1 = \mathbf{Z} - w,$$
$$p_2 = \mathbf{V} - k,$$

where values $\mathbf{Z}$ and $\mathbf{V}$ represent the settings of the first and second rotor from the key sent in the message header. Then, we choose the next headers starting with a letter we have not yet encountered. In the example above, the first letter

(continued)

that meets this condition is O. The selected header corresponds to the sheet with the designation

$$H + R - O = A + R - O = D$$

From the sheets corresponding to the assumed order of the rotors, select the sheet corresponding to letter D and place it on the one chosen in the first step with an offset in rows w_p and columns k_p equal to the difference between the settings of the first and second rotor rings in both keys:

$$w_p = V - K$$
$$k_p = Z - S$$

Repeating the described procedure for the following message headers we get one of two possible results. In the first case, after applying a few to a dozen or so sheets, it turns out that no hole is repeated in all the sheets. This means that the assumptions made regarding the rotor sequence or the left-hand ring setting were wrong and the process must be repeated for other values. However, if there is exactly one common hole left after the application of all the sheets, it represents the solution.

Due to the shifting of the first sheet, the coordinates of the common hole do not represent a straightforward solution and need to be adjusted accordingly. Let's assume that the common hole found has CF coordinates. The correct setting of the rings will be determined from the relationship:

$$p1 = Z - C$$
$$p2 = V - F,$$

where Z and V come from the machine setting transmitted by the clear text for the first sheet.

As can be seen from the example, the process of selecting the right headers, choosing the right sheets for the overlay and calculating their mutual offsets was complex. Therefore, the first step in the sheet stacking procedure was to draw up a set of input parameters, which cryptologists collectively referred to as a *menu*. Let us remember this term, as we will soon encounter it again, in new context.

Later fortunes of the BS4 team indicate that finding the right candidates to help in this work would not be a difficult task. Next to the team of the three codebreakers German section included also a group of operators, whose only task was current deciphering of *Wehrmacht* and *Luftwaffe* messages, their translation and distribution. Judging by their background all of them guaranteed the protection of the secret. Most, if not all, of the candidates were recruited basing on the recommendation from one of the earlier team members. This is certainly how Sylwester Palluth, Antoni's nephew, joined the group in 1935. Sylwester combined talent for the humanities and sciences. He polished his impeccable Greek and Latin in the best classical grammar school in Poznań, and supplemented his education with the sciences at college in the same town. Over time, he mastered also German, French, Italian, Spanish and English, becoming a natural candidate for a codebreaker. Kazimierz Gaca reached the BS4 in a similar way, although a little later. Born in Bydgoszcz in 1920, he was the youngest of four brothers, the oldest of whom was Zbigniew. They both studied at the same classical gymnasium in Bydgoszcz, which Marian Rejewski completed earlier. It seems that the familiar tone that Rejewski used in his later correspondence with the oldest of brothers could testify to their school contacts. Although Zbigniew was expelled from the high school for the excesses associated with his musical tastes, later on he earned degree at the Poznan School of Economics and joined the Polish intelligence service in 1930. His brother, Kazimierz, was studying mathematics at the University of Warsaw and engineering at the Warsaw University of Technology. Zbigniew remembered well that playing bridge and solving riddles together was their favorite family pastime and recommended him to the chiefs of the Cipher Bureau. According to a popular opinion combination of the mathematical and musical skills makes a good codebreaker. Among the three mathematicians forming the core of the BS4 team, both Różycki and Zygalski demonstrated unquestionable musical talent, entertaining their colleagues on various occasions with improvised concerts. Although Rejewski was mostly a passive connoisseur, he always spoke about music with passion and knowledge. Also, at Gaca's home there was a tradition of the family music performances; the candidate did not lack the competence to take on the new challenge. Kazimierz Gaca was engaged by BS4 in December 1938, becoming the youngest member of the team.

In the last months of 1938, the history of Enigma accelerated dramatically, as if trying to keep pace with the galloping events in the European political arena: the occupation of the Rhineland, the remilitarization of Germany, the Spanish civil war, the Anschluss of Austria and the Munich crisis. The number and nature of changes introduced in the Enigma system during this period might have puzzled Polish codebreakers. The modifications seemed to be aimed precisely at the methods developed and used by the Poles. Codebreakers' responsibility was limited to the recovery of the daily Enigma keys, they rarely had the opportunity to get acquainted with the content of the messages themselves. But even the few clear texts they read in passing heralded the coming storm. And precisely at this very moment, when knowledge of Germany's true intentions became particularly valuable, the most reliable source of information turned from a wide river into a narrow stream. It must have been a painful situation for the mathematicians. In the race of minds they kept on winning and maintained a safe advantage over their adversary. Their methods of breaking the cipher were still effective, even after recent changes. Theoretical foundations behind their attacks remained valid. It was mainly the limited economic potential of the country, that caused the crisis that reached Cipher Bureau in the first days of 1939. Interwar Poland and its army were not prepared for the intellectual success of the codebreakers. Paradoxically, it is likely that this weakness became an element of the Allies' future victory. Enigma breaking operation conducted in Poland on an industrial scale, comparable with later Allied effort, would be impossible to hide after the country was occupied by German troops. But in January 1939 Rejewski, Różycki and Zygalski despite their success on the ground of the cryptological theory, had to feel a little helpless and very lonely.

Their loneliness was not just an abstraction. At the time no codebreaking service in the world achieved successes comparable to theirs, in fact no one took up seriously the challenge posed by the military Enigma. Among the actual and potential allies of Poland, only France could count as a partner in the any conflict with the Third Reich. However, Polish-French relations became complicated since the military alliance was signed in 1921. According to the French, the historical role of their country, confirmed by their victory in the Great War, has entitled them to act as patron rather than partner. Polish-French contacts in the field of signals intelligence represented an isolated example of cooperation between armies of both countries. Bertrand showed a lot of patience in his relations with his Polish counterparts, but against all his hopes the flow of information was unilateral. The Cipher Bureau received a total of 38 months' keys out of 81 delivered by "Asché", without even a single decrypted message travelling in the opposite direction (although anonymized data from the deciphered messages were certainly included in the intelligence summaries exchanged between both countries). This situation was frustrating for Bertrand, who did not have the arguments to convince superiors of the value of the material purchased from his agent. Bertrand started to suspect that he was taking part in an intelligence game played by the Germans via "Asché". However, during a direct confrontation with "Asché" the agent invoked, somewhat ambiguously, his honor and honesty, convincing even the experienced and suspicious "Rex".

Contacts between Bertrand, Langer and Ciężki created an atmosphere of friendship and cooperation. On the other hand, Enigma itself provided arguments against open sharing with the French one of the most closely guarded secrets of Polish intelligence service. Rejewski mentions in his memories that among the broken messages there were some indicating at least the susceptibility of some of the French ministers to corruption. The person specifically mentioned in the messages was Paul Marchandeau, who during his long political career held many functions as minister of finance, interior and justice. Potential betrayal of one of the top French politicians was not a good topic to discuss with French partners, but it definitely strengthened the fears of Poles arising from the French authorities' tolerance of the numerous supporters of Nazism in their country. In view of all the doubts Langer was glad that decisions concerning sharing the information from Enigma decrypts were beyond his responsibility. For his personal use Langer interpreted the agreement with

Bertrand according to a somewhat Jesuit logic: he and his team were not using the materials provided, so they were not obliged to share the effects of their work.

Langer's reserve would have increased if he had known that some time ago his partner nearly compromised the Enigma's secret. During the conference on 5 November, 1937, Hitler presented his plan of conquests to the top Wehrmacht commanders. The conference's secretary was Hitler's military adjutant, Colonel Hossbach (from whose name the summary of the meeting went down in history as Hossbach's protocol). He shared information about the Führer's plans with his friend Rudolph Schmidt, and through him the news reached "Asché". Appreciating the importance of this information, "Asché" immediately forwarded the report, bypassing the usual channels, directly to the French embassy in Berlin. Ambassador André François-Poncet and local resident of French intelligence service, Maurice Dejean, considered it urgent and decided to send the report to Paris by telegraph, instead via secure diplomatic mail. The telegram was intercepted by one of the several German agencies competing with each other in the field of signals intelligence, camouflaged under the inconspicuous name of the research office—Forschungsamt. Forschungsamt was an early child of Nazism, established on 10 April, 1933, less than three months after Hitler took power, soon eavesdropping on most of the internal and external telephone, telegraph and radio networks in the Reich. Access to its reports, printed on characteristic brown forms, was reserved for Göring and Hitler himself. The Forschungsamt team was breaking the French diplomatic code without particular difficulty, so in one of November reports Hitler found clear confirmation of treason at the highest levels of the Reich's military hierarchy: a French report from a top-secret meeting. This discovery must have exacerbated the paranoid mistrust with which the Führer treated his army commanders. It also led to an investigation into the leak, which was supposed to find its finale only after six years. The telegraphic transmission of Hossbach's protocol was a reckless act considering that the documents previously delivered by "Asché" included the copies of a dozen or so French diplomatic messages broken by the Forschungsamt. German codebreakers were breaking not only French diplomatic code. In the early 1930s they managed to deal with the enciphered code used in communications between the French ministry of war and commands of the military districts. Looking ahead, it should be noted that after the outbreak of the war the French changed the code used between headquarters and army commanders, extending the scope of application of the cipher previously used between Paris and the department of Sabaudia, which German cryptologists had broken… a year earlier. In this situation the fact that the Enigma's secret was not leaked in the early days of the war must be considered an extremely fortunate coincidence. Langer was not aware of all the facts presented, but his caution in sharing success of his team was well justified.

Hans-Thilo Schmidt was also furious about the carelessness of his business partners. François-Poncet's telegram was broken by the Forschungsamt, with whom "Asché" worked occasionally at the time, so he quickly learned not only about French mistake, but also about the threat to his own security resulting from it. Less than two weeks after the fuss with the telegram, Schmidt met with Perruche, "Rex" and Bertrand in Switzerland. This time the meeting was turbulent: Schmidt demanded that his partners take seriously his personal safety. The problem resolved itself soon. On 28 September, 1938 "Asché" terminated his employment at the Chiffrierstelle. He remained an active French agent until few weeks after the outbreak of the war, providing some useful information, but Enigma keys were not included in his deliveries anymore (justifying thus Ciężki's earlier decision to keep the keys in his safe and not to pass them on to the codebreakers). The paradox was that the office he switched to was the Forschungsamt—the organization that was the first to signal his betrayal.

Chapter 7
Dress Rehearsal

When Rejewski, Różycki and Zygalski were busy with their apprenticeship in code-breaking, events were taking place at the opposite end of Europe, which were supposed to make the importance of the Enigma even more evident to the countries still indifferent to its challenge. After military disasters at the end of the nineteenth century, Spain was on the verge of political stability. General Primo de Rivera's coup d'état, the announcement of the Republic on 14 April 1931, attempted assassination of general Sanjurio in August 1932, finally suppressing of the leftist revolt in Asturias in 1934 marked the stages of the disruption. On 12 July 1936 the phalangists murdered Lieutenant José Castillio, a member of the republican military association. The colleagues of the murdered man set out in search of vengeance, and, unable to find Gil Robles as originally intended, murdered the leader of the monarchists, José Calvo Sotelo. Both murders have become a signal for action for the military conspirators: on 17 July military coup attempts were made in many regions of Spain. Successful in Burgos, Pamplona, Valladolid, Cadiz, Cordoba and Jerez, they failed in Madrid, Barcelona, Oviedo, Bilbao, Badajoz and Malaga. The next morning, Spain woke up divided.

The Spanish Civil War became the first arena where machine ciphers were practically tested. In November 1931, the Spanish foreign ministry instructed the embassy in Berlin to obtain information on the ciphering systems used in Germany. Within a week the minister received an offer from Chiffriermaschinen AG for three Enigma models. Two of them were well known commercial variants, one self-printing, the other one equipped with lamps. The third model, the Enigma Z30, represented a curiosity: equipped with three rotors and an adjustable reflector, it had only 10 keys and 10 lamps, marked with the digits 0-9. The rotors of the machine were similarly marked, suggesting that its purpose was the re-encipherment of messages in code. Despite the affordable price of 600 marks, the ministry chose the

M. Grajek, *Enigma Myth Deciphered*, History of Information Security,
https://doi.org/10.1007/978-3-031-65475-6_7

alternative offer of Kryha[1] machines. In April 1932 the embassy was authorized to purchase six copies, but the bargaining delayed the order; finally three copies of the Kryha machine were dispatched to the Spanish embassies in April 1934.

The success of the military coups in some regions of the country and their defeat in others led to Spain's division into a number of separate regions, with areas controlled by the republicans intertwined with domains of nationalists. In addition, nationalists controlled the country's overseas departments, both islands and north African. The coordination of activities on both sides required fast and reliable communication, which could only be provided by radio. The nature of the conflict has increased the importance of communication security, codes and ciphers. The bidding in cryptological poker was started by the government in Madrid, which in the early stages of the war summoned a 28-year-old engineer, Anselmo Carretero Jimenez, to Madrid, entrusting him with the republican codes and ciphers. Carretero arrived hastily from Mexico, but could not do much for his principals. The only ciphering machine at the disposal of the republic, the Kryha system, did not provide adequate security. Manual codes and ciphers were not much better; breaking them did not cause much trouble to the team of nationalist codebreakers working in Palma de Mallorca under the leadership of Lieutenant Balthazar Nicolau Bordoy. His team routinely read the messages enciphered with at least two republican keys—"Bocho" and "Victoria". Broken ciphers allowed, among other things, the cruiser *Canarias* to take over on March 8, 1937, the Republican freighter *Mar Cantábrico*, sailing with a cargo of war material from Mexico to Santander. On the opposite side of the front, communication security has improved with the arrival in Spain of Italians and German Condor Legion. British reports confirm that the first message sent from Spain and enciphered with the Enigma was intercepted on 5 November 1936. The German machines made such an impression on the Spanish hosts that in November 1936 the Burgos government decided to purchase 10 Enigma copies (according to an alternative report Hitler donated 15 machines to nationalists). Machines were placed under the supervision of Franco's staff officer, Antonio Sarmiento, thanks to whose meticulous notes we know the serial numbers of the machines, allowing us to identify them as a commercial model—Enigma K. One of the machines disappeared on December 7, 1937, in circumstances that were not explained by the following investigation. If the machine was stolen by Republican agents, they could save themselves the risk—Enigma K was available in the open market. What's more, its ciphers were already being broken at that time…

During Great War the British codebreakers contributed significantly to the victory over the Central Powers, reading various codes and ciphers of the Kaiserliche Marine and German foreign ministry, breaking in particular the famous Zimmerman telegram. After the end of the Great War, the British authorities considered that

---

[1] The encryption machine was patented on 16 January 1926 by Alexander von Krish from Berlin. Gears with a variable number of teeth ensured irregular movement, but the cipher period between 260-520 characters did not guarantee a reasonable level of cipher security. Despite this, the machine sold well, and later it was copied and implemented for use in the Soviet Army as "Kriga Malaya".

having an effective cryptological organization could also be useful in the peacetime. In 1919 the government decided to establish a new organization and, remembering the recent successes of the Navy codebreakers, entrusted the Admiralty with its supervision. The core of the new unit, which on 1 November, 1919 was given the neutral name of the *Government Code & Cipher School (GC&CS),* has been staffed by veterans of wartime Navy codebreakers, supplemented by Army signals intelligence staff. Thus, the GC&CS was joined by Alastair Denniston, Dillwyn Knox, Nigel De Grey and William Clarke representing Navy's Room 40, and Oliver Strachey representing the military intelligence, M.I..1b. The Admiralty was reluctant to accept responsibility for the new service. The Sea Lords and their subordinates have not been impressed by the codebreakers' civilian lifestyle, their lack of devotion to the Navy tradition, even the vocabulary of a common landlubber. The Admiralty's resistance was softened by the then Foreign Minister, Lord George Curzon, who pointed out that only the Navy commands a network of the intercept stations capable of providing codebreakers with the cipher material. This was not the final of the GC&CS's bureaucratic adventures. Just over a year after GC&CS was established, another bureaucratic storm broke out over the heads of its staff. In a conversation with the French ambassador Lord Curzon shared a bit too open his opinions regarding his cabinet colleagues. The ambassador dutifully quoted the Lord's views in a message to Paris, while GC&CS equally dutifully decoded the message and quoted its contents in a bulletin distributed to cabinet members, who were not amused by the indiscretions about themselves. The *faux pas* suddenly and dramatically changed Curzon's attitude to codebreaking. He suddenly discovered that in peacetime, the activity of the codebreakers focuses on diplomatic codes and ciphers rather than military ones, which justifies putting the organization under the supervision of the *Foreign Office.* Curzon found an ally in the Lord of Exchequer, who questioned the financing of the Admiralty's interception stations, noting that in the peacetime most messages are being exchanged by telegraph. The Navy did not defend its authority over GC&CS with determination; in a memorandum of February 1920, admiral Bentinck expressed his view that "*it is not impossible that in the near future other methods will replace the current functions of the School*". As a result, on 1 April, 1922, supervision over GC&CS was transferred to the Foreign Office.

However, a year later, the tide turned. The former head of naval intelligence, the charismatic Admiral Hugh Sinclair, has taken over as head of the influential but officially non-existent SIS. He was not only experienced in the intelligence matters, but equally passionate about secret services and operations. At the time one of London's theatres played a show entitled "The Gay Lord Quex", whose protagonist was described as "the most wicked man in London". Sinclair's lordly manners combined with his wickedness towards the King's enemies have earned him the nickname "Quex". Remembering the contribution of World War I codebreakers, Sinclair considered it necessary to regain influence on the functioning of GC&CS. As a result, a compromise has been made which, in any country outside the UK, would be a source of total paralysis of the service. GC&CS was formally subordinated to the Foreign Office, which financed the agency's activities from its own budget, but did not manage its operational activity, which were supervised by the head of the

SIS. Various services were to run their own intercept stations, with a certain geographical division. Inside the GC&CS, sections of Navy, Army, and Air Force were created. This complex history and structure allowed the British to avoid the postwar errors committed by France and Germany—dispersion of the codebreaking service into several agencies working independently.

In 1925, the relationship between GC&CS and the SIS was crowned by the transfer of the codebreakers to a building on 54 Broadway, the common headquarters for both services. The head of SIS remained the nominal director of GC&CS, but day-to-day operations of the center were managed by his deputy. The first Deputy Director of GC&CS was one of the veterans of Room 40, Alastair Denniston. It seems that his nomination was not received with enthusiasm by his colleagues, whose experience in the codebreaking was at least equal to his. Moving on to Room 40 from the post of a German language teacher in one of the naval schools, Denniston reasonably focused on the administrative rather than cryptological side of the service. Interestingly, his lack of managerial skills was common belief among his peers, qualifying him in the, undoubtedly malicious, opinion of one of his coworkers at most for "running a candy shop in the East End". It is difficult to decide whether his generally adopted nickname "Little man" referred more to his posture or rather his character. However, the Navy decided that Denniston would be a better candidate than his Army competitor, Major Malcolm Hay. Hay made no secret that he ruled out any service under Denniston's command. Turf wars between Navy, Army, Foreign Office and the SIS did not augur well for the future of the new institution that was supposed to reconcile their interests.

The interdepartmental nature of the School and the complications it entailed were only one of the sources of Denniston's problems. Like many services created primarily for war, GC&CS faced drastic financial constraints during the peacetime, which went so far that during the discussions on the name of the new service, Nobby Clarke was applauded by his colleagues for putting forward a proposal for "Public Benefactors". Despite all those problems GC&CS was successfully breaking codes and ciphers of several countries, making no special difference between potential adversaries and allies. Until 1926 British codebreakers were reading German messages encrypted using variations of the systems they learned during the past war. However, when machine ciphers appeared on the air, GC&CS codebreakers lost interest in German traffic to such an extent that they stopped intercepting German messages altogether. This decision probably explains the lack of interest in Bertrand's proposal to co-finance "Asché's" activity. During this period only one codebreaker worked on German systems at GC&CS. The British acquired a commercial model of the machine in Vienna as early as 1925 (most probably the Enigma D version), but it was not until two years later that Hugh Foss was commissioned to investigate it. Foss found that knowledge of at least 15 pairs of plain text and corresponding cipher message allows to determine which rotor was used as the fast one and its starting position, and 180 such pairs allow to determine the internal connections of the pair consisting of right and middle rotor. Further interest in the Enigma was not stimulated even in 1931, when the French intelligence provided the GC&CS with the technical description and instructions for the use of the machine. The situation was only changed by the outbreak of civil war in Spain. Suddenly abundance of radio traffic was taken over by British stations. The Germans with their toy fleet were of negligible interest for the British at the time, but the Enigma was also used by Italians fighting in Spain and by General Franco's forces. The Italians, a troublesome ally in the years of World War I, have meanwhile turned into a threat to British interests in Africa and the Mediterranean. The occupation of Abyssinia by the Duce soldiers threatened the shortest route to India, Mussolini's vanity made him speak of the Mediterranean as the Mare Nostrum... Franco's nationalists, with a little encouragement from outside world, might have been determined enough to try to get rid of the British from Gibraltar...

The accelerated pace of events resulted in 1936 in a return to British-French cooperation in cryptology. Its most superficial expression was the exchange of the intercepted messages. It was also agreed that both parties will undertake research on the machine itself and its operational procedures. A year later, Captain Freddie Jacob noted that the three-letter indicator played a special role in the messages exchanged between the German troops in Spain and Berlin. In 1938, the French and British again exchanged information on the frequencies and call signs used by the radio stations of the German Navy. The British, knowing that French intelligence had a well-placed agent at its disposal, have formulated a list of questions, focused on the nature of the keys used in the Enigma system. However, GC&CS was not limited to the passive role in the cooperation with the French. One of its assets was the ability to break the codes of the Italian fleet, largely due to the custom of the political indoctrination of its officer corps. The Italian command had the habit of encrypting and transmitting the texts of the Duce's speeches, widely available in the press. When British codebreakers learned this peculiar practice, breaking the code became easy and enjoyable. From the broken messages the British learned that the machine introduced to the Italian army and navy represents a commercial Enigma model. In 1936 Dillwyn Knox, known among his colleagues as Dilly, who was considered to be one of the most talented codebreakers in GC&CS, took over the work on the Italian Enigma. Dilly was known mostly due to his activities outside the world of secret services—the reading and editing of the heavily damaged Greek manuscript of "Mimes" by Herodas, torn from the sands of Egyptian desert. The exercise in completing the lacunae and correcting the mistakes of the poor copyist proved to be an effective preparation for purely cryptological challenges; over the next few years Knox, accompanied by Clarke and Strachey, broke the diplomatic codes of the USA and France. Dilly had reportedly the intention of abandoning his codebreaking career and returning to King's College and classical philology; Enigma stood in the way of these plans.

The machine posed a problem different from anything resulting from Dilly's education and experience. Despite this, after less than half a year's work, on 24 April 1937, Dilly made a breakthrough. The method he used was a skillful adaptation of techniques used so far in the attacks on the traditional manual ciphers. Given that most manual ciphers operated with a limited number of cipher alphabets, it was convenient to write them down on strips of paper or rods of erasable material, which were then juxtaposed in different combinations, looking for one where some order appeared in the chaos of characters. Role of rods in Knox's method gave it the name of *rodding*. Knowing the design and operation of the commercial Enigma Knox was able to write on each rod an alphabet generated in a specific position of one of the machine rotors. He then made assumptions about the clear text corresponding to the fragment of the ciphertext and the rotor used in a fast position. GC&CS was routinely reading most of Italian codes, and Knox could assume the same text he knew from the broken messages to be enciphered with Enigma. The name given later to one of the methods, "Percommandante",[2] confirms that Italians routinely put standard phrases in the message headers. If the method did not give a result under the assumed assumptions, the steps for the remaining possibilities had to be repeated. A detailed description of the method for those interested is set out below.

Rodding

The rodding method (strips method) developed by Dilly Knox refers to the methods used in the times of domination of manual ciphers. At that time, cryptanalysts quite commonly used the technique of writing all code letters on strips of paper or wooden blocks. By selecting successive stripes/pads and shifting one relative to the other it was relatively easy to see the position where the dependence or regularity of the cipher was revealed, suggesting the starting point for further analysis.

An obstacle in applying the strip method to the Enigma was the number of code letters, much higher than in the case of traditional ciphers: at three rotors, the code letters started repeating after 17.576 characters. However, if the analysis is limited to a piece of ciphertext and it is assumed that only the fast rotor of the machine moved during encryption, the number of ciphertext alphabets is at most 26. In order to determine the cipher alphabets associated with the selected Enigma rotor, three factors must be taken into account: the order in which the keyboard signals are fed to the first rotor, the structure of the internal wiring of that rotor and the effect of shifting the rotor after encoding each character. The strip method was used by Knox for the commercial model of the Enigma, which differed from a military machine in the absence of a plugboard, different entry drum and rotor wiring. However, assuming that no plugboard is used, the method also works for the military Enigma, which will be used to demonstrate the *rodding* below.

The connections of rotor No. I in the military Enigma correspond to permutation:

```
abcdefghijklmnopqrstuvwxyz
EKMFLGDQVZNTOWYHXUSPAIBRCJ
```

(continued)

---

[2] 'To the commander'.

Wires from the keyboard are fed to the entry drum in alphabetical order. Imagine that during the encryption of a text fragment, only rotor I, placed in the fast position, advances; the remaining, stationary rotors and the machine's reflector form a virtual reflector, while between the fast rotor and the virtual reflector there is an additional, stationary drum with contacts arranged in the same order as the entry drum (intermediate drum). We also assume that the position, in which the rotor realizes the permutation described above, will be defined as the rotor zero position. In the neutral position of the rotor I, the contact of the entry drum corresponding to letter A is connected to the contact of the intermediate drum corresponding to letter E. Moving the rotor I to the next position will cause letter A of the entry drum to be connected to letter K of the intermediate roller. By repeating the rotor rotation operation 25 times and determining all the connections of the input and intermediate rollers, we obtain a square containing 26 code letters:

```
    01234567890123456789012345

A   EKEUFNGWADHSKJCIDRNJXBHCKJ
B   KFLFVGOHXBEITLKDJESOKYCIDL
C   MLGMGWHPIYCFJUMLEKFTPLZDJE
D   FNMHNHXIQJZDGKVNMFLGUQMAEK
E   LGONIOIYJRKAEHLWONGMHVRNBF
F   GMHPOJPJZKSLBFIMXPOHNIWSOC
G   DHNIQPKQKALTMCGJNYQPIOJXTP
H   QEIOJRQLRLBMUNDHKOZRQJPKYU
I   VRFJPKSRMSMCNVOEILPASRKQLZ
J   ZWSGKQLTSNTNDOWPFJMQBTSLRM
K   NAXTHLRMUTOUOEPXQGKNRCUTMS
L   TOBYUIMSNVUPVPFQYRHLOSDVUN
M   OUPCZVJNTOWVQWQGRZSIMPTEWV
N   WPVQDAWKOUPXWRXRHSATJNQUFX
O   YXQWREBXLPVQYXSYSITBUKORVG
P   HZYRXSFCYMQWRZYTZTJUCVLPSW
Q   XIAZSYTGDZNRXSAZUAUKVDWMQT
R   UYJBATZUHEAOSYTBAVBVLWEXNR
S   SVZKCBUAVIFBPTZUCBWCWMXFYO
T   PTWALDCVBWJGCQUAVDCXDXNYGZ
U   AQUXBMEDWCXKHDRVBWEDYEYOZH
V   IBRVYCNFEXDYLIESWCXFEZFZPA
W   BJCSWZDOGFYEZMJFTXDYGFAGAQ
X   RCKDTXAEPHGZFANKGUYEZHGBHB
Y   CSDLEUYBFQIHAGBOLHVZFAIHCI
Z   JDTEMFVZCGRJIBHCPMIWAGBJID
```

(continued)

The columns of the square correspond to the cipher alphabets generated by the rotor I after advancing by r positions (values of the parameter $r = 0, 1, \ldots$, 25 are given in the column headings). As successive cipher alphabet letters are created by the rotation of the rotor by r positions, they are sometimes referred to as *r-rotated* alphabets in relation to the basic alphabet. The rows of the square correspond to the letters to which the letter indicated in the header of a given row is connected in all 26 positions of the rotor. The square of *r-rotated* alphabets shows some interesting property. Note that the diagonals of the square facing down and to the right of the elements of the first column contain successive characters in alphabetical order. The order of the diagonals of the square is the same as the order of the letters in the machine entry drum. For the commercial Enigma model, where the characters in the entry drum were arranged in the order of QWERTZ..., the diagonal elements would be arranged in the same order. The machine entry drum connections represent the key information for the construction of the square of the *r-rotated* alphabets. Knox used the term *diagonal* for as a synonym for entry drum connections.

After preparing the square of cipher alphabets we can proceed to prepare the stripes from which the method took its name. To do this, cut the square into rows corresponding to the individual characters of the entry drum. The application of the strip method requires knowledge of the probable plain text and its location in the ciphertext. Let's assume that we have reason to suspect that one of the encrypted texts, which has the content of **JFIDYVWZ**, represents an encrypted equivalent of **REJEWSKI**. We write pairs of plain text and ciphertext characters underneath each other (shifted by one position, corresponding to the movement of the fast rotor), and then to the right of each letter we put the corresponding rod with the cipher alphabet:

```
R  VLWEXNRUYJBATZUHEAOSYTBAVBVLWEXNR
J  QBTSLRMZWSGKQLTSNTNDOWPFJMQBTSLRM
E  MHVRNBFLGONIOIYJRKAEHLWONGMHVRNB
F  HNIWSOCGMHPOJPJZKSLBFIMXPOHNIWSO
J  QBTSLRMZWSGKQLTSNTNDOWPFJMQBTSL
I  ASRKQLZVRFJPKSRMSMCNVOEILPASRKQ
E  MHVRNBFLGONIOIYJRKAEHLWONGMHVR
D  GUQMAEKFNMHNHXIQJZDGKVNMFLGUQM
W  YGFAGAQBJCSWZDOGFYEZMJFTXDYGF
Y  ZFAIHCICSDLEUYBFQIHAGBOLHVZFA
S  CWMXFYOSVZKCBUAVIFBPTZUCBWCW
V  FEZFZPAIBRVYCNFEXDYLIESWCXFE
K  NRCUTMSNAXTHLRMUTOUOEPXQGKN
W  YGFAGAQBJCSWZDOGFYEZMJFTXDY
I  ASRKQLZVRFJPKSRMSMCNVOEILP
Z  WAGBJIDJDTEMFVZCGRJIBHCPMI
```

(continued)

We then analyze all the columns of hypothetical cryptograms and under-score letters that imply a contradiction with the nature of the Enigma code. Let us note, for example, that in the penultimate column, the **N** character was once encrypted as **R,** and otherwise as **S**. If we have eliminated all columns in this way, it means that one of the assumptions is false. However, in the example above, in the last column there are no logical contradictions (more-over—the repetition of the **R-M** pair additionally indicates the correctness of the alphabet used), so it is likely that the analyzed text was encrypted with the rotor I in the fast position, with rotor starting position 26 (**Z**).

If we are to believe the opinions recorded by his colleagues, the Enigma chal-lenge and first success changed Knox's lifestyle. Until now open and communica-tive, he suddenly became constantly embarrassed, refusing for the first time in his life to participate in a university celebration he had attended regularly for years. Much later his family members remembered that he gave up his social life for fear that during talks with his friends he might be able to reveal information about his work. The first slit in the veil of Enigma ciphers had to be welcomed with enthusi-asm at GC&CS. Its traces can be found in Foss' note, regretting that the French were completely indifferent to the British revelations. The enthusiasm of Foss, Knox and their colleagues would probably have been moderated if they had known that the machine constructors were aware of the weakness their attack was based on: the possibility of a rodding-like attack was a direct reason for adding a plugboard in the military model.

Soon after his first success Knox added a second one. Rodding was a practical method of breaking the code provided the machine rotor wiring was known. However, British stations were intercepting also messages enciphered using an Enigma with rotors obviously rewired in comparison with the standard model. Dilly proposed a method described as buttoning-up (cf. example presented in the box below). He assumed that only the rotors had been rewired, preserving the original structure of the entry drum, i.e. the *diagonal*. The application of the method required significant number of messages enciphered with the same key, whose open texts were known or at least could be reasonably guessed. It was estimated that the suc-cessful application of buttoning-up required at least 90 pairs of characters represent-ing open and cipher text, success was certain with about 140 pairs. Buttoning-up succeeded thanks to the earlier breaking of a number of Italian hand codes and ciphers, its application to a cipher in which no cracks were previously found, was problematic.

Buttoning Up

Suppose we intercept radiograms encrypted with the modernized Enigma commercial model. We know that its modernization consisted in replacing the wiring of all the rotors, while leaving the other details of the machine design intact. Our task is to reconstruct an unknown rotor wiring. Let's also assume that among the intercepted radiograms there is a set of messages known to have been encrypted at a common machine settings. In addition, we know the fragments of the clear text corresponding to the beginning of the messages (quoted in the table below along with the corresponding cryptogram fragments).

REJEWSKI	INFANTERIE
ZAFOGHGS	FVJWDWLVZO
ZYGALSKI	PANZER
RUNWVHGS	HEGFXJ
RISKS	ANGRIFF
ZITNZBP	TVNSFNC
LANGER	ABWEHR
CEGIXJ	TJCOSJ
CIEZYKI	MUNITION
LOVFME	KYGGYESP
PALLUTH	GEHEIM
HEXKQWW	YAYOFV
WEHRMACHT	FLOTTE
SAYSKPFWP	ICUQYI
LUFTWAFFE	TORPEDO
CYJQGPCOJ	AISBXZS
KRIEGSMARINE	ARTILLERIE
MPBOWHTTNMFP	TPZGVQLVZO
BEFEHL	GEWEHR
VAJOSQ	YACOSJ
OBERKOMMANDO	WAFFEN
EJVSMUTBXSTI	SEJZXF
MELDUNG	SOLDAT
KAXMQFK	WIXMBW

The tool we will use when reconstructing the wiring of the Enigma fast rotor will be presented when discussing the method using the table of *r-rotated* alphabets. Its zero column, corresponding to the neutral position of the rotor, together with the knowledge of the entry drum structure, describes unambiguously the rotor connections. The purpose of the analysis can be presented as a reconstruction of the zero column of the table of *r-rotated* alphabets.

In the table above, only the first characters of the cryptogram correspond to the zero column of the square, the others belong to the rotated alphabets. However, we assume knowing the structure of the machine's entry drum, i.e.

(continued)

the   diagonal   of   the   table,   ordered   in   the   sequence
QWERTZUIOASDFGHJKPYXCVBNML. Knowing the position of the character in the message and the diagonal, we can reduce each letter of the cryptogram into the form in which it would appear in the zero column of the table.
To illustrate this process, let's select the letters DU from the MELDUNG text.
Taking into account the diagonal structure the DU characters in the fourth and
fifth position of the cipher box correspond to the H and S characters in the
zero column:

```
MELDUNG
 FI
 GO
 HA
 S
```

The characters of the corresponding word MELDUNG of the cryptogram
can be transformed in a similar way:

```
KAXMQFK
 LW
 QE
 CM
 T
```

Dillwyn Knox, has defined the process by which a pair of characters in any
position in the cryptogram is reduced to the corresponding character pair in
the zero column of the square. The effect of the buttoning-up in the example
above is that if there is a HS pair in the zero column, there will also be a
CT pair.

Repeating the process for all the messages above and identifying the character pairs occurring in the first six positions of each message, we obtain the
following set of letter pairs:

```
01 02 03 04 05 06
==================
AT AE BI AW AB AP
BV BJ CW BP CZ BK
CL CL EV DM DN DZ
EO IO FJ EO EX EI
FI NV GN FZ FI FN
GY PR HY GI GW HS
HP UY LX KL HS JR
KM    OU NY KM LQ
RZ    RS QT LV MV
SW    TZ RS QU OU
            TY TW
```

(continued)

Then we convert the found character pairs by buttoning-up, receiving:

```
01 02 03 04 05 06
==================
AT SR MA FT GQ HB
BV NK BR LC MA WV
CL VQ TN HW JW KS
EO OA HP ZD UN IF
FI MB JL JO KD PE
GY YT KC KS PZ XJ
HP IX WV XE YH CO
KM SO QV CE TZ
SW UI UG TS GD
         OB AU
```

The set of character pairs thus determined is a material which allows for determination of the content of the zero column. The verification process begins with the selection of the hypothesis—a pair of characters, whose presence in the zero column we want to confirm. We will demonstrate the way of verification of the presence of the **AB** character pair in the zero column.

Note that in the first column of the above table **A** corresponds to **T** and in the second column **B** corresponds to **M**. This means that if in the zero column there exists pair **AB**, there is also pair **TM**. This relationship is recorded in the form **AB-1-TM** (the number between the pairs of characters identifies the column of the table that was used to establish the relationship). We can also establish further relationships:

```
AB-2-OR
AB-5-MH
```

Based on the available data, it is not possible to determine further relationships for the **AB** pair, so we analyze the **TM** pair, obtained in the first relationship:

```
TM-2-YA
TM-4-FA
```

The above relationships are contradictory because **YA** and **FA** pairs cannot be present in the zero column at the same time; the hypothesis of **AB** pair presence has been verified as negative. In the same way we can verify the hypotheses concerning the following pairs of letters: **AC**, **AD**, **AE**, … In all cases, except for one, the hypothesis verification procedure will result in a

(continued)

contradiction. Let's assume, however, that we are verifying the assumption that there is a pair of **AK** letters in the zero column. We get the following chain of relationships:

AK-1-TN	TN-2-YT	YT-1-GY	RJ-3-BO
AK-2-OC			RJ-4-IW
AK-3-MS	OC-3-SL	SL-2-RJ	RJ-5-LX
AK-5-MS	OC-4-JE	SL-4-KR	
	OC-5-BO		LX-1-CI
		JE-3-LX	LX-5-RJ
	MS-1-KR	JE-5-WP	
	MS-2-BO		VA-2-QM
	MS-3-AK	BO-1-VA	VA-4-QM
		BO-5-OC	
			NB-5-UH
		KR-2-NB	
		KR-3-CI	CI-1-LX
		KR-4-SL	CI-5-EF
		KR-5-DQ	
			IW-2-XV
			IW-3-UH
			LX-1-CI
			LX-3-JE
			UH-4-GY

There are no internal contradictions, and by reordering and combining the pairs we get three chains in form:

`DQMSLXVAKRJEFGYTNBOCIWPUH`

The above chains are an almost complete reconstruction of the first rotor of the enigma model used by the German railways, for which the zero column of the square of the code letters has a form:

`CIWPDQMSLXVAKRJEFUHZGYTNBO`

This success cemented Dilly's position as a leading authority on the machine ciphers at GC&CS, but in the context of the earlier work, not even taking into account the Polish Cipher Bureau, it did not represent a breakthrough. When Edward Hebern offered his machine to the US Army and Navy just after the end of World War I, samples of its cipher were evaluated independently by William Friedman for the Army, and Agnes Mayer Driscoll for the Navy. They both had broken submitted samples of the ciphertext, leading to the rejection of Hebern's offer. Driscoll exploited this experience later, breaking the ciphers of early versions of the ciphering machines used by the Japanese Navy in 1921 and in the mid-thirties. We do not know the reasons why in the second half of the 1930s the US Coast Guard became interested in the ciphers of the commercial Enigma used by the Switzerland. In any case, its codes had been broken by another codebreaker, Elizebeth Friedman.[3] Her job might may have been facilitated by the fact that she was married to William Friedman, who had already handled a different type of rotor machine and probably shared with her his experience.

A slightly postponed side effect of the British interest in ciphering machines during the Spanish Civil War was the development of their own model. When GC&CS acquired the commercial version of Enigma, a committee was set up to analyze the idea of implementing a similar device for its own use, but the results of seven years of its activity were null. One of the options the committee examined was an adaptation of the add-on for Enigma proposed by an RAF officer, Lieutenant Colonel O.G.W.G. Lywood. The acceptance of his concept may have been hindered by legal considerations: the Enigma's constructors obtained also a British patent. Meanwhile, the RAF pressed for the replacement of the traditional code books with a modern and, above all, faster system. In August 1934, without waiting for the committee's verdict, Lywood began working at the RAF training center in Kidbrook on the implementation of his concept. Effect of this work was originally named after the original, RAF Enigma with Type X attachments, and was presented on 30 April, 1935 to the Air Ministry for approval. In early 1937, 30 copies of the device were delivered to RAF units.[4] At this point, the history of the machine accelerated rapidly, probably as a result of Enigma use in Spain and Knox's success. In February 1937, design work began on the second model of the device and in June 1938 the result was presented to the committee, which approved it and ordered 350 further copies. This time there was no reference to the name of the original; the machine was named Type X or simply TypeX. TypeX remained in the service of the British and Commonwealth forces for the next 20 years and was manufactured in eight successive versions. Its designers have well identified commercial Enigma's weaknesses and introduced well designed measures to eliminate them.

All versions of TypeX used five rotors, with the first two remaining stationary during operation. However, they could be set in any position, performed thus a function similar to the plugboard in the military Enigma. Each rotor consisted of 2 parts: an insert with wiring connections and an external ring. The number of available inserts ranged from 10 to 14 (Mark VI version), each insert could be placed in the ring in two ways (heads or tails upwards) and in any of 26 positions. TypeX rings were also different from those used in Enigma. The Enigma ring had only one turnover notch; TypeX rings had several, distributed irregularly around the

---

[3] NARA RG457, HCC, NR1737, Box 705, Enigma Conferences, Theory'.

[4] The manufacturing of the device was commissioned to Creed & Company, a company specializing in the production of teletypewriters.

circumference. But the crucial difference between TypeX and Enigma was the setup of the rotors relative to the keyboard. In Enigma, keyboard was connected via an entry drum with the fast rotor. It was precisely this feature that allowed Rejewski to treat the aggregate of the middle and right rotor, and reflector as the fixed permutation. In TypeX, keyboard was connected with slow rotor. As a result, each keypress caused a complete change of permutation executed by the rotor battery. Interestingly, GC&CS, nominally responsible for the design of Britain's own codes and ciphers, made little or no contribution to the creation of TypeX.

In the meantime, the Spanish Civil War was slowly dying out amidst paroxysms marking the end of the great historical dramas. Disasters at the fronts were exacerbated by conflicts inside the republican camp. When, on the eve of Christmas 1938, Franco's troops launched a campaign against Catalonia, they did not encounter significant resistance from demoralized republican troops. On 26th January 1939 Barcelona has fallen, and tens of thousands of republicans went into exile in France. Among the refugees crossing the border was a group of republican codebreakers, who gained considerable experience in breaking mostly Italian ciphers. At the same time, Gustave Bertrand was scanning the internment camps at the Franco-Spanish border, looking for potential candidates for his own service. In one of the camps he met the head of the rather infamous republican agency, Servizio Investigacion Militar, by name—military intelligence, in practice—the communist-dominated political police, accused of numerous crimes and brutal torture. Bertrand's interlocutor did not know anything about the codebreaking, but his brother reportedly interned in Hendaye was a member of the republican codebreaking team. Bertrand managed to identify and recruit a whole team of five people, who were later joined by two former political commissaries. Team members have rescued from the chaos of disaster a passport to a better future—the reconstructed or captured code books of Italian, German and Franco's forces. The team was headed by Faustino Antonio Camazón Valentín, former president of the National Council of the Security Corps of the Spanish Republic, so the whole group became known as the " Camazón's seven". Bertrand proposed that they continue their work on the codes and ciphers of Italy and Spain, as part of the French intelligence unit under his control. For the survivors of Republican Spain, this was a proposal that could not be rejected. Seven Spaniards played the role of a link between the prologue and the first act of the historical drama.

Chapter 8
Transfusion of Hope

As early as February 1939, Admiral Hugh Sinclair, the head of the British secret service, claimed that the alarming rumors of the coming war were only being spread by Jews and communists for purposes known only to them. The opinion of a man who, by virtue of his function, should be the best-informed person in the British Empire was quite disturbing. Nazi Germany has done much to convince Europe and the world of its aggressive intentions, documenting them not only with slogans but above all with clear-cut actions. Perhaps the Admiral's statement should have been seen as an expression of the loyalty of a well-trained official to his chief executive… Didn't Chamberlain declare on his return from Munich that he was bringing peace to our times? On the other hand, it didn't take a subtle intelligence game to observe the rapid growth of the power of the German armed forces—Goebbels' propaganda machine itself made sure that the world learned about the power of the German army and freshly built air service. Perhaps Sinclair thought that if every army created in history were to find practical employment, humanity would still live in caves? Such a thought should, however, be accompanied by a reflection that if every treaty signed in the history of mankind had been kept, armies would have turned out to be merely an expensive extravagance. Fortunately, the Admiral did not draw too far-reaching conclusions from his misguided opinion, and his practical actions seemed to be based on the opposite assumption. Remembering the unexpected expansion of "Room 40" in the early phase of the First World War, he sought for its modern equivalent a seat capable of meeting the challenges of modern conflict. The psychosis of the air war that prevailed in the UK led to the search for a location outside London, which in the event of a war was to become the target of the enemy's devastating air raids. Sinclair has found a convenient location less than a hundred kilometers north-west of London, near the Bletchley railway junction. Between 1882 and 1883 Sir Herbert Samuel Leon built a complex of buildings there, located in a vast park. After his death in 1926, the estate was managed by Leon's widow, Fanny. When she too died in 1937, her son, Sir George Leon, decided to sell the whole estate. Originally, the buyer was to be a local building trust, which planned to

M. Grajek, *Enigma Myth Deciphered*, History of Information Security, https://doi.org/10.1007/978-3-031-65475-6_8

divide the property into smaller plots and sell them separately. When Admiral Sinclair's advisors were looking for a war seat for GC&CS, the buyers were preparing to demolish the mansion forming the center of the estate.

Bletchley Park mansion was a strange place for secret service headquarters. The building reflected the aesthetics of its upstart owner. It was described as a residence in the Tudor style by more delicate critics, and by others, less delicate, as a majestic work of a plumber. If one accepts the point of view of the latter, serving the war in such an environment may have seemed less inappropriate, especially when improvised barracks grew up next to the mansion. The extensive park provided isolation from prying eyes of passers-by and the possibility of expanding the complex. The location and distance from London reduced the risk of enemy raids. On the other hand, the location at the North Road provided an efficient connection to the services remaining in London. In Bletchley itself, North Road crossed another line—the non-existent at present railway link between Cambridge and Oxford, the country's most important academic centers. When the historians began writing the history of Bletchley Park, they somewhat ex post deduced that Admiral Sinclair, in a surge of brilliant intuition, anticipated the recruitment of key GC&CS staff from both universities and located the headquarters in a place easily accessible to them. The Admiral's earlier quoted opinion does not allow to see in him an aura of a visionary attributed to him later. But he was undoubtedly a patriot ready to make personal sacrifices to the country's security—when the Treasury began bureaucratic maneuvering around payment for the estate, the Admiral reportedly put the necessary £7.500 out of his own pocket.

The new residents of the mansion immediately began work on adapting it to the needs of the secret service. The installation of the telephone and teletype lines connecting Bletchley Park to London's GC&CS headquarters and the interception stations almost caused the center to be compromised before it was properly organized. The work has not escaped the attention of journalists from the local "Bletchley District Gazette", which at the end of May 1938 published speculations about new hosts and the function of residence. It is possible that Admiral Sinclair's purchase of the property in his own name helped to cut off this speculation: the public authorities could in good will deny to have anything to do with the transaction and the work in progress. The Admiral, on the other hand, learned a useful lesson of conspiracy. When in September of the same year, during the Munich crisis, he ordered a trial transfer of GC&CS personnel to Bletchley Park, the operation was presented to the public as a visit by a group of guests of a certain Captain Ridley (in fact, the SIS officer responsible for administrative matters), who decided to spend a few careless days shooting the birds. Given that the core of the team consisted of 24 people and about 150 members of the support staff were deployed in the local hotels, it must have been a loud hunting. Sinclair joined his subordinates, probably wanting to evaluate himself the value of the residence. His decision was enthusiastically received by the team, as the admiral, sensitive to the charms of life, took the chef from London's Savoy, one of his favorite restaurants, with him to prepare the dinners served at the common table. The presence of the chef became a problem after a few days. Shocked by the working conditions and the sudden democratization of his talents and services, the chef attempted a suicide. The admiral was forced to call the local constable, who had to be persuaded not to make the incident public. On 9 October 1938 the entire GC&CS team returned to London. The new wartime head-quarters of the British codebreakers remained far behind the center of the Polish Cipher Bureau in the Kabaty Forest in terms of organization, equipment and, above all, the effects of their work. But Bletchley Park was to become the main arena for upcoming events. It would be known as Station X, as the tenth property in Foreign Office's administration. For the hundreds and thousands of people who will work in this place during the war, it will remain in memory under the acronym BP, which, through respect for the veterans, we will continue to use.

Sinclair was aware that the upcoming conflict would require the expansion of GC&CS staff, and the change that has occurred in the nature of ciphers since the previous war required the employment of people with a fresh look and background different from "Room 40" veterans. As early as 1937, he managed to convince the Treasury that in case of war the team should be expanded by about 50 codebreakers with academic background and 30 young women with a secondary school diploma and the knowledge of at least two foreign languages. Having agreed terms and con-ditions of employment for future codebreakers (£600 per year for students and grad-uates[1] and £3 per week for support staff) Sinclair and his deputy set out to look for candidates representing, as they called them, "professor type". Their task was greatly facilitated by the fact that many veterans of "Room 40" returned to academic careers after the end of hostilities and occupied significant positions in the scientific community. Denniston later recalled that "in some university centers, positions were held by people who worked in our ranks between 1914 and 18. So it turned out that we were most successful in recruiting at these universities". Talent search in Cambridge and Oxford was to be done by professors of both universities. Both Hugh M. Last of Oxford and Frank E. Adcock of Cambridge were historians spe-cializing in ancient history. It seems that Professor Last preferred intellectual dis-course with the Greeks and Romans of the classical era, without exposing his students to the vulgar modernity. Adcock was supported by Frank Birch, who was his colleague from King's College in Cambridge before entering the world of secret service. Their knowledge of the environment, promising students and graduates helped to expand the GC&CS team more than the railway line connecting the uni-versity centers and BP. As a result, Oxford's representation at the BP was modest in comparison with its eternal rival, Cambridge. For Cambridge, the participation of his graduates in BP's successes was a partial compensation for the role played by another group from its milieu—Blunt, Burgess, Cairncross, MacLean and Philby; the famous five Soviet spies.

Unlike the Poznań course, Sinclair's recruitment was not focused on exact sci-ences—in the first group of twenty candidates there were art historians, Germanists,

---

[1] Taking into account the differences in the cost of living in both countries, British cryptologists were to receive a similar salary as their Polish counterparts. Rejewski and his colleagues were paid about 800 zloty a month, which at the exchange rate of the zloty to the pound 25.93 (data from 1938) corresponded to about 370 pounds a year.

traditionally strong representation of classical philologists and only four mathematicians: John Jeffreys, Peter Twinn, Alan Turing and Gordon Welchman. A group of
candidates for codebreakers gathered in the autumn of 1938 at the GC&CS headquarters in London for a course lasting just several days, during which their future
colleagues offered them a condensed pill of knowledge about signals intelligence,
main code and cipher systems and methods of breaking them. At the end of the
course participants were instructed to appear at GC&CS in case of a war and
returned to their universities, where they remained, with one exception, until the
outbreak of war. The exception was Peter Twinn, who was invited to GC&CS in the
early weeks of 1939. He was offered an interesting, but not very grateful role as a
pioneer of a new generation of British codebreakers. At the outset, he was made
aware that the employment of mathematicians at the School was the subject of long
debates and doubts, as they were "considered strange and totally impractical guys".
In internal discussions, it was argued that in view of the "regrettable need to employ
people of a professor type (…) it would be better to look for candidates among
physicists, given that they should show at least a little understanding of the real
world". Fortunately, Twinn also graduated in physics, so his candidacy was an
acceptable compromise between the unpleasant necessity and the expectations of
his new employer.

As Gordon Welchman recalls, during the training no attention was paid to the
machine ciphers, believing that "the chances of doing anything with the German
Enigma's messages were minimal, at least until the machine itself was accessible".
Pessimism about the Enigma dominated also the statements of other GC&CS representatives of that period. In a memorandum of June 1937 (thus made after the
ciphers of the commercial model of the machine had been broken), Nobby Clarke
outlined an optimistic picture of the School's ability to break most of the ciphers
and codes, but ended his considerations with a statement: "There's only one cloud
left on the horizon – the possibility of common use of ciphering machines. It can be
argued that this will mean a complete end to cryptanalysis. However, it can also be
argued that the use of the machine during the war will not be equivalent to the problem of this measure that it represents in the present conditions". Pessimism sometimes turned into defeatism: already after the war broke out Frank Birch noted the
statement of one of the BP heads that "(a)ll German ciphers are unbreakable. …
putting pundits to work on them represents a waste of time". The same Birch, writing in August 1940 to the one of the deputy BP directors, Commander Travis, considered that in his opinion "defeatism at the beginning of the war played a key role
in delaying the cracking of ciphers".

These quotations draw a very surprising picture of GC&CS. On the one hand, the
School was at that time the only cryptological organization in the world with an
uninterrupted tradition of functioning over several decades. Its core was made up of
people at whose eyes and with whose participation unveiled the greatest codebreaking successes of World War I. Thanks to foresight of its superiors, the School was
well organized and had at least the basic material resources. Leading universities of
the world assured its intellectual background. And yet Admiral Sinclair's deck was
missing some unknown card or cards to join the bidding. Their lack weakened the

GC&CS team's belief in success, and the prevailing pessimism acted as a self-fulfilling prophecy, preventing full determination to be unleashed. Interestingly, GC&CS had considerable record of successes in its attacks on Japanese machine ciphers. In 1934 Hugh Foss and Oliver Strachey managed to break the cipher of the machine named "Red" and used by the Japanese Foreign Office. In fact, the machine was named "Red" by the American codebreakers, who broke its cipher a year later, but their name was generally adopted by the historians. In the mid-1930s GC&CS was still concerned mostly with the Italian and Japanese codes and ciphers, practically ignoring German ones. Priorities must have completely changed sometime during 1938, most probably as a result of Austria's Anschluss and the Munich crisis. When the Japanese started using a new model of the machine (called Purple by the Americans) in February 1939, the British did not take up the challenge. Most historians claim that their indifference to the new challenge resulted from GC&CS' absolute focus on the Enigma problem.

In view of the stalemate in their own backyard, the British were more inclined to listen to signals from outside world. At the turn of 1938 and 1939 the first such signal came from Poland, through Paris. According to Hinsely, the first attempt to exchange experiences between Polish and British codebreakers took place in December 1938, but neither in Polish nor in French sources references to such a meeting were found. The historian of the British secret service most probably refers to the British-French meeting that took place in November 1938 in London. In his report Bertrand described the enthusiasm with which the British have finally welcomed documents delivered by "Asché". From the British source we know that they included four fragments of open text, the corresponding cipher texts and the machine settings used to encipher the messages. In spite of the British enthusiasm the documents did not open the door to the cipher, as the conclusions of the meeting included instructions for the agent to deliver pairs of open text and the corresponding cipher text of about 1.000 characters. Foss' account of the meeting indicates certain confusion regarding the documents delivered. According to Foss the open texts had been encrypted using the commercial Enigma model. Moreover, it was only from the documents provided in December 1938 that the British learned for the first time about the existence and role of a plugboard in the military Enigma. Question arises—what had happened to the documents delivered by Bertrand in 1931?

The first tripartite conference was the result of Bertrand's initiative, which he himself referred to in the discussion with Langer as Machiavellian. Bertrand suggested arranging an information leak to convince the Germans that the Enigma cipher was broken and make them give up the machine! Langer found himself in a rather awkward situation. He had to dissuade the author of the idea at all costs, but without revealing his real motives. Fortunately, Bertrand also had a second idea. It boiled down to bringing to one table the representatives of British, French and Polish codebreaking services in the hope that their discussion would yield new ideas. Cooperation in the intelligence matters, and in particular in the codebreaking, requires considerable capital of mutual trust. Franco-Polish and Anglo-Polish relations in the years directly preceding the outbreak of World War Two could hardly provide it. Franco-Polish military alliance signed in 1921 eroded since then and its practical dimension was reduced almost to null. Its last practical manifestation was Marshall Pilsudski's proposal of possible preventive war with Germany addressed to Paris in 1933. Negative response from France marked the practical end of the alliance. Less than a year before the outbreak of the war, British-Polish relations did not go beyond bare diplomatic correctness, and relations between the military, especially the intelligence services of both countries, were nonexistent. The Poles could not forget Lloyd George's behavior in the most difficult days of 1920, when the Bolshevik armies stood at the gateway of the country's capital and his own ambassador in Warsaw listed the objects of his aversion in a significant order: "I don't know who I hate more: Lloyd George, the Bolsheviks or the Poles." On their part the British were inclined to treat Pilsudski and his successors as fascist dictators of minor importance. In a nutshell, as early as December 1938, there was no sign of a future alliance between the two countries, so a meeting of the codebreaking services would act as the harbinger of its birth.

The meeting was organized in Paris, in January 1939. The instructions for the delegates of the Cipher Bureau, Langer and Ciężki, were precise: obtain as much information as possible about the work and possible achievements of the other services; do not reveal own success if the other parties could not demonstrate comparable level. Instructions to stick to generalities, presence at the table of so far unknown representatives of GC&CS, and finally language problems (German proved to be the only language common for the representatives of the three countries) turned the meeting into a total failure. Ciężki has reached or even exceeded the limit of his instructions, arguing that the key could be recovered using the Enigma operators' custom of choosing "childish" (as Denniston put it in his account) message keys. Despite these indiscretions, Knox described his interlocutors as "ignorant fools", adding the verdict "practical knowledge of the Enigma – zero". The participants decided that the work on the recovery of the machine's internals got stuck in an impasse, from which it could only get out using new information from the agent. A technical questionnaire was developed for this purpose and was to be submitted to the parties' intelligence services. The items of the questionnaire confirm that the fundamental problem of British and French codebreakers was the recovery of the rotor wiring. It was also agreed that the next rounds of talks would take place in Warsaw and London. Their dates were to be defined depending on the emergence of new circumstances. Such circumstances were soon provided by the Third Reich, occupying the Czech Republic and the Klaipeda region in March and making demands to Poland concerning the status of Gdańsk and the extraterritorial route through the Pomeranian corridor. These events have led the British government to take a decision unprecedented and contrary to the entire tradition of imperial politics, expressed in Lord Salisbury's motto: "Britain does not seek alliances, she accepts them". On 30 March, the Prime Minister of His Majesty's Government offered Poland a unilateral guarantee of its security.

At that time the BS4 team was going through the most frustrating period in the history of the service. It was not necessary to read the broken messages to recognize the seriousness of the impending storm; after Austria, the Czech Republic and Lithuania Hitler's gaze fell on Poland. And it was at this critical moment that the codebreakers, whose work for several years assured the most reliable information about the intentions of their neighbor, were almost out of business. As a result of changes in Enigma's operational procedures the river of information turned into a trickling stream. The real scale of the informational eclipse of the pre-war months remains unclear. British documents date the last Wehrmacht message broken by BS4 back to 14 December 1938. The report prepared by Poles in 1940 confirms that the last Enigma key before the outbreak of war was broken on 26 August 1939, the day of general mobilization in Germany. The introduction of two additional rotors did not invalidate all the Polish methods of attack. Thanks to the enemy's error the BS4 team managed to reconstruct the wiring of the new rotors. Before the outbreak of war Poles were able to complete the sets of Zygalski sheets corresponding to two rotor orders only. Their application should permit breaking the key in at least one day in a month. The blackout was not complete; BS4 team could read the Enigma messages to the very last moment before the outbreak of the war, although the percentage of broken keys fell dramatically in relation to the spectacular achievements of 1938.

Meanwhile, the political situation was heading towards a climax. On 28 April Hitler terminated the non-aggression pact with Poland. In the course of the staff talks, the French allies confirmed that in the event of an unprovoked German attack on Poland, France would strike with the majority of its forces on the fifteenth day of the conflict at the latest. British military mission continued talks in Warsaw, trying to find out how its country could meet the commitments resulting from its guarantees. At the same time Polish intelligence service assessed the international developments and the ability of the Polish army to resist the Germans. It was obvious that in the early stage of the hostilities Poland would lose control of a large part of its territory, the functioning of the state and the army would be disorganized, and in the ensuing chaos achievements valuable for the whole coalition could be lost. As a result, Cipher Bureau obtained the consent to reveal the Polish success with Enigma. On 30 June 1939 Langer dispatched the telegrams to Paris and London, containing the code word agreed during the January meeting, "Il y a du noveau", and scheduling a meeting for 24 July in Warsaw.

For Bertrand the invitation to Warsaw represented just another episode in the French-Polish cooperation. On the British side it caused some problems when Knox refused to participate in another meeting with "ignorant fools". Denniston convinced Dilly arguing that Poles specifically asked for his participation. As a result, on Monday 24th July, Commander Humphrey Sandwith, officer in charge for the Navy's listening stations,[2] flew to Warsaw directly from London. Alastair Denniston and Dillwyn Knox chose the train, taking, they thought, the last opportunity to take a look at Germany. Also, Gustave Bertrand and Henri Braquenié, codebreaker of the

---

[2] In Polish sources he is referred to as Professor Sandwich, however distortion of his name took place on the British side: in his memoirs Alastair Denniston wrote 'the Admiral suggested that Sandwich should go as well' (Robin Denniston, Thirty Secret Years, Polperro Heritage Press, Worcestershire, 2007, p. 118). The alternative version, according to which Stewart Menzies, then Deputy Head of the SIS, was hiding under that code name, is not corroborated by the sources.

French Air Force, came to Warsaw by Nord-Express train. Langer had to address a number of diplomatic problems. How to reveal to Bertrand the success of his service and hiding it for so many years? How to assure that his expected anger would not threaten the success of the meeting? He started accommodating the British in the Bristol Hotel and the French in the Polonia. This allowed him to arrange a private dinner with Bertrand on the eve of the conference, and share with him the news before anyone else could learn about it. Bertrand seemed pacified, but both officers knew that the day ahead is going to be long and difficult.

Next day in the morning everybody gathered at the BS4 headquarters, hidden in the Kabaty Wood. During the first, courtesy part of the meeting, Colonel Marian Smoleński, the then head of Polish intelligence, played the role of the host, but quickly passed the baton to the codebreakers. Major Ciężki delivered a three-hours long presentation outlining the story of Polish struggle with Enigma. Denniston in his account honestly admitted that he could not understand a word from it. He might have been distracted by Knox, who openly demonstrated his lack of interest, betraying with his body language that he found himself there against his will, considered the discussion a waste of time and has not changed his opinion about the hosts since the meeting in Paris. The atmosphere was slightly relaxed when, after the introductory presentation the participants were invited to a room in the basement of the building, where a more practical part of the show was prepared. There, with a dramatical sense of effect, one of the hosts took the veil off the device waiting on the table and, lo and behold, Enigma appeared to the guests' eyes. Langer answered Bertrand's inevitable question "Where did you get it from?" stating with satisfaction: "We made it ourselves". In the light of later accounts, not all of guests trusted this declaration.

More surprises awaited under the veil of black fabric, being revealed successively by the hosts. Further on the codebreakers demonstrated the process of breaking the key using Rejewski's bomb and Zygalski's sheets, using the German messages intercepted on the same day. This time demonstration was so convincing, that even Denniston "understood the reasoning". He wished to call British embassy in Warsaw to bring the engineers and draftsmen from London, to document the equipment. Langer assured that there was no need to bother the embassy, as the Poles prepared for each delegation one Enigma copy and full documentation of the remaining equipment and methods used by Polish codebreakers. At the latest at this point Denniston realized that in Paris it was the Poles who were vetting they potential partners, and they failed the test spectacularly.

Polish declaration of sharing the Enigma secret without any preconditions dispelled the fears of all the guests except Knox, who was still "silent as enchanted, clearly angry for some reason". He only opened up to Denniston in the car driving guests back to the hotel. Ignoring the presence of Bertrand and the accompanying Polish officer, he nearly screamed that "they are lying to us now, just as they did in Paris. (…) He repeated that it was a theft, they never worked it out, they stole it years ago and followed its development as anyone could do, but at first, they had to buy or steal it". Denniston was in an awkward position. Polish officer politely pretended to be disinterested. But even Bertrand, who did not understand English, started asking questions about sources of Knox's agitation. Denniston was afraid that his colleague could cause a diplomatic scandal and considered immediate return to London, without attending the second day of the meeting. Fortunately for the future Allied cooperation in the codebreaking Bertrand was able to dissuade him this idea over a dinner.[3] In his memories Bertrand remained discreet about the incident, noting only with obvious satisfaction that he witnessed the first situation when British codebreakers had to bow before the results achieved by Poles.

The second day of the meeting was dedicated to the technical questions, with Rejewski, Zygalski and Różycki playing the leading role. Rejewski's calm competence, supported by Zygalski's flawless English, pacified even Knox, who became a natural partner in this phase of the discussion. Dilly's first questions focused on the Enigma element on which his own attempts failed: the wiring of the entry drum. The discussion was not easy, as both sides used different terms referring to the same element. Rejewski was confused when Knox started asking questions about the "diagonal" of the machine, calling it alternatively "QWERTZU".[4] In his later report Knox expressed his concerns that material provided would not be properly understood, as the British use a different notation. However, in the constructive atmosphere of this morning an agreement was reached, permitting to proceed and leading directly to another crisis.

After Rejewski realized that the "QWERTZU" refers simply to the wiring of the entry drum, he explained with a smile that the wires are arranged alphabetically, which he discovered thanks to intuition and patience. Dilly found this information particularly frustrating. In his opinion machine constructors were clearly unfair reaching for the solution too obvious to deserve attention. Moreover, the Poles were successful in exploring the variant that he himself rejected a priori, even though its testing was suggested by an otherwise unknown Mrs. B.B.! In fact, Knox had a more important clue than Mrs. B.B.'s intuition. Scherbius had patented his machine also in the UK, and in the patent application the entry drum was wired strictly in alphabetical order, which suggested checking, even *pro forma*, this variant also in a military machine. Despite the temporary confusion further conversations went smoothly, permitting Rejewski to conclude in his post-war memories that the British "quickly figured out how we found the internal connections of the drums and quickly understood how our sheets and bombes worked". Contrary to initial concerns, the second day of the conference was a complete success. Even Dilly regained his good mood, and on the way back to the hotel hummed an improvised song whose chorus sounded "Nous avons le QWERTZU, nous marchons ensemble[5]"...

They all parted the next day. Bertrand and Braquenié flew to Paris. During their stopover in Berlin, they walked along Unter den Linden avenue and under the Brandenburg Gate. Near Reichstag they blended in with the crowd waiting for Hitler

---

[3] *In Warsaw, it was you who advised me to come back and made me think - you were right.* Denniston's letter to Bertrand, dated 03.08.1939.

[4] See description of *rodding and fastening on page xxx*.

[5] *"We have QWERTZU, we march together".*

to appear. Denniston and Knox went to London by train, but at the German border it turned out that Knox's passport was incorrectly stamped when entering Poland. Dilly had to go back to Poznan and get a new visa at the local German consulate. In this way, he unexpectedly gained the opportunity to visit the place, where the adventure of Enigma breaking began. Spending an evening in a local hotel, he continued writing report, which he started the previous evening in Warsaw Bristol. Judging by its tone irritation caused by visa problems and the loneliness of the hotel room caused another change in Knox's mood (although many people who knew Dilly well would probably say that he simply returned to his usual state of mind). In the report he attributed Polish success only to luck ('they read the machine [...] because they were lucky'). If Knox had listened to Mrs. B.B.'s suggestion in due time, now 'we would have instructed them'. He expressed doubts concerning knowledge and honesty of his partners ('I'm sure Schensky [Ciężki] has very little knowledge about the machine and tries to hide facts') and the validity of Polish methods of attack ('the military cipher from September 15 to April [...] has disappointed. [...] we must examine their system... with considerable skepticism. [...] Even the [rotor] wiring must be treated with mistrust'). Dilly did justice to Rejewski, Różycki and Zygalski's talents ('these young men seem to be very talented and honest'), and identified the crucial weakness of their methods—their dependence of the repeated message key ('all their successes were based on a factor that can be removed at any time').

Knox's report deserves particular attention, because it was not intended for the outside world and represents the most authentic reaction to Polish revelations. For many years it rested in the archives, unavailable to historians, but many of arguments presented therein keep on returning in historiography until today. It is therefore worthwhile taking a closer look at Knox's conclusions and assessing to what extent the future narrative was based on his first, very emotional reaction. Dilly blamed the Poles for concealing their success during the January meeting. Denniston in the report from Pyry confirmed that during the conference in Paris the British had not revealed their breaking of the commercial Enigma either. The Pyry meeting took place about a month before the signing of the British-Polish military alliance. Sharing the most precious secret with potential allies, without any preconditions, represented an unprecedented act in the secret service community. Over time, we will learn the history of British-American cooperation in the codebreaking and we will see that it represented a sharp contrast to the open attitude of the Polish Cipher Bureau.

Knox's second hypothesis concerned the sources of Polish success. He claimed that Poles had to steal or buy an early version of the machine and then watch its development. Knowing the history of breaking the cipher by Rejewski and his colleagues, we could have simply ignored Knox's argument if it hadn't been for the fact that it has determined the British perception of this story for many years and is still influencing the historical narration, even in the twenty-first century. Knox's claim that the Poles owed the cracking of the cipher only to luck requires comment. It obviously refers to Rejewski's guessing of the wiring of the entry drum. Dilly's emotional reaction was due to the fact that his own efforts broke down on this very element of the machine. Rejewski did not hide that in the first, decisive attack, he simply guessed the wiring of the drum, emphasizing immediately that he later worked out a method allowing to determine the wiring also purely mathematically. Unfortunately, he failed to convince of his honesty Knox, who claimed that the diagonal was obtained as a result of *Verrat* (betrayal) or 'through luck'. This part of the conversation must have been in French, because Dilly quoted Rejewski's words "mais nous l'aurions pu trouver par mathématique",[6] that referred to his purely mathematical method. Knox chose to pretend that he did not understand Rejewski's words and that they probably refer to the date the secret was purchased on!

After Knox's return to London, he passed on the information about the 'diagonal' to Peter Twinn, who tested it on the message delivered by Bertrand in December. It took Twinn just two hours to recover the wiring of the two rotors used when encrypting the message and confirm that the information brought from Warsaw is correct. When a few days later Josh Cooper congratulated Twinn on his achievement, he accepted congratulations indifferently, because with 'the information that Dilly brought from Poland, the task was not much more than a routine'. Knox presented this routine job as a bureaucratic alibi, trying to convince his colleagues that he was defeated on the home straight and only thanks to the rival's luck. Knox's report represents a model reaction of an official whose competence has been verified by the outsider. He travelled to Warsaw as an unquestionable British leader in the attacks on the machine ciphers. Passus in his report indicates that he was used to lecture rather than to be instructed. He questioned the very purpose of the journey, the knowledge and good faith of the meeting's hosts. Completely unprepared he had to face the success of people whose competence he seriously misjudged. His very non-British reaction was to minimize the scale of his rivals' success and attribute it to external (luck) or morally ambiguous (theft, bribery) factors to offset its influence on his own position within GC&CS In this situation, we can understand Denniston's words to Bertrand in a later letter: 'You must forgive me for caring about him, but (…) I will never take him to the conference if I can avoid it'.[7]

Knox's reaction addressed directly to Rejewski, Różycki and Zygalski, was completely different. He did not feel comfortable with mathematics and mathematicians. His classical education, his experience focused on the old-fashioned hand codes and ciphers, differences in terminology and trivial language misunderstandings made him unable to assess the real value of Polish revelations *ad hoc*, still in Warsaw. After return to London Dilly invited Alan Turing for the weekend to his country house in Courns Wood. There he presented the results of the meeting and, perhaps, a report provided by Polish codebreakers. Turing's verdict on them must have been favorable, as the following week Dilly dispatched to Poland the gifts, the meaning of which was not clear to the recipients. Each of the mathematicians received a silk scarf with, as Rejewski put it, 'a view from England', and a set of 'sticks with letters on them'. In fact, the scarves depicted a horse having won the Derby, while the 'sticks' were nothing else but rods—the main tool in Knox's attack

---

[6] "But we could also have found it through mathematics."
[7] Excerpt from Denniston's letter to Bertrand dated 03.08.1939.

on the commercial Enigma. The meaning of the gift was obvious: the image of the winning horse emphasized the Poles' priority in the race to the Enigma secret. Adding his tools Dilly acted like a medieval knight who, defeated in a duel, lays down his weapon at the feet of the victor.

Apart from the misunderstandings and emotions the conference brought also practical results. It was agreed that Poles would continue theoretical work on the machine and its ciphers. The most resourceful British would take care of the design and construction of the necessary devices. In practice, this meant perforating more than two million holes in new sets of Zygalski's sheets. The French were to continue cooperation with their agent attempting to obtain additional information. The parties also agreed to exchange the intercepted messages, with certain level of specialization: the British were to provide the Kriegsmarine messages, the French were to focus on the Wehrmacht traffic. On 7 August, Langer delivered to Paris in the diplomatic pouch two Enigma copies. He has spent two days instructing his hosts in the functioning of the devices, ending his visit with a dinner at Drouant's, celebrating with Rivet and Bertrand the strengthening of the cryptological alliance. On 9 August Langer boarded a plane to London, where he spent another week waiting for the French to deliver a copy of Enigma intended for the GC&CS. After Langer's departure, one machine remained at the intelligence service headquarters at Rue Tourville 2bis, the other one was repacked into the British diplomatic pouch and on August 16th it set off on its way to London, accompanied by a numerous committee. On the Golden Arrow train, Commander Wilfred Dunderdale, SIS resident in Paris, his deputy called "Uncle Tom"[8] and Bertrand himself took care of the precious luggage. The journey would have gone without noteworthy incidents if it wasn't for Dunderdale's imagination. He considered that the size of the pouch could attract the attention of German spies, whose presence on the train or at British customs in Dover he predicted with his sixth sense. On board the Calais-Dover ferry, he met his friends and neighbors from Paris, the actors couple Yvonne Printemps and Sacha Guitry. Considering the number and size of suitcases they were carrying on their way to tour in Britain, he decided that among them the pouch with Enigma would escape the attention of the Germans. In exchange for the promise to pass through customs without formalities, the French couple agreed to add another piece to their luggage. On the platform of Victoria Station, the Enigma's entourage was met by the deputy head of British intelligence service, Stuart Menzies himself. As he was just going to the party, he was in his evening dress, with a rosette of the Légion d'Honneur in a boutonniere: Bertrand misinterpreted his dress as an *accueil triomphale* of a secret delivery. In fact, Dunderdal's conspiracy could have been counterproductive. France-Germany Committee was one of the organizations through which the Germans tried to manipulate the French opinion. Its activities were being coordinated by Otto Abetz, later to become the Reich's ambassador to Paris. Both Abetz himself and a group of French activists were under close watch by the French

---

[8] The familiar name Bertrand used for Tom Greene, Dunderdale's long-time colleague and, after the war, his successor as a British intelligence resident in Paris.

counterintelligence service, SCR. Despite the Quay d'Orsay's strong opposition, the SCR was able to expel Abetz from France in July 1939. Sacha Guitry was among the French citizens on the SCR's watch list in connection with the activities of the France-Germany Committee. Entrusting him with a secret cargo in order to avoid the attention of German spies was thus somewhat grotesque idea.

Two weeks later real bombs were falling on Warsaw, where the Allies were recently watching the operation of codebreaking bombes. Polish intelligence service correctly identified most of the forces that Hitler gathered at the country's borders in preparation for the coming war. The Enigma's decrypts, provided by BS4 until 26 August 1939, might have contributed to this success. But even perfect intelligence cannot replace an armed force capable of exploiting the fruits of its work. At the end of the eighteenth century, just before the partition of the country by its neighbors, Jean Jacque Rosseau gave the Poles a practical advice: "If you can't avoid being swallowed, at least try to prevent them from digesting you". Over the next two centuries of their difficult history, Poles have made this advice into a useful motto. This time the role of the poisoned pill was played by the Enigma secret, discovered a month before Hitler came to power and handed over to the Allies a month before the war broke out.

Chapter 9
First Round

On September 1, 1939, not one, but two Polish-German confrontations began. The first, widely known, began with rounds of the *Schleswig-Holstein* battleship aimed at the Polish outpost on Westerplatte, bombs falling on Polish cities in the morning and tanks breaking through the border barriers. The second one was a very personal duel between a group of Polish citizens, fought away from their home country, but just like the first one, it attracted the attention of audiences around the world. On September 1, 1939 in the capital of Argentina the finals of the VIII Chess Olympics began. The Olympics, organized for the first time ever outside Europe, attracted so many participants that preliminary eliminations had to be conducted. Even in this phase of the tournament politics played an important role. Although the Czech representation appeared under the banner of the Protectorate of Bohemia and Moravia, the organizers rejected the demands of the German team to display only the swastika flag: the Czech flag waved in front of the Teatro Politeama. The German team was a great unknown of the tournament; it was strengthened by Eliskases and Becker, who had represented Austria so far. The Americans were a big absentee of the tournament; their financial demands exceeding the organizers' capabilities. In their absence, the Polish team became the favorite of the Olympics, having regularly ranked second in the previous few years. When news of the outbreak of war reached Buenos Aires, the tournament hung by a thread. Several participating teams represented countries which were or were to become involved into the conflict in the coming days. A couple of other teams gave up the competition for purely practical reasons. However, the Argentine hosts did not allow the thought that the long-awaited Olympics could be interrupted, so they pushed through the continuation of the competition during the team leaders' meeting. Doubts were soon confirmed. The matches between France, Poland and Germany did not take place and were entered into the minutes as a 2:2 draw. The jury faced a problem when the Palestinian team refused to play against Germany. The Germans demanded a default 4:0, which neither the jury nor the opponents wanted to accept. The problem was solved by the hosts, who offered to register their match against Palestine and the

M. Grajek, *Enigma Myth Deciphered*, History of Information Security,
https://doi.org/10.1007/978-3-031-65475-6_9

Palestine-Germany as a draw. All the participants considered it a generous gesture, except for the Germans, who also demanded that the Czech national team be disqualified. In their opinion the Protectorate of Bohemia and Moravia, participating under its own flag, was now a part of the German Reich.

In such an atmosphere, many duels on a chessboard had little to do with the royal game. The Polish team, dominated by players of Jewish origin (Tartakower, Najdorf, Frydman, Regedziński and Sulik) remained particularly sensitive to news coming from the country. They were gloomy, which caused sensational ties with Chile and Palestine and failures against Sweden and the Netherlands in the early stages of the finals. Finally, a victory 4:0 over Lithuania helped Poles to get together and start their climbing in the table. After a precious victory over the hosts, they got to the podium, and after the victory 4:0 over France even came to the fore, but the rivals again outstripped them, when the Polish team had a break in the next round. Eventually, winning 3.5:0.5 over Denmark in the last round did not allow to make up for the distance lost in the early stages and the points awarded to Germany as a result of political games. Nevertheless, the chess players defended their honor as convincingly as the army fighting in the country: they took second place, losing to the Germans only half a point, 36:35.5.[1] After the end of the tournament, many players from several teams, especially those of Jewish origin, decided to stay in Argentina, permitting the hosts of the 1939 tournament to win in the coming years five Olympic medals. Najdorf and Frydman were among them. Interestingly, all players of the German team also decided to follow their steps.

The British national team lost 1:3 in the elimination against Poland, but managed to qualify for the final sixteen. When the war broke out, the British, as the only participants of the finals, decided to quit immediately and boarded the first ship sailing to Europe. We shall soon meet three of its members—Conel Hugh O'Donnel Alexander, Philip S. Milner-Barry and Harry Golombek—working in BP.

As the chess Olympics in Buenos Aires came to an end, the main battles of the first campaign of the Second World War in Europe were over. Poland was a convenient target to confirm the effectiveness of the Blitzkrieg doctrine. Already in the first version, it proved that psychological effects are more important than purely military ones: paralysis of the will of the commanders, demoralization of the troops void of command, disintegration of the state structure and desperation of the civilian population, which became the target of the attack on an equal footing with the troops. Individual soldiers and units usually fought bravely, often effectively, but the Polish armies soon turned into a largely soldierly mass, without a coherent command. J.F.C. Fuller's principle, according to which "tactical success in war consists in turning an organized force into a disorganized one", was confirmed with a disastrous effect for the Polish army. The expected offensive on the Western front was never to come. The declarations of an attack with majority of forces on the fifteenth day of mobilization, repeated by France before the outbreak of the war and

---

[1] Individually, Najdorf took first place on the second chessboard, Frydman second on the third, and Regedziński third on the fourth. Tartakower, who played the first chessboard, had to give way to Capablanca and Alechin, among others.

confirmed in the interpretative protocol to the military alliance signed on 4 September, could be placed alongside the security guarantees offered to Czechoslovakia.

Officers of the Polish General Staff, educated in French military academies, undergoing internships in the units of French army, regularly reading French military journals and aware of the moods prevailing in France (*on ne va pas mourir pour Danzig*), should have soberly assessed ally's military doctrine. As early as 1935, in the course of a parliamentary debate, General Maurin, then Minister of Defense, rejected the arguments presented by supporters of the offensive strategy asking: "How can you think about taking the offensive, when we spent billions on building a fortified line? Are we supposed to be crazy enough to go beyond it looking for adventure"? Allied activity on the western front was limited to the occupation of several undefended villages. French soldiers posing under signs with their names written in Gothic letters presented themselves nice in the photographs in the French papers. Waiting for an allied offensive in the west, Poland was attacked from the east instead, which finally derailed any plans of defending the bridgehead at the border of allied Romania. Marshal Piłsudski's consistent, even if never openly stated, foreign policy principle was to keep peace with Germany. He realistically assessed that Poland could cope with other challenges in the region, while a conflict with Germany would have to end in disaster. Basing on this principle and considering the reality created by the victory over the Soviets in 1920 Polish military preparations were focused on the eastern border. Warsaw was a natural base for commanding the troops fighting in the east, at the same time the proximity of the German borders to the west and north posed a real threat to the capital and its headquarters in case of war with Germany. When this threat became a reality, Poland was unable to adapt rapidly the structure of the army, its command and logistics to the new challenge.

This applied also to signals intelligence. Its command center in the Kabaty Forest was approached by the German armored units within less than a week after the outbreak of war. When on September sixth and seventh, the Army headquarters was evacuated from Warsaw, the team of the Cipher Bureau was also boarded on one of the echelons leaving for Brest. Even before its evacuation BS4 was not able to support the fighting divisions. The intercept stations in the western part of the country were evacuated or fell into German hands during the first few days of war. Moreover, there was no secure way of passing information to the commanders of the field units. Field ciphers have been compromised in the first few days of fighting. Without any backup even at the highest level of command the officers communicated using the jargon codes based on facts known, at least in theory, only to both parties. A few years before the war, the Cipher Bureau developed and ordered several dozen copies of the Polish cipher machine Lacida. They were designed for the staffs from the division level up, but never reached their destination. Colonel Langer explained the reason in his report submitted in Paris: "cipher machines were a reserve for a later period of the war so as not to risk losing them in the first month of the war. (...) Without an express order, the head of the Cipher Bureau was not allowed to distribute the machines".[2] Contrary to the general opinion it seems that Lacida did see operational service during the September campaign, however, being used on board of Polish Navy ships. Bolesław Romanowski, famous submariner, wrote in his war memoirs that in the period immediately preceding the outbreak of the Second World War, " a new ciphering machine was introduced and Kuba (*one of the ship's officers*) was writing various sneaky texts, which one of us had to encrypt and the other had to decipher".[3] Polish Army HQ planned to organize a Lacida-based communication network once the fronts had stabilized, without risking the machine falling into the enemy hands as a result of unpredictable maneuvers opening the campaign. However, the most unpredictable element of the 1939 campaign was the lack of any stabilized fronts.

Unable to support the HQ with the fresh decrypts, Cipher Bureau's had to focus on protecting its secrets. This operation required obfuscating the tracks leading in several directions. The most important place requiring removal of traces was the BS4 site in the Kabaty Forest. The complex was basically designed as a communication center rather than the codebreaking facility, so it was easy to erase any traces of BS4 activity. The most sensitive components, in particular the rotors, bombes and Enigma clones were transported to the AVA workshop and melted in a blast furnace. Fokczyński and Betlewski buried less important devices in the forest surrounding the center. The remaining machines and documentation were carefully packed, transported to railway station and loaded onto the wagons. We know from the later German sources that the team left on the spot several copies of outdated codes and ciphers, possibly as an element of disinformation.

According to the diaries of the youngest member of the BS4 team, Kazimierz Gaca, September fifth was devoted to packing equipment and documentation and destroying unnecessary documents and equipment. On the next day, the staff and crates were loaded into the echelon "F", which departed from the Warszawa Wileńska station through Minsk Mazowiecki, Mrozy and Siedlce to Brest. The Brest fortress was originally chosen as a temporary seat of the Polish Army HQ after its evacuation from Warsaw. However, General Guderian's 19th Armoured Corps threatened the fortress from the north even before the C-in-C's staff was able to settle in the fortress for good, forcing the headquarters to evacuate further south on 11 September. The journey of train "F" to Brest lasted about three days, delayed by continuous air attacks. When the staff of the Cipher Bureau arrived on the spot, they were ordered to continue their evacuation immediately, following the staff of the C-in-C towards Włodzimierz Wołyński. The race between the evacuated team and German troops continued: the team left Warsaw two days before the Wehrmacht troops approached the city, and left Brest 3 days before the German attack on the fortress began. In Brest the team was divided into two parts. Parallel to the train, a truck column of the Cipher Bureau reached Brest. Key staff and equipment were

---

[2] Gwido Langer's report of 12.05.1940, p. 3 (from the collection of the Piłsudski Institute in London).

[3] Romanowski Bolesław, *Torpeda w celu*, Wydawnictwo MON, Warsaw 1981, p. 14.

transferred to incredibly crowded and overloaded vehicles. The column was ordered to proceed through Lutsk, Dubno and Krzemieniec, towards the Romanian border. Rejewski recalled an exhausting journey in a car packed with crates, during which he himself sat on a pile of spare fuel tanks. Despite the resupply in Brest, it quickly became obvious that there was not enough fuel for all vehicles. Langer and Ciężki decided to scrap some trucks and scramble the staff and equipment in the few remaining. Rejewski recalled how during the stops near Lutsk and Wlodzimierz Wolynski the unnecessary documents were burned and the equipment was buried. Somewhere in the land belonging today to Ukraine, remnants of Polish Enigma clones and several dozen Lacida machines are rusting.

The fate of the second group, evacuated from Brest by train, was more dramatic. The troop-train was travelling through Kowel, Rowne and Zdolbunow. Until then, the train was heading southeast, moving away from the attacking Germans. However, in Zdolbunov the railway line turned towards Lviv, to the southwest. On September 13th, during the stop at Brody, the convoy command had to receive messages that were interpreted as a threat signal, deciding to burn the most secret Cipher Bureau documents. The danger materialized soon. The train arrived in Busk, where the railway line to Tarnopol branched off—from that moment on, it was supposed to be distancing itself from danger with every kilometer. However, still at the outskirts of Busk, the train was heavily bombed. During the air raid at least 2 officers and about 40 soldiers were killed, one of the employees of the Cipher Bureau was badly wounded. The train was unable to continue its journey, so passengers travelled another 30 kilometers using improvised means of transport. In Zloczow they luckily found another military train, which managed to get through Tarnopol and Trembowla almost to Czortkow, from where it would be not far to Zaleszczyki and the Romanian border. But it wasn't the end of the team's adventure. Gaca's records show that on September 14th, near Kopyczynce, passengers tried to push the train in an attempt to roll it to the nearest station where coal could be found. Desperate attempt was doomed to failure; unable to continue the journey by train, the fugitives decided to look for horse-drawn carts. They decided also to deviate from the main route in the direction of Husiatyn, probably for fear that the wave of refugees cleared the neighborhood from any transport. Luckily in the nearby Wasylkowce they managed to find hospitality and some horses. The wagons travelling via byroads did not attract German aircrafts, but they were moving very slowly. It was not until September 16th that they managed to get out on the main road in Muchawka, bypassing Czortków from the south. When it seemed that the road to the Romanian border was straightforward, on September 17, the news of Soviet invasion spread in the crowd. For the team of the Cipher Bureau the prospect of Soviet captivity was just as menacing as the German one. In order to move away from the Soviets and use the corridor between the armies of both aggressors, the convoy turned from the road to Zaleszczyki to the west, to reach the border in Kuty. After a night spent in Siemakowce, the fugitives managed to cover only a dozen or so kilometers, when the German bombing surprised them again in Horodenka. They decided to continue the journey at night, covering a distance of 84 kilometers within a dozen or so hours. On 18 September at 15.46 the group was on the border bridge in Kuty.

Both echelons of the Cipher Bureau crossed the Polish-Romanian border on 17 and 18 September, the race that had started less than two weeks ago came to an end. Enigma's secret slipped through the last open gate, just a few hours before it finally slammed. The team was an attractive target for both aggressors, representing the sum of the Polish experience in the codebreaking. Its commanding officers knew all the details of the structure and operations of the Cipher Bureau. Researchers represented precise knowledge of the methods of breaking ciphers. The cipher clerks' memory could replace a big part of the documentation lost along the way. Besides, the team did not part with the most important items of its equipment: two Enigma clones and Lacida. For safety reasons, the equipment was divided between the teams evacuated by cars and railway. In the railway group, its youngest member, Kazimierz Gaca, took care of the valuable cargo. Cipher Bureau obviously hoped to be able to resume its operation at the Romanian bridgehead or in exile. The evacuated team included staff representing two crucial sections of the Cipher Bureau: BS3—Soviet ciphers, and BS4—German ones. Its seizure by the invaders meant a deadly threat to the people and their service's secrets. Falling into German hands meant compromising the Enigma secret, and the loss of its value for the Allies. Team members could not have known that when crossing the border, but more cruel fate resulted from the prospect of Soviet captivity; all of them would probably end up in the mass graves of Katyń or other places of the massacre ordered by Stalin.

For the team of the Cipher Bureau and selected AVA employees, September campaign ended when they crossed the Romanian border. The concern for the Enigma secret has temporarily disappeared, but the anxiety about families left in the country remained. Rejewski left his wife and two children in Warsaw; son Andrzej, was only three years old, and his daughter Janina, was born just a few months ago. Polish Radio reports from Warsaw gave an image of life in the besieged and bombarded city, and the imagination suggested what was absent in the news. Różycki's son, Jan, was born just four months before the war broke out. His father's anxiety was multiplied by the decisions he had made on the day of the evacuation. For the employees of the intelligence service it was obvious that during the campaign a large part of the country will be occupied by the enemy, and central Poland, including the country's capital, may become the scene of fierce battles. Hoping to protect his wife and four-month-old son from the horrors of war Różycki convinced them to seek safety with the family living near from Rzeszow. On his last visit to their Warsaw home before the departure, he was surprised having found his wife and son there. They were both unable to get through the crowd on the platforms of the railway station and reach the train. Desperate, he took his family to the train waiting to evacuate the staff of his unit. Everyone happily found a place on the wagons when the heavy bombing gave a foretaste of the further journey. In the evening the train set off, but the tracks destroyed by bombing caused it to reach Siedlce only two days later. On the way, several times they had to hide from the bombers in the fields near the tracks. During one of the raids, the bomb carved a large crater a dozen or so meters away from the hidden family, covering them with sand. On the same day, they watched the German plane landing in a field nearby and releasing a group of agents, including a few dressed up as priests. On 9 September Brest welcomed the train with beautiful weather and another air raid. Różycki and his wife, who went in search of milk for their baby, were just running for a hiding in the trench when the bomb hit it. After returning to the station, they found station buildings and train heavily damaged, but their son alive and healthy. Their joy was disturbed by a message brought by Zygalski: three codebreakers were to leave the train and continue their journey by road. This was Różyckis' third farewell in the few days of the war. Jerzy left his young wife and child alone, amidst the chaos of evacuation, bombing and food shortages. His anxiety was compounded by the later news of the Soviet invasion and occupation of the area where his loved ones found themselves.

Antoni Palluth was in a similar situation. On the eve of his own departure, he managed to get his wife and sons out of Warsaw, giving at their disposal a company Chevrolet and a driver. Predicting the direction of the evacuation, he agreed with his wife on several possible meeting points, and later, during his journey towards the Romanian border, he used every stop to reach the nearest contact point by motorcycle—to no avail. Meanwhile, his family, travelling only at night, was breaking through to the east among the increasing chaos and danger. The first part of the journey came to an end on September 18, when just before Równe they were stopped by the Red Army, which confiscated their car and luggage. Then came the night of horror, when the Polish population of Równe and the refugees gathered in several houses at the market square fearing that the local Ukrainians would meet their threat to kill every Pole in their reach, with the silent approval of the Soviets. The next morning a strenuous return to Warsaw began, the stages of which were paid for with the remains of property left by the Soviets. When, after nearly a month's journey, the whole family reached the capital, they were relieved to find the house untouched by bombs, bullets and fires. Maria Różycka and her baby were also able to reach their Warsaw home.

Sylwester's sister and Antoni Palluth's niece, Irena, was less lucky. She has just graduated from the gymnasium and finished her last peaceful holiday, and on 1 September she was supposed to start her first job ever, of course at the Cipher Bureau. The first days of the war were not a good time for the adaptation of the novice, so her cooperation with the Cipher Bureau ended on the same day it started. As a result of military automatism, Irena was evacuated by one of the trains in the direction of Brest. There, the refugees from Warsaw were seized by the Soviet army. Young girl would probably avoid the repressions the military and policemen were subject to, but the soldiers searching through her belongings had found a document confirming employment in the Cipher Bureau. On this basis, she was considered an enemy of the working people and sentenced to 10 years in the gulags. Thus, Irena Palluth's first practical work experience was the timber felling in the Siberia, from which she was saved only two years later by the Sikorski-Mayski agreement.

Pyry facility was identified by the Germans as the communication center only. Some reports indicate that during the war a visit to the center represented a part of training for the German signals staff. Another place requiring attention was AVA. Repeated shelling of the company's headquarters and warehouses during the siege of the capital indicates that the Germans might have been aware of the military profile of the plant, as well as the search for and questioning of the company's employees after the capitulation of Warsaw. All Enigma-related equipment and supplies had been destroyed beforehand. Edward Fokczyński, deeply involved in engineering work on the codebreaking equipment, was evacuated to Romania. Those employees who stayed in Warsaw were not only loyal to the company, but their fragmentary knowledge of the secret projects did not allow to determine their nature. Their interrogations gave Abwehr the impression that AVA was focused entirely on the wireless technology. Its peace was only disturbed by the elusiveness of Palluth and Danilewicz, whose search continued throughout the war.

Army HQ, and in particular its second Department—Military Intelligence, were the main recipients of the information from the signals intelligence, so the traces of the codebreaking operation had to be erased there too. Defense of Warsaw lasted for over three weeks, giving ample time to properly secure or destroy the crucial documents, therefore the events described below represent embarrassing and still unexplained secret. After the capitulation of Warsaw, Abwehr's officers combing through the capital, found empty safes and worthless documents in the Saxon Palace, earlier Army HQ. However, when Captain Bulang opened the gates of the Legions' Fort, he was surprised by the sight of shelves filled with files with signatures clearly indicating their intelligence background. They represented a considerable part of the archive of the Second Department, including files of sections working on Germany. Documents filled 6 trucks, and were immediately transferred to the Abwehrstelle in Gdańsk for analysis. Their content contributed to the compromise of over 100 Polish agents in Germany. Among the documents there were no files relating directly to signals intelligence or the codebreaking. The Abwehr officers have only identified three messages transmitted in 1937 from a German cruiser operating in the Spanish waters and originally enciphered using Enigma. Isolated character of the discovery suggested that Poles could have received their clear texts from the agents rather than the codebreaking; the operation of systematic breaking of German ciphers should have left more abundant traces.

But Abwehr was able to come across the circumstantial evidence of the Cipher Bureau. The archive included financial documentation of the Polish intelligence service. The analysis of payrolls confirmed that it included the section, codenamed "Wicher", whose other traces were absent in the seized archive. The section employed mathematicians, which might link its activities with the cryptology and the codebreaking. Their remuneration was relatively high, which might indicate some success of their work. Those traces of the Cipher Bureau's activity triggered an investigation whose code name was formed from the Polish codeword for signals intelligence—"Fall Wicher". The disturbing news quickly reached France. It was only in 1983, that Paul Paillole, former officer of the French intelligence, learned from his former associate, Henri Navarre, that the last meeting with "Asché" took place already during the war, on 10 March, 1940, in Lugano. During the meeting Schmidt was furious about the French indiscretions that put his safety at risk: "SD has obtained evidence in Warsaw that the Poles managed to recover the Enigma machine. They're screening the Chiffierstelle staff". The post-war interrogations of the German intelligence staff did not confirm their knowledge about the recovery of

Enigma by Polish codebreakers. Writing his memories Paillol was recalling grave errors committed by the French intelligence service, starting from the leak of Hossbach protocol. He might consciously or unconsciously distorted Asché's true words to shift the blame for endangering Enigma secret onto the Poles. But one thing was definitely true; German investigation into Enigma security was worrying enough.

Chapter 10
In the Dark

When thousands of Poles were breaking through to the Hungarian and Romanian borders in the hope of continuing their fight in exile, the Allies were considering the choice of an appropriate name for the actions undertaken on the western front. The British began by carrying out on the night of the 3rd to 4th September an operation called "Western Air Plan 14"; ten planes dropped thirteen tons of leaflets over the Ruhr, warning the Germans of the consequences of war. The action was vigorously continued, and on 1 October the planes with leaflets reached Berlin. Ninety-seven million leaflets were printed during the first month of operation, but fortunately for air crews risking their lives over Germany, the plan was abandoned after the first thirty-one million copies were dropped and the remaining leaflets were spent for milling. This episode was termed "confetti war".

On September 7, small French force crossed the German border at Saarlouis, Saarbrücken and Zweibrücken, entered the Reich for a few kilometers, and then retreated behind the Maginot line in a feeling of job well done. At that time all German units representing any combat value were involved in the fighting in Poland. On the western front the situation of the Germans was similar to the one described in a saying from the previous war: *Im Osten steht das wahre Heer, im Westen nur die Feuerwehr.*[1] After this incident, the western front froze for a few months in stillness, leaving aside the occasional artillery rounds, given occasionally in honor of the generals visiting the front line. Also, in this case care was taken to ensure that the shells did not hit any important target, which could provoke the enemy to respond. The mobilized soldiers praised the barbed-wire mesh, making easier drying the washed pants and footwraps. Veterans remembering the frontal realities of the previous war wiped their eyes in amazement and described the ongoing conflict as a "funny war" (*drôle de guerre*).

---

[1] "A real army is fighting on the eastern front. In the west, only the fire department."

M. Grajek, *Enigma Myth Deciphered*, History of Information Security,
https://doi.org/10.1007/978-3-031-65475-6_10

The British had less reason to treat war in terms of humor. The first British victim of the war, less than two hours after joining it, was a bomber pilot who took off from Hendon airport and crashed his plane on the street of a nearby town. Further developments in the skies also did not inspire optimism. On 4 September the first British air raid on the ports of Wilhelmshafen and Brunsbüttel turned into a tragedy. A third of the planes failed to find their target. On the way back to the base, the pilots, who had to get rid of their bombs, have found in the North Sea a potential target. Fortunately, at the very last minute before the bombs were dropped, they noticed that they were attacking Royal Navy ships. Pilots of two Wellingtons dropped the bombs on the port of Esbjerg in neutral Denmark, 110 miles away from their target. Half the planes that found the target were shot down by the Flak. The three bombs that hit the battleship Scheer did not explode. The author of the only German losses was an airman named H.L. Emden, whose plane was shot down and crashed into the cruiser... Emden. Fortunately, nothing that happened in the air could affect the mood of the British, focusing their attention mostly on the seas and oceans. But the beginning of the war didn't go well there either. On September 3 the German submarine U-30 sank the liner *Athenia*. On the day the Soviet troops entered Poland, the U-29 sank the carrier HMS *Courageous*. On October 14th, the U-47 slipped into the Royal Navy base at Scapa Flow and sank the battleship *Royal Oak*. The pride of the British Empire was challenged on the seas, which it used to treat as its domain. At the same time, the British divisions, gradually transferred to the continent, remained as active as their French comrades-in-arms. Expected, devastating German air raids on British cities and ports have not materialized. The British have coined for this particular war a term as adequate as the French: "phoney war".

The Germans also made their contribution to the new war terminology. The first practical attempt at Blitzkrieg doctrine in Poland ended in a spectacular success, but it was paid with significant losses and disorganization of the units. The reorganization of the battered divisions, their transfer to the western border and preparation for the upcoming operations were to take more time than planned. On 27 October Hitler issued instructions to strike in the West on 12 November. When the deadline proved unrealistic, he was postponing the date of the attack several times by another week. Finally on 20th November he issued a new directive in which he ordered the troops to remain on standby, but did not define a date for the start of the operation. To the surprise of the Germans, the Allies did not take any action in the period of their weakness, leaving enough time to prepare their own strike. In contrast to Blitzkrieg in the east, this phase of the campaign in the west has become known as "Sitzkrieg", or Sitting War.

This early phase of the conflict could nearly become a kind of a phoney war also for the GC&CS codebreakers. The knowledge and hope they had received a month earlier from their Polish colleagues has changed the situation dramatically. Around 1 August, the staff of the military sections received orders to prepare for their departure from London, in the middle of the month they installed themselves in BP, and in the third decade of August diplomatic section followed them. The start of work in the new place was difficult, mostly due to the chaos following the relocation and scarcity of space. SIS headquarters also moved to Bletchley occupying the upper floors of the mansion. The rooms on the ground floor had to house 186 people of the GC&CS staff. Malcolm Kennedy recalled that after arriving from London there was no lunch waiting (the cook from Savoy apparently did not allow himself to be commissioned again), the Elmer's School building, which was to become the seat of the Japanese section, was full of students, and the distance of the designated billets from the place of work convinced him and his friend to hire a room nearer from BP.

Luckily for the codebreakers, the July conference in Pyry already produced results before the team moved to BP. After return of the British team from Warsaw, the Exchequer allocated several thousand pounds to adapt the property to codebreakers' needs, even before they managed to break a single Enigma message. Captain Hubert Faulkner's construction company, which originally intended to demolish the mansion and build a housing estate in its place, took care of the construction of temporary wooden huts. Faulkner himself, an avid rider and hunter, was watching over the progress of the work from the height of his thoroughbred. Water tanks were removed from the mansion tower and the SIS wireless station moved in there from its previous location at Barnes High Street. Antennas were stretched between the chimneys of the building and the tops of trees in the park. Telephone and telegraph wires were connected to the former Sir Leon's billiard room, where the communications center was located. The park area was fenced with barbed wire, local population was discretely informed that the fence protects an anti-aircraft defense unit. This story did not explain the presence of a large group of civilians representing, to put it mildly, non-military behavior. The rumor has it that later it has been replaced by the one presenting the facility as a lunatic asylum for the particularly dangerous cases. It seems that the rumor was not too far from the opinions of the center's staff. When one of them, Angus Wilson, suffered later on a nervous breakdown, he was offered therapy at one of the asylums working with the Foreign Office. However, Wilson thought it best to undergo a period of convalescence in an asylum already familiar to him, i.e. the BP. The huts being built were soon to be populated with the new staff. Starting from September 4 previously selected candidates for codebreakers were appearing one by one at the center's gate. Facing the war necessity BP heads were forced to drop their preference for physicists and start to accept also the mathematicians. Denniston regretted that "for several days we have been forced to hire professors from the reserve list". Among the earliest mathematicians to reach BP were John Jeffreys, Alan Turing and Gordon Welchman.

Of all three of them Alan Turing attracted the most attention, both because of his position in the world of mathematics and his particular way of being. He came from a respectable family whose members were scattered around the world, from Canada to India. When entering King's College in Cambridge, Alan followed in the footsteps of his grandfather, who after completing his math studies chose a vicarage at the university. Alan represented a remarkable personality, which modern psychologists would probably describe using a number of academic terms, most of them starting with a- or dis-… He was brilliant in anything he treated as his passion, mostly in mathematics and long-distance running. In other areas of life his behavior oscillated between funny and embarrassing. He appeared in BP in a glory of a scholar who solved one of the classic problems in mathematics defined by David

Hilbert. It is doubtful whether his new employers were able to appreciate his scientific achievements. In their eyes Turing's personality certainly confirmed any earlier concerns regarding employing mathematicians in the secret service.

Despite the organizational chaos prevailing in the center in the new, wartime reality, all of the new recruits fell into a whirlwind of feverish work from the very day one. The least spectacular and at the same time the most important task was reserved for Jeffreys. He was supposed to complete a task the Poles were unable to finish before the outbreak of war: to prepare sixty sets of Zygalski's sheets. Knox considered it a risky venture. Even a slightest change in the connections of any of the rotors threatened to make this work useless. However, in September 1939 the Allies had no alternative to Zygalski's method. Jeffreys began by designing his own device, the design of which mimicked the Polish cyclometer. The device was to distinguish "males" from "females", so his team members called it a "sex cyclometer". The first version was constantly breaking down, so after some time it was decided to build new one. The first one, known as *Mouse*, put the mark on the sheet where the hole should appear (then perforated by hand). The second one, *Waterwheel*, made it easier to check the correctness of the sheets by lighting the bulbs in the positions of the holes and allowing to compare both patterns. Jeffreys and his colleagues must have felt the burden of their responsibility, which was almost palpable in the gazes and delicate allusions, the meaning of which could be reduced to just one question—when will be the sheets complete? Jeffreys' team only took one break when the first million holes were punched to celebrate the halfway point, very appropriately with the vase of punch. The completion of work was celebrated in the same way, in mid-December 1939. Zygalski's method was referred to in the slang of British codebreakers as the "Netz", and the Zygalski sheets were consequently called "Netz sheets". Later on, Jeffreys implemented another idea proposed originally by the Polish team—construction of a catalogue indicating not only the appearance of a "female" in certain position of rotors, but also defining the letter representing this female. This catalogue was known among the Poles as the "Catalogue F". Jeffreys' team was responsible also for the implementation of yet another catalogue, known in BP as the "Jeffreys sheets". That catalogue was used in conjunction with Knox's 'rodding'.

Welchman, formally assigned to Knox's group, failed to win his sympathy. When he arrived in BP, he must have been surprised by the brusque reception offered by his new boss, but later, on the basis of his own experiences and conversations with other team members, he came to the conclusion that "Dilly seemed to hate most of the people he was in contact with". However, Knox had an instinct allowing him to make the right decisions even on the basis of false prerequisites. Wanting to get rid of Welchman's presence, he assigned him an unspectacular task, which in time was to break out of the original scope and shape the structure of BP as an effective intelligence organization. Welchman and Alex Kendrick were placed in the Elmer's School over the piles of German messages intercepted in previous months. Their task apparently had nothing to do with codebreaking. The British stations intercepted thousands of encrypted messages forming a chaotic picture. German operators were in the habit of transmitting a reply to a message on a different frequency to the one they received it on, making it difficult to link both parts of the exchange. The frequencies used in each network were changed daily, as were the station call signs. The deciphered message reveals the identity of both the sender and the receiver, but in September 1939 the British were not ready to break the Enigma. The structure of the communication networks reflects to a large extent the organizational structure of the army. The identification of this structure without breaking ciphers, solely on the basis of the relations between communicating stations, is called traffic analysis *(TA)*. Welchman had no experience in this field, not even knowing about the existence and function of the call signs, but he received help from two sources. Colonel Tiltman offered the advice of an experienced sergeant, who in his team was archiving the messages and knew a lot about their identification. The second source was placed on Welchman's desk by Josh Cooper, head of the air section, although it originated from Poland. It was a package of messages and corresponding clear texts, covering only two or three days, probably handed over at the Pyry meeting or sent later with Enigma clone. Welchman decided to touch upon the experience of the people who intercepted the enemy's messages, and paid a visit to Chatham, at the Army intercept station. Some of the information obtained during the visit proved to be useful in solving the puzzle of the structure of German communication networks. Drawing the diagrams every day, on which he marked the call signs and frequencies of communicating stations, Welchman soon noticed that they were arranged in three independent structures, corresponding to different communication networks. After identifying each of them, Welchman marked their signals with crayons of a different color: red, green or blue. His selection of colors was generally adopted and since then German networks have been referred to as Red, Green and Blue. After the new German networks appeared on air, other colors were used until the palette of available crayons was exhausted. As during the war the production of school crayons has been suspended, British cryptologists ordered them in the USA. Several boxes of colored pencils were apparently the earliest American contribution to the code-breaking operation in BP.

At that time the listening stations were sending the intercepts to BP in bags delivered by the motorcycle dispatch riders. Only after the bag was delivered Welchman was able to start working on network identification and station assignment. On the other hand, until networks and stations were identified, listeners were working in blind, mixing messages from different networks. Welchman suggested that after receiving any message its header only (including the call signs) should be immediately forwarded from Chatham via a teletype to BP. With the call signs, Welchman was able to complete the network identification and pass this information back to the intercept stations, where work was distributed among the operators according to the network structure. This procedure was named "Welchman's special". In the first months of the war Chatham provided an abundance of Red and Blue, but relatively few Green messages. The subsequent breaking of the keys of these networks revealed that Red and Blue were used by the Luftwaffe, while Green by the Wehrmacht. During the "phoney war", Wehrmacht units rarely used wireless, sending messages by wire only. Moreover, most Wehrmacht transmitters remained out of range of the British intercept stations. In that period most Green messages were intercepted by French stations and reached BP through a liaison group in Bletchley, consisting of several French officers.

A technical digression will be useful. The creation of multiple communication networks has at least two objectives. Radio waves can be received by all receiving stations within the range of the transmitter, while the sender generally directs its message to a defined group of correspondents. For the stations beyond that circle the message not only has no meaning, but should remain secret. If all the stations were working in a common network, any break-in would have disastrous consequences. Dividing the communication system into separate sub-networks reduces the effects of an opponent cracking the ciphers used in any of them. The architects of the German communication system allocated a different set of Enigma keys to each communication network. In each message, before the relevant ciphertext, the sender provided a header containing information of an administrative nature: the call signs of the sending and receiving stations, the time of sending the message and the number of characters contained, a tag identifying the message as a one-part or continuation of a message sent earlier and a three-letter identifier indicating which of the network's Enigma keys were used to encipher the message. If the operator on the receiving side found an unknown key identifier, he could skip the reception of the message—without the right key he could not decipher it. In a system constructed in this way, there was a direct relationship between working in a specific network and the use of a specific Enigma key. As a result, codebreakers in BP used the terms "break a communication network" and "find the daily Enigma key used in this network" interchangeably. The economics of language usually dictated that this phrase be shortened to the concise statement "we broke the Red key" or, simply, "we broke the Red".

When Jeffreys and Welchman were busy creating technical and organizational foundation of the intelligence gathering process, Knox and Turing have focused on the pure codebreaking methods. Dilly was concerned about the fact that the methods of analysis developed by the Poles take advantage of a weakness in the way the Enigma is used, consisting in the repetition of the message key. Avoiding any repetition is a cardinal principle of cryptography, while the experience and professionalism of the enemy made him believe that the Germans would notice their own error and eliminate it. Knox was determined to develop cipher analysis methods that do not rely on the repeated message key. He installed himself away from the bustling mansion, in a former gardener's office, simply called the Cottage. Dilly faced a problem that required a completely different experience from the one he had gathered so far. After the news from Warsaw Enigma cipher, which so far represented for him a kind of a linguistic puzzle, turned into a purely mathematical, or rather logical, problem. The complexity of the challenge seemed to be beyond the capabilities of any single person. Even with the solution at hand, its exploitation required the teamwork in a dimension unprecedented in the history of the codebreaking. Finally, Polish experience indicated that the ciphering machine needs to be confronted with another machine, replacing a duel of minds with a competition of machines and their constructors. Dilly did not feel comfortable in any of those fields.

Fortunately, Turing could fill in at least some of the gaps. His mathematical interests included group theory—field of mathematics useful in describing Enigma. While at Princeton University, Turing met Godfrey Hardy, but never imitated his tendency to lock himself in an ivory tower of pure mathematics, undertaking frequent excursions into experimental and applied math. Before devoting himself to mathematics, he was passionate about the problems of experimental chemistry. His personality was also influenced by his youthful friendship with the prematurely deceased Christopher Morcom, who infected Alan with his predilection for astronomy. As a result, even if Turing's mind was focused on mathematics, he practiced it in his own special way. He based the proof of the thesis on computable numbers on a revolutionary concept of an abstract machine, capable of performing several simple operations. His idea proved to be a useful way of describing many theoretical problems and entered the world of mathematics permanently as a Turing machine. Soon he went from theory to practice and that's because of the ciphers. As early as December 1936 he wrote in an account from a party organized by two friends: "they have prepared (…) cryptograms, anagrams and other puzzles, completely incomprehensible to me. …I'm not good at such things." Less than a year later he developed the concept of a cipher, the use of which required the multiplication of large numbers. To make his idea practical, he designed and constructed a relay-based device, automatically multiplying the numbers represented in binary form. The outbreak of the war interrupted Alan's work on yet another device: he designed a machine whose task was to determine the approximate location of the zero point of Riemann's *Zeta* function. While working on it, he did not avoid getting his hands dirty while turning and grinding the gears and other parts of the device. It was Turing's non-gentlemanly, according to the standards of the time, interest in engineering that made him well prepared to take on the challenge of the Enigma, both in theory and practice.

His predisposition to teamwork was slightly worse. He usually worked alone, challenging his own mind and not paying attention to the research and publications of other mathematicians. As a result, he twice achieved his important research results soon after other scientists solved the same problems. While supporting Turing's application for a scholarship at Princeton, Max Newman wrote to Professor Church that one of the goals of Alan's journey to the United States is "for him to come into contact with leading researchers in this field, so that he does not develop into the type of declared solitary". It seems that time spent at Princeton did not meet Newman's expectations, possibly due to the fact that Turing did not hide his sexual orientation. Even if homosexuality was not the norm in British universities and public schools of that era, it was accepted or at least tolerated. Dilly Knox himself has found this out: if we can trust Lytton Strachey's memories, Dilly rejected his advances during his studies. However, according to the memoirs of the great economist, John Maynard Keynes, Knox couldn't reject his offers. Behavior accepted in the universities of the Old England was rejected in the puritan New World—perhaps in part because of this Turing was unable to enter the inner circle of scientists shaping the intellectual atmosphere of Princeton. Consequently, his tendency to live and work alone was rather strengthened during his stay in the USA. News of Turing's preferences must have preceded him, as when he came to BP, Knox made it clear that tolerance for homosexuality in the British civil service is not as high as at the universities, so it would be prudent not to advertise his preference too openly. This delicate suggestion later became an additional source of clash between Knox and Turing, when Alan found out about Dilly's youthful adventures and accused him of hypocrisy.

In the Cottage, the constant noise of equipment perforating Zygalski sheets did not allow for concentration, so Turing decided that the best place to work would be the attic. Although the new Turing's temple could only be reached by means of a ladder, and a jug of coffee or tea was pulled up by means of a block and rope, the whole hustle and bustle of the world did not get there. Turing's introduction to the Enigma's problems must have been quite hasty, but he certainly received a package of documents from Warsaw. In accounts prepared after the end of the war several BP codebreakers mentioned a document written in "stilted German ", in which Poles summarized all the methods of analysis of Enigma ciphers they had developed.[2] There is indirect but solid evidence that Turing has been deriving from it systematically from the first moments of his work. In 1940, Turing summarized the results of his studies on the Enigma in a document referred to by modern historians as " Turing's Treatise on Enigma", but known in BP under a more familiar name: "Prof's Book".[3] Both Prof's Book and other post-war accounts by BP veterans contain numerous and obvious quotations from and references to the original study by Rejewski and his colleagues, readable despite translations into two foreign languages on the way. These quotations refer not so much to the description of the machine itself or the cipher it generates, but to the theoretical foundations of its analysis. Turing also had a package of messages provided by Poles. This is indicated by the genesis and examples of some methods (e.g. **FORTYWEEPY**) and the details of the message interception (torpedo boat call sign **AFA**).

Despite obvious borrowings, Turing seemed to deviate from the direction of the Polish research so far. From the early success on the attention of the BS4 team was focused on the repeated message key. German cryptologists must have been aware that the repetition represents a potential weakness, but they probably thought that it is more than compensated for by the strength of the machine cipher. An inconspicuous feature of the cipher was used by Rejewski first to find a gap in the Enigma's cipher security, then to open the door wide. Keeping up with the changes introduced by the opponent, the Poles used the effects of his initial error to the farthest limit. An important side effect of grappling with the new challenges was their continuity of cipher analysis. This, plus adversary's minor mistakes, contributed to the fact that the Poles were watching the evolution of the cipher system rather than revolution. They were able to address the new challenges one by one, never having faced the

---

[2] This document has not been found in the Allied archives to date (2024). In 2017, the author found a document in the French archives representing probably an abridged version of the report.

[3] Turing was nicknamed '*Professor*' among his BP colleagues. It was partly due to his position as an intellectual team leader, and partly probably due to Turing's behavior, embodying the stereotype of the Professor's not very practical attitude to reality.

crisis of several changes happening at the same time. Knox probably could not match Rejewski's and his colleagues mathematical background and that put him at a disadvantage in the race to the first breakthrough, but he certainly had a longer and more comprehensive experience in practical cryptanalysis. Experience has taught him that the gradual evolution of the opponent's system will not last forever: at the least desirable moment, revolutionary changes may occur, which will require an equally revolutionary response. One of the natural and probable forms of such a change was the elimination of the repeated message key. Thinking ahead, Knox instructed Turing to develop a method of breaking the cipher that wouldn't rely on this feature. This assumption left no choice as to which attack method the new approach could be based on. When the message key was declared a forbidden fruit, only the message body remained as the subject of analysis, and this in turn entailed almost automatically the probable clear text attack method.

The guess that a specific text appears in the analyzed message is not a sufficient basis for an effective attack on the cipher. It is also necessary to guess the precise location of this phrase in the cipher text. In this job Turing had an important ally— the Enigma construction itself. Its reflector eliminated the need to switch the machine between the ciphering and deciphering mode, but at the price of an important cipher feature—no letter could be enciphered as the same letter. It was thus possible to reject many possible positions simply by juxtaposing the probable clear text with the message body. By moving the crib (as the British codebreakers used to describe the probable clear text) along the message, some positions were easily eliminated. Let's assume that we have intercepted the message in the form **XJARWIJSBEQZTASWRAT** and we suspect appearance of the phrase **MARIANXREJEWSKI** therein. Sliding the crib along the message we look for pairs of identical letters in the columns.

```
M A R I A N X R E J E W S K I
X J A R W I J S B E Q Z T A S W R A T

  M A R I A N X R E J E W S K I
X J A R W I J S B E Q Z T A S W R A T

    M A R I A N X R E J E W S K I
X J A R W I J S B E Q Z T A S W R A T
```

In the second pair above, the characters **A** and **E** of plain text have been transformed into the characters **A** and **E** of the message, respectively. In the third pair, the same happens with the letter **S**. Bearing in mind that Enigma construction does not allow to transform the letter of clear text into the same character in the cipher text, we can rule out both positions. Such a procedure does not guarantee neither confirmation of the crib's appearance in the message, nor its proper position, permitting only elimination of some obviously false guesses.

It was an interesting beginning. Turing imagined the construction of a machine similar to Rejewski's bombe, whose task would be to match fragments of the intercepted message and probable open text in order to determine the rotor positions at which their convergence is achieved. Unfortunately, Enigma's plugboard stood in the way of this idea. The plugboard became the crucial challenge, but Rejewski's notes gave hints as to how the problem could be dealt with. Most of the attacks developed by Poles started from the observation that the plugboard does not affect cipher's cyclical structure. The attempt to apply this approach to the message body, instead of its key, was full of difficulties. Rejewski's original approach required some 80–100 messages enciphered with the same key in order to determine its cyclic characteristics. Ciphering procedure introduced in September 1938 caused each message key and body to be enciphered with a different key. Turing had to base his attack on the content of a single message body.

Fortunately, Rejewski's basic principle retained it validity even when applied to the single message. In order to trace the reasoning that led Turing to the solution of the problem, let us assume that we have an enciphered message and the corresponding crib in form:

position in message:	12345678901234567890123
crib:	REJEWSKIROZYCKIZYGALSKI
cipher:	FJLJJQBJEKPAPFNAORCNGXR

Let us note that in the third position of the message letter J has been transformed into L, in the eighth position I into J, in the fifteenth position I into N, and in the twentieth position L into N. Analyzing in the same way the successive transformations of the characters of crib into cipher letters we can construct the following graph (with characters representing the nodes, and positions in the message, in which transformation takes place the edges)[4] (Fig. 10.1).

This graph contains two disjoint sections. The section at the top of the picture does not contain any cycles and is of no interest for the codebreaker, but the bottom one is interesting. Let's take a look at the cycle consisting of the letters J-L-N-I and consider the logical consequences of its presence in the message. To do this, we shall analyze the transformations to which the clear text characters in each of the four nodes are subject.

Let's assume that the letters J and Z are connected in the plugboard (which we denote as J-Z). This means that the letter J in the third position of the crib is replaced by the plugboard with the letter Z, which is input into the rotor battery. The rotors convert Z into an unknown letter z_1. Since from the message we know that the final product of the transformation in position 3 is character L, it means that the character z_1 is connected in a plugboard to letter L (z_1-L). In position 20 of the crib, the input letter is L. From the previous step we know that L is connected to the letter

---

[4]The structure of the strings indicates that the graph should be orientated—the nodes/signs present in it should be connected by arrows indicating the direction of the characters' transformation. However, we also know that the Enigma code has a reciprocal feature; if in a given position of the machine X is transformed into Y, then Y in the same position goes into X. This feature of the cipher allows you to replace a graph oriented graph with a non-oriented graph.

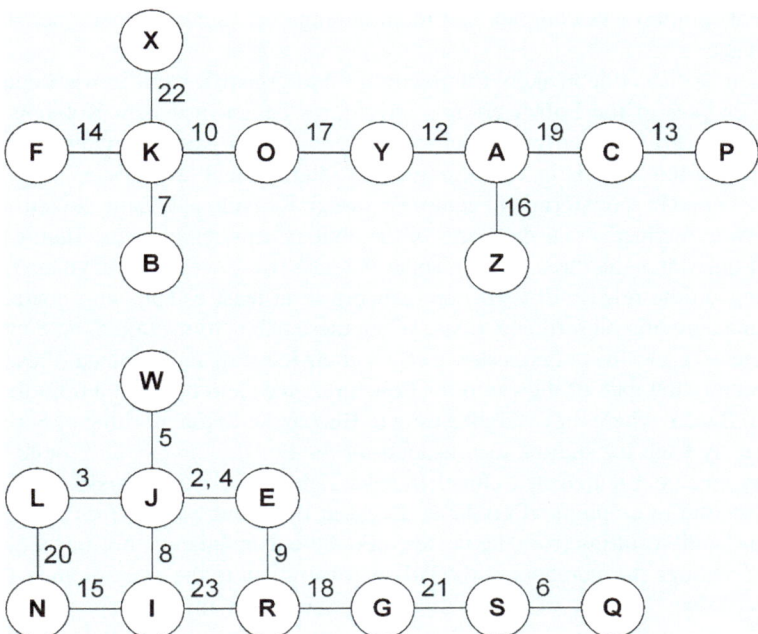

Fig. 10.1 Cycles resulting from the crib above

z_1, so at the input of the rotor battery appears the character z_1 and at the output an unknown letter z_2. However, we know that the final product of encryption in position 20 is **N**, which means that z_2 is connected to **N** in the plugboard (z_2-**N**). Consequently, in the position 15 of the crib z_2 will be fed into the rotor battery, at the output of which an unknown letter z_3 will appear. However, we know that the product of the encryption in step 15 is the letter **I**, which means that z_3 and **I** are connected in the plugboard (z_3-**I**). This means that in position 8 the letter z_3 will be fed to the input of the rotor battery, and its output will be marked with z_4. We also know that the letter z_4 must be connected to the letter **J**. Let's assume, tentatively, that $z_4 = $ **C**, which also means that the letters **C** and **J** are connected in a plugboard. However, at the start, we assumed that the letters **J** and **Z** were connected, which implies that the letter **J** would be connected with both **Z** and **C**. The construction of the Enigma rules out such a scenario, so we must reject the assumption $z_4 = $ **C** as leading to contradiction. In a similar way other hypotheses can be verified; we get a logically correct result only with $z_4 = $ **J**. Without any knowledge of the intermediate values we are able to verify the hypothesis concerning the plugboard settings! The only condition permitting the application of this approach is the presence of the closed loop in the crib. Within the loop electrical signals pass through the plugboard twice, each time in the opposite direction, cancelling its influence on the cipher. Turing has got a basis for an attack, however, the path leading from theory to practical solution was complex and required engineering rather than mathematical experience.

When British codebreakers frantically tried to organize work in war conditions, a new chapter of the Polish group's adventures began in distant Romania. After crossing the bridge over the border river Czeremosz in Kuty, the team of the BS4 was divided into two parts. According to the international law officers and soldiers were detained in the internment camp. Rejewski, Różycki and Zygalski, despite the raging war, retained so far the status of the civilian army employees. Taking advantage of this status the three of them spent the next two days in Starożyniec, perhaps counting on the release of their commanders or at least establishing contact with them and agreeing on further actions. When these calculations failed, on September 19, Rejewski and his colleagues set off on their journey. In Adâncata, they passed indifferently the gate of the camp for Polish refugees and boarded a train that took them to Bacău, where they caught a train to Bucharest. In the morning of September 21, directly from the station, they headed for the Polish embassy and found there a military attaché, Lieutenant Colonel Tadeusz Zakrzewski. Zakrzewski could not do much for the three nominal civilians. Right at the same time a storm was rolling over his head, resulting from the maneuvers of the Romanian authorities around the right of passage for members of the Polish government to the allied France. He was assigned also the task of organizing the transfer for military specialists, most urgently awaited in the Polish army being reorganized in exile. It seems that the codebreakers did not belong to this category, because Colonel Zakrzewski immediately referred his guests to the allied embassies.

The direction of mathematicians' next steps illustrates their assessment of the Allies. Langer and Ciężki would probably choose the French embassy, basing on the tradition of contacts with Bertrand. Despite the fact that the Polish-British cooperation was short and despite Knox's whims during the meetings in Paris and Warsaw, Rejewski and his colleagues went straight to the His Majesty's embassy. Had they stopped in one of the Bucharest cafés along the way, their personal fate, and to some extent the history of Enigma, would probably have taken a different turn. The codebreakers reached the embassy almost simultaneously with a convoy of vehicles carrying staff evacuated from Warsaw. Sir Howard Kennard, the ambassador, and the members of his staff certainly had much to say about the events of the recent days. On their way from Warsaw to Romanian border they learned the new face of war. They saw the bombed cities and columns of refugees; they were forced to travel at night seeking in the darkness protection from Luftwaffe bombers. Their local colleagues had to die of curiosity for news so different from the official reports.

Under these conditions, the fact that Rejewski managed to exchange a few words with the Ambassador bordered on a miracle. The Ambassador asked the Poles for patience, promised to contact London and get instructions. However, the arrival of a group from Warsaw caused such a mess, that the mathematicians were not sure that anyone would remember their case. They politely thanked for the words of encouragement and proceeded straight to the French embassy. There, the very mention of Bertrand's code name "Bolek" caused a hurricane of activity indicating that the codebreakers were being expected. During the first days of the war, Langer maintained contact with Paris via the radio network BLR. When the Polish station went silent on 10 September, Bertrand began to worry about the fate of his partners and the secret they were protecting. His anxiety must have increased when the news of the Soviets' aggression reached him on 17 September. As soon as a call from Bucharest confirmed the safety of the Cipher Bureau staff and secret, Bertrand exploded with activity. It took only two days for a local resident of French intelligence, Colonel Neuhauser, to equip the three codebreakers with the documents necessary to continue their journey (resistance of the Romanian official was overcome by several banknotes tucked into the passports). Officer made sure that the Poles found their seats on the train that set off through Belgrade, Zagreb, Trieste, Milan and Turin to the Modane border station, already on the French side of the tunnel under Mont Cenis. There, an officer of the Polish evacuation post picked them out of the crowd, equipped them with tickets for the further journey and placed them on the train leaving for Paris. Rejewski recalled that they arrived to Paris around 25 September. All three were accommodated by the hosts in Hôtel Sèvres, where for some time they could recollect the events of the past month. On their way to Paris, they passed Bertrand, who immediately took a plane to Bucharest to get the rest of the Cipher Bureau's team into security.

He was determined to use the strong French influence in this country to get the officers out of the internment camps, in his own words "through friendship, in equal measure for own interests, and finally, because (...) I[ntelligence] S[ervice] was striving for the same thing". The deep respect of the Romanian officials for everything French combined with skillfully distributed bribes, did the job. Already on 2 October Bertrand was able to return to Paris with a sense of duty well done: he won his "race with the Intelligence Service, which pressed for their [Polish] seizure". A group of Cipher Bureau officers were found in an internment camp in Câlimâneşti, a small resort on the River Olt in the Southern Carpathians. The camp was improvised in local hotels, where the supervision of the interned officers was at that time symbolic. It took only a few days to get the group out of the camp, transport them to Bucharest and arrange the documents necessary for the journey to France. Bertrand mentions in his memoirs that part of the team arrived in France by a plane hired especially for this purpose, the rest by train. Customs control stamps in the preserved Ciężki's passport show that he was travelling to France by train. What's more, his journey didn't go entirely according to plan: the stamps confirm that on 6 October he tried to enter Switzerland from Italy, but was turned back and only reached Paris on 12 October. Ciężki and his company might try to get to Switzerland to seek help from Major Michałowski's family living in that country. It seems that Bertrand was the only one to take a plane back to Paris, while the Polish embassy in Bucharest did not issue passports allowing officers to continue their journey until 3 October.

Part of the team evacuated by train was still in Romania. The difference of several hours in crossing the border or a difference in the military rank meant that they landed in an internment camp improvised in the barracks in Balş, which they reached on 6 October. Leaving them in the country freely penetrated by agents of Nazi intelligence service would be dangerous. Germany's increasing pressure on the Romanian authorities translated into tightening of supervision, which complicated the escape. Although the first attempt to escape from the camp, made at the beginning of December, ended already in Piatra Olt, the first town on the way to Bucharest, the second attempt, in early 1940, was successful. On 9 January the group including Kazimierz Gaca and Sylwester Palluth, appeared at the Polish embassy in Bucharest to collect passports with the necessary visas and tickets to Belgrade for 13 January. The fugitives entered the route to the Polish army in France that many Poles were following in those months. So many, that German propaganda gave them the collective name of "Sikorski's tourists". The journey through Yugoslavia and Greece took some time—exactly one month later, on February 13th, they boarded in Piraeus the Polish ship "Warszawa", which after a week-long journey reached Marseille. The Enigma secret was safe again.

Meanwhile, political turmoil broke out over the heads of a group of Polish codebreakers. It had modest influence on the further story of the Enigma's secret, but its results turned out to be decisive for the personal fates of the Cipher Bureau team members. Bertrand was the first to take action. After the outbreak of war, he was promoted to the rank of major. The Service de Renseignement became the fifth department of the Army General Staff (Etat Major de l'Armée); Bertrand assumed the position of the head of one of its four sections—Section d'Examen. Looking at the preparations of the French army in the field of cryptology, he was more than skeptical about army codebreakers, and, on the other hand, painfully aware of the advantage of fast and safe communication for German troops. It seems that neutralizing this advantage has become something of a mission for Bertrand. However, the mission itself did not change the simple fact that he found himself in the situation of a commander without an army. Guards, drivers, messengers and typists assigned to his section did not compensate for the total lack of codebreakers. The seven Spaniards he hired some time ago might have rendered services in case of conflict with Italy, but it was Germany to be the main opponent in the upcoming campaign. In this situation the cooperation of Polish codebreakers represented the raison d'être of his unit. This fact heavily influenced his position in the discussions on the future of the group.

Some letters exchanged between the heads of British intelligence and the management of BP indicate their concern about the fact that the last keys of the military Enigma were broken on 14 December 1938. This is a somewhat surprising statement, as during the conference in Pyry, in July 1939, Polish codebreakers demonstrated the process of breaking the key using a newly intercepted message. It is also known that the last day on which Poles broke the key to the military Enigma was 25 August 1939. Regardless of what caused this discrepancy, the British codebreakers were unable to break the keys in the first months of the war. Modifications of the ciphering procedure undertaken during the Munich crisis allowed to suspect that the next wave of changes followed after the outbreak of the war, rendering Polish methods ineffective. If that were the case, the task facing BP would not be an update to the known problem, but a completely new challenge. Direct help and participation by the authors of the earlier breakthrough would be of paramount importance, bringing to them an element of continuity, an almost intimate knowledge of the nature of the cipher, which could not be conveyed in a formal description. As late as 9 January, 1940, Denniston wrote to Menzies: "These young people have ten years' experience and their short visit could be extremely valuable".

This opinion must be interpreted within the context of the inter-Allied discussion regarding the organization of the codebreaking service. The British proposed establishment of a joint center, where the best British, French and Polish codebreakers could work hand in hand on breaking the enemy's ciphers. This proposal naturally reflected the integrated nature of the British and Polish codebreaking services, which in time was to prove to be one of the key factors in the Allies' success in the signal intelligence. Langer's account suggest that the proposal was rejected due to Bertrand's resistance. Some mechanisms of coordination were put in place; in BP a French liaison mission was installed, and a British liaison officer appeared in the French equivalent of GC&CS. Both centers were connected by a direct teletype line and were to exchange information about broken German keys in the future. Exchange of intercepted cipher material was instituted, with the German land forces messages flowing from France to BP, and Kriegsmarine messages in the opposite direction. Subordination and location of the Polish team represented the limit of compromise in the French-British cooperation in signals intelligence.

The dispute over the fates of Polish codebreakers was surprisingly bypassing Polish government in exile. Both, government in exile and the army being reconstructed in France, enjoyed considerable autonomy. One might expect that in view of the controversies regarding Polish team, the voice of the Polish authorities would prove decisive. It seems, however, that the fate of the codebreakers has never become an object of serious interest for the Polish army HQ. This development was probably the effect of several factors, of which only one was objective in its nature: there was no justification for creating independent signals intelligence service within the Polish army in exile. The second probable factor was the way Prime Minister and simultaneously C-in-C of the Polish army, General Sikorski, understood the Polish interests. Sikorski seemed to assume that during the war the measure of the importance of Polish raison d'état is the number of divisions in the field. Every pair of Polish hands were directed to the infantry divisions being formed, the only exceptions being pilots and sailors. It is hardly surprising that Sikorski, burdened with the responsibilities of both Prime Minister and C-in-C at the same time, adopted a somewhat simplified vision of reality. Remembering his experiences from the war of 1920, we can be sure that he appreciated the importance of the codebreaking. However, the General had spent several years before the outbreak of war in the exile, separated from the current developments in the Polish army. Consequently, he probably did not know anything about the codebreakers' success and its potential importance for the Polish cause.

The disaster of September 1939 put an end to the structure of the Polish state formed as a result of Piłsudski's coup in 1926. Some historians speculate about Piłsudski's political last will, recommending his successors to remove the divisions in society caused by his own policies. This will, if it ever existed, was not implemented, which resulted in Marshal's epigones taking full responsibility for the military disaster and the preceding it democratic deficit. The leaders of the former opposition, who took the helm of Polish politics in exile, did not make a gesture and started a poorly masked campaign of revenge. The environment of the pre-war General Staff, especially its intelligence service, represented for the new policy-makers the embodiment of evil, a place where the concentration and influence of the old policy was particularly strong. The Cipher Bureau without any doubt repre-sented an element of the intelligence service, so its members were classified as untrustworthy, without entering into the detailed analysis of individual fates, atti-tudes and sympathies. We don't know much about Langer and Ciężki's political orientation. They have both started their military careers in the armies of the parti-tioning countries, disliked and distrusted by Piłsudski. Their careers do not show the rapid acceleration after 1926 coup, characteristic for the supporters of the new regime. The public and political attitudes of the codebreakers were formed during their studies at the University of Poznan, which at that time was dominated by the National Democracy, opposing Piłsudski. It seems that the views of the core of the Cipher Bureau were far from the pre-1939 regime. Despite this, members of the Cipher Bureau team were initially denied commission in the Polish Army. Available documents confirm that the key members of the Cipher Bureau team were subject to the humiliating accusations and interrogations. From Langer's report of May 12, 1940, we learn that the Cipher Bureau and Langer personally were accused for all the problems of the Polish army communications during the 1939 campaign.[5] Even Antoni Palluth was subject to interrogations. Military prosecutors were accusing him, and the heads of the Cipher Bureau, for the collusion during the tenders for the wireless equipment. These accusations might point to the real source of the prob-lem. Awards granted by the Chief of the General Staff for breaking the Enigma and the salaries of the Cipher Bureau employees raised their living standards above the average of the intelligence service. Even Różycki's mother, trying to find out about the sources of her son's visible wealth, suspected him of criminal connections... Newly built or purchased villas and apartments, holidays spent in fashionable resorts—all this attracted the attention. At the same time, the real cause of this pros-perity had to remain secret even inside the service itself. This contrast between the external attributes of wealth and the mystery surrounding its causes must have caused strong emotions within their environment. In Poland they were protected by their success. In exile the codebreakers fell victims of campaign of revenge they were completely defenseless against.

The Cipher Bureau team started to break up, and the news of its problems must have become a bit of a public secret at the intelligence market. The Japanese came forward first. Thanks to many years of cooperation with the Cipher Bureau, officers of the Imperial Army and Navy knew quite a lot about the successes of Poles in breaking Soviet codes and ciphers and offered their authors a switch to the Japanese

---

[5] *If current research shows that not all armies or operational groups have received (...) ciphers, this shortcoming cannot be blamed on the Cipher Bureau, because (...) it did not know these units, which have always been surrounded by great secrecy. (...) Without an express order, the head of the Cipher Bureau had no right to issue* [ciphering] *machines* [Lacida]. From Gwido Langer's report to Ministry of War, May 12, 1940.

pay. Five codebreakers from the Bureau's Soviet section accepted the proposal. Part of them were installed in a signals intelligence center operating at the Imperial Embassy in Sofia, most were transferred to Manchuria, where they kept breaking Soviet messages for the Kwantung Army. After Japan joined the pact of steel in September 1941, and the Empire and Poland found themselves in opposite camps, Polish codebreakers terminated their service. The Japanese only managed to convince the leader of the group to continue his work. Remaining codebreakers were transferred to the area under Allied control and their further fate is so far unknown. The rest of the team was transferred under the French command. For the Polish authorities, Bertrand's offer was a convenient way out of an awkward situation. At the end of October 1939, the codebreakers were commissioned only pro forma in the Polish Army and immediately seconded to the French service. As a result of the political and personal tensions within the emigration, Polish government with some relief got rid of one of its most important assets.

In this way Bertrand acquired a missing piece of his puzzle and energetically started to organize his service. Colonel Louis Rivet's and his Fifth Department found the wartime headquarters near Gretz-Armainvillers, at Château Péreire, which on this occasion gained the name of the command post "Victor" (*Poste de Commandement Victor, P.C. Victor*). Bertrand's intercept and codebreaking section was placed in the nearby Château Vignolles, renamed P.C. "Bruno". The core of the section of about 70 people were two foreign teams; "L'Équipe D", comprising seven Spaniards, and "L'Équipe Z", i.e. Poles. Later, just before the German attack in the west, the French Grand Quartier Général realized the uselessness of its own codebreaking unit. Bertrand was pleased to learn that most of its members were transferred to "Bruno" as reinforcements. By April 1940 virtually all French activity in the field of cryptanalysis passed into the foreign hands.

The Polish " Équipe Z" started its activity in "Bruno" on October 20, 1939, even before the formal transfer under the French command. The internal organization of the team remained the same as in the pre-war Cipher Bureau: Langer remained the commanding officer, Ciężki his deputy. Bertrand isolated the two foreign teams locating the Spaniards in a separate building. This resulted partly from the need of the secrecy, but partly from certain distrust with which members of the Polish team treated their Spanish colleagues. Poles might hear about the reputation of the SIM— the secret police from whose ranks the Spaniards were supposed to come. Besides, only a few weeks ago they had been fleeing Soviet tanks themselves, so their reaction to the word "commissar" seemed understandable. Subsequent joint experience has contributed to the reduction of the original reserve. While "Équipe D" was at Bertrand's disposal immediately, the Poles gathered gradually. Langer, Ciężki, Palluth and Fokczyński arrived in early October. The remaining members of the team were getting through the evacuation channels from Romania, which took time. The organization of Polish Army in France gained momentum, so the newcomers were qualified as infantry immediately upon their arrival, and transferred to battalions being formed, from which Langer had to extricate them with a great deal of effort. On March 7, 1940, Kazimierz Gaca and Sylwester Palluth were assigned to an infantry company stationed near the castle in Mauron. It was only in the second

half of April that Langer managed to locate them and enforce their assignment to
"Bruno". The lack of staff was accompanied by a lack of equipment. Bertrand
proudly demonstrated a copy of the Enigma received in August, which had been
perfectly preserved, but nothing was done to manufacture its copies. As the team
brought a second copy, Anthony Palluth disassembled this machine and started to
prepare, with Fokczyński's help, its blueprints. Based on this documentation
Bertrand contracted the local company, Etablissement Edouard Belin, to manufacture forty copies of Enigma as soon as possible. As part of the wartime preparations
"P.C. Bruno" was connected to the French intercept stations by teletype lines.
British liaison officer, Kenneth McFarlan, joined the team and arranged "Bruno's"
connection with the field headquarters of the British expeditionary corps in Arras.
According to Captain Braquenié, the deputy commander of "Bruno", security of
communications between "Bruno" and BP was protected in somewhat perverse
way—the messages were enciphered using Enigma. He himself, when preparing the
clear texts for encryption, attempted to make life difficult for his opponents, adding
at the beginning and the end a phrase unrelated to the content of the report and
rather unexpected for German codebreakers—*Heil Hitler*. The members of "Bruno"
team lived an isolated life. Bertrand used to joke to be the abbot of the Order of the
Brunists. The functioning of "Bruno" at the end of 1939 resembled the tuning of
instruments by an orchestra awaiting the concerto.

Its score was at the same time being written in BP. Jeffreys and his team forgot
the taste of punch drunk when they celebrated halfway point of their work, when on
1 November they finished perforating two sets of Zygalski sheets for their own
codebreakers and started making copies for the Poles and the French. It would be
incredible if British codebreakers had not tried out their new tool immediately, but
their attempts to break the cipher all failed. Initial optimism started to give way to
concern. In a letter of 9 January 1940, Denniston wrote referring to Polish team: "If
there's a change in the Enigma system when the war broke out (as we're starting to
suspect), the experience of these people could shorten our work by months ". A few
months earlier news from Poland turned the defeatism regarding the prospects of
success in the struggle with Enigma prevailing at that time within GC&CS into the
wave of optimism and activity, but Knox remained skeptical about Polish methods.
He emphasized that since 14 December 1938 Poles had not read a single message
encrypted with the military Enigma (which was not true), while the Germans certainly kept an ace up their sleeves in case of war—a change that could invalidate
some or all the methods developed by Rejewski and his colleagues. Turing's work
on the new methods of attack could not produce practical results during the next few
months. The fresh failure of Zygalski sheets seemed to confirm Dilly's concerns.
Closer cooperation with the Polish team seemed to offer the best prospects of
progress.

The first opportunity to negotiate secondment to BP of the Polish team arose during Colonel Langer's visit to the UK in December 1939. Langer was no longer the
master of even his own destiny. Knowing Bertrand's negative stance, the British
decided to seek help at the higher levels. On 10 January, 1940 Sir Stewart Menzies,
who took over as head of the SIS following the death of Admiral Sinclair, sent a

request to his French counterpart, Colonel Rivet, to delegate three Poles to BP. We know Rivet's answer from the effects only; none of the Polish codebreakers appeared at BP. On 7 January 1940, Dilly Knox sent a memorandum to Denniston recalling that during the Pyry meeting they made a promise to deliver the complete set of Zygalski sheets both to Poles and French when it is available. Although the work of Jeffreys' team was completed in mid-December, the sheets have not yet been delivered to France. Dilly offered himself as a courier who could deliver the sheets to "Bruno" announcing intention to tender his resignation if the promise is not kept. Knox's tone suggests that Denniston might try to use the sheets as a commodity that could be traded for the temporary or permanent loan of Polish codebreakers. Denniston, resigned, must have acknowledged it did not really matter where the Poles would verify the effectiveness of their methods or its lack, and in mid-January 1940 decided to send Alan Turing as a courier to Paris. When sending Turing to the continent, he probably hoped that his direct contact with Rejewski, Różycki and Zygalski could be a substitute for the visit of the Polish trio to BP.

The January meeting was the only personal contact between the two people, whose knowledge and talent were crucial to the history of Enigma: Rejewski and Turing. Rejewski remembered that Poles treated their interlocutor "(…) as a younger colleague who specialized in mathematical logic, but is only taking his first steps in cryptology. Our discussions, if I remember correctly, were about plugboard and its connections, which were Enigma's strength". Turing was not known as the great communicator and his particular manners somewhat deceived Rejewski's intuition, who usually accurately assessed the character and qualifications of his interlocutors. But the meeting brought the expected results. On 17 January Poles, in Turing's presence and using the set of Zygalski sheets he delivered, have broken the key of a message from 28 October last year. The reason of problems encountered by British codebreakers turned out to be trivial: they have swapped the turnover points for rotors IV and V. Bertrand decided to celebrate the success with dinner in one of the Parisian restaurants. Both the host of the evening and Langer remembered an episode seemingly meaningless, but prophetic in the context of Turing's later fate. During the dinner, someone noticed on the table a bouquet of pale lilac flowers, and Langer threw in their Latin name—*Colchicum autumnale*. Turing remarked that the colchicine contained in the flowers was highly poisonous; a few petals are enough to get rid of all the problems of the temporal world.

For the time being, however, there have been celebrations both in Paris and BP, where the news from Paris caused real euphoria among the staff; codebreakers were back in the business. The first Enigma key broken during the war represented the network referred to in Welchman's color code as Green.[6] The content of the decrypted messages indicated that Green is a Wehrmacht's network of administrative and logistics nature. After Turing's return to BP the British managed to break between 20th and 24th January several Green messages. Interestingly, the key which

---

[6] In time, the *Green* network was renamed *Greenshank*. Since it served for communication between German military districts, it was also known as *Wehrkreis*.

was the first to yield during the war, later proved to be a difficult opponent. By the end of the war the Allies managed to break the Green key for only thirteen days, including some broken thank to the captured keys. Even breaking of the first messages intercepted during the war did not dispel all fears. Green messages were intercepted still in 1939; the British feared initially that a change might occur at the beginning of the new year. Soon, however, they managed to break the key of messages sent in the Red network on 6 January 1940. The first success was soon followed by others, fears have been dispersed, the time for tuning the instruments was over. Time was ripe for the concert.

Chapter 11
Concerto Without Listeners

Captain Ridley proved to be as effective an administrator as the organizer of the "hunting". After moving GC&CS to Bletchley, he was able to master the chaos of the first days and launch projects that were to give BP the final shape. Captain Faulkner's company managed to construct in August four huts. Already in September the staff of the center grew to 186 people and it became clear that the existing and constructed structures could not accommodate all the sections. In mid-September it was decided to build three more huts for the Army, Navy and Air sections, and their construction started at the end of the month. It was decided also to connect all buildings with a pneumatic mail system, which was to improve internal communication and increase security. Before the system began to function, two of the barracks were connected by a wooden tunnel, in which the basket with the documents was pushed in both directions with a brush stick. During this period of operation, conditions in BP did not match the relative luxuries that surrounded the French-Spanish-Polish team at Château Vignolles. The problem of the space was gradually solved as the next huts were finished. They weren't a pleasant place to work. Their interiors were divided inside with wooden walls that did not isolate from the noise of typewriters and conversations in the neighboring rooms. Buildings were heated by military peat or oil stoves, which filled the interior with pungent smoke during bad weather. Night shift was usually the worst time; cipher keys being changed at midnight caused the work to be most hectic, and the tightly closed covers on the windows contributed to the dense atmosphere. In all of this, several people in BP have become increasingly aware that its organizational structure, mostly unchanged from the interwar years, resembles the rough walls of the huts being constructed around, requiring the planning of dividing walls, structuring and filling with furniture.

Among the frantic activity the room at Elmer's School, where Welchman was working on the structure of enemy networks, represented an oasis of peace. In just a little more than a month, Welchman managed to hand over most of the routine work (and the colored crayons) to his assistants, Patricia Newman and Peggy Taylor. Having some spare time, he started to study deciphered Enigma's messages. He

M. Grajek, *Enigma Myth Deciphered*, History of Information Security, https://doi.org/10.1007/978-3-031-65475-6_11

quickly noticed the subtleties associated with the repeated message key and the resultant cycles. It took him just over a month to design a method that exploited the weakness of Enigma's *females*. He intended to use the photographic film in which a bright field represented the female. When he rushed excited into the Cottage to report his idea to Knox, he was disappointed: "Dilly was angry". He instructed Welchman that his colleague from Cambridge, Jeffreys, was just busy working on the production of such sheets, and advised him to return immediately to the Elmers School and his proper assignment. Writing down his memoirs many years later, Welchman was not sure whether the method he developed was identical to the sheets produced by Jeffreys' team. In fact, he probably has reinvented Zygalski sheets. However, Welchman was wrong in claiming that he had also invented the term "female" for a 1-cycle. This term obviously has caused interpretation problems in BP. Another codebreaker wrote about the 1-cycles "called, for no apparent reason, females". Welchman thought having coined the colloquial name for the "females" from their representation in the form of a hole in the sheet. The confusion around the name indicates that as early as in October 1939 elements of knowledge about the Enigma were being passed on inside BP without indicating their true source.

The reprimand that Welchman received from Knox became the foundation of Bletchley Park's future success. In an empty room of Elmer's School, far from the events that absorbed his colleagues, Welchman saw problems escaping their attention. Denniston and other members of the BP management seemed to live with one question—will it be possible to break the Enigma ciphers? Welchman was concerned that BP in its current form would not be able to take advantage of the effects of a likely breakthrough. Common sense told him that the ultimate goal of the whole operation is not to break the enemy's ciphers, but to provide the leaders of the country and the military commanders with the resulting information. Knowing the start and the end of the chain, he designed the intermediate links in a way ensuring maximum efficiency and included the results in a memorandum, which he finished writing in November/December 1939. It is worth noting that he decided to share this memo not with Denniston, but rather his deputy, Commander Edward Travis. Quoted earlier opinions about Denniston indicate that he did not enjoy the full confidence of GC&CS staff when it came to resolving organizational issues. His deputy represented, reportedly, a different personality. Travis joined the Navy in 1906. During the First World War he served as a signalman and cipher clerk on board the battleship *Iron Duke*, Admiral Jellicoe's flagship during the Battle of Jutland. The position of a cipher clerk determined the direction of his future career. From 1925, when he joined GC&CS, to the outbreak of war Travis has been working in the section designing British codes and ciphers. Having joined the navy at the age of eighteen, he did not have a chance to complete his studies, so he must have felt a bit uncomfortable in a group dominated by graduates of the country's best universities, cultivating the academic ethos also in the reality of the secret service. He compensated the lack of formal education by common sense, practical approach to problems, energy and the dose of risk he was willing to take in a case that he thought was worth it. It seems that Welchman's memorandum belonged to that category. Having appreciated the importance of the arguments presented, Travis was able to gain the approval of the other members of GC&CS management—Welchman's project was accepted on 5 December.

A few weeks of frantic activity at the turn of 1939 and 1940 led to the formation of an organizational model, the core of which survived until the end of the war. The starting point of Welchman's plan was to ensure effective management of the intercept stations. While it was possible to centralize the codebreaking, each service maintained its own intercept stations. There existed a body called the *Y Committee*,[1] consisting of representatives of GC&CS, Naval Intelligence (NID9), Army (MI1(b)) and Air intelligence (AI1e) to coordinate interception. Each station specialized in intercepting traffic of a particular type or in particular geographical area. The system functioned well during the peacetime, but after the outbreak of the war the enemy perversely ignored the existing habits and arrangements. In March 1940 War Ministry, supervising the Chatham station, was shocked by the directives emanating from GC&CS, requiring station to focus its attention on the Luftwaffe networks. At that time only Luftwaffe messages were being intercepted in Britain in numbers permitting the codebreakers to attack the cipher, and the Luftwaffe networks were the only ones whose keys BP was able to break. However, according to the Ministry, its station should focus entirely on the Wehrmacht messages. Welchman argued that the RAF station at Cheadle did not have sufficient experience in intercepting Enigma messages to provide valuable material. Aware that breaking the Enigma key requires a considerable number of intercepted messages Welchman planned to focus limited interception resources on most important or most promising networks. His plan called for the creation of the unit called the Intercept Control Room. Its task was to convey to the intercept stations information about the discovered enemy networks, their frequencies and call signals, as well as interception priorities resulting from the codebreaking needs. The section was established in December 1939, under the direction of Dr. John Coleman.

Welchman's second idea was to separate the research from the operational codebreaking activity. The first issue was to be dealt with by the research section functioning in the Cottage under Knox's direction. Next to it, a new organization had to be created, whose tasks Welchman described as production or operation. This aspect of his plan was in line with decisions already taken: construction of a separate hut for the section focusing on the operational codebreaking was approved on 18 November. This building, along with the organizational structure it was to house, will go down in history as Hut 6. But Welchman's plan went one step further, setting out the structure of operations inside the production hut. Registration Room registered all the incoming messages, so that they could be searched according to many criteria. Its second function was a continuation of Welchman's old job: identification of the network structure, frequencies and call signs used. In the early stages of the hut's activity, it was responsible also for searching out the females in the message headers and passing them out to the Records section. The Records section,

---

[1] Any wireless interception in Great Britain was referred to as Y. There were two reasons for this. Firstly, Y was pronounced identically to the first letter of "wireless". Y resembled also the graphic symbol describing the antenna.

which since 3 January was managed personally by Welchman, has been distributing information in three directions. The intercept coordination section received data about the enemy's networks and passed them on to the intercept stations. Registered messages went to the Decipherment section (more about which in a moment), and females found—to the sections directed by Jeffreys: Sheet-Stacking Room and Machine Room. The Sheet-Stacking section was responsible for the first stage of the key breaking—identification of the rotor order and the ring settings using Zygalski method. Zygalski sheets only rarely provided an unambiguous solution. Usually they were able to reduce the number of possible solutions to a few, which were then transferred to the Machine Section, whose task was to determine the remaining elements of the key. When the Machine Section was able to recover all the elements of the key, it was passed on to the Decipherment section, which was responsible for deciphering intercepted messages. Welchman took care of a sufficient number of Enigma equivalents, permitting deciphering many messages in parallel. Unlike the Poles and the French, the British have not constructed their replicas of Enigma. Instead, GC&CS ordered a number of modified TypeX machines. Deciphered messages were forwarded to the neighboring Hut 3, where a team of intelligence analysts translated, interpreted, paraphrased and finally passed them on to the recipients.

Hut 6 introduced a novelty to the functioning of BP—continuous work in three shifts. In proposing such an organization Welchman assumed that the hours immediately after midnight, when the keys to the cipher, call signs and frequencies were routinely changed, were decisive for the effectiveness of the center. The three-shift system was later gradually introduced in most BP units, but at the time of its original implementation arouse some concerns at the Foreign Office, appearing as a payer, which considered it a costly extravagance. During the first months of 1940 the Red network became a training ground for the codebreakers. By the end of March both Allied codebreaking centers had managed to break the keys for a total of about 50 days. During this period experience of Poles, who were working in a team of just three people, contributed about 40% of that number. Soon, BP's organizational superiority contributed to a reversal of proportions: by 1 May 126 keys had been recovered and the BP's share rose to 83%. This was not a bad result for "Bruno", given its staff of only 70 people and the BP team having grown from nearly 200 people immediately after the outbreak of war to over 700 in May 1940.

The bare numbers suggest that both centers have achieved almost 100% success in attacks on Luftwaffe keys, and a stream of information started to reach Allied commands. The reality looked less optimistic. The keys were broken effectively, but often with a delay of several days to weeks. The first reason for the delays was a system of delivering the messages to BP in bags, by motorcycle dispatch riders. Welchman appreciated in his memories the dedication of those soldiers, who during the winter months, often in bad weather and at night, covered tens and hundreds of miles on winding roads to deliver their cargo. Despite their dedication, many messages were reaching BP a day after they had been received. Nature of the intercept station's job contributed additionally to the delay. Enigma message had to be intercepted without error to be deciphered. Operators were trained to omit rather one or more characters than write them down with error. They were indicating the approximate number of missing letters, or writing them down in variants (e.g. U or V, written usually as U/V). When several stations were intercepting the network, the interception coordination section was usually able to get a correct copy from one or more sources, but at the cost of time. The speed of breaking the cipher depended also on the available traffic. During the "phoney war" the traffic was scarce. The intercept stations were usually able to collect several dozens of messages necessary for Zygalski sheets, but they rarely had them soon after midnight, after the key change. However, the most important factor was the nature of military operation at that time. During the "phoney war" most messages concerned the banalities absorbing the attention of an idle army: current weather and forecasts for the coming days, fuel stocks in the warehouse and beer in the canteen, holiday leaves and duty transfers, at best—training or reconnaissance flights. From information of this kind Hut 3 could not infer any conclusions about the German plans. Message content was not the only problem with the traffic being intercepted. In Hut 3 a group of linguists and air intelligence officers were awaiting results of Hut 6 work. Reading the early deciphered messages must have been a shock for both groups: the linguists could have doubted their knowledge of German and the intelligence analysts shared their concerns. The language of the deciphered messages represented a specific jargon of Enigma traffic, where fragments of the legible German text intertwined with the terms specific to the Luftwaffe or Enigma. The rules of transcribing numbers, diacritics and other elements absent in the Enigma character set, combined with military jargon and numerous abbreviations, could put the Oxbridge philologists, educated on Goethe and Heine, into the corner. Their intelligence colleagues did not fare better. Luftwaffe, just like any military organization, used specific jargon, whose terms were absent in any dictionary available. Its character resulted partly from the Luftwaffe organization, different from RAF, partly from technical solutions that have not yet settled in the common language, and finally were the result of Goering's buffoonery. The exegesis of Enigma messages appeared to be far more difficult than editing papyruses torn from the sands of the desert. The difficulties did not discourage Welchman, who assumed that the number, character and content of German messages would change once the "phoney war" turned into a war for real.

Building up BP's credibility among the recipients of information it provided was not a fast and easy process, either. This process was disturbed in the early stage by the warning, given in good faith, but based on uncertain premises. When Travis introduced John Coleman into the duties of head of the intercept coordination section, he hoped that Coleman would be able to draw useful conclusions from the structure of the German networks and intensity of traffic. Coleman got to work with energy, and traffic analysis soon permitted to identify a new network—Blue. Fixing the positions of its transmitters revealed that they are located mostly in the northwestern Germany, unlike the Red network, whose transmitters were scattered all over the German territory. This was the area where the troops preparing for the invasion of France would concentrate, so watching the activity of the new network was considered to be one of the priorities of signals intelligence. Welchman recalled that the intensity of traffic of the new network changed often, and British staffs responded to every fluctuation announcing an alarm for the troops in France. However, when on 29 January 1940 Knox and his team broke the Blue key for the first time, it turned out it was only a Luftwaffe training network. But the harm has already happened; unnecessary alarms have eroded the commanders' confidence in the information emanating from BP.

During the Pyry meeting a division of work had been established, in which the Polish team was to continue its research on the theory of Enigma ciphers, while the remaining partners were to focus on exploiting their achievements. Practice has quickly made these arrangements obsolete. Without the adequate support from their French commanders, Polish mathematicians were forced to concentrate on the strictly operational activities. When still in Poland they were passing on the recovered keys to the clerical staff, who deciphered and translated the messages. In France considerable part of the Polish team was still busy practicing the drill and cleaning their rifles in the infantry barracks; order transferring Kazimierz Gaca and Sylwester Palluth to "Bruno" was dated 1 May 1940. Facing permanent lack of the support staff the codebreakers had to take over most of their tasks, which left no time for the research work. At the same time new GC&CS employees created a special blend of knowledge, permitting them to look at the problems from many angles. They discovered that there were many roads leading to their goal. One of them started in the Air section working under the leadership of Josh Cooper. The RAF station in Cheadle provided BP with the messages exchanged by the German aircraft with their ground controllers. While working on the three-letter code for transmitting meteorological data to the ground, the Air section staff noticed that the code words are superenciphered by the letter substitution changing daily. For example, the code word YBO could one day take the form of ZVT, and another DFO. Analyzing this cipher, British codebreakers noticed that if, for example, the letter Z in the code was enciphered as D, then D was simultaneously transformed into Z. When this information reached Knox's group, its members must have looked at each other in disbelief. This behavior corresponded to the characteristics of the Enigma plugboard—were the Germans so reckless as to tie the low-level field code to the machine settings? This suspicion could not be verified until Enigma was broken, but already the first success confirmed the guess: the letters in the Luftwaffe meteorological code were enciphered using current Enigma plugboard settings for the day! Breaking a simple weather code did not require much effort, and as a bonus revealed the plugboard settings. A mistake made by the Luftwaffe cryptographers converted the military Enigma into the simple commercial model, and the cipher key could be broken using already forgotten rodding method.

The second breakthrough was the work of a mathematician, although it was not of a mathematical nature. When Welchman and Jeffreys started organizing their teams in January 1940, they used a network of personal connections. Their escapade to Cambridge resulted in the recruitment of a group of their recent students, including John Herivel. Apart from mathematics he was interested in logic and psychology of a scientific discovery. In BP he came to conclusion that if there was logic in scientific discovery, one might find it also in the codebreaking. He tried to put himself in the situation of a German cipher clerk, who has to set the new Enigma key at midnight and then, under pressure of time and superiors, encipher and send a package of pending messages. Herivel imagined how an operator sets the current position of the rings for each rotor, puts them on the shaft in the right order, inserts the whole rotor battery into the machine and closes the cover. After this operation, the letters visible in the machine windows should be close to the current setting of the rings. Of course, German procedure assumed that the operator would turn the rotors by selecting an individual message key. Herivel suspected that an operator working under stress might consider the three letters visible in the machine windows after closing the cover to be a great starting position for the first message of the day. Verification of his hypothesis required recording the headers of the first messages transmitted after midnight. The rotor starting positions were given in clear, so comparing them was trivial. If among the registered headers there are several with identical or similar first three letters, it would mean that the operators had fallen into Herivel's trap. Practical application of Herivel's approach required use of Herivel square, listing 26 possible settings of the left rotor on one axis and 26 possible settings of the middle rotor on the other. At the intersection of row and column corresponding to the left and middle rotor position, letter corresponding to the position of the right rotor was entered for each intercepted key. Let's assume that the following rotor starting positions were recorded just after midnight:

ASD UAN PBL KTC GIJ TLX GHK RCO HIK QSB IHJ LFC MWO FSN

Inserting them into the Herivel square we easily notice the cluster of characters corresponding to the coordinates **G-I/H-I**. This cluster indicates that the ring settings might be in the range **GHJ-HIK**. Number of possible ring settings has been reduced from 17.576 to just 12!

	A	B	C	D	E	F	G	H	I	J	K	L	M	N	O	P	Q	R	S	T	U	V	W	X	Y	Z
A																			D							
B																										
C																										
D																										
E																										
F																			N							
G								K	J																	
H									K																	
I								J																		
J																										
K																				C						
L			C																							
M																							O			
N																										
O																										
P		L																								
Q																			B							
R			O																							
S																										
T												X														
U	N																									

(continued)

	A	B	C	D	E	F	G	H	I	J	K	L	M	N	O	P	Q	R	S	T	U	V	W	X	Y	Z
V																										
W																										
X																										
Y																										
Z																										

The codebreaker trying to recover the key usually did not know the selection and order of rotors, but trying all possible combinations required testing at most 60 * 12 = 720 hypotheses, allowing the key to be found in a reasonable time. The method of attack was known as the Herivel tip, or, in the pseudoscientific jargon that Oxbridge scientists enjoyed using, *herivelismus*. Both the method and its author had to wait for widespread recognition. Herivel came up with the idea in February 1940, only after three weeks in BP. In the following months, the Machine section worked hard to fill in the Herivel squares, but without any result. German cipher clerks obviously refused to act in the way Herivel considered likely.

Even if Herivelism worked as expected, the rotor order was found by exhaustive search, and plugboard settings were determined using the Luftwaffe weather code, codebreakers were still missing an important element of the key; rotor starting position. At the very beginning of work on Enigma the codebreakers observed that cipher clerks often select predictable and repeatable message keys, such as **AAA**. Hut 6 codebreakers noted that even in 1940 rotor setting sent as clear text occasionally formed particular patterns. Let's assume that among the intercepted messages the following headers were found:

Message 1:	**WSX VXIUZH**
Message 2:	**RFV PEHKJG**
Message 3:	**ZHN CSJWIT**

A glance at the Enigma keyboard shows that the rotor settings (**WSX**, **RFV** and **ZHN**), correspond to the characters selected diagonally from the keyboard. It is also clear that subsequent messages use the triplets selected every second keyboard diagonal. It is tempting to check whether the encrypted message keys represent the remaining diagonals. If that was true, the three messages had been enciphered using message keys **EDC**, **TGB** and **UJM**, respectively.

```
Q W E R T Z U I O
 A S D F G H J K
 P Y X C V B N M L
```

In this case the sequence of **VXIUZH** in the first message corresponds to the clear text **EDCEDC** enciphered at the starting rotor position **WSX**. If this assumption is correct, remaining elements of the key may be recovered, including the plugboard settings (an example of the procedure is described in the box below).

In practice, scenarios just described and similar were observed much more often than German communication security officers would like to admit. The repertoire of the predictable message keys was quite rich. BP veterans remembered most fashionable structures. Some were easy, like **HIT-LER** or **BER-LIN**. Hut 6 was sometimes surprised discovering also the pair **LON-DON**. There were keys referring to the current events. If later on, during the siege of Tobruk, the codebreaker found letters **TOB** in the message header, it was worth checking the message key in form **RUK**. Similarly, the starting position **ROM** often indicated the message key **MEL**. Another popular practice required selecting for the starting position and the message key two triplets 3 letter apart, like **BER** and **BEU**. After enciphering the message key, the rotors were thus exactly in position corresponding to the message key. This custom was known in Hut 6 as "nearness". Extremely lazy operators were selecting the same triplet for both the starting position and the message key. There must have been quite a lot of them, since this practice got in Hut 6 its own name, **JABJAB** (reportedly from the first time this error was noticed by Dennis Babbage). Mavis Batey recalled that during her time at Hut 6, she became a walking encyclopedia of German six letter curses and profanities, extremely popular as the message keys. Another operators' habit was to choose as a message key the position of the rotors left of after the encryption of the previous message, practice particularly popular for the multipart messages. Some operators used fixed keys so consistently that they became their trademark. One of them unwittingly contributed to naming the whole method. By constantly using his girlfriend's name, **CIL-LIE**, as the key of the message, he made repetitive keys known in BP as cillies. The ambiguity of the English pronunciation made Welchman remember cillies as sillies rather. His version describes the consequences of the German operator's thoughtlessness more accurately.

Key Recovery Basing on Cillies

Let's assume that we have intercepted a package of messages whose headers written in the Herivel square show a cluster around **OKA**, **PKA**, **OKB** and **GDP** ring settings. In addition, we have intercepted the following messages, the structure of which indicates the presence of *cillies*:

QAY AHG	(alleged **WSX** key)
EDC VCS	(RFV)
TGB SUP	(ZHN)
WSX SYZ	(QAY)
RFV YTH	(EDC)

Based on the relationships between the presumed plain text letters and the corresponding cipher text characters the following dependecies can be established:

(continued)

```
                        R F
                        | |
                E       V C
                |       | |
                W-A-Y-Z-S-H-U      X-G      N-P      D-T
                |
                Q
```

The dependencies defined above allowed to define the menu, which was passed on to the machine section in order to check and determine on its basis the plugboard settings. The menu consisted of the order of the rotors and the arrangement of the rings (determined by Herivel's hints), one or more-character strings, determined in the step shown above, and a list of starting positions of the rotors in which the individual characters of the string have been encrypted. Let us assume that Herivel's method indicates the order of rotors I-II-III and the ring setting OKB. Moreover, the knowledge of the rotors and ring settings allows us to confirm that when encrypting probable open texts, the central rotor has not advanced. The menu analyzed later in the example will be based on the string W-A-Y-Z-S-H-U. Its first character, W, is derived from the alleged open text WSX, corresponding to the starting position of the rotors QAY. According to the Enigma principle, after pressing the key, the rotors are first moved by one position, only then an encrypted character is displayed. This means that the W has been effectively encrypted when the rotors are set to QAZ. The next character in the chain, A, comes from the probable open text QAY, corresponding to the starting position of rotors WSX. This means that it has been encrypted in the position of the rotors WSZ. In a similar way, we determine the remaining starting positions of the rotors that will be included in the menu. The menu in the analyzed example takes form:

The order of the rotors:	I-II-III
Ring settings:	OKB
Test string:	W-A-Y-Z-S-H-U
Starting positions of rotors:	QAZ WSZ WSA TGC QAA TGD

Having received the menu in the form above, the operator in the machine section starts his job by encrypting the whole alphabet at the first starting position of the rotors from the menu and without a plugboard. For this purpose, the machine section used a device, called in BP jargon Letchworth Enigma, which, in comparison to the standard Enigma model, had no plugboard and no rotor movement mechanism, encrypting successive characters at their fixed position. The result of the operation is presented below:

(continued)

```
ABCDEFGHIJKLMNOPQRSTUVWXYZ
XCBGSMDYZVUQFONTLWEPKJRAHI
```

Then the operator encrypts the result of the above operation at the next rotor start setting from the menu, receiving:

```
XCBGSMDYZVUQFONTLWEPKJRAHI
PUETKLJNQRCZAHYGMIBXSDVFOW
```

By repeating the operation for each rotor start setting in the menu, we get the following list:

```
W    ABCDEFGHIJKLMNOPQRSTUVWXYZ
A    XCBGSMDYZVUQFONTLWEPKJRAHI
Y    PUETKLJNQRCZAHYGMIBXSDVFOW
Z    XWHINVRKZJGQOEFCSTDPMBLYAU
S    DIOWGAPJEKNBHZSLFYXRUQCTVM
H    GHJMDLROYCXZIBUATENPSVKFQW
U    SUNPFXORZBLYQCHVETJMGAWDIK
    ●●●●●●●●●●●● ●●●●●●●●●●●●●
```

The lines of the above list represent permutations, implemented by rotor battery in their individual starting positions. The start positions of the rotors in the menu correspond to the positions in which the test string characters indicated on the left-hand side of the list have been encrypted. The list columns can be interpreted as characters corresponding to the letters of the test string after passing through the rotor battery. For the data thus determined to correspond to the probable texts observed, the individual characters in the columns must be combined in a plugboard with the corresponding test string characters.

Starting the analysis from the letter **A** in the first column of the first row, we notice that to be consistent with the test string, it must be connected to the letter **W**. However, the **X** character in the first column of the second line requires a logical connection to **A** for consistency, which implies a contradiction with the Enigma (the contradictions are indicated by underlining the letters to which they relate, while the columns in which the contradictions occurred are highlighted with ●).

Continuing the analysis in the outlined manner, we identify the column in which no contradictions were found. Its content indicates that **OW**, **AN**, **FZ** and **HU** pairs are connected in the plugboard, letters **S** and **Y** do not belong to any connection pair (i.e. according to BP terminology, they are *self-steckered*). By analyzing the remaining test string characters in a similar way, the full set of plugboard settings can be reconstructed:

AN CT EK FZ HU IX JV LP MR OW

GC&CS staff members with some codebreaking experiences were assigned to either the Knox's research group or the sections dealing with ciphers used by particular German services. The team, that was to play a major role in the struggle with the Enigma in the coming months, represented a squad of scientists, students, bankers and young (mostly very young) women from good British families. Welchman's first acquisition was his old friend from Cambridge, Philip Stuart Milner-Barry. When they first met at Trinity College in 1925, they became friends despite their diverging interests: Welchman chose mathematics, Milner-Barry focused on classical philology. A common problem for the BP's headhunters was the complete ban on informing the potential recruits about the nature of their future job. We will never know whether Milner-Barry decided to accept Welchman's offer just because he trusted his friend or because he was bored with the stock broker's job, not quite suitable for his classic education. The number analysis skills resulting from his employment in the stock market were probably an asset in the codebreaker's job, but Milner-Barry had a better handicap; in 1923, at the age of sixteen, he won the British Youth Chess Championship. We met him already in Buenos Aires, as a member of the British team in the Chess Olympics. In January 1940, he appeared in BP, where Welchman assigned him a task, that was soon to become of fundamental importance. Milner-Barry, who spoke a perfect German, was entrusted the analysis of the deciphered messages in search for any stereotypical phrases—potential *cribs*.

A year after having won his first chess title Milner-Barry failed to defend it. At the tournament in Hastings he lost to Conel Hugh O'Donel Alexander, born in Irish Cork. Alexander graduated in mathematics from King's College in 1931, but sharing his time between mathematics and chess he did not become a member of the college. Hardy commented on this with regret, writing about his student as "the only real mathematician I knew and who did not devote himself to mathematics professionally". Hardy's standards of professionalism must have been pretty high, considering that Alexander started teaching mathematics in Winchester, although he was actually better known as a chess player. In any case, it was chess that made him leave Winchester and move to London: the owner of the chain of department stores John Lewis Partnership, a chess lover, offered him the position of head of corporation's research division. Milner-Barry, who had the opportunity to get to know Alexander well during his trips across the Atlantic, assessed that he does not feel comfortable in the business and deserves a job allowing him to make the most of his mathematical talents. In February 1940 Alexander also joined BP, early enough to become one of the founding members of Hut 6 (during 1941 the youngest member of the British team from Buenos Aires, Harry Golombek, also joined the team). Ties within the chess community permitted to recruit James Macrae Aitken, the chess champion of Scotland in 1935 and captain of the Oxford chess team. Welchman did not have to recruit another friend from Cambridge, Dennis Babbage. Mathematician from Magdalene College appeared in BP wearing officer's dress in the early stage of war, and after a few weeks spent in the Cottage he had switched to Hut 6, where he was taking the duty of the watch head, alternately with Welchman and Alexander.

When Travis and Welchman were busy recruiting new team members, Langer systematically pulled his associates out of the barracks and training grounds. Both teams were finally completed just in time, in the spring of 1940. In the morning of 9 April, German infantry disembarked in several Danish and Norwegian ports, and German paratroopers landed at several airfields in Norway. The end of the "phoney war" was not a complete surprise for the Allies, although the warning this time did not arrive through the broken Enigma messages. A large part of the German campaign plans could be found at the Allied headquarters, where they had been created a few months earlier, were created in relation to a conflict in which neither France nor the UK participated. On 30 November 1939 Soviets invaded Finland. The observers expected the Finns to save the honor rather than protect the integrity of the country. However, the Finns fought bravely and found an important ally in the winter, which at least once in the history was not on the Russian side. An unexpected Finnish success made the Allies see the conflict as an opportunity; Neville Chamberlain announced that Britain would deliver to Finland 30 airplanes and the French declared nearly 150 planes, 500 guns, 5.000 machine guns and other war material. Hundreds of volunteers headed towards Helsinki to join the Finnish army fighting the Soviet invasion. The true reason for the Allies' warm feelings towards the Finns was the Swedish iron ore. In summer, ore extracted in Gällivare County was exported through the Baltic port of Luleå, but in winter, when the Baltic froze, the only road led through the Norwegian port of Narvik. Whoever controlled Narvik was controlling also ore export and could reduce its delivery to Germany by 80%. Narvik's occupation would probably cause the Scandinavian countries to enter the war on the German side. The fighting in Finland and the sympathy that the Finns gained in Sweden and Norway gave the opportunity to achieve the same effect while maintaining the appearance of coming to the aid of a small and brave country. However, the Finns' defense collapsed and the war came to an end before the Allies managed to carry out their plan. Nevertheless, in the first days of April teams of British ships set off from Scapa Flow to mine Norwegian coastal waters and in this way block Narvik's ore exports. However, German ships carrying Wehrmacht troops, weapons and ammunition sailed two days earlier. BP provided a warning regarding German plans. Its Navy section managed to crack one of the Kriegsmarine's hand ciphers and correctly determined the route of the convoy leaving the Danish Straits for Norwegian ports. However, the Admiralty seemed more interested in carrying out their own mine-laying operation than in intercepting ships carrying German troops. The news arriving on April 8 left no doubt—the German invasion of Norway was in progress and the Admiralty had to recall its ships. The next morning, German troops disembarked in Denmark and several Norwegian ports. The Norwegian campaign has brought the Allies shreds of glory and good fortune, a lot of bitterness, and in its final the evacuation of the Norwegian royal family and the victorious Allied troops, which have just entered the Narvik, captured after the bloody battles.

Although the Norwegian campaign ended in a tactical defeat for the Allies, the codebreakers recorded a result closer to a draw. Enigma kept silent about the German plans, but BP was able to deduce the outline of the German plans from the broken lower-level ciphers. Unfortunately, the alarm signal from BP passed unnoticed among many warnings disregarded by the Lords of Admiralty. Disregard for warnings provided by the signals intelligence was to bring another tragedy in the Norwegian waters. But before they happened, codebreakers from both Allied teams enjoyed their day of triumph. The next day after the Germans landed in Denmark and Norway, British and French intercept stations identified a new communication network, known in BP as Yellow. Its key was first broken on April 15, less than a week after the campaign started. The first broken messages permitted to establish that Yellow serves as a coordination network between the German land forces and the Luftwaffe. Wehrmacht troops were accompanied by the Luftwaffe liaison officers, commonly known as *Flivos* (*Fliegerverbindungsoffizier*). They were relaying information about the position of the troops, directions of their march and the encountered resistance to the Luftwaffe units. The number of messages intercepted daily by the Allied listeners went into the hundreds.[2] The Yellow network key was usually broken within a few hours after midnight, and for the first time since the beginning of the war, codebreakers from BP and "Bruno" had an abundance of information about the enemy. What's more, the vast majority of this information concerned aspects, which could prove decisive for the outcome of fighting. By getting ahead of the Allied operation Germany gained a temporary, tactical advantage in Norway. It seemed that Royal Navy quickly took the lead at sea, destroying in a series of clashes in the Narvik fjord a squadron of German destroyers and delivering the Allied landing parties to their destinations. On land the English, French and Polish troops merged with the Norwegian ones and started pushing the Germans out of their positions. Unfortunately, on the first day of the invasion German troops captured the most important airfields in the country, from where the Luftwaffe started to operate. The Allies, strong at sea and fighting well ashore, were helplessly watching German planes send ships carrying their reinforcements and supplies to the bottom, destroy their warehouses, and support German land forces. At the start of the campaign the Luftwaffe created a command center in Trondheim, where all the information about its own and opponent's troops was gathered and where the operational orders were coming from. At the same time, a weather station installed in Oslo distributed the reports transmitted in the three-letter code broken by BP. The messages exchanged between Trondheim, Oslo and Luftwaffe units offered codebreakers an abundance of material to work on. The information from BP could have changed the outcome of the campaign, but it never or only sporadically reached the commanders in the field. From the British point of view, the Norwegian campaign was a Navy operation and the Allied forces fighting in Norway were under Navy officer's command. Only the Navy had a communication network allowing for the transfer of information between the UK and Norway. Unfortunately, the Admiralty showed only limited interest in the information provided by BP and a lofty indifference to all matters on the ground and in the air. BP managed to build an intelligence factory, which with great sense of timing delivered its product to the market. However, there was no time to build an effective distribution network and marketing.

---

[2] In May 1940, Hut 6 was processing about 1000 messages per day.

The episode closing the Norwegian campaign epitomized disregard for intelligence by the Admiralty. At the end of May 1940, the Allies decided to withdraw from Narvik the units, that had just captured the port and pushed the desperately fighting Germans towards the Swedish border. One of the ships participating in the evacuation was the carrier *Glorious*. As the Navy squadrons crossed the North Sea on their way to Norway, British intercept stations registered a wave of traffic transmitted by Kriegsmarine stations in the Baltics, Danish Straits and Heligoland. As BP was still unable to break the Kriegsmarine Enigma, codebreakers tried to infer some information about the German plans basing of traffic analysis only. When Welchman was learning the basics of traffic analysis for the Wehrmacht and Luftwaffe networks, the same task for the Kriegsmarine messages was entrusted to an equally inexperienced, just twenty years old Cambridge historian, Harry Hinsley. Switching from the archive queries and the study of regests and palimpsests to the analysis of call signs and frequencies, after a few weeks Hinsley turned into an experienced signals intelligence expert. His fresh experience allowed him to deduce from the sudden increase in radio traffic, that a strong group of enemy ships is heading towards the North Sea. When he tried to draw the attention of the Navy's intelligence officers to the new threat, he found out that he was missing several tens of years of service or several stripes on his sleeve to be credible in their eyes. Admiralty not only neglected to change the course the convoys, but had not even issued a warning. In fact, Hinsley detected the first signs of Operation "Juno"—a sortie of a group of German battleships (*Scharnhorst*, *Gneisenau* and *Admiral Hipper* with the escort of four destroyers) against convoys returning from Norway.

Disregarding Hinsley's warning didn't have to lead to tragedy. Originally *Glorious* was to join a convoy whose course led far enough away from the danger. On the morning of 8 June, however, its captain asked for permission to sail to Scapa on his own, accompanied by two destroyers, and received permission. Under normal circumstances, the officer commanding the deck squadrons would order an aerial patrol, which would probably detect enemy vessels early enough to take evasive action, but this officer was missing during the operation. During the previous mission he had a row with the captain, who left him ashore at Scapa, threatening to put him under the court martial upon his return. Reportedly it was captain's desire to speed up the court session, that caused him to separate from the convoy. In the afternoon of 8 June, German ships surprised the *Glorious*, sending it to the bottom with both escorting destroyers. The sacrifice of three ships and 1515 souls might have saved Britain from equally painful disaster. During the *Glorious'* last battle cruiser *Devonshire* sailed only a few dozen miles away, with the Norwegian royal family and government on board. The tragedy opened the minds, eyes and ears of the Navy intelligence staff to information coming from BP. At the Admiralty's invitation Hinsley visited their bases and command centers, learning the specifics of the Navy and its people. His monthly trip was supposed to bring dividends in the future, but for the time being the Kriegsmarine Enigma resisted all the codebreakers' efforts.

To make matters worse, both the Admiralty and BP were not aware to be losing another duel with the enemy. Germany's improvised operation in Norway just a day before Navy own mission, followed by an almost successful attempt to intercept the evacuation convoys, should have made the British think. If any intelligence analyst had written a report explaining these events by chance only, he would have been immediately transferred to the paperwork. The facts spoke loud and clear: the Germans must have known the timing and objectives of British operations and responded with own effective improvisation. The Admiralty's messages were coded using two systems; Administrative Code and Naval Cypher. Both systems were designed and implemented in the quiet pre-war times,[3] when the circumstances allowed them to be used in a somewhat relaxed way. The Administrative Code had to be superenciphered only for the messages qualified as secret. Such a usage turned the reconstruction of the code dictionary into a more extensive exercise in crypt-analysis. While the Germans sealed their communication systems in the advent of war adding new Enigma rotors and changing the operational procedures, the Admiralty persisted in using the old and proven ways. Cracking of a large part of the British codes by the cryptology unit of the Kriegsmarine, Beobachtungsdienst (more commonly referred to as B-Dienst), was not a real challenge. During the Norwegian campaign, B-Dienst was breaking 30–50% of the messages in Naval Cypher, and since March 1940 also messages in merchant navy code BAMS. The codebreakers of both parties have achieved a balance of power. The Navy played the crucial role in the campaign on the Allied side, its messages being broken by German codebreakers. On German side the major role was played by the Luftwaffe; BP achieved its first operational success breaking its keys. The tragedy of *Glorious* demonstrated the lack of understanding for the signals intelligence and inefficiency of the information flow at the British side. Against that background, the functioning of German communications represented a strong contrast. Welchman recalled being "impressed by the speed with which the Germans were organizing communication on the battlefield. Clearly well-trained and equipped with Enigma, communication teams accompanied the first assault troops wherever they landed".

When at the turn of April and May the fighting in Norway petered out, it seemed that codebreakers could sum up the lessons of this campaign. And it was precisely then that Knox's prophecy came true: the Germans dropped the repeated message key. The change was not completely unexpected; it was announced by some hints in the previously broken messages. Moreover, the Germans made a mistake introduc-ing the new system—some Luftwaffe stations started using the new procedure pre-maturely, in parallel to the old system whose key was broken. This permitted the Polish team in "Bruno" to work out the structure of the new procedure and to pass on its details to BP.

Circumstances indicated that the change meant the end of the "phoney war". At the beginning of May, there were also warnings from sources other than Enigma. The head of German intelligence service, the Abwehr, Admiral Canaris, never belonged to the supporters of Nazism, and having learned about the way the war was conducted in Poland and the subsequent crimes committed in that country, he

---

[3] *Administrative Code* came into use in 1934.

considered them unworthy of German soul. Having learnt Hitler's decision to attack Belgium, France, the Netherlands and Luxembourg, he decided to offer the Allies a warning. To do this, he sent a lawyer, Joseph Müller, to Rome to notify the Pope of the imminent invasion. As he expected the Vatican decided to send a warning to the Hague and Brussels, but the messages were intercepted by the Forschungsamt. Over the past months its specialists have recorded several messages or conversations indicating leakage of information at a high level of the Nazi hierarchy. On the eve of the invasion in Norway, the Forschungsamt overheard a conversation with a Danish military attaché in Berlin, who demanded an urgent meeting with the ambassadors of his country and Norway on the matter of utmost importance. On 9 May the Forschungsamt also overheard a conversation between the Dutch military attaché in Berlin, Colonel Sas, who was informing his government that the attack would take place the next day. Vatican's messages to the Hague and Brussels were broken and their fragments concerning the arrival of a German citizen in Rome with a warning of an imminent attack were quoted in a report to the Führer. The evidence of betrayal in his internal circle must have triggered an attack of fury in Hitler, contrasting with the joy of success in the ongoing campaign. He mistakenly assumed that the source of the Vatican leak was the same person or group behind the message from the French embassy that the Forschungsamt had broken 1937. As a result, he ordered the reopening of the then suspended investigation, which was to be conducted in parallel by the Abwehr and the Gestapo. Canaris was probably a little embarrassed to supervise an investigation in which he himself was a wanted person. For the time being, however, he has managed to divert attention from his own person and to confuse the Gestapo.

Signals of a near invasion depressed the codebreakers at BP and "Bruno". Turing was still working on an attack method independent of the repeated indicator, but by the beginning of May his work did not yield practical results. His achievements to date have aroused skepticism among his colleagues. Frank Birch judged that "Turing and Twinn are like people waiting for a miracle, but rejecting faith in miracles". The changes introduced by the Germans, putting an end to the application of Zygalski sheets, meant that in the course of the upcoming operations both GC&CS and "Bruno" would remain almost blind and deaf, unable to provide useful information. A slight consolation in the new situation was that the procedure change did not concern Yellow key, the only source of information at that time. The Germans probably decided that the troops involved in the fight should not be bothered by changes in procedures, and the Yellow key kept its former structure until the fighting in Norway petered out, when the key disappeared from the ether. The atmosphere of pessimism was aggravated by BP's first wartime loss. Until May, the sheet stacking section at Hut 6 was headed by Jeffreys, whose team played a major role in the reconstruction of the Enigma keys during the period of first successes. Change of the ciphering procedure invalidated the attack, to which he devoted so much effort. After this disappointment, Welchman noticed a hitherto unfamiliar note of irritation and impatience in his behavior. Thinking it was the result of exhaustion, he offered Jeffreys to spend a few days of rest in his apartment in Cambridge. Soon the news reached BP that Jeffreys was transferred to the hospital with a diagnosis of advanced tuberculosis and diabetes. The physical effort and mental strain associated with his work exhausted young mathematician, who did not return to BP and passed away in 1944.

If anyone in BP had any doubts about the meaning of the change in the message key structure, they were dispelled by total radio silence, which fell on the German side of the front two days before the storm began. The blow fell on the morning of 10 May. German paratroopers have seized the key bridges and airfields in Belgium and the Netherlands, as well as the key to the Belgian defense system—considered to be impregnable fort Eben-Emael. Armored columns set off along the roads paved by the paratroopers. Belgium, scrupulously observing the duties of a neutral, refused so far to allow Allied troops to enter its territory and prepare lines of defense. The attack dispelled the scruples, British and French troops set off towards the line of river Dyle, hoping to reach it before the Germans and prepare defensive positions there. The Allies did exactly what the enemy expected. The attack on Belgium and the Netherlands, despite its horrors, was only a bait to distract attention from the impact prepared further south. When the Allied divisions marched north and east in Belgium, the German armored divisions moved through the Ardennes in the opposite direction to reach Sedan on the third day of the campaign. Not long ago, the French smiled watching the snapshots of the chronicles, sparked by the Nazi propaganda, showing Polish lancers attacking German tanks. Now, to slow the progress of the armored columns through the Ardennes, they sent forward their cavalry divisions. The result was predictable; when the first German tanks crossed the Meuse near Sedan in the morning of 14 May, the fate of campaign was decided. At least that is what the French Prime Minister, Paul Reynaud, considered it to be the next day, declaring on the phone conversation with his British counterpart: "We are defeated. We are beaten, we lost the battle". Reynaud's interlocutor was no longer Neville Chamberlain, but Winston Churchill, who had only been in office for five days. His predecessor repeatedly, before and during the war, delivered evidence of a failed assessment of the political and military situation. In his speech on 5 April, he was mocking at Hitler's failure to attack in the west when the Allies were unprepared, stating that the Fuhrer had missed a bus. Less than a week later the German invasion of Denmark and Norway proved it was Chamberlain's watch that was hopelessly late. On 10 May, under the influence of reports from Belgium and the Netherlands, the House decided to designate Winston Churchill as the new head of the government.

Churchill was brought up in the atmosphere of the Great Game, a bit like Kipling's Kim. When the Secret Service Office, the predecessor of SIS, was established in 1909, he was among the first candidates for agents trained by the new service. Since then, his contact with the world of intelligence have not been loosened for a moment, even during periods when he did not hold official functions. As the First Lord of the Admiralty, at the beginning of World War I, he was receiving the copies of the German Navy code, extracted by the Russian navy from the wreck of *Magdeburg* and delivered to Britain. It was Churchill who created the rules of operation of the famous "Room 40". His post-war indiscretions, which was an important catalyst for the use of Enigma by the German services, confirms the passion with which he treated the secret messages. In his person, BP gained an insatiable recipient of information from the broken messages, as well as an enthusiastic supporter, capable of meeting the needs of the developing service. In one of his mailings from the Boer War Churchill wrote that "the intelligence service is starving for lack of money and lack of minds". When assuming the post of prime minister, he was determined to ensure that the contemporary version of "Room 40" would not run out of either. When the French struggled with their own despair, as the British Expeditionary Force marched back and forth through Belgium, along the trail of the old glory of the British army, BP and "Bruno" had little to offer. Moreover, on 20 May British MI5 seized the cipher clerk of the US embassy in London, Tyler Kent. A large collection of messages exchanged between Churchill and Roosevelt in previous days was found in his apartment. Kent did not consider himself a traitor, but rather an ideological supporter of American isolationism. He intended to expose the contacts between both leaders, who in his opinion conspired to engage the USA in the war. The very fact that treason was detected at such a sensitive point was depressing, especially when combined with the attitude of the American ambassador to London, Joseph Kennedy, who in his reports to Washington painted a bleak picture of Britain's survival under Hitler's pressure.

When BP desperately attempted to resume the Enigma breaking, the Poles were on the move again. "Bruno" proved to be well prepared for the "phoney war", but less so for a real conflict. After the German attack, its crucial staff was moved back to Paris. The Poles were placed in the Hôtel Vauban and worked in the French military intelligence headquarters at Rue Tourville 2 bis. Accommodation in a hotel was a bit pro forma, because in his memories from that period Rejewski emphasized the hectic work lasting day and night. The work on the Enigma naturally branched out into two threads these days. The first one, the easiest and gradually losing its importance, was the routine breaking of Yellow messages. From the list in Bertrand's memoirs, it appears that the attack on France caused a rapid loss of interest in the developments in Norway: the last Yellow message was read by the Poles on 12 May. The second thread boiled down to the search for a new method of attack, effective after the change of the ciphering procedure. The Poles did not know at that time that the continuation of the attack required an engineering rather than a mathematical approach, involving financial, technical and organizational support, which France was unable to provide. The fact that the hour of silent desperation translated into the beginning of new hope was due to the work that had so far kept on bringing disappointment. When after midnight on 1 May Hut 6 failed to break the Red key with Zygalski sheets, an attempt was made to attack the cipher with the only methods available under the new conditions. Over long weeks the codebreakers kept on recording the rotors' starting positions just after midnight, without effect. Fortunately, the fighting made the working conditions of the German cipher clerks closer to Herivel's assumptions. The number of messages increased in relation to the period of the "phoney war". The Enigma operators accompanied the forward Wehrmacht units, working under stress, sometimes under fire. In those circumstances operators should be committing the mistakes predicted by Herivel. Finally, on 22 May, the codebreakers filling in the Herivel square noticed a clear concentration of settings, which in the morning gave away the key to the cipher. When Herivel appeared at Hut 6 at the beginning of the morning watch, he was greeted by the appreciative looks of his colleagues and Welchman's words: "Herivel, you will be remebered".

The news of the success was immediately passed on to France and both centers were able to restart their work. Unfortunately, the time when the Enigma secrets could change the fate of the campaign was over. The pace of Blitzkrieg in France exceeded German achievements from Poland. On 16 May, the Wehrmacht command was so shocked by the progress of its own armor, that it decided to keep them waiting for the infantry. Generals Guderian and Rommel, in an unprecedented act of insubordination, pushed their troops to further attack, thus determining the fate of the campaign. On 21 May, the Guderian corps reached the coast, cutting off the northern Allied army group. On the same day Maxim Weygand, who had just become the C-in-C of the French armed forces, tried to coordinate a simultaneous attack from the south and north, which could cut off the German armored spearhead, allowing the Allied troops to restore the front line. The attack of several dozen British tanks at Arras caused a momentary panic in the German ranks. It was stopped by a deadly line of German 88-millimeter guns. The French only attacked from the south the next day, when the British were already in retreat, and they also failed. By chance it was also a great day for Herivel and his method.

A week earlier, the pessimistic assessment expressed by Reynaud described the morale of the French command rather than an real situation. Seven days later, the situation was as serious as his conclusion: the Allies were defeated. There were many reasons for their failure, but the decisive factor for the German success and Allied defeat was communication. The German wedge driven into the Allied positions was only 40 kilometers wide. It was attacked several times alternately from the south and north, but the Allies could not properly coordinate the operations of their two army groupings. At the same time, the Wehrmacht's divisions were operating in vast areas, effectively coordinating their activities, using Luftwaffe support to break through points of resistance and regularly receiving supplies of fuel and ammunition, delivered by air directly to the armored spearheads. The role of Enigma in this campaign is well illustrated by the famous photo of General Guderian in his command vehicle, leaning over two soldiers operating the cipher machine.

On the Allied side, things looked different. The French made a fundamental mistake at the very outbreak of hostilities changing their ciphers to the system used earlier and already broken by the enemy. The ability to read the French messages did not have a significant impact on the early stage of operations, as the German plan was based on the initiative and speed, without paying attention to the opponent. However, when the German troops turned to the south, the information from the signals intelligence helped to consolidate the victory at little cost. Any French attempt to mount a counter-attack or consolidate the defense was broken up by the Luftwaffe or armored troops before the French units managed to concentrate. It is surprising that in the midst of the chaos the Enigma secret has been protected. The French adopted a risky system of passing on to the army headquarters the original texts of the enemy's messages, without paraphrasing them of distributing their summaries. These texts, printed on characteristic yellow sheets (*feuillets jaunes*), were delivered to the headquarters three times a day, directly by Bertrand or a liaison officer (in urgent cases by teletype). Rejewski mentioned that the *feuillets jaunes* were distributed in many copies, and their distribution and use did not seem to be properly supervised. There was a serious danger that a staff officer could quote a fragment of the Enigma message verbatim. Given that the Germans were reading French communications, this could have had incalculable consequences. The French command was fully aware of the vulnerability of its own signals. Conclusive proof thereof was delivered on the night of 21 April, when the Polish team deciphered Enigma message confirming that "the Germans read instructions sent by the [French] C-in-C by radio".

During the Norwegian campaign the Allies were taken by surprise, and the Navy's command structure hampered the operational use of intelligence resulting from the codebreaking. Some steps to institute the system of its secure distribution were undertaken as early as August 1939, when a SIS transmitter was installed in the BP mansion tower. In November 1939 section responsible for SIS communications started its move from BP to neighboring Whaddon Hall, where Colonel Richard Gambier-Parry organized the headquarters of the service known as the Special Signal Units (SSU), to be renamed later as the Special Communication Units (SCU). The beginnings were modest; decrypted and properly processed messages were transmitted by teletype from BP to Whaddon Hall, from where, after another encoding, this time in British cipher, they were dispatched by the powerful transmitter located on the nearby Windy Ridge.

Their activity during the French campaign was somewhat improvised, but it was the starting point for an efficient and secure system, which from 1940 onwards was being constructed in two interdependent threads. Colonel Gambier-Parry's service represented the first one. The SCU1 station at Whaddon Hall communicated with its outstations installed at the headquarters of major British land and air force formations around the world (at the end of the war their number reached forty). SCU units were usually composed of NCOs of Royal Signals. Their only role was to assure reliable wireless communication with their headquarters at Whaddon Hall; they were not aware of the content of the messages transmitted and received. The second component of the system was the Special Liaison Units (SLU), constructed and supervised by the SIS officer, Colonel Frederick W. Winterbotham. They were usually staffed by RAF and consisted of several NCOs trained as cipher clerks, and a RAF officer, usually a Captain or Major, commanding entire complex SLU/SCU. SLU was responsible for deciphering messages from Whaddon Hall, and occasionally coding messages transmitted back to the HQ. In the early days most messages exchanged by the SLU were enciphered using one-time pad cipher, but as time went by, SLUs were gradually equipped with the British TypeX.

The commander of each SCU/SLU unit enjoyed a delicate status. He was usually the only person knowing the full context of the messages passing through his hands. He was both a provider of intelligence and supervisor of the of security of its source. He enjoyed the privilege of direct access to a highest-ranking officer, to whose staff he was affiliated. He usually presented to him and only to him the information resulting from the received decrypts, and outlined its context. The special status of the SCU/SLU commander usually did not permit him to gain sympathy at local HQ. The information he provided was out-of-bounds even for the members of C-in-C's close staff. They had to accept the exceptional position of an officer, usually several ranks lower than their own. To make matters worse, the SLU/SCU commander was also entitled to rigorous control of orders based of the information he provided. It was strictly prohibited to act exclusively on the basis of information obtained from the decrypts. In every case it was required to create a legend indicating that the information was obtained from an alternative source; interrogation of a PoW, air reconnaissance, etc. The SLU/SCU commander had the right to suspend the execution of any order issued in violation of this rule. It can be guessed that this role of the SLU/SCU commanding officers required selection of people of exceptional integrity.

A part of the security veil for the decrypts was taken care of at BP. Raw decrypts from Hut 6 were passed on to the intelligence staff working in Hut 3 nearby. There the messages were translated from German, indexed, collated with other information sources, and finally paraphrased. Paraphrasing was designed to obliterate the structure and vocabulary of the original, assuring at the same time that no important information from the original message is lost or distorted. Usually, this process required replacing the real source of information with a fictitious one. In the early period messages were attributed to the human sources, agents denoted as CX/FJ and CX/JQ.[4] During the Norwegian campaign, a fictional agent codenamed "Boniface" was born in Churchill's lush imagination, who soon achieved outstanding results in his activities. The camouflage of Enigma's decrypts in the form of agent reports had its disadvantages. Many recipients of the reports were distrustful of human sources of information. Others took the cover so seriously that they directed instructions to a fictional agent about the information he should focus his attention on. On the other hand, this form of camouflage justified restrictions protecting the source of information. The ban on undertaking any action exclusively on the basis of the "Boniface's" report was to be motivated by the concern for the safety of the precious agent.

Most of the described precautions were nonexistent in the first half of 1940, when the system was only taking shape. It is difficult to assess the real role of Enigma decrypts during the French campaign. Timing of events related to the construction of the SLU/SCU system indicates its limited role. The SCU component reached the British staffs in France already in March 1940, in time to assure the transfer of information from BP to the staffs of the British Expeditionary Force and the RAF Strike Force in the continent. The second part of the system, the SLU units, had been organized at the very last moment. It was not until May 24 that SIS officers, Humphrey Plowden and Robert Gore-Brown arrived at the headquarters of Lord John Vereker Gort, C-in-C of the British Expeditionary Force, and a little later that Major F.W. "Tubby" Long joined the RAF Strike Force HQ. Some historians

---

[4] The term CX meant in the SIS code a personal information source—an agent, the initials FJ and JQ were fictitious, individual identifiers of agents.

express their view that despite the resumption of decryption already after the decisive battle, BP managed to make its mark providing information confirming the need to evacuate British troops from France. This view does not seem to be accurate. Churchill ordered the Admiralty to make preparations for the evacuation on 19 May, still before BP managed to resume decryption. It was from Hitler's famous "stop order" of 24 May, that the British learned being given time to consolidate the defense of the bridgehead. But this order was transmitted in an open text, making BP's intermediation superfluous. German breakthrough on his eastern flank convinced ultimately Lord Gort to give up participation in Weygand's planned offensive. However, information about this threat was not delivered by BP, but found with German staff officer, whose car got entangled between British troops on 25 May. In order to protect his exposed wing, Gort had to reposition there the divisions initially assigned to Weygand's operation. The prospect of the Belgian surrender prompted the BEF commander to start the evacuation on the same day. The next day his decision was accepted by the government, and on 26 May an order was issued to start operation "Dynamo". On the other hand, when Hitler's order stopped the German armor, the Luftwaffe was entrusted the destruction of the northern Allied army group. This decision enabled BP to support the evacuation of troops from Dunkirk. The evacuation could not happen without gaining, even local and temporary only, air superiority over the embarkation area. The Luftwaffe's operational network, Red, was being broken at that time with hourly delay. Air intelligence officers working in BP certainly noticed a mistake by Luftwaffe command, assuming that the evacuation would be carried out through the Belgian ports and focused its action on Ostend. This information allowed the Navy to act with unprecedented courage, contributing to the success of operation Dynamo beyond its original assumptions. The initial plans assumed saving some 45,000 soldiers during the operation lasting two days. In fact, the evacuation lasted until 4 June and permitted to save over 330,000 British and French soldiers.

During the French campaign the Poles from "Bruno" did not experience even a symbolic satisfaction. They quickly mastered the subtleties of the new Enigma breaking technique and, according to Bertrand's notes, between 22 May and 14 June, they have broken more than 3000 messages. However, Rejewski noted his feeling that the French staff used the decrypts mostly to locate the positions of their own troops. One case of spectacular disregard for the information provided was particularly memorable: "Operation Paula". On 26 May broken messages brought for the first mention of the planned German operation under this code name. Over the next few days, the extensive message exchange provided gradually full information about the planned bombing of the Renault and Citroën factories. "Bruno" provided the HQ with a complete information on the time of the operation, the airfields from which groups of German aircraft would depart, their routes over France and the numbers of participating aircraft. According to Rejewski's recollections, on 3 June the Poles were watching from the courtyard of Château Vignolles the German squadrons, whose course led right above their heads. Bertrand went to Paris to verify the precision of the data provided by Enigma. German bombers arrived on time and did their job without any obstacles; 254 people were killed, including only 59 soldiers. The commander of the French Fighter Command argued later, that he could not detach units protecting the troops at front line.

After "Paula" the Poles could experience a crisis of motivation. Bertrand's sarcastic remarks showed that the decisions of his own commanders were incomprehensible even for him. This state of confusion wasn't supposed to last long. The air raid on Paris was the first sign that German troops were resuming their operations, this time heading south. On the morning of 5 June, they started a new offensive from the line of Somme and Aisne. German armored divisions broke through the front and headed for Paris. After the next five days, the situation of the French army became so critical that General Weygand suggested surrender. On the same day Italy declared war on Britain and France. The British responded bombing military targets around Genoa and Turin and interning Italian residents of the UK, which had a negative impact on the quality of kitchen and the waiter service in the City restaurants. The Italian action improved the mood of the French command for a while: at last, they had an opponent they could successfully face.

Bertrand was as much aware of the approaching end of campaign as he was concerned about German troops heading towards "Bruno". At the beginning of June, John Tiltman arrived from BP to Gretz-Armainvillers suggesting the withdrawal of the liaison officer representing GC&CS. In his conversation with Tiltman, Bertrand did not try to camouflage the facts ("it's over here"), but he assured his interlocutor that the secret of the Enigma would remain safe ("please assure your bosses that none of the secrets would fall into the hands of the enemy"). Concerned by the lack of instructions from his superiors, Bertrand paid a visit to French HQ on 10 June, asking for orders. However, the Chief of Staff had no orders for him, informing Bertrand only of the C-in-C's resolution to die rather than surrender. Bertrand, whose trust in the competences of his superiors must have reached the bottom at this point, decided to appoint himself his own boss and to act according to his common sense. Since "Bruno" was not provided with a transport worthy of the name, he drove to the abandoned military depot in Vincennes. There, he requisitioned a Parisian bus, putting his personal driver behind the wheel of a truck filled to the brim with cans of fuel. On his return to "Bruno" Bertrand loaded entire staff into the vehicles and formed a convoy which on the same day reached La Ferté-Saint-Aubin. There, for the next three days, the Poles tried to continue their work. It was there that they learned the details of the next Luftwaffe raid, planned for 12 June. When Bertrand communicated their findings to the headquarters of the French intelligence service, then in Briare, he was asked to stop bothering anybody with his revelations when everything around is plunging into chaos and the front and army command practically do not exist.

For "Bruno" this meant the practical end of the French campaign. For the Poles the situation inevitably brought to mind scenes from the recent past; roads crowded with refugees, bombardments, overcrowded cars, and the difficult way towards an unknown future. On 15 June, the convoy reached Vensat, where its staff was still trying to contact the HQ, but its station did not respond. On 17 June, their vehicles reached Larches, and on 19 June Agen. On their way the Poles were able to receive the news about the ongoing truce negotiations and General Sikorski's order for the

Polish troops to disregard French orders to surrender and to evacuate to Great Britain. The news about signing the truce on 22 June reached the Poles already in Agen. The quarters in the buildings of the Bon Encontre Seminary were a very appropriate place to meditate on the events of the recent days and the uncertainty of tomorrow. Bertrand, however, was not in a mood for the meditation. On 23 June he received a message from Stewart Menzies, in which the head of SIS expressed concern for the safety of his liaison officer and the team of the Polish codebreakers. If anyone of them fell into the German hands, the Enigma secret would be in danger. Menzies suggested an immediate evacuation of the entire team to the UK. Bertrand rushed to place MacFarlan on board of the last RAF plane leaving Cazaux airfield for Britain, but he had different plans for the Poles. At the airfield near Toulouse he managed to find three planes with just enough fuel available to reach North Africa. Two of them on the morning of 24 June departed with Polish team on board heading to North Africa. According to Zygalski's memories flight to Oran was quite an adventurous undertaking. His plane ran out of fuel just before reaching the African coast. Its pilot had to land on the fumes, ignoring the airfield control's refusal to land and risking being shot by its AA defense. With regard to the "Équipe D", the flight over the Spanish territory was considered too risky, so they were first flown to Marseille, from where they flew to North Africa on board of a seaplane. Having arranged the evacuation to Africa Bertrand could inform Menzies on 26 June that "as far as our Polish friends are concerned, they were evacuated by plane before receiving your message, so it is no longer possible to send them to UK". Bertrand's message was one of the last Enigma related episodes during the French campaign. On 28 June the wireless link between France and Great Britain got silent.

The period of the Norwegian and French campaigns became for BP and "Bruno" a time of their own "phoney war". During the fighting in Norway, both Allied codebreaking centers kept on breaking Enigma keys used by the enemy's network conveying information most relevant to the Allied operations. Deficiencies of the improvised command structure prevented the information from the codebreakers from reaching the local Allied commanders and staffs. German attack in the Benelux and France surprised them when the shortcomings of the previous campaign were only partially eliminated. In addition, the change in the enemy's operational procedures blinded the Allied codebreakers during the critical early phase of the campaign. On the other hand, break in the decipherment lasting just three weeks confirmed the maturity achieved by the BP team during the first months of war. Both campaigns brought a valuable experience for the codebreakers. Close cooperation of the Allied codebreaking centers and intercept stations paid off. Even if it proved impossible to organize a common cryptology center, the speed of structuring and the scope of cooperation between BP and "Bruno" contrasted with the bargaining, which we will be watching commenting on the birth of the British-American codebreaking cooperation. Exchange of liaison officers, division of roles, direct communication lines linking the centers, coordination of the key breaking, exchange of intercepts—in few other fields was the British-French-Polish military cooperation so close and brought such tangible results.

The most important Allied success during the French campaign went mostly unnoticed: despite the defeat the secret of Enigma breaking remained secure. Its security was not obvious. In the early spring of 1940 Bertrand recorded an event representing most probably a test by one of the German intelligence agencies of how the Allies were dealing with Enigma. Certain Raffali approached reportedly the French army command with an offer to develop a device permitting to break the German machine ciphers, requesting 200,000 francs for this service. Bertrand convinced his superiors to pay a large advance on his work, which would convince the bidder that the French are still looking for a solution. This unsophisticated manipulation could have been a private initiative of a shrewd dodger, but it could equally well represent enemy's plan to disclose the Allied knowledge of the German ciphers. Josh Cooper once expressed his belief that the Enigma would become less mysterious in case of war, when his service would certainly, sooner or later, get a copy of the machine. But this mechanism also worked in the opposite direction—in the chaos of warfare, the Germans could also intercept messages betraying the Allies' knowledge of their own, enciphered orders. This scenario was all the more likely, as the German side was breaking at least two Allied cipher systems; the British naval and the French army code. The secret of Enigma, however, rested primarily in the minds of those involved. The Norwegian and French campaigns did not pose a direct threat to British participants. Probably closest to danger was Lord Gort, commander of the British Expeditionary Forces in France. On the 30th of May, four days before the end of Operation "Dynamo", he was ordered to hand over the command on the continent and evacuate to England. Later on, the British will not take even such a limited risk—none of the recipients of Ultra could find themselves in a place or situation contagious, even potentially, with ending up as a PoW. The Poles found themselves in a more difficult position. Hardly any army in the world prepares in advance a plan of action in the event of a catastrophe, especially one as sudden and unexpected as France's collapse in June 1940. It must have led to a collapse of morale and paralysis of will. In effect, the evacuation of the "Bruno" was an complete improvisation. Nevertheless, people, machines, archives, and with them the Enigma secret, have reached safe haven in French North Africa. However, the Allies were to pay a high price for the lack of planning, responsibility and chaos of the disaster in the future.

Chapter 12
Bombes

Napoleon used to say that waging war requires three things: money, money and more money. Since the Napoleonic campaigns the reality of armed conflicts has changed. Money preserves its position as the engine of the modern warfare, but information became its nerve. The value of information may vary for different participants of the conflict. Strength usually compensates for the lack of information: it is the stronger player who imposes initiative, creating rather than reacting to the facts. Even an accurate information does not represent considerable value for a party unable to join the game due to the limited resources. Information represents thus a natural weapon of the weaker side, which intends to continue fighting using its limited resources. After the fall of France, Great Britain became such a participant in the war. During the French campaign the British army did not suffer catastrophic number of casualties, but left on the beaches of Dunkirk practically all the equipment of the British Expeditionary Force: 475 tanks, 38 thousand vehicles, 1.000 guns, communication equipment, even soldiers' personal weapons. At the beginning of June 1940, three combat ready divisions were stationed in the UK. The next fourteen divisions were in various stages of training. The crowds of soldiers evacuated from France required reequipping and reorganization from scratch. The Home Guard consisted largely of veterans of the Great War, so its members marched smoothly. However, troops equipped with pitchforks, hunting guns and knives would not be a match for German soldiers. Divisions transported from Australia, New Zealand and India came to an emergency stop in the Middle East, when it turned out that Egypt became an enclave squeezed between Italian armies in Libya and East Africa. With the news of Hitler's victories in Europe Japan saw it fit to take up its own game and requested the Vichy government to hand over French colonies in Indochina to Japan. Thereafter the Japanese demanded that the British close the Burmese Way—the only artery permitting delivery of supplies to their Chinese adversaries.

For several critical weeks, Britain was defenseless on land, counting only on defensive lines in the air and at sea. According to traditional British doctrine, the

M. Grajek, *Enigma Myth Deciphered*, History of Information Security,
https://doi.org/10.1007/978-3-031-65475-6_12

islands' outer defense line runs on the mainland, along the Belgian and Dutch borders. Now, however, the Germans not only controlled the ports of both countries and the Atlantic coast of France, but also installed Luftwaffe squadrons at local airfields, a dozen or so minutes' flight from the coast of England. The extension of German control over Norway meant not only that Swedish ore could be imported into the Ruhr uninterruptedly. The Kriegsmarine ships could hide in Norwegian fjords, ready to go out to sea at any time, threatening British shipping in the North Sea and Atlantic. The British leaders were aware that their defense lines are dangerously stretched. In that difficult moment they could only hope that their intelligence service, still considered to be the best in the world, would allow them to concentrate modest resources in the critical points and throw them into the battle at the right time.

But in mid-1940 British intelligence service was far from its best shape. Its recently deceased long-time head, Admiral Sinclair, was a charismatic leader and a great organizer. However, his passion for the secret work was matched by his tendency to say what the politicians wanted to hear. In the weeks leading up to the Munich Agreement, Sinclair submitted a report suggesting the need to appease Hitler. A few months before the outbreak of the war he kept on discounting the signs of the coming conflict, attributing them to Soviet propaganda. SIS reports dismissed the German-Soviet negotiations, claiming that their only function was to disrupt the parallel talks between the Allies and Russia. Sinclair's death in November 1939 destabilized SIS even further. Sinclair anointed as his successor his deputy, Stewart Menzies, whose candidature was strongly opposed by Claude Dansey. Churchill complicated the matter adding a third candidate. Finally, Menzies won, but as part of the compromise, Dansey assumed the position of his deputy, clearly demonstrating his disappointment. And in May 1940, he was probably amused, watching the relations of his superior with Churchill, who not so long ago was opposing his appointment.

Churchill's relationships with Admiral John Godfrey, head of the naval intelligence service, were not much better. The present chief of NID brought to his memory Admiral Hall, with whom he had cooperated during the First World War. Hall used to have his own vision of his service's role, going far beyond the tasks defined by the politicians. Churchill, acting as the Admiralty First Lord, needed Godfrey's cooperation, but he did not feel comfortable about Godfrey's seeking often Hall's advice and even moving his own office to Halls' private apartments. What's worse, a conflict of attitudes developed between both leaders. Churchill saw the war as a continuation of politics. The conduct of war required building up the morale, so the First Lord welcomed certain window dressing in the reports. Godfrey assumed that his service's role was providing the leaders with reliable information. Churchill's journalistic temperament was fully in line with the requirements of war propaganda, but was putting Godfrey's loyalty and standards to a hard test.

Personal conflicts of attitudes and interests represented only part of SIS's problems. During the first months of the war most of its stations and posts in Europe were annihilated. SIS had to roll up its posts in Berlin, Prague, Warsaw and Bucharest. Shortly after the Soviet invasion of Finland, station in Helsinki was relocated to Stockholm. As a result of the French campaign, stations in Paris, Brussels and The Hague were closed down. The Dutch station and agent network had been compromised as a result of the Venlo incident. Soviet occupation of the Baltic republics forced the SIS to leave Riga and Tallinn. In mid-1940 SIS's network in the continental Europe was limited to the capitals of three neutral countries—Bern, Lisbon and Stockholm. At a crucial time, the SIS could only cover up its weakness with a cloak of the glorious tradition.

In that gloomy situation BP's increasingly effective operation represented one of the few signs of hope. After the fall of France, the Wehrmacht occupied the northern part of the country unopposed. Before the German Army managed to install the wire network, it had to rely on the wireless communications. The proximity of their new garrisons to the British listening stations allowed to intercept most of the German transmissions, and the growing skill of Hut 6 facilitated monitoring German preparations for the invasion of the British Isles. At that time BP was only able to break the messages exchanged in Luftwaffe networks, but the air attack had to be the first stage of any operation over the English Channel. In the coming months BP was to become the main and only source of information about the German plans. Starting from the first days of June, Hut 6 was able to decipher up to thousand messages a day. Early on, on June 1, one of the deciphered messages brought the information that Germany did not intend to undertake any operations against the British Isles before the final victory in France. However, on 23 June, the day after the signing of the armistice, another message brought a warning about Luftwaffe's preparations for the operations against the UK. Five days later another message revealed an order to complete the aircraft maintenance, deploy the units to the new locations in France, and supply the airfields by 8 July. Next to the new threats Enigma occasionally communicated also good news. On 6 July, Hut 6 deciphered the message containing Luftwaffe's current *ordre de bataille*, informing that the enemy could field 1.250 bombers, carrying 1.800 tons of bombs, contrary to the earlier British estimates of over 2.500 aircraft with a capacity of 5.000 tons. Four days later, attacks of German bombers on the shipping in the English Channel started the air Battle of Britain.

As the waiting time ended the British could switch to more effective methods of recognizing the enemy's plans than Enigma. A chain of radar stations provided information about approaching aircraft, their course and altitude as soon as the enemy left the airfield. German scientists also worked on the radar technology and undertook an investigation of the British preparations in that field, sending over the English Channel one of their Zeppelins filled with wireless receivers. As their equipment was tuned to the frequencies used by the German, but not the British radar devices, operation of the existing Home Chain Low radar network was not discovered by the mission. Consequently, German plans disregarded the role of the radar, which was to cost the Luftwaffe lives of many pilots and, in the end, defeat in the Battle of Britain. However, contrary to the widely adopted opinion, deciphered Enigma messages have hardly affected its course. It seems that even the commanding officer of the Fighter Command, Air Chief Marshall Hugh Dowding, was not among the recipients of Enigma related information during the battle.

But the Germans also had some surprises in the field of wireless technology up their sleeve, and it was in the struggle with their innovations that Enigma was to make the crucial contribution. In the early phase of war, the Allies received a warning concerning the new weapons systems being developed in the Third Reich. On 4 November 1939 someone dropped a package to the British Consulate in Oslo containing a prototype of a proximity fuse and a letter describing several research programs being implemented in Germany. Some of the projects mentioned in this letter included jet and rocket engines, jet and rocket propelled aircraft, radar, wireless night navigation and guidance systems, guided bombs, ballistic missiles, magnetic mines and acoustic torpedoes. The first reaction of the recipients was disbelief. The mysterious shipment arrived only a month and a half after Hitler's September speech, in in which he threatened the enemies of the Reich with terrible and mysterious weapons, against which they would be completely helpless. The Oslo package seemed to confirm Hitlers warnings, giving them the appearance of scientific truth. Moreover, it seemed unlikely that a single person could be able to gain access to so many secret and probably well-guarded projects. Subsequent events confirmed most of the information included in the report, which for several decades after the war was considered one of its greatest riddles. It was also thought that its author would remain anonymous, but the British scientist, at whose desk the Oslo report landed in 1939, revealed many years later[1] its author—the German scientist Hans Ferdinand Mayer. He was persecuted during the war for the statements about the Third Reich considered dissident by the Nazis, but his persecutors did not know nothing about his report.

The first confirmation of the report's credibility arrived the same month, while examining a sea mine dropped by a German plane on the coastal shallows. The next one had to wait until March 1940, when in the wreck of the German bomber notes were found referring to one of the systems mentioned in the report. The documents have been placed on the desk of Reginald V. Jones, who at the time acted as a one-man scientific intelligence unit at the Air Ministry (*Scientific Information Branch, SIB*). Jones was a graduate of Balliol College, where he received his doctorate in physics. He had to sacrifice the planned journey to the United States, where he was offered a job at the Mount Wilson Observatory, to accept the job at SIB. When presented with the responsibilities linked with this position, he reportedly responded briefly: "A man in this position could lose the war. I take it!" Jones who was among the first people in Britain to have read the Oslo report and one of the few to have taken its warnings seriously. His analysis indicated the possibility of developing a wireless navigation system, but the evidence available so far did not corroborate this statement. Luftwaffe's use of a precision night-navigation system represented a huge threat to Britain. RAF was fighting desperately to repel air strikes during the day, but its defenses against night bombing were just as improvised, as ineffective.

On 5 June, BP has deciphered a Luftwaffe message mentioning the name of the system, *Knickebein,* known from the earlier POW's interrogations, and revealing the location of its transmitter and the direction of the radio beam: the existence of the system was finally confirmed. A flight of an aircraft equipped with the proper

---

[1] Reginald V. Jones, in his book '*Reflections on Intelligence*', Heinemann, London, 1989.

receiver allowed on 21 June to determine the nature of the system and to develop the countermeasures. The SIB referred to the *Knickebein* system using the code name "Headache", so it was logical to describe the method of its jamming as "Aspirin". *Knickebein* settings were usually communicated to the system's operator by an Enigma message. The same message broken by BP identified the targets for the next night and permitted to administer a proper dose of the "Aspirin". BP was able to monitor the developments almost in the real time. When the Luftwaffe started on July 27 installing new transmitters in the Brest and Cherbourg area, the codebreakers learned about it the same day.

Broken Enigma messages combined with "Aspirin" forced the Luftwaffe to abandon the application of the *Knickebein*. But it turned out that they have at least one more surprise in store. Already during the investigation of the *Knickebein* system, one of German PoWs was overheard mentioning the mysterious *X-Gerät*, whose mode of operation was supposed to be similar, but not identical to the *Knickebein*. Confirmation arrived again in the Enigma message. Around September 10, a message was deciphered informing about the construction of five radio transmitters in the area of Calais and Le Havre. The same message mentioned the cooperation of the new stations with the Luftwaffe unit, *Kampfgruppe 100,* whose usage of the wireless guidance systems was identified earlier, in connection with the *Knickebein*. Successive deciphered messages of the Red network permitted to catch the enemy red-handed on the night 19/20 September. While the Luftwaffe was testing the system, British scientists were able to fix the locations of its transmitters, the frequencies they were using and the characteristics of the signals. After several weeks of feverish work, they had prototypes of jamming devices at the beginning of November. As the challenge posed by *X-Gerät* was considered more serious than 'Headache', the cure was accordingly referred to as 'Bromide'. Its first application was, unfortunately, only partially successful and created a real headache for the future historians. On 11 November Hut 6 deciphered a message ordering the preparations for operation against the targets marked with numbers 51, 52 and 53. A message addressed to the *Kampfgruppe 100* revealed the code name of the same operation—"Mondschein Sonata". At this point the codebreakers have concluded their role in the operation, which was to become the first practical test of "Bromide" effectiveness. Alarmed by news from BP, the RAF reconnaissance aircraft detected *X-Gerät* wireless beam over central England. All "Bromide" transmitters in the area have been activated. Despite the countermeasures, during the night of 14/15 November over 500 German bombers dropped their load on Coventry, destroying the city, the historic cathedral and nearby aircraft factories, and killing about 500 people. After Enigma breaking was revealed many years later, a legend developed that Churchill had sacrificed the city to protect the codebreakers' success.

In the light of our present knowledge such an interpretation is not justified. Broken Enigma messages allowed to identify the new system in time and to design the countermeasures. The same source provided precise information about the date of the operation, but its targets were hidden behind the code names. In fact, the extraordinary number of bombers participating in the planned raid made RAF intelligence guess that London was its target. The proper question regarding the Coventry tragedy was not so much about Enigma's, but rather Bromide's failure. As far as we know today, its failure resulted from the competition between the Navy, the Army and the RAF. Bromide did not work because its constructors overlooked certain feature of *X-Gerät* receivers.[2] The receiver was found in the wreck of a German bomber, which crashed in the shallows off the British coast on 6 November. The place of its crash on the border of land and water turned out to be unfortunate, starting the dispute over whether the wreck investigation should be carried out by the Army or the Navy. It was finally settled in favor of the RAF, which did not present its report until 27 November, almost two weeks after the Coventry bombing. Bromide, corrected basing of the report's conclusions, performed later on correctly, although never as effectively as Aspirin. The Germans had another system in store, codenamed *Wotan* or *Y-Gerät*. RAF intelligence service learned about its existence also from the Enigma message intercepted in July 1940. When the intercept stations recorded new radio transmission in December, engineers quickly identified its source and purpose, proposing an efficient jamming method. The system, theoretically most technically sophisticated, proved to be the easiest to defeat.

The period of struggle between the German night navigation systems and the British signals intelligence and engineers is known in history as the battle of beams. During this battle British intercept stations identified the new Luftwaffe wireless network, code named Brown by BP. Its key was first broken on September 2, 1940, and the deciphered messages have brought interesting information. It turned out that the Brown network was operated by the *Kampfgruppe 100*, a special unit of German bombers, whose aircraft were equipped with radio navigation devices and whose crews were trained in blind navigation. The main task of the group was precise bomb drop before the arrival of the main forces, so that the aircraft not equipped with the blind navigation devices could see the fires of their target. The British remedy for this tactic was somewhat theatrical, but effective. In the vicinity of probable targets, particularly the aircraft factories, in the open field, installations were constructed from perforated pipes and fuel tanks, which could be turned into a flames inferno by pressing a button. Once the objective for the night was identified, on the approach of the German aircraft its fake target was ignited, providing German pilots with a clearly visible aiming point. The Brown network became the Hut 6 code-breakers' favorite due to the exceptional lack of discipline of its German operators: "Brown's ciphers have demonstrated that the Enigma is only as safe as its staff make it. (...) it is difficult to indicate a rule of Enigma usage or wireless communication that was not violated by the cipher clerks and radio operators of these units".[3] On the day of the operation *Kampfgruppe 100* was regularly informing system operators about the current target sending the message starting with fixed plain text **VORBEREITETXBETRIEB** (prepare operation). The operators confirmed the beam

---

[2] A narrowband filter with a central frequency of approx. 1.5 kHz is installed in the input circuit of the German receiver. The British interference beams were modulated by a 2 kHz signal, which was eliminated by the filter, so they did not reach further steps of the German device and did not interfere with its operation.

[3] *The History of Hut 6*, Volume I, NA HW 43/70, s. 198.

activation with a message starting with **FERTIGXEINGERICHTET** (ready) and were waiting for the end of operation in form of the message starting with **BERTIEBSSQLUSS**. Over time Hut 6 codebreakers confirmed that no other network offered comparable quantity and quality of cribs and cillies. Choosing the message keys in form of the commanders' and cipher clerks' initials was a standard practice, and the stability of the unit's staff meant that they remained unchanged for months. In December Hut 6 detected repeatable nature of the plugboard settings, which facilitated breaking the key every other day. Codebreakers speculated that unit's secret and experimental nature must have prevented proper security training of its staff. The Brown network was intercepted by British listening stations longer than other Luftwaffe networks. Most German garrisons in France were rapidly switching to the wire communication. But the *Kampfgruppe 100* resided at Vannes airfield, one of the westernmost and consequently the latest to be connected to the wired network. Its messages dried up only in 1941, but then the battle of the beams was already over. Even then it was not a last meeting with the Brown network, which was to return in 1944 in the service of experimental units involved with V1 and V2 programs.

Final assessment of BP's role during the Battle of Britain and battle of the beams remains inconclusive. BP encountered considerable problems converting the broken messages into usable intelligence. The messages were full of abbreviations and terms absent in any dictionary available. In time BP would build up an index of the German military dialect, but in August and September 1940 this work was in its infancy, and the clear text of many deciphered messages remained a puzzle. The Luftwaffe used maps of Great Britain with a grid from the original British map, which had been withdrawn years ago and all its copies destroyed. Hut 4 was thus unable to convert the target coordinates revealed by the broken messages into the present system. Even without the participation of Hut 6 the British would have been able to cope with night raids, perfecting radar equipment and introducing new types of night fighters. Enigma played a decisive role in identifying the threat, but it was the wireless scientists and engineers who made the most important contribution to defeating the threat of *Knickebein*, *X*- and *Y-Gerät*. The nature of this clash was important from the point of view of the history of modern warfare. For the first time in the history the military operations depended almost entirely on the results of the duels of mind in several fields of knowledge, science and engineering. The battle of beams represented the earliest episode, when the Second World War came so close to the modern conflicts, its crucial episodes unveiling in the university lecture halls and scientific institutes rather than on the battlefields.

From July to the end of September 1940, life on the British Isles was largely shaped by the invasion threat. This danger was also felt in Bletchley, far from the battlefields. Reports found in the literature about the locomotives standing under steam at a nearby railway station and the trains ready to evacuate BP to the ports of the West Coast, from where the ships were to take the codebreakers and their equipment to Canada, are exaggerated. In July and August 1940, the entire BP employed less than 200 people,[4] and booking entire train to transport existing staff, equipment and documentation would be extravagant. Preparations for the evacuation were limited to combining the private vehicles of the team members with center's means of transportation to create the Mobile Column. In the course of events the selection of staff members qualified for evacuation "(w)ith a very small group of girls, one hundred percent volunteers, triggered the most serious crisis in personnel relations at *Hut 6*".[5] Another token of preparations for the invasion was assigning the mobile wireless units provided by the SIS's Section VIII to the staff headquarters in the British Isles. The military planners were seriously considering that London might be overrun by the invaders, with all its communications centers. The only tangible sign of the ongoing war that reached BP was a few bombs, that a stray German bomber accidentally dropped on and around the park on the night of 20 November, 1940. One of the bombs ruined the stables and laboratories of Elmer's School, where Welchman used to start his adventure with cryptology. The second bomb fell a few meters from Hut 4, and the explosion displaced the whole structure from its foundations. The third one fell in the park without exploding, and had to be disarmed and removed by a team of Royal Engineers. None of the staff were hurt, although everyone was very proud to be so close to real danger. After this incident, the wooden huts were surrounded by the brick walls protecting from the shells.

BP's early successes in breaking the Enigma's ciphers during the French campaign and later, during the Battle of Britain, allowed the center to gradually build up its authority. For the time being codebreakers' success was based on the very fragile foundations. Both Herivel tip and *cillies* exploited bad habits of the German cipher clerks. In his role of Cassandra Dilly could point out that, in line with his earlier prophecy, the Germans had already eliminated the mistake making Zygalski sheets so efficient. Their radio security service would certainly notice current mistakes and put an end to them. Any hope for further success was linked to the work carried out in the Cottage attic by Alan Turing. He immediately grasped the importance of Rejewski's major breakthrough—the independence of the Enigma's cyclic structure from the plugboard. However, guided by Knox's remarks, he intended to apply this principle not to the message header, but the message text itself. At this point Turing decided to abandon mathematics and focus on his second passion—practical engineering. He intended to develop a device representing a more flexible version of the Rejewski bombe. Rejewski had been comparing three pairs of characters, which required a set of six ciphering machines. Testing the cycles resulting from hypotheses concerning the message texts would require many more Enigma equivalents. Enigmas in the Rejewski bombe were permanently connected with each other. The cycles based on the message texts had to include characters in any position. Turing's design had to permit modification of the connections between the Enigma equivalents. Basing on the observation of given day's traffic Rejewski knew which letter to

---

[4] The most numerous was the crew of *Hut 6*—89 people, followed by *Hut 8* (fleet codes)—32 people, *Hut 7*—10 people and *Hut 7* people (as of August 1, 1940).

[5] *The History of Hut 6*, Volume I, NA HW 43/70, s. 5.

search for. Turing only knew that in the correct rotor position one of the possible characters would generate the solution, but he did not know which one. It was possible to test each letter separately in the subsequent runs of the machine, but it would take too long. Therefore, Turing proposed a design that solved two problems at the same time: a variable machine configuration and simultaneous scanning of 26 hypotheses. The essence of his idea was to "straighten out" the Enigma rotor. In the real Enigma an electrical signal from the keyboard passed through a plugboard, an entry drum and three rotors, reached the reflector and returned via three rotors, the entry drum and the plugboard: inputs and outputs of the electrical circuit were mixed with each other. Turing's answer was simple: for the price of doubling the number of rotors in each set he achieved a complete separation of inputs and outputs, allowing him to connect the output signals of one Enigma to the inputs of another one and test all 26 hypotheses at the same time (a more systematic description of Turing's bombe is presented in a box below).

Knowing the history of Turing's earlier designs, one could suppose that he would be eager to transform an idea into an engineering design by himself. BP heads decided that it would be better to entrust the job to professionals. GC&CS had been using for some time business sorters and tabulators, very useful in searching for repeats in coded messages. Their manufacturer, the British Tabulating Machine Company (BTM), could provide solid references. The company was established in 1902 under the name of Tabulator Limited when its founder, Robert Porter, obtained the exclusive rights to distribute in the UK and the dominions equipment manufactured by the American company—Tabulating Machine Company. Neither the terms "tabulator" or "sorter", nor the names of the BTM and Tabulating Machine Company are currently linked with the microcomputer and mobile phone era, although they should.

Their common history began in 1884, when Herman Hollerith patented in the USA the first version of his invention, which he described as a tabulator. The tabulator was designed to facilitate processing of statistical data, in particular the results of population census. Its operation principle was simple: data was perforated on a paper card, with each column of the card corresponding to one of the categories into which the census divided the population. The cards were then passed through a device, which was set up to count cards corresponding to a given condition, for example all women under the age of eighteen who lived in New York. Basing on the processing speed of his design Hollerith was able to compile US census results in 1890 in a quarter of time and at a cost of five million dollars less than traditional technology required—his tabulator gained widespread recognition. In 1896 Hollerith founded the Tabulating Machines Company, which was renamed in 1924 as International Business Machines Corporation, more easily recognized by its abbreviation—IBM. Robert Porter's British company, renamed BTM in 1907, was originally distributing machines manufactured in the USA by Hollerith. However, after the First World War, the demand increased so much that it was decided to build a modern factory at Icknield Way in Letchworth, which in 1939 had more than 1.200 employees. After the next World War, the license agreement with the IBM became a barrier, so in 1949 the relationship between the two companies was loosened and subsequent mergers and transformations resulted in the adoption of a new name—International Computers Limited (ICL).

The tabulator quickly found new applications beyond the areas for which it was designed. In the 1930s its potential has been recognized also by the codebreakers. Breaking codes required finding among the intercepted messages any repetitions, the convenient starting point of attack. Sorters and tabulators were able to dig tirelessly and much faster than humans through the piles of cards, and liberated the codebreakers from this routine job. On the eve of World War II, the leading codebreaking services of the world were using a modest number of sorters and tabulators. Close contacts between GC&CS and BTM have made the British tabulator manufacturer a natural candidate for turning Turing's ideas into an engineering design. BP found a competent and enthusiastic partner in the person of Harold "Doc" Keen, head of BTM's R&D department. Keen started his career at BTM immediately after having received his degree in electrical engineering in 1912, and spent the rest of his professional life in the company. He was the author of over sixty patents and a leading designer of business machinery of his time. In the coming years the collaboration between Turing and Keen was to take a similar form as that between Rejewski and Palluth in Poland. Keen was facing many challenges, but two of them came to the fore. The first one was related to the mechanism detecting the stop condition. The shorter its reaction time, the faster rotors could rotate, shortening the search time. Fortunately, Keen could find a simple analogue with his earlier designs. Both sorters and tabulators included a relay circuit detecting a defined combination of holes. The second problem was beyond his engineering experience so far. The stop condition could be represented both by the presence (hot scanning) or absence (cold scanning) of electrical signal in just one of 26 circuits. We will probably never know whether this problem was solved by Turing or Keen, but the relay based logical circuit one or both of them proposed was functional.

Discussions between Turing and Keen on the concept and construction of the machine had to be concluded in October or November 1939, as GC&CS placed with BTM an order for 30 devices in November or December. BTM considered it wise to give the project a label suggesting manufacturing just another type of the tabulating machine, so the whole project was internally named 6/6502, and known among the employees as Cantab. In BP, where there were no reasons to hide the purpose of the machine, it was named after its Polish original—the bombe. Keen had to solve several organizational problems related to its production. The factory at Icknield Way couldn't handle the order without hampering the main production line, being equally important. The production of components had to be divided between the BTM suppliers. These included the Spirella company, also located in Letchworth, manufacturing the most secret part of the machine—its rotors. For the Spirella staff it was a completely new experience, as in the peacetime its main product were women's corsets. During the war the ideal waist shape of the British women was less important than victory, and Spirella was employing almost exclusively female staff, guaranteeing patience and accuracy in the tedious work. BTM had to solve the problem of bombe production secrecy. Keen did so in a slightly perverse way, giving up any visible attributes of secret and pretending the new device to be just another type of the business machine. The BTM's suppliers and partners manufactured components that did not reveal the nature of the final product. The final assembly of the bombe took place in the company's main plant, after which it was loaded onto an ordinary truck, driven by a lone unescorted driver, who drove the vehicle to a predetermined location, where he was passing on the car keys to another driver, responsible for delivering the cargo to BP.

In early 1940 the serial production of bombes was the melody of the future, and whether they would meet Turing's hopes depended on the results of the prototype tests. The first bombe was delivered to BP on March 14, 1940. The name it was given reveals the high hopes for the prototype, as the device was christened Victory. Its early tests were disappointing; Victory was unable to find the correct rotor settings. There could be many reasons for its failure. The most important was probably the selection of the test case. The codebreakers attempted to use the prototype to break the Kriegsmarine key, which has not yet been broken by other methods. Effective use of the bombe required having a reliable crib. Thousands of broken Luftwaffe messages gave the Hut 6 a good understanding of the structure and language of its messages, but Kriegsmarine ciphers represented so far a closed world. The bombe did not announce the reason of its failure; it could be a false crib, wrong crib position within the message being tested, or the machine's malfunction. The lack of reliable cribs undermined confidence in the bombe's effectiveness, and the codebreakers from Hut 8 had heated discussions with intelligence staff from Hut 4 on the crib selection, without effects.

The second potential problem with Victory was the required crib structure. The old hands kept on suggesting cribs, which in their opinion were more likely to appear in Kriegsmarine messages than the ones preferred by Turing and his team. They probably did not understand that a crib with no closed cycles was worthless for Victory. If the number or length of the cycles was limited, the bombe gave an ambiguous result, generating a lot of the false stops, which had to be manually checked. The device designed and manufactured within just four months was a masterpiece of engineering and, after the infancy problems were removed, worked smoothly. So, its initial failure must have resulted from the error in the test concept rather than the engineering. When the bombe project was at the verge of collapse, it was rescued by Welchman. He noticed that the current concept properly exploits the cyclic structure of the Enigma cipher, but ignores its other feature; Enigma's inability to encipher any character as the same letter. Welchman suggested using this property to construct the so-called diagonal array. His solution was extremely simple; it was a board on which a tangle of wires was fixed. In spite of its simplicity, it turned a technical failure into a brilliant success. Turing, to whom Welchman first presented his idea, could not believe that such a simple trick could be effective. But after reconsidering the importance of Welchman's proposal he enthusiastically incorporated it into his concept. The first mention about the diagonal board appears in source documents around July 1940, indicating that its concept was born as a reaction to a series of *Victory's* failures. The first bombe equipped with a diagonal board reached BP on 8 August, 1940 and was named Agnus Dei, usually shortened as Agnes. This time the name turned out to be prophetic, and Agnus Dei soon started to preach the good news. The bombe's earliest success was achieved still in the testing phase in Letchworth, when a codebreaker sent to the manufacturer's plant somewhat unexpectedly managed to break the first German message (Fig. 12.1).

The bombe must have impressed even someone who was not aware of the intellectual effort put into its construction or its significance for the conduct of war. The device weighed about a ton, was over two meters wide, almost two meters high and more than half a meter deep. Inside a tangle of multicolored wires, with a total length of several kilometers, connected about a million soldering points. The

Fig. 12.1 Turing-Welchman bombe front view. Each vertical column of 3 rotors corresponds to one Enigma machine. The bombe contains 3 rows of 12 machines each, for a total of 36 Enigma equivalents. (CCS-SA 2.0, photo Ian Petticrew)

device's front panel was occupied by the rotors emulating Enigma[6] machines. Unlike in the real Enigma the rotors representing a single machine were not mounted on a common shaft, but one under the other; in the highest row there were fast rotors, in the following ones—middle and slow. In order to facilitate quick change of the rotor order and to prevent the mistakes, each type of rotor had a different color (rotor I—red, II—chestnut; III—green; IV—yellow; V—brown; VI—blue; VII—black; VIII—silver). In the standard Enigma the electrical signals ran from the keyboard to the lights as shown in the picture below (Fig. 12.2).

Turing's idea required the separation of inputs and outputs of the rotor battery, which he achieved "straightening" the rotors. As a result, signals from the keyboard passed through 3 standard rotors of the machine, then through a modified reflector (with contacts on both sides) and finally through three subsequent rotors, whose connections represented a mirror image of the standard rotors (Fig. 12.3).

In the reality, both standard rotors and their mirror images were placed in a common casing, which facilitated their synchronous movement. Moreover, their contacts were all placed at just one side; four rings of contacts represented the separate inputs and outputs of the two rotors (Fig. 12.4).

Four concentric contact rings of the rotor's surface corresponded to four stationary rings at the front of the bombe. The two central rings were internally connected with the contacts located in the row below (which corresponded to the connection to the next rotor in the real Enigma) (Fig. 12.5).

Fast rotor   Middle rotor   Slow rotor   Reflector

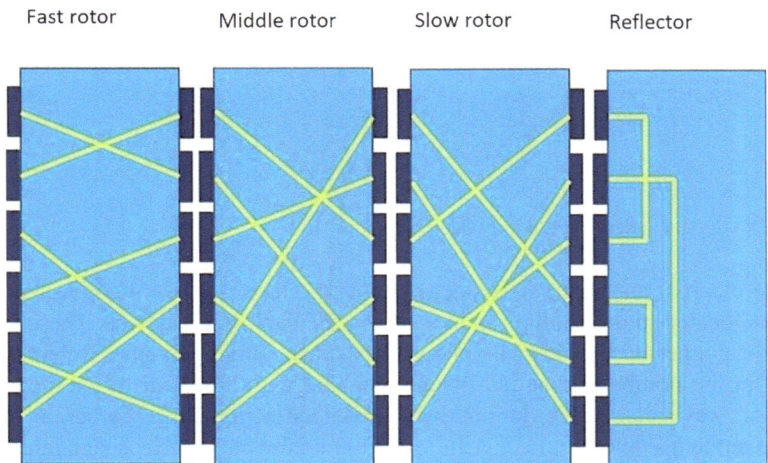

Fig. 12.2 Signal flow in a standard Enigma

---

[6] The prototype copy of '*Victory*' *was* constructed as an equivalent of 30 Enigma machines. The next copies, starting with '*Agnus Dei*' were built in a 36-machine version, to which the '*Victory*' bombe was also rebuilt over time.

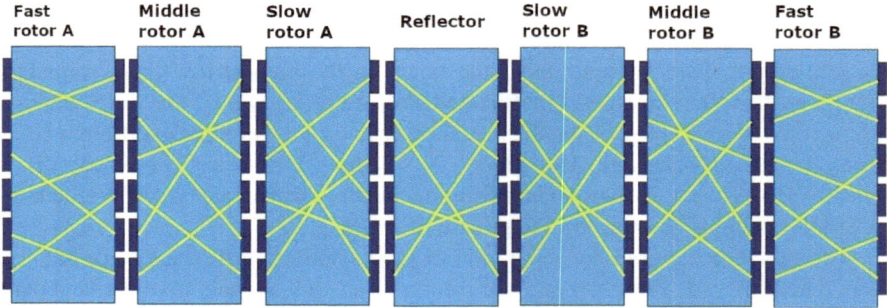

Fig. 12.3 Signal flow in the Turing-Welchman bombe

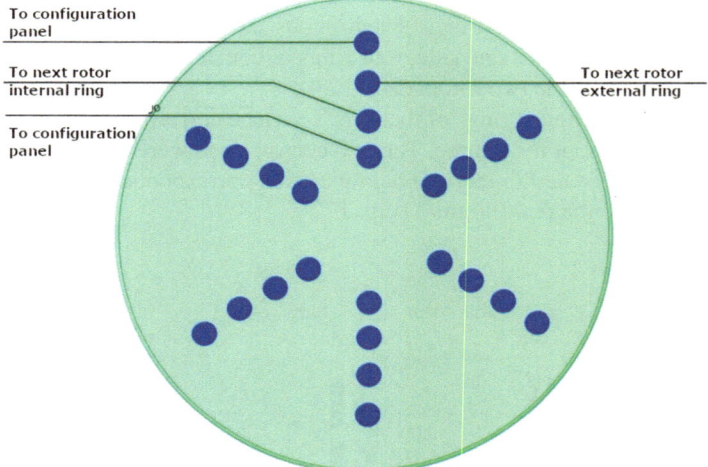

Fig. 12.4 Six-character bombe rotor (side view)

The external and internal contact rings of the fast rotor were connected to the sockets located in the configuration panel on the back of the device. These sockets could be connected with 26-wire plugs to sockets belonging to other Enigma equivalents in the bombe; the output of one machine with the input of another. Entire rotor set was called in BP jargon a double-ended scrambler.[7] The scrambler was equivalent to a single Enigma machine, excluding the plugboard. A standard bombe contained 3 rotor banks, each of which consisted of 12 scramblers, and was equivalent to 36 machines (Fig. 12.6).

Knowing, at least in the outline, the bombe's construction permits us to return to the theory and describe its application principle. The diagram above illustrates the situation, when in the second and fourth position of the Enigma message, the letter

---

[7]When Americans later began building bombes, they used the term '*commutator*' to describe a set of rotors.

Fig. 12.5 The described structure of Enigma's 'straightened' rotor can be seen in the photo of the machine's front panel. (CC0 1.0 UPD)

E of the plaintext is transformed into the letter J of the ciphertext, in the third position—J into L, and so on. The above diagram is one of the forms in which instructions were given to the bombe operators, usually referred to as the bombe menu. Implementation of this menu requires 7 scramblers connected as in the diagram below (Fig. 12.7).

Fig. 12.6 Identification of cycles in the probable text of the encrypted message

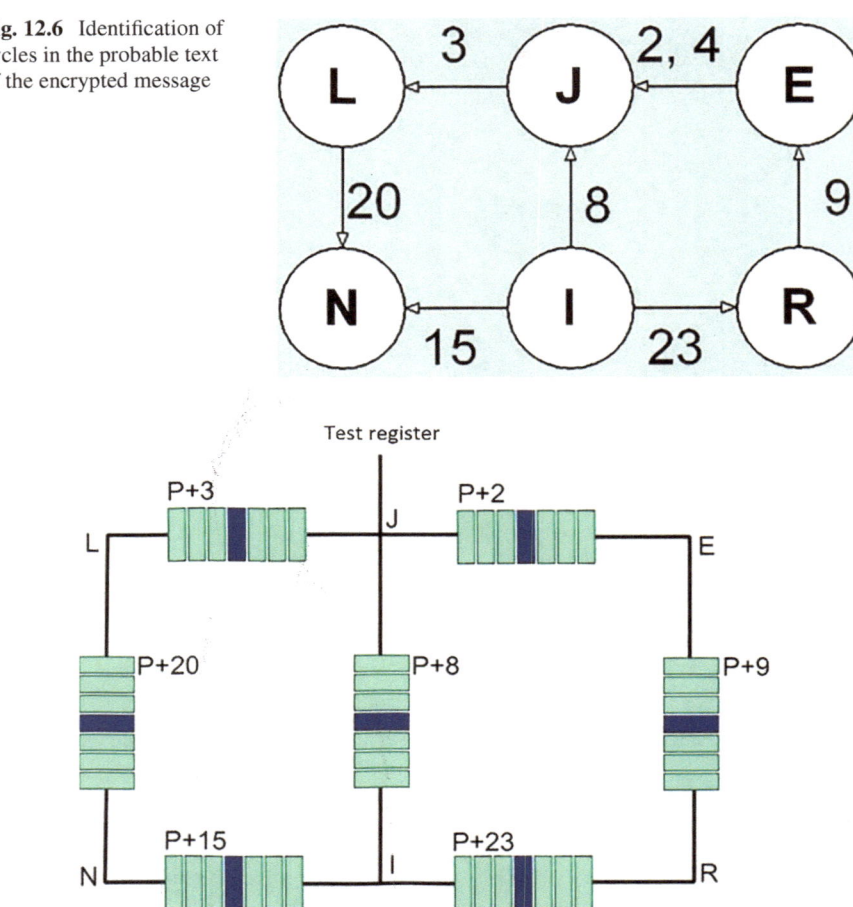

Fig. 12.7 Transition from identified cycles to machine settings in the bombe 'menu'

Setting the bombe according to the menu started with the selection of rotors. The structure of the cycles does not provide any guidelines for selecting the right rotors and determining their possible order. Over time, mathematical methods were developed for selecting the most probable rotor combinations, but in the worst case, it was necessary to check the menu for all 60 possible rotor orders. The search could be accelerated by using the three independent scrambler banks contained in the bombe; the example menu requires only 7 scramblers, fitting into one bank. The other two banks can be used in parallel to test the same menu with a different rotor order.

After selecting the rotors and placing them on the shafts they had to be set in the right position in relation to each other. From the menu we know that the rotor placed between the letters J and L had to be shifted by three positions forward in relation to the reference starting point, the rotor between the letters L and N—by twenty positions forward, and so on. The next step was connecting the scramblers. Example menu requires that the output of scrambler between J and L is connected to the input of the scrambler between L and N, the output of the latter to the input of the scrambler between N and I etc. Interested reader will find more details of the bombe functioning in the box below (Fig. 12.8).

After setting the menu the operator could start the bombe. Rotors rotated synchronously running through all the possible positions. The fast rotors rotated at about 50 RPM (increased to 120 RPM in later bombe models). In each rotors' position the test circuit checked whether the bombe stop condition was met. If a cycle structure corresponding to the menu was detected, the machine was automatically stopped. The inertia of the mechanical parts caused the rotors to move after the stop condition was detected; the operator read the position of the rotors corresponding to a possible hit by reading the state of the circuit relays. In the majority of cases, the menu structure did not guarantee an unambiguous solution—during most runs the bombe stopped several times, so the operator was noting the stop position, moving the rotors back to the position just read and restarting the device. The registered stop positions were tested using the special checking machine. If none of them gave a solution, the same menu was used with a different set of rotors or their order. When only few bombes were available, testing of the single menu could take several days, without any guarantee of success.

Fig. 12.8 A view of the bombe's configuration panel illustrates its complexity. Prior to the bombe run, the Enigma equivalents included in the menu were connected by wires, which are missing from the photo. (CC-BY-SA-4.0, Mike Peel (www.mikepeel.net))

Principle of Operation of the Turing-Welchman Bombe

In the general description of the bombe two elements were only mentioned, without presenting them in the detail: Welchman's diagonal board and the test register. Their role is so vital to the functioning of the bombe that we need to make up for this omission. We will be using a simplified example—a hypothetical bombe used to break the cipher of the six-character Enigma. A crib suitable for this kind of a bombe could look like:

position	012345
open text	**ABECAD**
ciphertext	**CADEEB**

The following cycle could be constructed on the basis of the above texts (Fig. 12.9).

The essential part of the bombe's construction, hidden from the operator's eye, are the bundles of cables, each consisting of as many wires as there are letters in the Enigma alphabet (for the six-character Enigma we have 6 bundles of 6 wires). Scrambler inputs and outputs are accessible via the sockets at the back of the unit. Suppose we mark the bundles with the capital (**A**,..., **F**) and the wires in each bundle with the small letters (**a**,..., **f**). Each wire can be tagged as **Xx**, where **X** designates the bundle, and **x** the wire therein. All the wires in all the bundles are interconnected in the diagonal board in such a way, that, for instance, wire **Ab** is connected to wire **Ba**, wire **Bc** to wire **Cb** and so on. When preparing the bombe for menu checking the individual scramblers are being plugged between the bundles (**A**...**F**) according to the structure of the cycles included in the menu (Fig. 12.10).

In the example menu, the transition from the character **A** to **C** occurs at the start of the ciphertext (position zero). This situation corresponds to plugging of the scrambler in the reference position between the bundles marked **A** and **C**. The transition from the letter **C** to **E** takes place in the third position of the ciphertext, so the scrambler shifted by three positions in relation to the previous one (**AAD**) is plugged between bundles **C** and **E** ; other scramblers are configured in a similar way. Before starting the bombe, one of the wires in one of the bundles included in the menu is connected to the battery—let's assume that we select for this purpose wire **b** of the bundle **A** (**Ab**). Simultaneously the same bundle is connected to the test registry, detecting the bombe's stop condition. The figure below illustrates the state of the bombe in one of the stages of the menu testing (Fig. 12.11).

Fig. 12.9 Cycle resulting
from the crib above

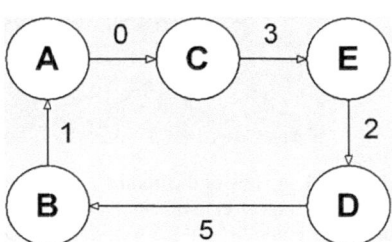

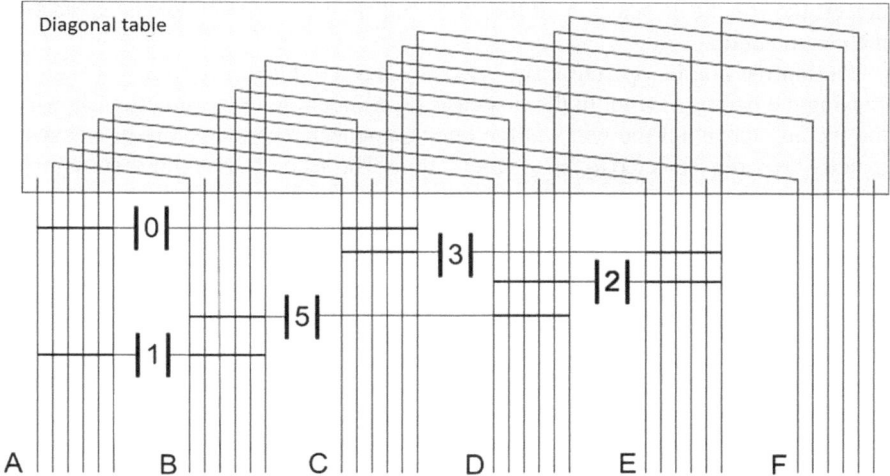

Fig. 12.10 Simplified bombe configured to run a menu corresponding to the cycle above

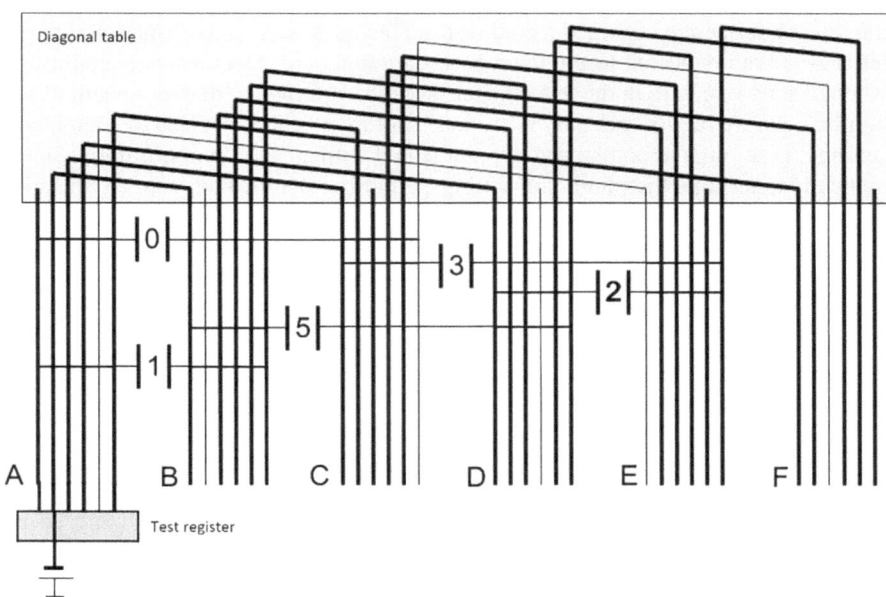

Fig. 12.11 Simplified bombe, having found a possible solution

A thick line marks the wires which, in a present bombe position, are connected to the selected **Ab** wire and, consequently, to the voltage source. The situation described in the above diagram is a bit unusual: In most cases **all the** wires in **all the** bundles are connected to the voltage source. In such a situation the test register detects the voltage presence in **all the** wires and allows the scramblers to advance to the next position.

During the bombe operation two special cases are detected by the test register, causing the bombe to stop. In the first case, corresponding to the situation shown in the picture above, all the wires in the bundle connected to the test register **except exactly one** are connected to the voltage source. The second case represents a mirror reflection of the first one: exactly **one** wire is active. This behavior is easily explained. With rotors in a position that does not correspond to the solution, the voltage is scattered by the scramblers' connections between all the wires of all the bundles. If the current position of rotors corresponds to the solution, one and only one wire in all bundles represents an electrical circuit isolated from the other wires. If the battery is connected to this circuit, the signal will only reach one wire in all the bundles. If the battery is connected to any other wire, the voltage is scattered among all wires except exactly one, isolated from the rest. In the jargon of bombe operators, the first case was described as a 'hot test' and the second, by symmetries, as a 'cold test'.

Translating the language of electrical signals into the language of logic, applying voltage to one of the wires was equivalent to the assumption that the Enigma plugboard connected the letter denoting the bundle with the letter denoting the wire under voltage. In the example presented in the picture above this not the case, as all the wires but one are activated ('cold test'). This represents a confirmation that the letter **A** is not connected to the letter **b** in the plugboard. Hypothesis is confirmed when exactly one wire in the test register is active. It turns out that as a result of the bombe's run we recover not only the rotors' starting position, but also the plugboard settings! Later on, the bombe was supplemented with an add-on permitting to automatically determine the plugboard settings. After the bombe had stopped, the add-on started feeding the voltage individually to the consecutive wires in the test register. Whenever the test register detected a 'hot test', the letter pair was printed as an element of the plugboard setting. The diagonal board deserves an additional comment. Its functioning was based on Enigma's non-reciprocity. The diagonal board included $(26 * (26{-}1))$: 2, i.e. 325 wires. This simple device permitted considerable reduction of bombe's dependence on the number and length of cycles and the reduction of the crib length.

Turing bombe in its original version, without a diagonal board, required cribs containing cycles. The more and the longer cycles occurring within the crib, the less false stops generated by the bombe. Turing estimated that the number of false stops during the bombe run was equal to $26 * (4{-}x)$, where x – number of cycles in the crib. An unambiguous solution could be expected with at least 4 cycles. This large number of cycles could be expected in the long cribs only, subject to another restriction. Bombe run result was correct when only the fast rotor moved during the crib ciphering. For the long cribs this condition was rarely met. The cipher property being exploited by the diagonal table was not dependent on or related to the cyclic structure. This allowed for the use of cribs with fewer or even with no cycles at all. In the practice, a crib without any cycles, at least 16 characters long, could give an unambiguous solution (i.e. a bombe run with only one stop). The cycle-free crib of 12 characters generated about 820 false stops in one run. However, it was enough for the crib to contain two cycles, in order to achieve an unambiguous solution for cribs about 12 characters, and for three cycles—even 9 characters long.Device just described is usually referred to as the "Turing bombe". But its construction was clearly based on the theoretical and practical foundations provided by Marian Rejewski. Turing was not hiding this fact. He gave his construction a somewhat ungrammatical name "bombe", in an obvious reference to Rejewski's original. But the bombe's success was only made possible by Welchman's contribution. The diagonal board was probably one of the simplest appliances in history of the codebreaking, but it was this device that saved Turing's project, transforming Victory's costly failure into success. Intellectual honesty suggests that the product of their intellect should rather be named the "Rejewski-Turing-Welchman bombe".

When the first bombes were delivered to BP, no one could predict the long-term effect of their appearance. "Room 40", which significantly influenced the course of the First World War, was a kind of gentlemen's club. This old-fashioned but effective model of signal intelligence had been outdated by the appearance of a ciphering machine. Its application had democratized the cryptography and the codebreaking in the same way as mass production had democratized the demand for the goods earlier considered exclusive. Training an Enigma operator required less effort and time than developing, printing and distributing traditional code books. The widespread use of radio combined with the simple machine encryption has triggered an avalanche of messages, that no traditional signals intelligence service could handle. When Churchill visited France a few days before its collapse, Prime Minister Reynaud complained of his helplessness against the power of German aviation and the armor, that were destroying not so much the country's armed forces, as their will to resist. Churchill's response was probably a poor consolation for Reynaud, but it turned out to be more prophetic than its author could have imagined. The British Prime Minister predicted that "(…) in the end the regime that owes its victories to machines will fall. The machines will be defeated by other machines". Indeed, the German Blitzkrieg doctrine relied on three types of machines: a tank providing firepower and speed; an aircraft assuring fire support, reconnaissance and supplies; and finally Enigma combining the previous two into seemingly irresistible force. The Allies had to wait months or years to develop tanks and aircraft capable of defeating their German counterparts, but Enigma found an equal opponent only two months after Churchill's prophecy. The use of the bombe has changed the nature and organization of the signals intelligence work. The natural environment for the machine is the factory; we will be watching the codebreakers' gentlemen club being replaced by the organization based on the industrial manufacturing standards. Men "of professor type" accepting the new roles of the production managers. Signals intelligence factory organizing subsidiary plants and building a framework allowing for the optimal use of its machinery and the transfer of production capacity between plants. With the appearance of the first bombes in BP, the era of the codebreaking seen as a romantic crusade of a handful of gentlemen came to an end. In a way, it was an act of historical justice that leveled their status with their environment. The war itself had lost its gentlemanly character at least one global conflict earlier, and the codes and ciphers were ever less used for purposes associated with the concepts of glory and honor.

Chapter 13
…Rule the Waves?

In the autumn of 1940, victory at the Battle of Britain and the impending season of storms put the danger of a Nazi invasion away, giving Britain at least a few months to mobilize its forces. In the end of the same year, and in the first months of the following one, the British conducted two brilliant campaigns in Africa, in which their modest forces crushed the Italian forces holding several times the advantage. It began on 9 December, when General Richard O'Connor, who with only 31.000 men at his disposal smashed in two months of fighting the approximately 215.000-strong Italian army of Marshal Graziani, forcing it not only to retreat from Egypt, but to withdraw to Tripoli. One element of the success of this campaign was the functioning of the A Force disinformation team under the command of Colonel Dudley Clarke, appointed by the British commander in Egypt, General Archibald Wavell. The team, installed in a hastily adapted brothel in Cairo, received information from BP about the intentions of the Italian side from decrypts of Italian codes. Knowing the plans and concerns of the Italians, the A Force team was able to suggest to O'Connor actions to strengthen the Italians in their false opinions, or to encourage them to abandon activities unwelcomed by the British. While fighting continued in the deserts of Egypt and Libya, in the third decade of January 1941, two arms of the British attack on the Italian army in East Africa, led by Amadeus II of Savoy, Duke of Aosta, were launched from Sudan and Kenya. 9.000 soldiers under the command of General Platt entered Ethiopia from the Sudan, and less than 6.000 commanded by General Cunnigham set off from Kenya heading towards the Italian army of some 200,000 soldiers. Despite the power imbalance on 18 May, Duke of Aosta capitulated at Amba Alagi, and two weeks earlier, Emperor Haile Selassie was able to descend from the mountains and return to the capital of his country. In this campaign too, the decryption of the Italian messages played a key role.

Victories on land, growing confidence in the air, and the discreet but growing help from the United States have changed the mood in Britain. The change of atmosphere was also felt in BP. The regularity in breaking Luftwaffe messages, breaking of a number of minor enemy code systems, and the breakthrough, which was the appearance and effectiveness of the first bombes, offered codebreakers some confidence in their own strength. The British, however, had good reason to treat their successes as half-hearted. The victories in the Battle of Britain and in both African campaigns have shifted the burden of battle, at least for the time being, almost entirely to the seas and oceans. There, the Royal Navy has maintained its supremacy with the utmost difficulty. The annihilation of German destroyers at Narvik made it impossible for German capital ships to act against the British navy as a squadron. However, the possibility of taking refuge in Norwegian fjords threatened at any time with raids into the waters frequented by the British merchant fleet. Shipping was a nerve of the British Empire, and its global nature forced the Royal Navy to protect the routes in spaces covering half the globe.

However, that threat from the Kriegsmarine which most kept British leaders awake at night was submarine warfare. From the second half of 1940 the German navy had at its disposal European coasts and ports, stretching from Nord Cap to the Bay of Biscay. German submarines could attack allied shipping at any time and place of their choice. The U-Boot Commander, Admiral Karl Dönitz, came to the conclusion, based on his own experience of World War I, that the actions of a single submarine are not a form of combat proper for the submarine and developed his own method, called, not quite properly, the wolf pack tactics. A true wolf pack is an independent unit guided solely by the experience and will of its leader. Dönitz, however, decided that independence is an attribute of the romantic past and imposed on his captains a paternalistic system with himself in the center. The U-Boot setting out to sea was not aware of the purpose of his mission; it received instructions by radio only when leaving the waters of the Bay of Biscay or crossing the 60th parallel on Arctic missions. Dönitz's wolf pack had no commander; each of the captains received instructions directly from headquarters. The procedure of direct communication between the ships was only an emergency measure used exclusively when contact with distant headquarters was not feasible. It was not acceptable to change the orders received, under any circumstances. Obtaining permission to return to the base required giving reasons for the request (for instance ship's damage list), specifying the route and expected time of crossing checkpoints. In each communication session with the U-Boot HQ, the U-Boot was obliged to provide the current location, weather report and information about remaining fuel and supplies. Some of this information was justified in maritime practice, other was an expression of Dönitz's personality, for whom the relationship with the crews was one of the ways to motivate them. If to the elements just described one adds messages to and from crew members, remaining at sea for long weeks, the avalanche of messages necessary to assemble the wolf pack and the signals sent by the U-boats participating in the attack, Dönitz's tactics entailed filling the ether with a flood of enciphered signals. Their security was ensured by Enigma, thus becoming an equally important element of the naval tactics as it was for the Blitzkrieg's land-air battle concept. Like almost every type of arms, the cipher machine was a double-edged weapon. The German rules of submarine combat stressed that "the characteristic feature and greatest strength of a submarine is its invisibility". The U-Boot "cannot be seen, heard or located" either before or during the attack. When detected by an enemy, it "loses almost all chances of success". The safety of the Enigma was the condition of the ship's invisibility, its violation threatened to reverse the roles—an invisible attacker would become an expected target. Assuming the security of communications is maintained, the information transmitted allowed the submarine command to direct the actions of forces dispersed over vast areas of the ocean. The cracking of ciphers provided the enemy with an extremely detailed knowledge not only of the actions of specific ships, but of the strategic plans of the entire navy. Enciphered messages about fuel and ammunition stocks could allow to identify the type of reporting ship. Identifying the ship's captain gave an important indication of the combat tactics it used.

As early as 1937, Poles noticed that the use of Enigma by Kriegsmarine differs from the practices adopted in the Wehrmacht and Luftwaffe. The difference in terms of equipment was seemingly modest: machines used by army and air force had three (later five) rotors at their disposal, from which they selected three. Navy cipher clerks selected three rotors, but from among eight. Five of them were identical to the machines used by Wehrmacht and Luftwaffe, three were specific to Kriegsmarine. The information provided in Pyry did not include the wiring of navy rotors. British codebreakers failed in the early months of the war to apply Rejewski's methods to the reconstruction of unknown rotors. Fortunately, the hope of seizing a machine in the event of war was confirmed fairly quickly. The first opportunity happened in winter on the west coast of Britain. On the night of 12 February 1940, the U-33 under the command of Captain Hans-Wilhelm von Dresky tried to get through to the Firth of Clyde in order to place mines on one of the busiest shipping arteries. Navigation conditions in this area are so complicated that the ship had to sail on the surface using the darkness of dawn. Captain Dresky was unaware that the noise of his ship's propellers was heard by the hydrophones of the HMS *Gleaner*, patrolling nearby in an attempt to locate the intruder. Around 4 AM her captain noticed a cloud of fog coming out of the submarine's tanks during the dive, and forced Dresky to lay his ship on the bottom with a series of depth charges. The U-33 was trapped; surrounding waters were infested with underwater rocks and troubled by fast currents, so a submerged escape was out of the question. After an hour spent at the bottom, among the explosions of depth charges and facing increasingly heavy damage to the ship, her commander decided to rescue the crew. In preparation for the inevitable captivity, the second officer distributed the Enigma rotors among three sailors, ordering them to throw them into the sea as soon as they left the ship. Not everything went according to plan: when around 5.20 AM a ship surfaced, the crew went aboard, and the ship's engineer was too hasty to set off the detonators, the explosion of which sank the ship. In the darkness, it took *Gleaner* two hours to fish the survivors out of the icy waters, so in order to save their lives, the sailors were immediately stripped and placed under a hot shower. When the warm bath restored their ability to think, radio operator Kumpf realized that he had not thrown the rotors into the sea the moment he saw the British sailor removing them from the pocket of his suit. Among the equipment that was thus brought to BP were two previously unknown rotors of the Kriegsmarine Enigma, VI and VII.

The circumstances surrounding the acquisition of rotor VIII have not yet been confirmed. The most probable version is that it was captured aboard the U-13, sunk by HMS *Weston* near Lowestoft on 31 May 1940. The U-13 was assigned the task of sneaking through British minefields and attacking convoys evacuating soldiers from Dunkirk. In the opinion of several surviving sailors, her commander, Max Schulze, overreacted after the British attack with depth charges and decided to surface the ship prematurely. On the surface, ship's engineer Hans Grandjean quickly opened the kingstones; the ship sank with him still onboard. Fortunately for Grandjean, the sea was shallow there: he managed to free himself from the sunken ship and get to the surface. The U-13 wreck, lying in the shallow waters, was searched by British divers, who in early August managed to recover a lot of equipment and documents from it. The version repeated in some sources that the British captured the eighth rotor during operations in North Africa is a mistake or disinformation: in August 1940 no German submarines were involved in operations off the coast of North Africa.

As a result of these events at the same time as the first functional bombe passed its maturity test, BP also had a full set of Kriegsmarine Enigma rotors. This satisfaction had to be dampened to some extent by the discovery of a specific feature of the new rotors. On the ring of each Enigma rotor there was a turnover notch, the position of which determined when the rotor would advance to the next position. In five "old" rotors the notch was located at different letter of the alphabet: in rotor I at R, rotor II at F, III at W, IV at K and V at A. In order to make it easier for codebreakers to remember the turnover positions in BP, the not very sensible but very patriotic acronym RFWKA, expanded as Royal Flags Wave Kings Above, was coined. Different turnover position was intended to make the Enigma's movement more irregular and thus less predictable. Jerzy Różycki had a different opinion on this issue and used the described feature to design a method that would allow for the identification of the right rotor (clock method). BP codebreakers must have been disappointed when examining the rotors VI—VIII. They found that each rotor has not one, but two turnover points, located identically around the circumference, at a distance of 13 letters. The constructors of the Enigma understood their mistake and decided to complicate the life of a potential adversary.

Seizure of the rotors didn't solve the problem of reading Kriegsmarine's messages. The instructions coming from U-13 only confirmed what Turing knew from the Poles or managed to deduce. The source of the problems was the way the German navy used Enigma, different from other services and more complicated. The techniques of breaking the Enigma ciphers that dominated BP's work during the second half of 1940—herivelism and *cillies*—were based on the mistakes of German cipher clerks. The essence of both types of errors boiled down to a predictable choice of those elements of the key of the message, the determination of which the procedure gave to the cipher clerk himself. The second potential weakness in the procedures of the army and the air force was the encryption of the key of the message using the same machine and the same procedure that was used to encipher the message body. This did not pose a significant risk as long as the cipher itself remained secure. However, when the first layer of its security was breached, the other elements fell in turn like domino stones. Breaking the key in the network meant that all the messages exchanged in the network during the key validity period were open to the adversary. Theoretically, each of them was enciphered with an individual message key, but after breaking the network key, reading the individual message key represented a trivial task. The designers of procedures in force in the navy were determined to eliminate precisely this weakness. A cipher clerk in the navy was not supposed to choose the message key himself. Instead, he received a ready-made book, from which he chose one of the possible keys for a given message and additionally processed it in a complicated way. The book and the procedure itself were developed by professional cryptologists, aware of potential threats, to ensure the security of the code.

The Kriegsmarine message encryption procedure required several additional documents in addition to the Enigma machine. The first one was the so-called *Kennbuch*, whose name was usually shortened to K-Buch, and in BP it was known as K-book. The individual key of the message consisted of three characters; the choice of three letters from Enigma can be made in 17.576 ways. German cryptologists used all possible three characters in the key book, spreading them randomly in 732 columns, each of which had 24 trigrams. Since the K-book was used for both encryption and decryption of the messages, it had to take the form of a bilateral dictionary. The second part of the K-book was a list of 17.576 trigrams in alphabetical order, with each trigram being accompanied by a pair of numbers indicating the column and row in which a given trigram is included in the first part of the dictionary (see examples below).

	51	52	53	54	55	56
1	WDP	IKZ	QDX	ZBU	OML	DXB
2	PCF	GFG	JVA	UVT	DIJ	IUT
3	UWZ	OUV	YTT	JFQ	FYY	LPQ
4	IJG	LBQ	XMR	CYY	LUA	XFK
5	NBX	TND	SHC	GNE	RNQ	TQG
6	SYC	YPH	MBJ	OJA	XCK	NKK
7	FEV	RZT	FWW	XZG	JOP	ADX
8	BHD	AWC	PLD	AUO	MXE	FTK
9	KIU	FLS	CZO	MHI	ULW	QGR
10	RTJ	KRX	IQY	EUF	PZI	ZIO

GHA 098 03	GIA 515 12	GJA 161 22	GKA 425 21	GLA 535 02
B 129 12	B 238 25	B 701 12	B 532 15	B 123 01
C 654 22	C 712 02	C 513 14	C 424 17	C 132 12
D 423 15	D 593 10	D 614 08	D 174 03	D 532 19
E 513 08	E 333 12	E 482 15	E 524 17	E 422 19
F 544 19	F 611 24	F 283 21	F 442 11	F 525 20
G 281 21	G 208 11	G 466 23	G 534 11	G 671 13
H 619 03	H 634 14	H 164 02	H 141 18	H 322 05
I 011 14	I 212 09	I 462 04	I 522 02	I 015 11
J 720 20	J 099 10	J 625 11	J 432 24	J 093 14

The K-book was common to all Kriegsmarine units, regardless of the cipher and communication network they used. The division of the German navy's communications into separate networks was done by allocating column blocks in the K-book, as defined in a separate document called the allocation list (*Zuteilungsliste*, see below).

Geheim!		Prüfnr. **996**
Kennwort: Haifisch		

Zuteilungsliste für Kenngruppen
zum K. Buch – M. Dv. Nr. 98.
Teil B.

Schlüsselkenngruppe	Verfahrenskenngruppe	
	Schlüssel M.	R.H.V.
Spalte	Spalte	Spalte
1-100 M. Triton (M. Tri)	1-733 Allgemein	1-564 Allgemein
101-130 M. Neptun (M. Nep)		656-733 Offizier
131-180 M. Hydra (MH)		
181-200 M. Aegir (MA)		
201-225 R.H.V.		

The contents of the first column of the list above indicate that the network named Triton uses K-book columns 1–100, Neptune network 101–130, etc. (the procedure description below will explain the meaning of the other columns). The third document necessary for the encryption was the so-called bigram tables, which consisted of nine tables, marked with the letters A-J (excluding I). Each of them contained bigrams, randomly selected equivalents for each of 676 possible pairs of letters. The transformation of the bigrams was reciprocative; if within a single table a pair of **NB** was transformed into **QG**, then also the pair of **QG** was transformed into **NB** (see the fragment of the table below). A separate document, called *Tauschtafelplan*, specified which of the nine bigram tables was valid on any day (Fig. 13.1).

When encrypting or decrypting the message, the cipher clerk started with setting the Enigma in the base position (*Grundstellung*). Then he selected from those columns of the K-book that were allocated to his communication network two arbitrary trigrams, for instance **CGA** from part of the table indicated as *Schlüsselkenngruppe* and **KDB** from the part indicated as *Verfahrenskenngruppe*. The first three characters were enciphered in the base position of Enigma, receiving **VOD**, then the rotors were moved to position **VOD** and the message was enciphered starting from this position. The **VOD** trigram represented the individual message key; the Kriegsmarine procedure required the key to be additionally encrypted before it was passed to the recipient. To this end, the cipher clerk would write the **CGA** and **KDB** trigrams selected in the previous step in two consecutive lines, adding one randomly selected character before the first trigram and after the second one (in the example below F and M were added).

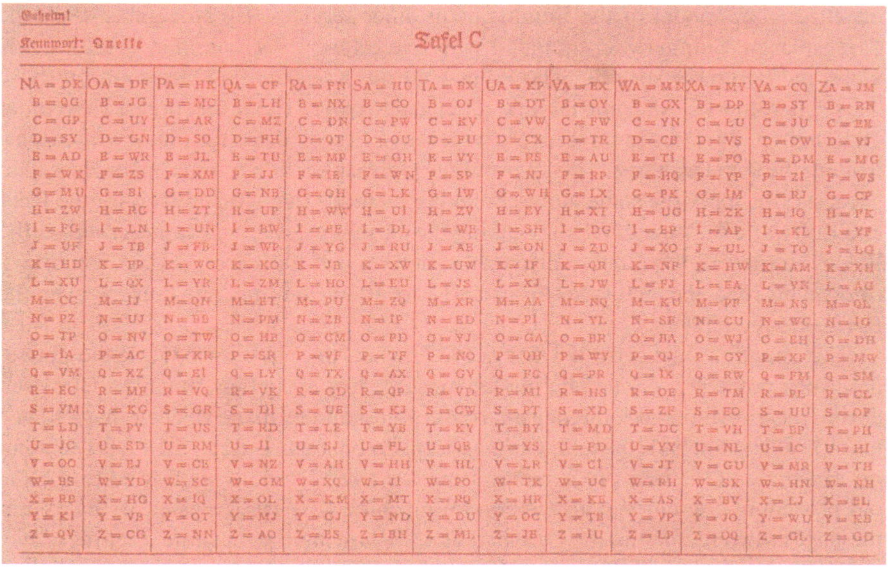

Geheim!

Kennwort: Quelle

Tafel C

NA = DK	OA = DF	PA = HK	QA = CF	RA = FN	SA = HU	TA = BX	UA = KP	VA = EX	WA = MN	XA = MY	YA = CQ	ZA = JM
B = QG	B = JG	B = MC	B = LH	B = NX	B = CO	B = OJ	B = DT	B = OY	B = GX	B = DP	B = ST	B = RN
C = GP	C = UY	C = AR	C = MZ	C = DN	C = PW	C = KV	C = VW	C = FW	C = YN	C = LU	C = JU	C = EK
D = SY	D = GN	D = SO	D = FH	D = QT	D = OU	D = FU	D = CX	D = TR	D = CB	D = VS	D = OW	D = VJ
E = AD	E = WR	E = JL	E = TU	E = MP	E = QH	E = VY	E = RS	E = AU	E = TI	E = FO	E = DM	E = MG
F = WK	F = ZS	F = XM	F = JJ	F = IE	F = WN	F = SP	F = NJ	F = RP	F = HQ	F = YP	F = ZI	F = WS
G = MU	G = BI	G = DD	G = NB	G = OH	G = LK	G = IW	G = WH	G = LX	G = PK	G = IM	G = RJ	G = CP
H = ZW	H = RG	H = ZT	H = UP	H = WW	H = UI	H = ZV	H = EY	H = XT	H = UG	H = ZK	H = IO	H = FK
I = FG	I = LN	I = UN	I = BW	I = EE	I = DL	I = WE	I = SH	I = DG	I = EP	I = AP	I = KL	I = YF
J = UF	J = TB	J = FB	J = WP	J = YG	J = RU	J = AE	J = ON	J = ZD	J = XO	J = UL	J = TO	J = LQ
K = HD	K = EP	K = WG	K = KO	K = JB	K = XW	K = UW	K = IF	K = QR	K = NF	K = HW	K = AM	K = XH
L = XU	L = QX	L = YR	L = ZM	L = HO	L = EU	L = JS	L = XJ	L = JW	L = FJ	L = EA	L = YN	L = AG
M = CC	M = IJ	M = QN	M = ET	M = PU	M = ZQ	M = XR	M = AA	M = NQ	M = KU	M = PF	M = NS	M = QL
N = PZ	N = UJ	N = BB	N = PM	N = ZB	N = IP	N = ED	N = PI	N = YL	N = SF	N = CU	N = WC	N = IG
O = TP	O = NV	O = TW	O = HB	O = CM	O = PD	O = YJ	O = GA	O = BR	O = BA	O = WJ	O = EH	O = DH
P = IA	P = AC	P = KR	P = SR	P = VF	P = TF	P = NO	P = QH	P = WY	P = QJ	P = GY	P = XF	P = MV
Q = VM	Q = XZ	Q = EI	Q = LY	Q = TX	Q = AX	Q = GV	Q = FC	Q = PR	Q = IX	Q = RW	Q = FM	Q = SM
R = EC	R = MF	R = VQ	R = VK	R = GD	R = QP	R = VD	R = MI	R = HS	R = OE	R = TM	R = PL	R = CL
S = YM	S = KG	S = GR	S = DI	S = UE	S = KJ	S = CW	S = PT	S = XD	S = ZF	S = EO	S = UU	S = OF
T = LD	T = PY	T = US	T = RD	T = LE	T = YB	T = KY	T = BY	T = MD	T = DC	T = VH	T = EP	T = PH
U = JC	U = SD	U = RM	U = II	U = SJ	U = FL	U = QB	U = YS	U = FD	U = YY	U = NL	U = IC	U = HI
V = OO	V = EJ	V = CE	V = NZ	V = AH	V = HH	V = HL	V = LR	V = CI	V = JT	V = GU	V = MR	V = TH
W = BS	W = YD	W = SC	W = GM	W = XQ	W = JI	W = PO	W = TK	W = UC	W = RH	W = SK	W = HN	W = NH
X = RB	X = HG	X = IQ	X = OL	X = KM	X = MT	X = RQ	X = HR	X = KL	X = AS	X = BV	X = LJ	X = BL
Y = KI	Y = VB	Y = OT	Y = MJ	Y = GJ	Y = ND	Y = DU	Y = OC	Y = TB	Y = VP	Y = JO	Y = WU	Y = EB
Z = QV	Z = CG	Z = NN	Z = AO	Z = ES	Z = BH	Z = ML	Z = JE	Z = IU	Z = LP	Z = OQ	Z = GL	Z = GD

Fig. 13.1 Sample page of the bigram table "Quelle". (cryptomuseum.com)

```
F C G A
K D B M
```

Then he read the resulting vertical character pairs (**FK**, **CD**, **GB** and **AM**), selected the bigram table valid for that day and replaced each pair with its equivalent from the table (using the example bigram table above he received **FK**→**ZC**, **CD**→**VT**, **GB**→**AQ** and **AM**→**GO**). By combining the pairs received in this way, he created a message indicator, which he added at the beginning and end of the cryptogram; in our example, the indicator takes the form of **ZCVTAQGO**. The decryption procedure was symmetrical, with some steps using parts of the tables complementary to those used in the encryption. Note that the encryption of an individual message key with the use of bigram makes it impossible to read all messages in a given network even in the case of a reconstruction of the day's base position (*Grundstellung*). Reading the text of the cryptogram requires the reconstruction of the individual message key, which is encrypted without Enigma, using bigram tables. The procedure described reveals that already at the time of its introduction in 1937, the authors envisaged the possibility of implementing a four-rotor Enigma model over time.

The Poles took over relatively few Kriegsmarine messages. Moreover, Poland was not a naval superpower, so the ability to read the Kriegsmarine ciphers was less important to the Cipher Bureau than the communication networks of the army, air force, and SD. Nevertheless, Rejewski and his colleagues closely observed the development of German navy communications, which allowed them to spot the moment of its weakness and work out the procedures used. Since 1934, Kriegsmarine used the same model of the Enigma as in other services. By November 1936 the Kriegsmarine Enigma was equipped with five rotors; three common to all the services and two navy-specific. The Poles read the Kriegsmarine messages on those days when the key did not include additional rotors. However, on 16 November 1936 both navy-specific rotors were withdrawn from service for unknown reasons, leaving the standard three. This allowed the German section of the Cipher Bureau to successfully apply the same cipher breaking methods that worked well for the Heer and Luftwaffe messages. When breaking the current messages became a routine, the Poles returned to the archive dispatches, determined the wiring of the withdrawn rotors and read the messages from 1934 to 1936, gaining full control over the Kriegsmarine ciphers.

The situation changed on 1 May, 1937, when the navy introduced the above-described procedure of enciphering the message key using bigram tables. However, during the transition period Kriegsmarine cryptographers made an error that allowed to unravel also the new system. German torpedo boat using a call sign AFA was not informed about the new procedure in time. As a result, she was allowed to use the old procedure combined with the current machine key. The Poles easily broke her messages, thus recreating the *Grundstellung*. Knowing the machine's settings, they were able to break the keys of the messages sent by other ships, using the new procedure, between 1 and 8 May. With a packet of over a hundred broken messages from that period, the Poles determined the nature of the new procedure, although without knowing the bigram tables they were not able to continue decryption after 8 May. The results of their pre-war work and a package of archival copies were handed over to their British colleagues during a meeting in Pyry.

Despite the fact that the Poles have worked out the structure of the system, its complexity originally gave rise to pessimism in BP about the possibility of overcoming it. Alexander later recalled that originally in the whole of BP only two people believed in the possibility of breaking the Kriegsmarine Enigma. In the case of Frank Birch, the head of the Navy Section at Hut 3, this was an official optimism: naval Enigma had to be broken because it was necessary for victory. The second optimist was Turing, who started working on the Kriegsmarine ciphers, as they were the biggest challenge available. The material that helped to make the next step was a package of about 100 messages broken by Polish Cipher Bureau immediately after the system change. However, at the turn of 1939 and 1940 unknown wiring of rotors VI, VII and VIII and the lack of bigram tables presented an insurmountable obstacle.

Turing quite quickly liberated himself from the care of Knox, who assigned Peter Twinn to assist him, and tactfully withdrew to his own challenges. The Turing-Twinn duo was joined by graduates and students from British universities coming to BP, giving rise to an organization that went down in history as Hut 8, the equivalent of Hut 6, working on the naval ciphers. Turing took over as head of the new hut, without seeming to have been formally nominated for this position—solely by virtue of his knowledge and competence. Watching his activities at the head of a rapidly growing team, coping with diversified tasks, operating in a semi-military, hierarchical environment, must have been an interesting experience. He had no problem with establishing authority in the group. His mathematical fame preceded his arrival in BP and accompanied him at every step. When his future subordinates crossed the doors of Hut 8, he was also a recognized authority on issues related to Enigma. In this role, his authority was based largely on the knowledge he had previously acquired from his Polish predecessors: Turing was one of few people who had access to all documents provided by Poles or have met them in person. It seems that he took the need-to-know principle very seriously, rarely and enigmatically indicating the sources of his knowledge. However, he quickly added his own contribution, the scale and importance of which justified the respect of his colleagues. The newcomers brought to BP the unpretentious custom of calling their colleagues by their first name, but Turing represented an exception to the rule. The vast majority of the team addressed him or referred to him by the nickname of Prof.

Turing's mathematical genius was complemented, however, by equally significant portion of eccentricities and peculiarities of character. He did not care about his appearance; unwashed and disheveled hair and dirty nails matched with not very fresh collars and shoes chosen from different pairs. From the tavern "Under the Muttons' Shoulder", where he lived, he commuted to BP on a tattered bicycle, which regularly dropped the chain. Instead of adjusting its tension, Turing counted the revolutions of the pedals, then got off the bike just before the chain fell and moved it forward, starting a new cycle. A large alarm clock hanging from the chain around his neck helped him to stay on time. Before the BP guards got used to his extravagance, he had to shock his surroundings by parading in a gas mask to relieve his allergy. Even when he tried to show some common sense and practical attitude, the results were unexpected. When, in 1940, in recognition of the first results of his work on Enigma, he and several of his colleagues received a substantial cash prize, he decided that in the event of the Nazi invasion, pounds would lose their value, so he exchanged them into silver bars, which he carefully buried in a place known only to him. When the danger of the invasion was definitely over and he was able to extract his treasure, it turned out that he forgot where he had hidden it.

Even if his minor extravagances were terrifying for the military part of BP staff, they passed without much echo in the tolerant atmosphere of the Turing's team. It remains a mystery how he managed Hut 8, given that he was, in the unanimous opinion of his colleagues at the time, a rather uncommunicative person. According to Mahon, who as the last head of Hut 8 wrote down its official history, "his ability to communicate was regrettable (…) and he was not a practical man". Frank Birch lamented that "Turing and Twinn are brilliant, but as many brilliant people are not practical. They are messy, they lose things, they can't rewrite anything properly, and they are still torn between theory and cribs". Turing's preserved manuscripts seem to confirm these opinions. Dotted with strokes and blots, which are like Turing's trademark, definitely unorthodox when it comes to spelling and grammar, make us think that if their author were to work in a team consisting of less talented and tolerant people, his genius would not find recognition or application in BP.

When Turing and Twinn hesitated over the choice of tactics for further attack on Enigma, and a group of students and fresh graduates integrated into the Hut 8 team, a campaign was unleashed in the Atlantic that was to become Churchill's greatest fear. He knew this kind of conflict from many sides. During the First World War and early Second World War he was the first Lord of the Admiralty. Reports of losses in merchant shipping were sent to his desk. He knew that at the beginning of the conflict the Royal Navy's ability to resist the underwater offensive was lower than at the end of the previous war. There was a shortage of escort ships, aircraft, trained sailors and modern combat tactics. The disadvantages were not compensated for by the invention of the Asdic and the 50 old American destroyers, which Britain received in exchange for providing the U.S. with bases in the western hemisphere. Later, from the Prime Minister's seat, he saw how dramatic the relationship between Britain's survival and its ability to bring people, raw materials and food from across the oceans had become. Meanwhile, the losses suffered by the British navy threatened to produce a gloomy scenario from the time of the WWI Zimmerman telegram: the loss of Britain's ability to continue the war effort. During World War I, U-Boots sank 4.837 Allied ships of 11 million BRT. At the beginning of the Second World War, all shipyards in Great Britain and its dominions were able to launch about 200 vessels with a total displacement of one million tons per year. If the German submarine fleet managed to repeat its achievements from the previous conflict, Britain was threatened by famine, lack of fuel and weapons.

Although the first months of the war brought losses, their amount did not herald disaster. By the end of March 1940, the fleets of the Allied countries had lost over 400 ships of 1.3 million BRT. The balance of losses better than expected was influenced by two factors. The outbreak of the war surprised Dönitz, who had about 30 U-Boots on stand-by during the first months of operation, being able to send about 15 of them out to sea. The British were unaware that they owed their breath to Hitler, who initially forbade an unrestricted submarine war in the hope of convincing Britain to conclude peace. After the fall of France, however, the mirage of quick peace was irreversibly dissipated, and the defeat at the Battle of Britain made Berlin realize that the only option available to bring England to its knees was a naval blockade. Hitler gave the orders for a total blockade of the Islands on 17 August. Taking advantage of controlling almost the entire northern coast of continental Europe, the Third Reich threw a murderous combination of weapons against Allied shipping. From April to December 1940, the Allies lost 878 ships of 3.441 thousand BRT. When the British improved the protection of coastal shipping, Dönitz moved the hunting grounds of his ships to the west. The Atlantic quickly became the largest tomb of Allied ships and sailors; losses on western approaches to British ports reached during this period 1.4 million BRT. The shift of the campaign to the Atlantic also meant that U-Boots advanced to first place in the hunting statistics (363 sinkings). Italy's accession to the war added the country's submarine fleet, then more numerous than the German one, to the list of threats to Allied shipping. The U-Boots operating in the Atlantic were soon joined by 26 Italian submarines operating from Bordeaux base.

In 1940, the United Kingdom lost four times as much tonnage as the speed at which its shipyards could replace losses. At the same time, intelligence service reported a frantic expansion of the Dönitz's submarine navy. Winter storms temporarily hid the Allied convoys from the eye of U-Boot captains, but the British knew that they would be faced with a trial at sea again in the spring. In the upcoming struggle, they were to have at their disposal similar weapon that helped Dowding to win the Battle of Britain; a modern intelligence and command coordination center—the Operational Intelligence Centre (OIC). The Spanish Civil War had convinced the Admiralty's bright minds that peace would not last forever. The waters surrounding the Iberian Peninsula suddenly became crowded with ships with war contraband, and all parties to the conflict attempted to paralyze supplies destined for the enemy. The British were concerned about the activity of the Italians, who tried to support their troops fighting in Spain also from the sea. It was less important when they limited themselves to handing over a few freighters to Franco's fleet. The Italians, however, began sending their own ships to the Spanish shores, which used the anonymity of a submarine and pretended to be nationalist units, firing torpedoes towards the patrolling British ships. It turned out then that the information they had about the Italian fleet did not allow the Royal Navy to counteract it. The situation has alarmed Admiral William James, who convinced the then Chief of Naval Intelligence, Vice Admiral James Troup, to create an organization that could integrate intelligence with operational directives. The beginnings of the unit, established in June 1937, were a modest announcement of further development. Its first task was to follow the movements of the Italian navy, especially its submarines. Already at that time, one of the sources of information were reports coming from GC&CS as a result of breaking the ciphers of the Italian Navy Command—Supermarina. The second source was a network of radio direction finding devices, capable of locating a working radio transmitter. Over time, the network became known as the Huff-Duff, which is a slightly distorted abbreviation of the name "High Frequency Direction Finding, HF-DF". Not only did the Huff-Duff allow to determine the approximate position of the transmitting ship, it was also a great method of camouflage of breaking ciphers. During the war, orders directed to their own ships, based on the content of broken enemy dispatches, pointed to the Huff-Duff as the source of the information. In this way, even if the other party should break the British ciphers, the secret of the source remained secure.

Turing and his colleagues at Hut 8 were paying a high price for the long-time neglect of British signals intelligence. When the German navy first used machine ciphers in 1926, GC&CS was so deeply pessimistic about the possibility of breaking them that they stopped intercepting Kriegsmarine messages for about ten years. When the insight into the communication of German U-boats became a matter of life and death, it was too late to follow in the footsteps of the Poles solving the problems arising from gradual changes in procedures—one had to break into the mature system. An unexpected gift from the Polish Cipher Bureau, the development of an effective bombe model and the acquisition of three unknown Enigma rotors at sea were handicaps. However, there were three reefs left, by which the attacks on the naval Enigma were broken up: the order of the rotors, unknown bigram tables and a lack of good cribs. The choice of three rotors from among eight gave a much larger number of possible combinations than in the Wehrmacht machine: 336 versus 60. In the most pessimistic case, where all 336 possibilities had to be examined, it took a week of bombe's work to confirm or reject the correctness of a single crib. In the early days of BP's work, when bombes were still in short supply even for attacks on keys that Hut 6 knew how to break effectively, it was difficult to carve out such significant time for a precarious undertaking such as attacking Kriegsmarine ciphers.

The second problem was related to the encipherment of the message key. In the Wehrmacht and Luftwaffe communication networks, the key was given in open text in the message header. This seemingly reckless solution took advantage of the fact that the key did not represent a value for a listener who did not know the current Enigma settings. The situation changed radically, however, when the opponent managed to determine the base setting of the machine. From that moment on, breaking each cipher required encryption of the three letters of the key sent in an open text message at the base setting of the Enigma, turning the rotors of the machine into the received position and reading the rest of the message. In the case of Kriegsmarine the problem was more difficult. Even knowledge of the machine settings in force on the day did not offer information about the individual message key, which was enciphered using the bigram tables. In order to break the ciphertext of the German navy, it was necessary not only to determine the Enigma's base setting, but also to know the bigram table valid for the day.

The third problem faced by Hut 8 was the lack of good candidates for effective cribs. The success of the bombe in the search for a valid Enigma base setting depended entirely on the accuracy of the hypothesis regarding the open text of the message. When the bombe concluded its run without stopping, it did not announce the cause of its failure. The failure could have been due to the wrong choice of rotors, placing them in the wrong order, shift of the middle rotor while encrypting the crib or an incorrect hypothesis about its text. The erroneous order of the rotors was, as it were, inherent in the nature of the attack; this is why Rejewski at one time pushed through the ordering of six copies of his bombe, investigating all six possible rotor orders simultaneously. The increase in the number of possible rotor configurations from 6 to 336 basically ruled out the possibility of testing all of them at the same time—even by the end of the war BP did not have the necessary number of bombes to do so. The problem of the accuracy of the cribs was much worse, as the number of theoretically possible combinations in advance ruled out the sense of an exhaustive search. The task of the neighboring Hut 6 was less complicated. The Polish methods with which the Enigma cipher was broken in the early stages of the war were independent of the content of the message. Provided by the Poles knowledge of the typical structure of German messages, updated already during the war, guaranteed a wealth of high-quality cribs, even if the previous methods lost their validity. Meanwhile, practically everything that the BP knew in this phase of the war about the content of the Kriegsmarine messages came from a handful of radiograms from May 1937 broken by Poles. As a result, two dividing lines have emerged inside Hut 8. The first ran between theoreticians and the crib supporters. Theoreticians postulated that in the absence of good cribs, work on purely analytical methods of attack should continue. The supporters of cribs cited their effectiveness against Luftwaffe networks, claiming that after the first break of the cipher, things will turn out easier. They also suggested that a well-conceived manipulation may lead the Germans to send out messages of predictable structure, providing the desired crib.

Birch's biting remark, quoted earlier, confirms that even Turing and Twinn were torn between both approaches.

The group of the crib supporters was also not uniform; the division was roughly between the codebreakers from Hut 8 and the intelligence analysts from Hut 4. The intelligence officers argued that their knowledge of events at sea allows them to determine with considerable probability the open text of many of the enemy's dispatches. If it is known that an Allied tanker was torpedoed in a convoy, a U-Boot message intercepted afterwards should contain the words "tanker" and "sunk". In response, codebreakers tried to explain that a good crib requires not only a probable text, but also an appropriate internal structure, containing at least one or two cycles. The subject of the dispute between the camps was the working time of the bombe. In the period when BP had one or two machines at its disposal, and it took several days of work to confirm the accuracy of the hypothesis, discussions on the quality of the cribs and the priority of checking them gave rise to strong emotions.

Turing started his attack on Kriegsmarine Enigma at the point where the Poles stopped working. Thanks to the materials received from them, he had at his disposal a set of about one hundred messages, for which he knew both the keys encrypted with the bigram substitution and the starting position of the rotors recovered by the Poles. Among the messages from 5 May 1938 Turing noticed a group that allowed him to reconstruct the method of encrypting the individual key of the telegram. He was thus able to confirm the correctness of the conclusions reached by the Poles even before the outbreak of war. In the next step Turing continued to follow Rejewski. In November 1939, the interrogation of the prisoner, mat-radio operator Meyer, made it possible to establish that Kriegsmarine changed the way numbers were presented in Enigma ciphers. Instead of using the top-order equivalents of the machine's keyboard, the numbers were presented in words; the number 2330, which is remembered from the *Fortyweepy* method, was represented as ZWODREIDREINULL in the new system. The attack based on the structure of multi-part messages, so effective in the hands of the Poles, deserved another chance, but Turing couldn't apply it to the current intercepts. At the beginning of the war, Kriegsmarine stopped using the call signs, which previously allowed for the identification of messages representing a continuation of the previously intercepted ones. However, he could confirm the effectiveness of the method for messages previously interceptted. In the first days of 1940, Turing and Twinn began analyzing the messages from 28 November 1938. This was one of the last days when Enigma used only the 'old' rotors, and the exchange of only 6 pairs of letters in the plugboard gave hope for success. The *Fortyweepy* cribs once again confirmed their effectiveness; after two weeks of painstaking work, the first message was broken. The recovery of the base setting for the first day made it easier to widen the gap. One of the observations that the Poles shared with their British colleagues was that the same letter of the alphabet is never subject to change in the plugboard for two consecutive days. With 6 pairs of connections in the plugboard, 12 letters could be ruled out on the following day. This error by German cryptologists made it easier to determine the Enigma's base setting for consecutive days.

Restoring the base machine setting based on the updated *Fortyweepy* method did not allow other messages from the same day to be read. A different method of breaking the message key had to be developed, assuming knowledge of the base machine setting. The **ANX** method previously used by the Poles was based on the stereotypical structure of the beginning of a message. Unfortunately, the regulations in force in Kriegsmarine forbade the use of stereotypical phrases in easily recognizable parts of the message. This excluded the occurrence of **ANX**, or anything similar, at the start of message. Turing's team indicated a worthy successor to the phrase used by Poles. The solution to the problem was provided by a change in the way numbers were represented in the Enigma messages. It has become clear that the words **EINS**, **ZWEI**, **DREI**, etc., must appear frequently in the messages. The most common number became **ONE**; the statistics developed later on confirmed that the phrase **EINS** was present in 80–90% of Enigma messages. Turing encrypted the word sought at all 17.576 start rotors positions, and began browsing through the messages for occurrences of any encrypted variant. This was a drudgery job. Just listing all the equivalents of the **EINS** word required many hours of tapping the TypeX keyboard. Searching in the texts of the intercepted dispatches for one of the more than 17,000 variants of the word was downright murderous work. Therefore, when the method was used in an operational way, it was automated. Its first stage was designing the *Baby*—a device combining four Enigma units. The rotors of each were moved one position forward relative to the previous one, corresponding to the situation during the encryption of a four-letter word. Electrical signals corresponding to the word **EINS** were fed into the device and the *Baby* produced perforated cards corresponding to all its possible encrypted forms. Since the device appeared in Hut 8 in June 1940, "minding the Baby" has become part of the teamwork routine. The second part of the task, searching in the messages for possible occurrences of **EINS**, also found a quick and patient performer—electromechanical sorter. It has become a practice at Hut 8 to transfer the content of all analyzed dispatches onto perforated cards. From that moment on, the search for occurrence of one of the searched phrases could be implemented quickly and without any negative impact on team morale. The method gained a quite natural name of *Einsing*.

A considerable number of historians echo the opinion that even if the breaking of the Wehrmacht and Luftwaffe Enigma was possible thanks to the pre-war achievements of the Poles, the breaking of the Kriegsmarine ciphers represented the sole achievement of Turing and his team. Even in the light of the scanty testimonies available, this is not obvious. There is no doubt that the first successful attack on the ciphering procedure used by the Kriegsmarine was also carried out, soon after its premiere, by the Poles. It is difficult to determine why after the first success Poles lost interest in Kriegsmarine ciphers. Probably the number of messages intercepted by Polish stations was too modest to undertake a systematic attack, as a consequence, Rejewski treated the break in the marine Enigma as an incident. Alexander writes that the Poles "were not able to come to a more concrete conclusion than to say that the system was some form of bigram substitution". Mahon's account shows that they went one step further and "pointed to another possibility – that the bigrams are not chosen at random. They suggested that the starting position of the rotors, encrypted in the machine's base position (...) is selected from a list of predetermined combinations". Scant mentions in the British reports, combined with the contents of a report by Polish codebreakers declassified in 2015, confirm unequivocally that the Poles not only correctly reconstructed the German procedure, but also managed to break more than a hundred messages that later became the starting point for Turing. The achievements of the Hut 8 team presented so far have been limited to confirming data previously obtained from the Poles. Even *Baby*'s design and concept of *Einsing* were purely mechanical developments of methods previously used by the Poles.

At this point, however, Turing sailed into uncharted waters. A true monument to his mathematical genius became the method of attack developed in late 1939 and early 1940, known by the somewhat cryptic term *banburismus*. Its main purpose was to circumvent the too modest number of bombes at BP's disposal. The attempt to check one crib for all 336 possible rotor orders required about a week of bombe runs. The cure was to increase the number of bombes available, but it was not an immediate cure. Turing was aware that the successful use of the bombe required a reduction in the number of rotor orders to be checked. He also remembered that a few years earlier Różycki had used the statistical properties of the language to develop his *clockwork* method, which permitted to identify the fast Enigma rotor and thus limit the number of possible combinations of the other two rotors.

The starting point was the observation that all the individual message keys during a given day are encrypted at the same Enigma base setting. Although the base setting was unknown, even without knowing it, it was possible to draw conclusions about their structure. Turing, like Różycki did before, was searching for messages that had two common characters on the first key positions, for instance **CEG** and **CEN**. Not knowing the base machine setting, he did not know what form both keys would take after encryption. However, the fact that encryption of both trigrams begins in a common machine setting meant that also the encrypted message keys will probably have the first two letters in common, for instance **TPD** and **TPK**. The distance separating the third letters of the key could be determined by shifting mutually the texts of the two messages in search of the position corresponding to the highest value of coincidence index (see the description of the *clockwork* method). If the texts of the messages corresponding to the keys **CEG** and **CEM** showed the highest coincidence index value at a shift of 7 characters, the third letters of the keys of the two messages also differed by 7 characters.

In fact, nearly all the messages were reviewed for coincidence. If the aim of the method was to determine the identity of the fast rotor only, it would have been sufficient to compare the messages over a range of offsets of up to 25 characters to the right and left. Turing, however, set himself a more ambitious goal; he also intended to determine which rotor occupied the central position. In theory, this required an analysis of shifts between -26^2 to $+26^2$ characters. Fortunately, the codebreakers' task was somewhat simplified by the opponent, who limited the maximum length of the Enigma message to some 200 characters. Searching for coincidences in such a large range of shifts required a specific approach. The text of each message was perforated on a long sheet of paper, whose 26 lines consisted of repetitions of 26 letters of the alphabet, and columns were numbered in the top line. The sheets were perforated with a hole in the column corresponding to the position of the letter in the message text and the line corresponding to its position in the alphabet. The example below shows text *Wem Gott bestrafen will dem erfuellt er seine Wuensche* registered in the described form (character ●represents a hole in the sheet).

```
01234567890123456789 0123456789
AAAAAAAAAAAA●AAAAAAAAAAAAAAAAAA
BBBBBBB●BBBBBBBBBBBBBBBBBBBBBBB
CCCCCCCCCCCCCCCCCCCCCCCCCCCCCCC
DDDDDDDDDDDDDDDDDDDD●DDDDDDDDD
E●EEEEEE●EEEEE●EEEEEEE●E●EEE●E
FFFFFFFFFFFFFF●FFFFFFFFFFFF●FFF
GGG●GGGGGGGGGGGGGGGGGGGGGGGGGG
HHHHHHHHHHHHHHHHHHHHHHHHHHHHHH
IIIIIIIIIIIIIIIII●IIIIIIIIIIII
JJJJJJJJJJJJJJJJJJJJJJJJJJJJJJ
KKKKKKKKKKKKKKKKKKKKKKKKKKKKKK
LLLLLLLLLLLLLLLLLLL●LLLLLLLLL●
MM●MMMMMMMMMMMMMMMMM●MMMMM
NNNNNNNNNNNNNNNN●NNNNNNNNNNNNNN
OOOO●OOOOOOOOOOOOOOOOOOOOOOOOOO
PPPPPPPPPPPPPPPPPPPPPPPPPPPPPP
QQQQQQQQQQQQQQQQQQQQQQQQQQQQQQ
RRRRRRRRRR●RRRRRRRRRRRRR●RRRR
SSSSSSSSS●SSSSSSSSSSSSSSSSSSS
TTTTT●●TTT●TTTTTTTTTTTTTTTTTTT
UUUUUUUUUUUUUUUUUUUUUUUUUU●UU
VVVVVVVVVVVVVVVVVVVVVVVVVVVVVV
●WWWWWWWWWWWWWWW●WWWWWWWWWWWW
XXXXXXXXXXXXXXXXXXXXXXXXXXXXXX
YYYYYYYYYYYYYYYYYYYYYYYYYYYYYY
ZZZZZZZZZZZZZZZZZZZZZZZZZZZZZZ
```

The sheets were printed in Banbury, from which they took their name—*banburies*, in turn lending their name to the method. When the texts of the two messages were registered on *banburies*, it was easy to identify their coincidences by shifting the sheets on an illuminated table. Particularly interesting were the coincidences of pairs, triples and the rare occurrences of longer repetitions. The result of the search for coincidences was recorded in a conventional notation, including the open form of the keys of both messages, the shift of their texts, the length of the overlapping text and the number and type of coincidences found. For example, the notation **FGH** = **FGX** − 24:7^{3x}/54 meant that the texts of messages with keys **FGH** and **FGX** were compared with each other, the beginning of the second message was shifted 24 positions to the left of the first one, and after shifting 54 characters of both messages were overlapping. Seven coincidences were found, including one trigram (marked as 3 in the exponent) and one bigram (marked as x).

As a result of comparing the texts of many messages a large number of coincidences were obtained; a tool was needed to facilitate the search for those indicating a probable relationship between the two messages. Turing found the tool in Bayesian statistics—an almost forgotten branch of statistics developed in the eighteenth century by a presbyterian priest, Thomas Bayes. Turing did not intend to give lectures on statistics to the staff members of Hut 8, and limited himself to developing tables in which he assigned probability measures to individual coincidences. According to the logic of Bayesian statistics, the probability was determined by a logarithmic unit, which the author called a *ban*. Turing created his measure of probability for a strictly defined and practical purpose, probably without considering the theoretical implications of his concept. However, almost at the same time, on the opposite side of the Atlantic, another mathematician, Claude Shannon, worked on a similar measure of information, basing it on a binary system and calling it a *bit*. War stimulates research, but does not always encourage the exchange of results; Shannon and Turing met during the war, but it is doubtful that they would talk about their parallel pioneering work on information theory.

Originally, Turing's tables used a unit ten times smaller, *deciban*. Later on, Jack Good noted that one could save about 50% of the time needed to prepare data by using a unit expressed as an integer. As a result, the tables were converted into half *decibans*, rounding the values to the full unit, which from its abbreviation, hdB, gained the jargon term *hubdub*. This incident illustrates the industrial character of BP's functioning in the second phase of the war; as if in a well-managed factory, each operation was analyzed and subject to modifications if this saved time or effort. Perhaps not all mathematicians were wholly impractical people, after all…

The use of *banburismus* tables required the consideration of many factors. Basically, they were not overly complicated: on one axis the length of the common fragment of both messages was written, on the other—the number of letters of repetitions, at their intersection the probability expressed in half-decibans was noted (below is an example of a table fragment).

	Length of overlapping fragment										
Number of repetitions	29	30	31	32	33	34	35	36	37	38	39
0	−4	−4	−4	−4	−4	−4	−5	−5	−5	−5	−5
1	−1	−1	−1	−1	−1	−2	−2	−2	−2	−2	−2
2	1	1	1	1	1	1	1	1	0	0	0
3	4	4	4	4	4	4	4	3	3	3	3
4	7	7	7	7	7	7	6	6	6	6	6
5	10	10	10	10	10	10	9	9	9	9	9
6	13	13	13	13	13	12	12	12	12	12	12
7	16	16	16	16	15	15	15	15	15	15	15
8	19	19	19	18	18	18	18	18	18	18	17
9	22	22	21	21	21	21	21	21	21	20	20

Let's assume that we want to determine the significance of a coincidence FGH=FGX–5:6^{xx}/37. The table above shows that 6 repetitions over 37 characters corresponds to 12 hdB. The obtained result should be adjusted taking into account additional factors. Both messages were shifted by 5 positions. The greater the shift of the two texts, the more likely it is that the middle rotor has advanced during their encryption. The obtained result should be corrected by a factor reflecting the probability of the middle rotor movement; a separate table indicates that for a shift of 5 characters the obtained result should be reduced by 5 hdB. On the other hand, among the repetitions, there were 2 bigrams (marked as xx). Another rule was that each bigram corresponds to a shift of the result to the next row of the table (and the trigram—by two rows), so the result is equal to 13 hdB.

Another adjustment is due to a complication that was purposefully applied by the German cipher clerks. They were obliged to transmit, among the actual ciphertexts, dummy dispatches, consisting of random characters but retaining the external format of the real Enigma messages. Such cryptograms were described in BP as dummy. The content of dummy messages did not respect the statistical properties of language, and their inclusion in the calculations disturbed the results of *banburismus*. As a result, a discipline was developed that aimed to eliminate the influence of dummy messages on *banburismus* results, described as *dummyismus*. According to BP veterans *dummyismus* was an art rather than a science. If the circumstances of the message transmission indicated that the message could be dummy, several points were deducted from the result read from the *banburismus* table. As a result of the described operations, for each pair of messages the characteristics of coincidence were obtained and the corresponding measure of probability that the observed coincidence is not a result of chance. From among them, pairs with a probability higher than 1:1 were selected and ranked in order of highest probability. So called fit list (see example below), was the starting material for the next stage of work. Let's assume that an excerpt from the fit list drawn up in the previous step is in the form:

VRK=VRS+5	28/95	100:1
RWN=RWA+6	27/100	50:1
STZ=STS+3	15/98	15:1
BAA=BAZ+8	13/122	10:1

The notation above indicates that for a pair of messages enciphered using keys **VRK** and **VRS**, with a 5-character shift, 28 coincidences over 95 overlapping characters were obtained, which is equivalent to a 100:1 chance that the coincidence is not accidental. Taking into account relative shifts of characters occurring in the third position of the keys above, it is possible to arrange them into a string (the occurrence of unknown characters is marked with asterisks):

<u>S</u>**Z**K****<u>A</u>*****N

With the information about the distance between some of the letters of the right rotor's cipher alphabet, the codebreaker can attempt to reconstruct the entire alphabet. For this purpose, the above string will be moved along the alphabet in search of positions generating contradictions with the nature of the Enigma cipher. The table below illustrates this process. Character pairs implying a contradiction are underscored and the combinations of alphabets in which they occur are highlighted by a lighter colour. Note, for example, that in the first pair of the left column, the letter **A** would be encrypted once as **S** and once as **L**, which the Enigma does not allow. In the second pair of the right column, the letter **A** would be transformed into itself, which is also excluded in Enigma.

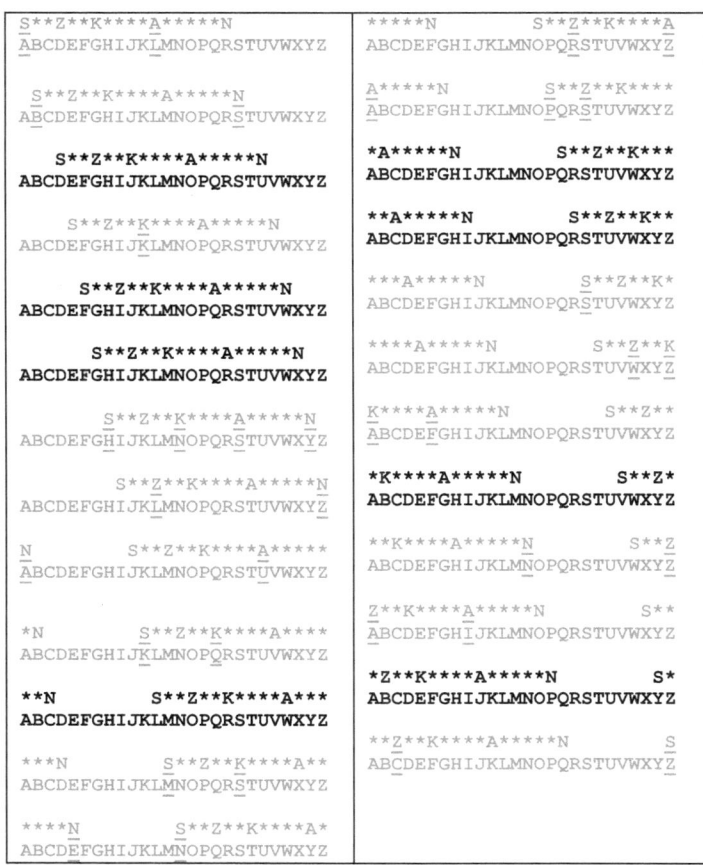

After eliminating mutual positions containing contradictions, 8 combinations remain. In our example, however, we have used a fit list containing only four items. Had we had a longer list, it would probably have been possible to eliminate all but one item. This process was called scratching out the contradictions. The position where the contradiction occurs was said to be *screeching*. As a result, the method was also referred to as *scritchmus*. At this point there are only two steps left to the finale of the *banburismus*—the determination of the right and, possibly, middle rotor. Let's assume that adding a few items to the fit list we were able to reduce the number of possible reciprocal text positions to one:

S**Z*LK*W**A*F***N
ABCDEFGHIJKLMNOPQRSTUVWXYZ

Banburisums assumes that the central rotor has not advanced during the encryption of the message key (i.e. between the characters **F** and **W**, corresponding to the characters **S** and **N** of the original string). If rotor I were present in the fast position, the central rotor would advance between **Q** and **R**, resulting in a contradiction to the assumption. Similarly, rotors II—IV (with turnover points at the letters **F**, **W** and **K**) can be eliminated, leaving rotor V as the only possibility.

The identification of the fast rotor was a significant achievement, but Turing set himself a more ambitious goal—to identify the middle rotor. To this end, he extended the scope of the analysis to shifts of more than 26 characters, which corresponds to comparing messages where only the first letter of the key is common (for instance **WAC** and **WTG**), and even keys that do not contain common letters. For such pairs of cryptograms, a different procedure was applied for the coincidence test. Giving up looking for high-probability pairs, he looked for practical certainties—repeats of four or more letters. For this purpose, the texts of the messages were copied onto perforated cards and forwarded to the section of Hollerith's machines (headed by Freddy Freeborn and therefore commonly referred to as Freebornery), which produced the so-called tetra catalogue, i.e. a list of repetitions of four and more characters. When the fast rotor was already known, by recalculating the relative shifts modulo 26 and repeating the *screeching* procedure, it was possible to identify the middle rotor as well, or at least to narrow down the list of choices.

Banburismus became for Hut 8 alternately a source of deep frustration, a lifebelt, a sense of life and rejected ballast. The initial frustration stemmed from the practical impossibility of applying a method in the development of which considerable effort had been invested. Turing and his team developed several methods of attack that were interdependent. The bombe could only find the solution in a reasonable time if *banburismus* had previously limited the scope of its search. However, the application of *banburisums* required prior knowledge of the bigram tables. The bigram tables could be reconstructed using the *Einsing*, but only on the condition that a large number of messages were previously broken using the bombe. In the closed circle of dependence of the three attack methods there was no gap through which one could slip into the German cipher system.

Breaking the closed cycle of impossibility required an external impulse. Fortunately, Hut 8 only had to wait until the Norwegian campaign for such a boost. Royal Navy, which allowed itself to be overtaken by the Germans by just 24 hours, energetically set about making up for lost time. Its ships seemed to be everywhere and patrolled every square mile of the waters around the Norwegian shores. In the morning of 26 April, at the entrance to Romsdalsfiord, the destroyer *Arrow* came across a fishing vessel flying the Dutch flag. However, when *Arrow* tried to send a group of sailors aboard the alleged Dutchman, it hoisted the Kriegsmarine flag on the mast, moved full steam ahead and rammed the British destroyer by puncturing its hull just above the waterline. The fake trawler was in fact a trawler *Schleswig* armed by the Kriegsmarine. Several volleys from *Arrow* and the accompanying cruiser *Birmingham* sent her to the bottom, and the British ships returned to patrol duties. Soon after 10 o'clock *Arrow* noticed another unidentified ship, but probably because of its damage, the destroyer *Griffin* was ordered to carry out reconnaissance. His commander, Captain John Lee-Barber, stopped his ship at a safe distance from the target on which the Dutch flag was flying and the name *Polares* was displayed on the side. Lee-Barber sent a boat with a group of sailors under the command of First Officer Alec Dennis towards the *Polares*. As they approached their destination, their confidence began to give way to anxiety—under dikes of fishing nets on board they recognized the contours of the deck cannon and torpedo launchers. The result of the inevitable confrontation was decided by chance. When one of the British sailors was boarding the ship, he accidentally pressed the trigger of his gun, and a series directed at the sky convinced *Polares'* crew to surrender the ship. Before that, however, one of its members managed to throw overboard two bags containing Enigma and secret documents. The bag with Enigma went down, but the one containing the documents stayed on the surface long enough for *Griffin*'s gunner, Florrie Foord, to be able to fish it out. The alleged Dutch fishing vessel turned out to be another Kriegsmarine unit, *Schiff 26*, carrying ammunition for the German units fighting in Norway.

Griffin's crew was close to turning their success into failure. The documents obtained aboard the *Schiff 26* were useful, provided that the capture was kept secret. Meanwhile, *Griffin* arrived at Scapa Flow in the middle of the day, triumphantly towing her prey, which was decorated with all the Nazi paraphernalia found on board. She also found a grateful audience for her entrèe: the arrival of the ships was filmed from the shore by a team from Universal. The naval intelligence officers were shocked when they heard about the show. Just six months earlier, the U-47 broke through the obstacles defending access to Scapa Flow, torpedoed and sunk the battleship *Royal Oak*. The British were convinced that assistance in identifying the route leading to the Royal Navy anchorage must have been provided by secret agents located in the area. If that was the case, the spectacle directed by *Griffin*'s crew must have been a real attraction for them. Someone conscious took care to change the decoration, confiscate the film and professionally search the prize. This proved to be a troublesome task: during the voyage from Norway, *Griffin*'s sailors spent a lot of time searching through their prize for souvenirs, turning its rooms and deck into a real dump. At Scapa Flow even Captain Lee-Barber had to return two pairs of excellent Zeiss binoculars, which his sailors took as a souvenir of success.

Although the events just described took place in the waters surrounding Ålesund, far from the Narvik fjord, the incident went down in history as a "Narvik pinch". Florrie Foord would have been disappointed to learn that the bag he saved at the risk of his life was not the most important trophy from Schiff 26. Alexander's notes show that it included, among other things, instructions for the use of Enigma, which confirmed the guesses about the construction of the message indicator. Memory failed Aleksander somewhat, for the report of the Polish codebreakers shows unequivocally that even before the German attack on France, a copy of the instruction captured on board of the sunken U-Boat had reached them.

Among the rubbish found in cabins ransacked by sailors, the cipher clerk's notes containing the settings of the plugboard and rotors valid on 23 and 24 April, as well as the texts of the messages sent on 25 and 26 April and the corresponding ciphertexts were found. Reconstructing the missing machine settings for the 23 and 24 of April was not a complicated task. It was probably on this occasion that Hut 8 discovered that the naval Enigma keys are being changed on a regular basis; the internal settings (rotor order and ring settings) every two days, the remaining (plugboard settings and starting rotors position) every day. The knowledge of this fact made it easier to find the missing elements on the second day of the pair. The BP codebreakers were lucky—the notes from Schiff 26 included 2 days belonging to different pairs. As a result, the keys valid on 22 and 25 April were also easily broken.

For Turing, a discovery more exciting than the machine's settings were the notes giving an open and encrypted text of the messages from two days of April. They formed a perfect crib; the captures from the German ship were to become a fire test for a bombe that had been undergoing tests for over a month. Of the two available messages, from 25 and 26 April, the latter was selected for the bombe test; 25 April was broken based on the pairs rule. One can imagine the tension with which the codebreakers were watching subsequent bombe runs. Finally, after about two weeks of work, the bombe stopped in a position corresponding to the correct solution. This success included a stroke of luck; in May BP did not yet have the rotor VIII. Luckily, the key of 26 April did not use the unknown rotor. Knowing the base position on 26 April, codebreakers easily broke the key for the second day of the pair and found that the message indicators on 27 April were encrypted using the same bigram table that was used on April 24. Breaking a considerable number of messages from both days allowed for the reconstruction of part of one of the bigram tables.

"Narvik pinch" enabled Hut 8 to break into the Kriegsmarine network, which was henceforth referred to as *Dolphin*, giving rise to the tradition of naming the Kriegsmarine networks after fish names (in the German nomenclature, the network was called *Heimische Gewässer*, or briefly—*Heimisch*). The breaking of the first messages has poured into the hearts of the Hut 8 team the optimism that had been so lacking in the preceding months. The reading of numerous dispatches allowed work to begin on the reconstruction of the bigram tables. It seemed that the codebreakers would manage to reconstruct them in their entirety, which would open the way for the application of *banburismus*. It must have been all the more frustrating to return to earlier failures. Soon, the war moved from Norwegian waters to the waters surrounding the British Isles, but new prey, on which the future of *Dolphin* breaking depended, was not coming. Tensions were growing in BP between codebreakers from Hut 8 and intelligence analysts from Hut 4. The essence of the conflict was the quality of the crib. Frank Birch waved his hands over the mathematicians' attitude to the proposals provided by his team: "Sometimes we provide a 90 percent probability of a crib. Turing and Twinn insist on adding a word about a probability of less than 50 percent (…)". The communication between both sections was obviously failing: Birch was clearly not aware that the bombe efficiency depended entirely on a cyclic structure, without which even an absolutely certain crib was useless. Codebreakers did not lose hope that they would manage to return to *banburismus* under more favorable conditions. They needed a day when at least 200 *Dolphin* messages were intercepted. The second condition was a little more difficult to meet: on that day the same bigram table had to be in force, which was partially reconstructed thanks to "Narvik pinch". On 8 May, 1940, both conditions had to be met, as in the coming months the effort of Hut 8 concentrated on the messages intercepted on that day. But even this narrow alley ran uphill, and weeks and months of work went by without a result.

In this tense atmosphere, surprising ideas began to emerge in the codebreakers' environment. The auction was started by Knox, who in early September 1940 was reported to suggest playing *va banque*—sending an Enigma message containing a request to send the current key by radio. If the reports correctly conveyed the essence of Knox's proposal, his colleagues must have rubbed their eyes with amazement: a cryptologist known for his caution made a proposal that was purebred adventurism. To make matters worse, Knox's proposal was short on detail. Sending a message encrypted with one of Enigma's known keys and requesting the key of another network was doomed to failure. Its only effect would be to betray that the British can not only read the Enigma messages, but are also able to encipher them correctly. It seems that Dilly's real intention was to send an open text with a request to respond with the key. It would probably make the adversary smile on the naivety of the British, who, helpless against the Enigma ciphers, reach for desperate means.

Whatever Knox's true intentions were, he was soon outbid. Lieutenant Ian Fleming was at the time a personal secretary of Admiral Godfrey, head of naval intelligence service. The news of the Hut 8's problems had to circulate in the naval intelligence, since Fleming learned about their nature without having direct contact with the BP. The solution he proposed in September impressed with its imagination. According to his scenario, a captured German bomber with a British, but German-speaking and dressed up in Luftwaffe uniforms crew was to be used (Fleming himself was, of course, to participate in this operation). The airplane was to ditch in the Channel, purely accidentally near the Kriegsmarine rescue ship. Then things were to go straightforward: as soon as the airmen are on board the vessel, the bomber's crew, quoting Fleming's proposal, were to kill the German crew, throw the corpses into the sea and escort the ship, together with the equipment and documents on board, to the British port. The strangest thing was that the superiors took the proposal seriously. Codenamed "Ruthless", the operation was assigned a captured Heinkel-111. Completing the aircrew proved to be a slightly more complicated task. Experienced pilots indicated that ditching this type of aircraft in the rough Channel waters meant a certain disaster. Even if the crew had survived the ditch, the wounded or at least exhausted after being pulled from the icy water airmen could not be expected to take on the enemy ship. In fact, the only airman who volunteered to take part in the mission was suffering his own spiritual torment due to a love disappointment, so surviving the mission was not his main priority. Fortunately, the trial hour never came. The plane and crew waited for several days at the airport near Dover for a suitable ship to appear on the Channel. The problem for those in charge of the operation was to find a ship large enough to have Enigma on board, yet small enough for the 'aviators' to have a chance of defeating its crew. On 16 October the operation was suspended, after which no one ever returned to the idea. Maybe with the exception of the plan's author himself, but it happened much later and under completely different circumstances. After the end of the war, Ian Flaming focused on a career as a writer and author of screenplays describing the adventures of Agent 007, James Bond. It was probably a good thing that he did not try to test his ideas in practice.

Meanwhile, the Hut 8 team focused its attention on the 8 May dispatches with such intensity that its members became averse to dispatches analyzed for weeks without effect. Over time the team was joined by one of GC&CS veterans, Hugh Foss, an eccentric Scotsman who attracted attention with his red beard, his predilection for wearing the traditional kilt and sandals and for Scottish folk dancing. It was Foss who carried out the first GC&CS security analysis of the Enigma in the 1920s and proposed the so-called geometric method of attacking its cipher. He also had another success in the struggle with machine ciphers—in 1934 he broke the cipher of a machine used by Japanese military attachés. This was GC&CS's first ever success in grappling with machine ciphers, made all the more valuable by the fact that it was achieved 15 months before the identical success of US codebreakers. After a break due to illness, Foss returned to BP in summer 1940. Perhaps the impasse in the attack on the naval Enigma caused his temporary assignment to Hut 8.

Foss belonged to a generation of cryptologists who did not expect immediate results, and so he began to work with a regularity and patience that his younger and more impatient colleagues could not afford. After several months of work, in November, the success came unexpectedly; Foss and *banburismus* simultaneously confirmed their value. The rule of pairs made it possible to break the key that was in force on 7 May, and the large number of intercepted messages made it possible to almost fully reconstruct the bigram table used in those days. For a while it seemed that the Hut 8 team could look to the future with optimism. Knowledge of the bigram table made it possible to find among the archived messages the days on which it was used, and then, at the cost of several weeks' work, to recover the keys for 14 April and 27 June. The second of these days helped allay codebreakers' fears that the adversary had altered the bigram arrays. The outcome of the work helped to confirm the effectiveness of *banburismus*. And it was at this point that another blow was struck—on 1 July 1940, the Germans changed the bigram tables in force. In the second half of 1940, the attack on the ciphers of the naval Enigma came to start again, practically from scratch.

Chapter 14
Sikorski's Tourists

Even though the fall of France in 1940 did not surprise attentive observers, the circumstances of June's defeat made the world astonished. The defeat of the Polish army, attacked in September '39 on two fronts by much stronger aggressors, was met with comments tinged with protectionism reinforced by Goebbels's propaganda; what could cavalry charges have done against tanks and planes? With France, history was to turn out completely different—the victors of World War I, safe behind the Maginot line, having an advantage in almost all weapons, were to show this funny corporal his place in line. But when Marshal Pétain, the triumphant from Verdun, reached for power not as a symbol of past victories and further struggle, but as a spokesman for surrender, the world was astonished.

Among the members of the Polish team of codebreakers, astonishment was mixed with concern for tomorrow. The notes in Kazimierz Gaca's diary show that when the members of "Équipe Z" learned about the negotiations around the surrender of an ally, differences of opinion about the future of the team emerged. The dividing line seems to have run roughly between the military and civilian parts of the team. Civilians with Rejewski at the helm argued that the fall of France would destroy their usefulness for their current ally, while their experience may prove useful to Britain continuing the fight. Officers stressed obedience to orders directing the team under French orders. However, on 19 June the BBC broadcast General Sikorski's order in three languages, ordering Polish troops in France to ignore orders to surrender trying to break through to Atlantic ports and further to the UK. The team conducted continuous radio monitoring also during the evacuation. This interception included BBC frequencies, as confirmed by a note in Bertrand's memoirs about the impression made by General de Gaulle's appeal broadcast from London the day before. Under these conditions, it is inconceivable that an order from the Commander-in-Chief of the Polish army would pass unnoticed.

From Agen, a road winding along the banks of the Garonna led directly to Bordeaux, not yet occupied by the Germans, which was then the seat of the French government. Only a few dozen kilometers from Bordeaux Libourne housed Polish

Springer Nature Switzerland AG 2025
M. Grajek, *Enigma Myth Deciphered*, History of Information Security,
https://doi.org/10.1007/978-3-031-65475-6_14

government agencies and a center for the evacuation of Polish soldiers. Between the 16 and 24 of June, operation *Ariel* was underway, a post scriptum to the evacuation of the British from Dunkirk. During this operation, over 160,000 British, Canadian and Polish soldiers were evacuated from the ports of the Atlantic coast of France. Sikorski himself returned from London to Libourne to finally leave the continent on 21 June, and the last representative of the Polish authorities, General Marian Kukiel, did not leave the small port of St Jean de Luz until 25 June.

The Poles did not know that an international intrigue was unfolding around their further fate. In the face of French defeat, the heads of British intelligence were understanably concerned not so much about the fate of the Poles as about the secret of Enigma, of which they were the depositories. On 23 June, Denniston sent a messge to his French counterpart suggesting the redeployment of all members of the team, French, Spanish and Polish, to Britain, away from danger. His suggestion unleashed a geyser of activity from Bertrand. In the first instance, he had the British liaison officer to P.C. Bruno, Kenneth McFarlane, delivered to Cazaux airfield, from where the Briton departed aboard the last RAF aircraft leaving the base. He then used his contacts in Air Force intelligence service to place Poles and Spaniards on board three planes taking off from Toulouse airport on 24 June. Two of these, with members of the Polish team on board, were to go directly to Oran in French North Africa. The direction of flight and the range of the planes required flying over Spain, which was considered too risky for the Spanish part of the team. The plane with the Spaniards on board therefore flew to Marseille, where the passengers transferred to a flying boat and only thus reached Africa. Zygalski's notes indicate that the evacu-ation was an improvised and risky undertaking. His plane ran out of fuel before crossing the African coastline. In desperation, the pilot landed despite the lack of permission from airfield control, which could have ended with the plane being shot down by the French. After Bertrand ascertained that both foreign teams were in Africa, out of reach equally of the Germans and the British, he replied to Denniston on 26 June, that evacuation of the Polish team to the British Isles was no longer pos-sible, as they had already been flown to Africa before receiving his invitation. Bertrand himself remained in France. The day after the departure of the Poles, events took place in the seminary complex at Bon Encontre, near Agen, which determined the fate of the Polish team for the next over two years. A significant number of personnel from various sections of French intelligence service were gath-ered there, including the commander of French intelligence and counter-intelligence service, Colonel Loius Rivet. On 25 June those present took an oath to continue fighting also in the new realities resulting from the Armistice Agreement.

After the experience of the lost campaign and the evacuation, Polish team was recovering at the Royal Hotel in Oran. While Bertrand oversaw the demobilization of his French team in France, Captain Henri Bracquenié took charge of the Poles in Africa. In his role as guardian of the group he could also be supported discreetly by Captain Emilé Bertrand, Gustav's brother and, before the war, head of the French counter-intelligence post in Lille. After the defeat of France, Emilé Bertrand was evacuated to Algiers, where in time he was to become head of the local counter-intelligence section (TR121). The role of the group's caretakers was not limited to paying the hotel bills, as history soon knocked again on the door behind which the Poles found temporary refuge. On 3 July, in the early morning, the entire team was transported unexpectedly to the local train station and, after a long and arduous journey, found themselves in Algiers the same day. Circumstances indicate that the unexpected decision to leave Oran was the result of the panic that seized the French hosts at the sight of a squadron of British ships in the roadstead of the nearby port of Mers el-Kebir. Operation "Catapult", of which the action under Mers el-Kebir was part, had its roots in the events preceding the fall of France. The British-French alliance contained a clause providing for the joint waging of war to victory and prohibiting any partner from entering into a truce or peace agreement on its own. When on 16 June Marshal Pétain deemed the military situation to be hopeless and demanded a cease-fire, Prime Minister Reynaud desperately asked Churchill to release his country from its treaty obligation. The UK gave Reynaud's note a two-fold answer. In the first reaction, she agreed to surrender, provided that the French fleet is assembled in British ports. When the French government did not accept this proposal, the British made a final, somewhat desperate attempt to save the alliance and continue the fight together, and submitted to the French government an *ad hoc* project to merge the two countries into a British-French union, capable of continuing the fight with the resources of the British Empire and the French colonies. Reynaud picked up the idea, but it did not find recognition in the eyes of his colleagues; having played the last trump card without success, the French Prime Minister resigned. The government of Marshal Pétain, formed on the same evening, did not feel obliged to respect the provisions of the alliance agreement with Great Britain, as a result of which the fate of the French fleet was to be settled in a way dictated by war necessity rather than the letter of the treaties.

While performing Operation "Catapult", British teams in most cases did not encounter any resistance. In the early hours of the morning of July 3 British sailors boarded French ships moored in British ports. A British and French seaman lost their lives on the *Surcouf* submarine as a result of a misunderstanding rather than hostile intentions. In Alexandria, Admirals Cunnigham and Godfroy peacefully agreed to immobilize the eleven French ships in this port. The most difficult task was put before Admiral James Somervill, who was to solve, peacefully or by force, the problem of the French squadron in the port of Mers el-Kebir near Oran. On the morning of July 3, the British ships appeared in its roadstead and gave the French commander an ultimatum covering several options; from his squadron taking up the fight at the side of the Royal Navy, through its transfer to ports far from the European theatre of war in Martinique, to the scuttling of the ships. Having received no satisfactory response by late afternoon, Admiral Somervill ordered his ships to open fire. Within a quarter of an hour, most of the French ships were destroyed and nearly 1300 French sailors lost their lives. The action of the British fleet at Mers el-Kebir and the blood of the dead had far-reaching consequences. They have aroused a psychologically understandable wave of anti-British sentiment in French society. Most of the French have not yet recovered from the defeat of their homeland. The perception of the recent ally as today's enemy facilitated the interpretation of France's defeat as a result of the desertion of the British from the beaches of Dunkirk. On 5

July the French government in Vichy broke off its diplomatic relations with the United Kingdom. Mers el-Kebir became the first step on the path of escalating mutual hostility. It later led to the Dakar, where the combined forces of the British and the Free French tried to take over another part of the French fleet in vain. In retaliation for Mers el-Kebir and Dakar, French aircraft bombed Gibraltar, and World War I aviation ace René Fonck collected a French contingent of about 200 pilots, ready to participate in German attacks on England. Later in the war, the British fought against Vichy government troops in Syria and Madagascar, finally during the American-British invasion of North Africa in 1943. Since July 3, 1940 between Great Britain and the government of Vichy there was a strange state of affairs, not an unspoken war, not an armed peace.

The records of Polish codebreakers do not contain any references to the tragedy that took place right under their side. Rejewski only mentions that after an exhausting train journey due to the heat, the whole team arrived in early July in Algiers. More precise was Kazimierz Gaca, who wrote that the whole group left Oran on 3 July at 11.26 a.m. It seems that the intended destination of 'Équipe Z' in Africa was to be Oran, not Algiers. A glance at the map shows that from Toulouse, where planes with Poles on board took off, it is easier to get to Algiers than to Oran. Given that the planes transporting the team arrived in Africa on leftover fuel, this was not an irrelevant circumstance. Moreover, the route to Algiers did not require a flight over Spanish territory, which forced a team of Spanish cryptologists to take a detour through Marseille. The plan was spoiled by the British, appearing at dawn on 3 July in the neighborhood. Worse still, part of the French fleet was based in Oran itself. The British ultimatum also concerned her—any moment now the Royal Hotel could be within range of the British guns, or maybe even landing of British Royal Marines. The news of the British ultimatum alerted the codebreakers' guardians, and uncertainty about the British intentions made them evacuate the Polish team to Algiers in an emergency. The evacuation was carried out so smoothly that when the guns spoke at Mers el-Kebir, the Poles were already sitting on the train, away from Oran and the British.

The real threat to the team posed by the Royal Navy's attack on the French fleet did not justify such a strong response. Combining the scattered sources and opinions expressed, it is hard to avoid the conclusion that the real reason for the evacuation was a crisis of trust in relations between the French authorities and members of the Polish team. Doubts about cooperation with the French appeared among Poles even before their departure to Africa. They had numerous and important reasons to question the sense of further work under Bertrand. When the command posted them to serve in the allied army, they expected to cooperate with their French colleagues, whose competence they evaluated through the achievements from the previous war. They quickly realized that they, with the addition of the Spanish group, constituted the entire cryptanalytical service of France. Even the banal manufacturing of several copies of the machine grew to be a problem. The following months spent surrounded by French colleagues made the Poles learn to value them as caring, cordial people, but they did not find the expected proficiency in cryptology among them. This state of affairs was contrasted with information coming from behind the Channel. The growing number of keys that "Pinky" was handing over indicated that an effective cryptologic center had begun to operate in the UK. Occasional contacts with British colleagues made it possible to assess their professionalism better than in the official atmosphere of the Pyry meeting. If one adds to the list the obvious problems of the French command in making use of the information provided by the cryptologists and General Sikorski's unequivocal order for Polish military personnel to move to British-controlled territory, one cannot avoid the conclusion that, by the summer of 1940, the possibilities for Polish-French military cooperation had shrunk practically to zero.

In fact, Bertrand stresses in his memoirs that when he started organizing the further work of his team, he did not experience any difficulties on the part of the Spaniards, but "some (Poles) felt remorse and decided that before returning to work for the French, and consequently for Vichy, they had to get some cover from the Polish government in London". The testimonies of Poles show that they perceived the problem differently. Colonel Stefan Mayer, one of the heads of pre-war Polish intelligence service, confirms that "(…) Langer tried to contact the Polish authorities and evacuate his team and the Polish soldiers remaining under his command to England, but he did not receive proper help from his French patrons". During the few critical days in France that decided the direction of the evacuation and the future of the team, its members were divided. Then the authority of Langer's emphasis on obedience to orders may have prevailed. In Africa it was Langer who insisted on evacuating his team to Britain, "remorse" was no longer for part of the group, but for the group as a whole.

In his memoirs, Bertrand seems to have distorted the true motivations of the participants in the events. Everything we know about the crew members of the former BS4 precludes them from trying to provide themselves with a bureaucratic alibi. They were patriots in the most traditional sense. If they were carrying out any calculations, they were only concerned with one matter: the fight against the Germans occupying their country. France, an ally in this fight so far, could not or did not want to continue it, and its achievements in this field so far were not impressive. As a result, the soul of the anti-Nazi coalition moved to England, and was followed by the Polish government in exile. The codebreakers recognized that their place was where they could most effectively serve the liberation of their homeland. Bertrand and his subordinates were perfectly aware of the mood in the Polish team and that is why the appearance of British ships near Oran caused a panic reaction. The officers accompanying "Équipe Z" were not sure of the British intentions or of the Poles' reaction to the news that the Royal Navy ships were nearby. They thought it would be safer to evacuate the team from Oran before the news of the events taking place nearby reach its members.

Unaware of the confusion, codebreakers reached Algiers, where, as befits "Sikorski's tourists", they were accommodated in the Touring Club hotel. In the following days, they were gradually being informed of the crisis at Oran, the retaliatory strike by the French on Gibraltar and the severance of diplomatic relations between the Vichy and the British governments. They were aware of the impact of these events on the fate of the entire team. The dilemmas they faced were synthesized in Kazimierz Gaca's diary, who annotated this entry with several question marks and exclamation marks: "Our team…!!!???" By this time, however, the stakeholders themselves no longer had any significant influence over their own destinies. Therefore, the next few weeks went down in their memory mainly as a period of idleness in an exotic environment, learning French language, trips and sunbathing.

At that time Bertrand was trying to tie together the threads broken by the defeat of his country and find himself in the reality created by the capitulation agreement. Those attending the Bon Encontre meeting received support for their plans from Marshal Weygand, Minister of War in the first weeks of Marshal Pétain's cabinet. The implementation of the plans required the setting up of several overt, semi-secret and completely secretive organizations. Terms of the Armistice Agreement provided for the possibility of organizing a residual intelligence and counterintelligence services. The official intelligence service functioned from 25 August 1940 under the command of Lieutenant Colonel d'Alès under the name of the office for combating anti-national activity (*menées antinationales*, MA) and included the Russian and English sections. However, Colonel Rivet added the underground sections MA1 (information acquisition), MA2 (former Bertrand's section) and MA3 (counterintelligence) to the openly operating ones. The counterintelligence service was camouflaged under the cover of the Rural Works Company (*Travaux Ruraux*, TR), operating under the auspices of the Ministry of Agriculture. In the first decade of July, the headquarters of the "enterprise" were installed in Marseille and the field network expanded over time to eight sites, employing more than 400 "agronomists". Bertrand concluded that the structures of Vichy counter-intelligence service offer sufficient facilities for his cryptology center to resume operations. Starting on 8 July he began to convince Colonel Rivet to allocate the necessary resources for the organization of the successor of P.C. Bruno. The achievements of the Polish team to date had to be highly appreciated by Rivet, as the proposal was accepted without deliberations. Just two days later, Bertrand set off in search of a new seat. Having found a villa in a remote and quiet area at the end of July, Bertrand bought it with General Weygand's permission, under the assumed name of Barsac and under the pretext of investing capital in real property in uncertain times.

Obtaining funding and a material base for the planned operation proved to be an easier part of his plan. Bertrand faced a more complex problem when it came to finding the executors. He had extremely modest French personnel, limited to his deputy, Captain Honoré Louis, and chauffeur, Maurice Isambert. He could not count on the French members of the former "Bruno" team because, according to his own words, "they were all demobilized and had no desire to expose themselves again". The Spanish team accepted his offer of cooperation without discussion; for the epigones of the Spanish republic, the world had shrunk immeasurably in 1940. However, Langer insisted so consistently on the evacuation of his team to the UK that Bertrand had to use extraordinary diplomatic maneuvers.

We do not have direct accounts describing the course of negotiations on the further fate of the group of Polish cryptologists. We can only reconstruct the arguments that both sides could hypothetically have used. The team represented the greatest value to those involved in the conflict determined to continue the fight. The de facto surrender of France at least temporarily excluded this country from the anti-Nazi coalition. The Vichy government accepted the directives coming from Berlin, partly because it could not object, partly because it was convinced that such actions were compatible with French interests. Vichy had neither the will nor the strength to take an active stand against the occupant, so the experience and qualifications of the Cipher Bureau's team were of little value to it. The previous order delegating the team to serve under French orders became voided and replaced by General Sikorski's instructions ordering Polish soldiers to advance into British-controlled territory. Langer must have been familiar with the text of the Franco-German Armistice Agreement of 22 June. Potential breaking of the Enigma's ciphers was a violation of several provisions of the agreement at once, so the activity of the Polish team could only be conducted in deep conspiracy. In this light, Bertrand's proposal was an act of recklessness or even perversion unprecedented in the history of cryptology, requiring the location of a codebreaking center in the territory under the actual control of the enemy. The Poles were ready to put their own safety on the line, however, the possible compromise of the planned center would bring about irreparable losses for the Allies. The concept presented by Bertrand signaled that he did not see the secret of the Enigma as the common good of the anti-Nazi coalition any more, thinking of it in the somewhat parish categories of his service's interests, if not his personal ones. If one rejects the option of evacuating the team to Britain, Algiers or any other place in French North Africa was an infinitely better place to continue the mission of the Poles; the local French authorities could spread an effective protective umbrella over the project. In the autumn of 1940, it was difficult to predict that the next episodes of the war would take place in the Mediterranean basin; in Yugoslavia, Greece, Libya and Egypt. It was clear, however, that southern France would at least temporarily remain in the periphery of the conflict. Thus, the availability of the Enigma dispatches would become symbolic.

Bertrand's arguments had to be based on three pillars. The first was tradition; pre-war cooperation, numerous personal meetings between Bertrand and Langer and a period of working together in France formed the basis for the understanding. The second pillar was the organizational and financial issues. Bertrand could have argued that the French government was covering the costs of the Polish team's stay in Africa. The fact that its members were not left to fend for themselves even after the French defeat was due to his loyalty. The third problem that had to emerge in the discussions was the tension in Franco-British relations, complicating the prospect of team's evacuation to Britain. British recognition of de Gaulle's committee set up in London as the representation of the French people constituted an assault on the Vichy government's position. The blood of the sailors shed at Mers el-Kebir allowed anti-British sentiment in French society to reach a boiling point and led to the severance of diplomatic relations between the two countries. Bertrand could credibly argue that the possible evacuation of the Poles to London would have required the cooperation of many people, and no one on the French side would have engaged in a venture benefiting the perfidious English.

In fact, Bertrand did not try to persuade Langer to change his position and negotiated the future fate of the team over his former partner's head. To this end, he contacted the residual Polish diplomatic representation to the Vichy government. It was headed at the time by a modest embassy counselor, Feliks Frankowski. According to Bertrand's memoirs, he presented Langer's position to Frankowski in a slightly distorted form. He omitted to demand that the team be transferred to the UK, signaling that Langer was only seeking a purely bureaucratic approval for continuing his cooperation with the French. Frankowski's position did not allow for a decision to be taken, and the inquiry submitted to London was stuck among the more important issues at the time. Frankowski's mission in Vichy soon came to an end; when the Vichy government announced in October the appointment of its own representative to the Polish government in London, the annoyed Germans demanded the immediate closure of both delegations.

In this situation, decisions were influenced more by events than by rational argumentation. And events were not conducive to even the appearance of cooperation between Vichy and Britain. Between 22 and 25 September the Royal Navy and the forces of the Free French tried to implement operation *Menace*. Its aim was to establish a bridgehead from which the de Gaulle Committee could start the process of taking over Vichy-controlled territories in Africa. A Royal Navy squadron was to suppress the resistance of French ships based in Dakar. The naval operation was to be accompanied by the landing of 8000 Free France troops. De Gaulle had hoped that the Dakar garrison would offer no resistance, but he miscalculated. The duel between the navies ended in a draw. When Free French troops were trying to get ashore, they encountered resistance and a fire from well-established positions. Making virtue of necessity, de Gaulle declared that he would not allow the French to fight against the French, after which on 25 September the expedition set off on its way back to England. The next day, 26 September, members of the Polish team were on board the liner *Lamoricière* sailing from Algiers to Marseille, where they arrived two days later. The coincidence of dates again points to the next impulsive reaction of the French hosts to the British operation off the African coast. The alleged threat appears to have prompted a unilateral decision to move the team to the continent, without waiting for talks to conclude. In the light of this reconstruction of facts, later events become understandable.

On 5 October all members of the team except Langer left Marseille for a place that was to become their home for the next two years. They found their way to a villa located in a small park, known as Château des Fouzes. Its nearby vicinity—the Roman monuments of Nîmes, the Pont du Gard aqueduct, the souvenirs of Avionion's papal glory—all encouraged to forget about the war and turn the thoughts to the past. The nearest town, Uzès, was as if it was transferred from the Middle Ages, the Cathars, the Albigenses and the Crusades. The environment, in which little has changed over the years, was both favorable and unfavorable to the conspirators installing themselves in the villa; any change in its surroundings was visible against a landscape that had remained unchanged for centuries. Still before getting his foreign legion from Africa, Bertrand adapted the building to the new role. This required the installation of radio equipment rescued from the June disaster. In the attic rooms four radio stations have been installed, and their antennas skillfully masked. In the walls of the villa, recesses were prepared, in which documents and equipment could be immediately hidden and bricked up in case of danger. The size of the building did not allow for separate rooms for work and rest. Each of the sections of the center received a room at their disposal, in which its tenants were to live and work. The villa's tenants could meet for meals in three dining rooms arranged in the basement of the villa, separately for each nation, during walks in the park and the surrounding area, or at an aperitif and a billiard party in a makeshift club improvised in the greenhouse. The idyllic estate of Château des Fouzes has developed into an heir to the tradition of P.C. Bruno, a military facility codenamed P.C. Cadix.

While the adaptation of the building went smoothly and quickly, the reconstruction of the team went with some problems. The relations between Bertrand and Langer and, more broadly, between the Polish group and its French hosts at that time contained a shadow, a mystery and, at the same time, a harbinger of future events. It was difficult for Bertrand to persuade Langer to leave Africa, and the doubts, which were not clarified on this occasion, came up again as soon as the Polish team arrived in France. Recalling the discussions that preceded his departure from Algiers, Bertrand writes that "soon afterwards the same doubts returned (…)". The crisis of confidence already felt in Algiers has only been subdued and has already erupted again in France. It has been contained only superficially, on formal grounds. After Frankowski's departure from Vichy General Juliusz Kleeberg remained there, the last Polish military attaché to the French government. Working under the guise of the "General Interpreter", he took care of the fate of demobilized Polish soldiers in the areas administered by Vichy. Less officially, he was the commander of the Polish Army in France and the head of the underground organization dealing with evacuation of Poles to Great Britain. He also took on the role of Bertrand's partner in negotiating the future of the Polish team. No information survives as to what rationale prevailed during the discussions in London. What we do know is that Polish intelligence headquarters forwarded orders to Langer instructing him to remain with his entire team in France. A copy of the order which Gen. Kleeberg gave to Langer has been preserved. Both officers were operating in clandestine conditions that are not conducive to issuing written orders. The custom adopted in most armies in the world, however, allows an officer who disagrees with an order received to request that it be issued in writing. There must have been a profound difference of opinion between Kleeberg and Langer about the point of further work by cryptologists in France. As a result, one or both of them found it necessary to issue a written order. Langer signed in a shaky handwriting, indicative of strong emotions, but this was not the end of his problems. When the next day he communicated the content of the order to the members of the team, he had to meet with some form of opposition. He found himself in a highly uncomfortable situation. He had to discipline the team to carry out an order the sense of which he himself did not accept. He reached for the purely formal mechanisms of discipline. Under his own signature, he added a clause acknowledging the order, to which all team members signed. On the way from Marseille to Uzés, there must have been an overwhelming silence in a group of codebreakers.

Surviving correspondence confirms that opposition to the order never ceased. Langer reported back in February 1941: "(…) I consider it inappropriate for our group to stay here… I did not sign any commitments to the French either personally or on behalf of the group, so we can be transferred to a different work at any time". By giving Langer an order, Kleeberg declared that he was doing so with the knowledge of the Commander-in-Chief in London. To some extent, this is confirmed by the structure that the Polish-French cooperation has adopted since then. The team was subordinated directly to the Polish intelligence headquarters in London, bearing the code name "Ekspozytura 300". London was to manage the operational priorities of the team's work. At the same time, for purely practical reasons, the team was to submit to Bertrand in administrative terms. The operational subordination to London was to be kept secret from the French hosts. One can only speculate on the reasons for implementing such a complex and not very functional structure. Bertrand's motives are most understandable. His position in French intelligence community depended on the competence of the Polish team. He was prepared to go far to ensure the continuation of its service under his own orders. But did the work of the Poles represent value to the organization under which the P.C. Cadix operated? Until the French defeat, Bertrand's section functioned within the structure of French military intelligence, for which the Enigma decrypts certainly represented a significant value. After the defeat, the P.C. Cadix center was placed in the structure of *Travaux Rureaux*, a counter-intelligence organization whose operational priorities were located in a completely different sphere. We will therefore see the activity of codebreakers move away from Enigma and into areas of real use to the Vichy counter-intelligence service.

The motivations of the heads of Polish intelligence service must have been more complex. Chief among these seems to have been the desire to keep the cryptologic team and the officers in charge of it away from London. After two military defeats during one year, emotions boiled over in Polish London and there was a ruthless struggle for influence and power. These disputes were particularly intense within the intelligence community. The codebreakers' team and its commanders represented the greatest success of Polish intelligence before the outbreak of war. Their eventual appearance in London would have made them significant contenders in the political jigsaw and tussle. Keeping the group on the sidelines, away from London, seemed a far better option.

Bertrand got the "cover" he wanted. Whatever the motives for the decision made by the Polish HQ, Langer could not discuss with the received order; the organization of work in the "Cadix" could move on at full speed. As before, the core of the center was made up of two teams of codebreakers; seven Spaniards and fifteen Poles. Poles still in Algiers were equipped with new identities and documents adapted to the reality of underground existence in France. Marian Rejewski transformed himself into Pierre Raneau, a math teacher from a high school in Nantes. In creating the new identity for Antoni Palluth, no imagination was shown. He was called "Black" by his colleagues and became Jean Lenoir in his French edition. The other members of the team were also given new identities, constructed in such a way as to justify their poor knowledge of French. To complicate matters further, everyone was also given a pseudonym under which they were registered in the records of the Polish intelligence service in London. And so Rejewski became "Oksza", Różycki "Rola" and Zygalski "Bemol".

Under clandestine conditions, it was not possible to count on the cooperation of services that had previously provided, for example, radio interception and communications with headquarters. Over time, Bertrand was able to secure a supply of captured ciphertexts to some extent, but at least in the early days the center had to rely on its own resources. Moreover, the dual subordination of the Polish team meant that abundant communication with London had to be maintained. It was necessary to train in the role of radio operator team members who have hitherto fulfilled engineering and technical functions. Correspondence with London had to be secured, so other members of the team became cipher clerks out of necessity. Considering that the most important task of the team was to attack German ciphers, it was natural that Major Ciężki took over the duties of Langer's Deputy. During his periodic absence from P.C. Cadix, this role was taken over by Major Michałowski. German cipher section represented the most numerous group. Apart from Rejewski, Różycki and Zygalski, it also included Major Michałowski (who specialized in transposition ciphers), Antoni Palluth (who dealt with ciphers used by German agents) and Kazimierz Gaca (who focused on ciphers used by German Armistice commissions). The Soviet Cipher Section was composed of Stanisław Szachno and Jan Graliński. Two of the four radio sets were assigned to maintain center's own communications, the other two were used to intercept enemy signals. Edward Fokczyński took care of the maintenance and repair of the equipment. Unfortunately, little is known about the structure, tasks and results of the Spanish team. The available reports do not contain information on the results of its work, although in his post-war book Bertrand precisely enumerated the German dispatches broken by the Poles. The operation of the center has been organized in such a way as to keep contact between members of the two groups to a minimum. As a result, the accounts of the Poles contain virtually no references to the results achieved by their Spanish colleagues.

With the exception of Bertrand and his wife, the French played purely auxiliary roles in P.C. Cadix. The maintenance of the property was handled by certain David, whose wife prepared meals for the residents. Bertrand's driver, Maurice Isambert, described by him usually as *mon fidèle Maurice,* also lived on the estate with his wife and their daughter. It seems that his task was primarily to care for the morale of foreign teams, which both Bertrand and Maurice understood as ensuring an abundance of alcohol. Maurice reigned in the bar improvised in the greenhouse, where he prepared aperitifs for the codebreakers. Bertrand himself was a rare visitor in the early days of the center; he appeared once a week to provide intercepted materials and receive decrypts. He lived with his wife in a villa rented in nearby Nîmes, which was, moreover, the official headquarters of the center in the documentation of the

French secret service.[1] In Château des Fouzes, however, a room was reserved for him, in which he became a more frequent guest from 1942 onwards. Theoretically, Bertrand's permanent representative in the Cadix was Captain Louis, acting as an adjutant to the center commander. Bertrand, however, did not value his talents very highly, so he tended to entrust the role of deputy to his wife in his absence. The list of French employed at the center was to include a total of 10 names. One of the missing names was probably an old acquaintance of the entire team, Air Force reserve captain Henri Braquenié. If that had been the case, Braquenié would have been the only French codebreaker employed by the P.C. Cadix. He would have been, because Bertrand himself does not mention him in his memories of that period of work, while Rejewski merely notes in passing that Braquenié lived with the Bertrands in Nîmes. The last missing person was probably the somewhat mysterious Françoise B., a secretary in the mayor's office of Uzès, who assisted Bertrand in organizing the stay of the entire group (in particular by providing food ration cards), and later was to play an important role by participating in hiding the equipment and documentation of the center.

The origins of P.C. Cadix's work must have been a contrast to the earlier functioning of P.C. Bruno. At the time, especially after the German attack on France, cryptologists were already trying to cope with the avalanche of intercepted dispatches. In the first weeks and months of the new center's operation, virtually no ciphered messages reached it that could be attacked. Improvised center's own listening service was not a solution either; operators had to be trained and German communications networks worthy of attention had to be identified. For weeks, the codebreakers were virtually unemployed for lack of material. Bertrand attempted to counteract the decline in morale in two ways. The first belonged to a classic of the functioning of any bureaucracy—the drafting of reports. He demanded a written report on the operations to date with Enigma in the background, which would capture the contributions made by each of the three countries involved. The report produced by Polish cryptologists in late 1940 and early 1941 did not win his approval. In a conversation with Langer, he stated that he would have to rework the document, as its current form shows that France had not made a significant contribution to breaking Enigma. Arguably, he did not manage to fulfil the announcement. The report was preserved in Bertrand's private archive and was declassified by the French secret service in 2015, adding some new elements to the knowledge of the history of breaking the cipher. The story of the second way to keep cryptologists busy belongs in the next chapter.

---

[1] Villa at 14 Rue de la Lampèze.

Chapter 15
Relatives Are Coming Forward

The first successes of Turing's bombe came at a special moment. With the end of French campaign, the number of messages intercepted by British signals intelligence significantly decreased. German troops installing themselves in France have quickly switched to wire communications. After the Battle of the Beams came to an end functioning of Hut 6 has taken on a somewhat routine character. The Herivel method and the cillies still worked well, allowing to break the Luftwaffe keys within a few hours. The codebreakers were prepared for the elimination of errors both methods were based on; bombes were waiting on standby. However, the proficiency in breaking German messages achieved by BP in the second half of 1940 was becoming less and less applicable. Contrary to the scarce material provided by Wehrmacht and Luftwaffe, British listening stations were intercepting a growing number of Kriegsmarine signals, as German ships installed themselves at bases on French and Norwegian coasts. Their abundance has become more of a source of frustration than satisfaction; at that time Hut 8 has not yet mastered the subtleties of the Kriegsmarine Enigma. At the turn of 1940–1941 most frontlines were either closed or have been stagnating. In this situation the secondary until then theatres of war were gaining importance, and it turned out that the military Enigma had a bunch of relatives. They remained in the shadow of the more famous sister, but when her dominant tone was missing, their less resonant volume could be heard.

Shortly after the defeat of France, between the 15th and 25th of July, British listeners intercepted the messages exchanged by stations located in Germany, France and the Benelux countries. The messages were of a different nature to those previously known, making it possible to identify a hitherto unknown communication network. The new network served the German railways, one of the main factors of German victories. Helmut von Moltke the Elder once formulated a thesis that a battle in which soldiers and supplies have to travel more than a hundred kilometers from the final railway station, is lost by definition. During World War Two the truck allowed to increase this distance to about three hundred kilometers, but the railway still carried over 90% of the soldiers, equipment and supplies of the German army.

© The Editor(s) (if applicable) and The Author(s), under exclusive license to
Springer Nature Switzerland AG 2025
M. Grajek, *Enigma Myth Deciphered*, History of Information Security,
https://doi.org/10.1007/978-3-031-65475-6_15

As a result of the defeat of France, all the distance limits between and the new Wehrmacht garrisons and their logistics bases have been exceeded. Deutsche Reichsbahn (DRB) had to undertake a gigantic task of transporting the troops to their new garrisons all over France, deliver to the coasts of the English Channel the units involved in the preparations for the Operation Sea Lion, finally supply the Luftwaffe units participating in the air Battle of Britain and the following Blitz campaign. Its managers needed a reliable means of communication to coordinate the activities of the new logistics centers located in the Benelux and the occupied part of France. It was no surprise that one part of the solution turned out to be Enigma.

DRB used a commercial Enigma model, which has undergone a standard modification, i.e., rewiring of rotors. The problem faced by BP codebreakers seemed to resemble the one solved by Knox during the Spanish Civil War. Unknown rotor connections had to be reconstructed using the *buttoning up*, which would allow *rodding* of the message content. However, the starting point for both methods was the reliable crib. Later successes in breaking the railway Enigma's proved that the content of her messages differed from everything that codebreakers knew from their previous experience. The messages contained long series of numbers defining the train schedules: departure and arrival times, station codes, number of wagons and their load. The non-stereotyped content of the messages and the consequent lack of effective cribs made the railway Enigma safe, despite the use of a simple, commercial machine model. BP owed the breaking of the railway network to the elementary error of German operators. The letter of 17 August, 1940 mentions that a few days earlier John Tiltman managed to break the railway Enigma setting in depth nine messages enciphered with the same key. When BP realized that the network served the German railway, it was given the name *Rocket*, referring to Stephenson's first locomotive. After the first break, Tiltman noticed that the cipher characteristics corresponded to the commercial Enigma, with rotor rewired and asked Turing for help in recovering their connections. Turing offered the help of his assistant, Joan Clarke. Armed with Turing's manual, known as "Prof's Book", and description of Knox's classic methods, Joan Clarke was able to recover the rotors' wiring. Despite her success, the codebreakers still had problems interpreting the clear texts of broken messages. Their interpretation became possible only after the employees of British railways were invited to cooperate. But the luck didn't last long: on 27 August, last messages were recorded in the *Rocket* network; Deutsche Reichsbahn had switched to wire communication. When it seemed that Tiltman's and Clarke's achievements would only play an episodic role, on 23 January, 1941, British listening stations intercepted the messages broadcast with a similar cipher, but this time transmitted by stations located in Eastern Europe. The experience from the previous break permitted to break the new key already on 7 February.

The decisions leading to BP's renewed contact with *Rocket* were made in the summer of 1940. On 21 July, Hitler announced to his army commanders gathered in Obersalzberg his intention to attack Soviet Union. At that time, plans for Operation Sea Lion and attack in the east were discussed in parallel. When the fate of Sea Lion was sealed at the end of October, Churchill predicted that Germany would turn to the Caspian Sea and after the Baku oilfields. His prophecy was based more on brilliant intuition than informed analysis. Circumstances and available information indicated rather Middle East as the next German target. Both Wehrmacht and Kriegsmarine commanders tried to persuade a strike in this direction; nevertheless, on 6 December Hitler ordered General Warlimont to develop a plan of attack on the Soviet Union. The outline of the plan soon submitted envisaged the use of three army groups. The northern one was to advance towards Leningrad, the middle one towards Smolensk and Moscow, and the southern towards Kiev. Vast expanses of the theatre of operations indicated that the campaign would represent a logistical nightmare, therefore the representatives of Deutsche Reichsbahn were included in the planning process from the very beginning. In the first half of 1941, three field railway directorates (Feldeisenbahndirektionen, FBD) were established, each for every of the army groups. The messages intercepted by the British stations in January probably came from the FBD1, earliest to be organized. The directorate originally created to serve the Army Group North, was transferred suddenly to the Balkans, where since October 1940 Italians were attacking Greece. The Greek counter-attacks pushed Italian troops into Albania, and the situation of the battered ally required urgent support from the Third Reich.

It seemed that the preparations for the German intervention in the Balkans did not interfere with the plans of attack against Soviet Russia. FBD1 was immediately replaced by FBD4 organized in Danzig; FBD2 established in Dresden was shifted closer to the future front line, to Warsaw, where it took over the premises after FBD3; FDB3 was meanwhile moved to Krakow. Operation "Barbarossa" was to involve nearly 150 divisions. The transport of one infantry division required about 70 trains, the armored division required more than 100. The transport of all units to the Soviet border required more than 12,000 trains, and transport of the supplies required approximately the same number. The number of trains passing mostly through the occupied Polish territory during that period was to reach 33,000. Four field railway directorates were commanding nearly 300 trains every day, exchanging information via the *Rocket* network. Its broken messages provided BP with a wealth of extremely precise information about Germany's strategic intentions. In retrospect, it is difficult to resist the conclusion that this mass of information had been wasted, some of them having led Churchill to false strategic decisions.

As early as World War I, Churchill saw the Europe's soft underbelly in the Balkans. After the Italians had been defeated in North Africa and Abyssinia, British Prime Minister sought an opportunity to return to the continent through the Balkans. Churchill was probably alone planning a campaign in this part of the world. The king and government of Yugoslavia tried to ensure the security of their country by declaring an alliance with the Axis countries. The Greeks were satisfied with pushing the Italians away from their borders. British commanders in Africa were busy planning operations to rout the Italian army from the coasts of North Africa; they needed every soldier and every tank to ensure their success. The German commanders were so focused on the attack on Soviet Union that every disturbance in the Balkans represented the distraction from the main goal. Even Mussolini seemed to treat the Greek adventure as an *hors d'oeuvre* to his involvement in Africa. And yet, it was Churchill who managed to impose his will. The British SIS had arranged a military coup in Yugoslavia; the new government had abandoned its alliance with the Axis. The Greeks, though reluctantly, finally had accepted British military assistance. Knowing from the broken railway Enigma messages the scale of German threat to Yugoslavia and Greece, British PM pushed through the transfer from Egypt to Europe of an armored brigade and the experienced Australian and New Zealand infantry divisions. These forces could not influence the course of the upcoming campaign, but their lack was felt Libya, where the British, instead of pushing the enemy out of the last ports on the African coast, had to stop their depleted troops.

Railway Enigma permitted the British watching closely the concentration of German forces at the Soviet Union's borders. Churchill did not mention in his memoirs the source of his information, but knowing the context leaves no doubt as to its origin: "(…) I learned at the end of March 1941 from one of the most reliable intelligence sources that large movements of German forces were observed on railway lines from Bucharest to Cracow". On 3 April, 1941 the British Prime Minister decided to warn Stalin of the imminent threat. To this end, he ordered Stafford Cripps, the British Ambassador to Moscow, to convey to Stalin, in person only, the following message: "I know from absolutely certain sources that when the Germans thought they had Yugoslavia in their grasp (…) they ordered three of the five armored divisions to leave Romania for south-eastern Poland. (…) I ask His Excellency to assess the importance of these facts himself". The warning was not accepted: Stalin did not trust anyone. He ignored Churchill's message, depriving the railway Enigma of one of the greatest opportunities to play a historic role.

The day of railway Enigma's unfulfilled glory passed unnoticed, but the history has reserved for this machine some episodes in subsequent acts. The victories of summer and autumn 1941 scattered German troops all over the vast areas of Russia. Practical lack of wire network there forced the field railway directorates to rely mostly on the wireless communication, offering the BP insight into the backyard of the eastern front. Information from this source was only incidentally useful to British commanders, especially when it concerned the transfer of units from the eastern front to Italy or Africa. However, the vast majority of messages referred to the operations on the eastern front and would probably be useful for the Soviets, should they cooperate with their allies in intelligence matters. On some occasions information about German logistics found its way from BP to the Russian HQ; in some cases, at the will of the British, in others without or even against it, due to the activity of the Soviet spy in BP, John Cairncross. The end of the *Rocket* network came in the autumn of 1944, when the front moved west enough for the Germans to revert to the wire communication.

Earlier, however, in September 1942, a second railway network appeared, this time covering centers in Western Europe: Paris, Bordeaux, Rouen, Toulouse, Liège and Brussels. This prompted BP to rename the existing *Rocket* network to *Rocket I*; the new one becoming *Rocket II*. Earlier experience with the Reichsbahn networks allowed the new key to be broken still in the same month. Broken *Rocket II* dispatches did not bring interesting information; stations mostly exchanged training messages. In May 1944, in the final weeks before the planned landings in Normandy, the intercept stations identified still another railway network, named *Rocket III*. The occurrence of numerous multi-part messages and messages marked as urgent indicated its fully operational nature, but its keys initially did not yield to an attack. In the period immediately prior to the invasion, any information regarding German military transport was of critical importance to the Allies, so more resources were allocated to attacks on *Rocket III*. Some messages were forwarded to American codebreakers, who had a specialist device at their disposal permitting to determine whether the network was using the railway Enigma model. The negative test results did not reach BP until the end of August, and only a few days after a German deserter handed over the cipher keys to the British with the information that *Rocket II* and *Rocket III* were now both using the standard military Enigma. The railway Enigma was used in the *Rocket II* network only during the first month of its operation, after which it had been replaced by a military model. The further history of the *Rocket III* network confirmed that Enigma properly used represented a difficult problem for the codebreakers. Reading its messages for several days of August BP realized that the unusual structure of the messages did not provide a good crib for bombes. *Rocket III* was broken once again in October, when a message was identified, the open text of which was known from another broken network.[1]

Another adventure with the Enigma variant had more spectacular effects. GC&CS' first success with the machine concerned the commercial Enigma model used by the Italian Navy. When, on 10 June, 1940, Italy declared war on France and Great Britain, the messages appeared again on the air enciphered using the system known for several years. Dilly Knox took up his old challenge: while staying at the Cottage, accompanied by a few girls assigned to help, he started to break the Italian messages. A complication in his work was the indecision of the Italian navy in choosing the main cryptographic system. Even before the war, the British broke the super-enciphered codes of the Italian Supermarina, but a change in the superencipherment tables in the first days of the war cut BP off from this source of information. To make matters worse, the Italian codes proved to be resistant to attacks until the country capitulated in 1943, except for a few days in the summer of 1941, when the code was broken on the basis of documents captured in Africa. For some time, the second ciphering system used by the Italian navy was Enigma. The Italians had few copies of the machine, installed in high-level staffs. They were only used to communicate with the Navy Command, which meant that the number of intercepted messages was negligible; often only one or two a day. When Dilly started the attack on the Supermarina ciphers he didn't have to look for new solutions: the old *rodding* technique was still working well. The scant number of messages meant that the first break into the cipher was not recorded until September 1940, but since then Supermarina's dispatches have been read in the Cottage quite regularly. However, at

---

[1] In the autumn of 1944, another railway network appeared, baptized as '*Stevenson*'. Its functioning seems to prove that the Germans were already experiencing serious supply problems; the '*Stevenson*' network used the military Enigma model, but the choice of rotors was limited to 3. Over time, '*Rocket II*' was renamed '*Blunderbuss*' and '*Stevenson*' was renamed '*Culverin*'. However, these were purely administrative changes that did not change the fact that the decryption of the railway communication networks no longer played a significant role in the Allied war plans.

the turn of 1940/41, the Italians changed the rotor wiring. Mavis Lever's perceptiveness proved to be a key moment in breaking the ensuing crisis. Among the intercepted messages, she noticed a long cryptogram in which the letter L was not present. Bearing in mind that Enigma does not encrypt the letter as itself she assumed that the lazy Italian cipher clerk decided to send a training message by pressing the L several dozen times. She was right, and her insight allowed her to identify new rotor connections. Despite the successes of Knox and his team, the Admiralty was skeptical about the information they were gathering. The very use of a somewhat archaic machine was a controversial operation. Weighing down the modest number of messages, the Admiralty concluded that the Enigma could be used by Supermarina to disinform the opponents, as Nobby Clarke noted on 24 March 1941. Just a few days later Their Lordships had to reject that opinion and give Dilly full satisfaction. In March a series of messages revealed that the Italians were going to take the fleet out to sea to intercept British convoys transporting troops from Egypt to Greece. The warned Admiral Andrew Cunningham, commander of the British squadron in Alexandria, kept his ships in port until the enemy's intentions were fully explained. On 25 March, Supermarina sent a message to the Rhodes garrison stating: OGGI 25 MARZO E GIORNO X-3 (*Today 25 March is day X-3*), which indicated the beginning of the Italian operation on 28 March. Dilly's broken message got into Admiral Cunnigham's hands in time to become the source of victory in the naval battle off Cape Matapan. The Italians lost three heavy cruisers and two destroyers, and with them, the naval initiative in the Mediterranean.

But Dilly's satisfaction was only temporary. The Italian command used Enigma less and less frequently, gradually replacing it with Hagelin C-38m machines, and finally, in the summer of 1941, the Italian Enigma messages disappeared from the ether. Fortunately for BP at this point the number of messages in the new system made it possible to launch a systematic attack. In April 1941 the problem of ciphers of the C-38 m machine was transferred from the section of Italian naval ciphers to the research section of *Hut 8*, and already in the second half of June the cipher was broken. The work on the Supermarina codes was divided between two teams. The day keys were reconstructed at *Hut 8* by a team led by Colin Thompson. The individual message keys were recovered at *Hut 4* by by an almost all-female team working under the leadership of Professor R.E. Vincent. From the name of the machine whose ciphers they were attacking, the girls earned the entirely undeserved but very Shakespearean nickname the *Hags*. The C-38m ciphers were broken just in time for BP to support the British forces fighting, with varying degrees of luck, in North Africa. A measure of BP's success in this sphere was the percentage of Axis military supplies sunk *en route* from Italian ports to Africa. If in June only 4% of the cargo was sunk, in November it was already an impressive 62%.

Before we can finally say goodbye to the commercial model of Enigma, we have to return for a while to Château de Fouzes, where we left the Polish codebreakers when Bertrand delivered a package of Swiss cryptograms in an attempt to keep them busy. Some of them turned out to be a simple commercial code, which was immediately broken by the Poles, but the content of the messages did not arouse anyone's interest. The structure of the remaining dispatches indicated encryption using the machine, but the somewhat naive way in which the Swiss used it did not even require the use of the instrumentation developed by the Polish team for Enigma. Analysis of the message headers indicated that all radiograms on a given day were encrypted with the same machine settings. If such a hypothesis was correct, all the first letters of the messages intercepted during the day represented a monoalphabetic substitution, the second letters another monoalphabetic substitution, and so on. Methods for breaking monoalphabetic substitutions were first described by the Arabic scholar Al-Kindi in the ninth century, so the messages should not have caused trouble for Polish codebreakers. However, attempts at interpreting the character frequency tables that were produced did not yield meaningful results. When the package of dispatches from one day finally brought a solution, the Poles understood the source of the problems. The open texts of the attacked dispatches represented a mixture of the official languages of the Helvetic Republic; French, Italian and German, so the statistical features of the ciphertexts were distorted. Already in the first broken messages Poles noticed a well-known reciprocity of cipher; if the letter A was encrypted as X, in the same position X was encrypted as A. This relationship indicated the use of the Enigma variant and encouraged codebreakers to recover the rotor wiring. The old methods passed the test and allowed for a quick and complete reconstruction of the machine—it turned out to be a commercial Enigma with rotors rewired.

The contents of the dispatches were of no particular interest, so the cryptologists considered the task completed and Langer passed on information to London about the weakness of the Swiss cipher, perhaps with a suggestion to warn the Swiss. We will return to the appalling consequences of the warning given. For now, let's just note that indirect indications are that the warning has reached its destination; the Swiss began enciphering each message with an individual key, appended at the beginning and end of each message. This did not make the cipher materially more secure, but the Swiss apparently lost faith in the security of Enigma and proceeded to work on another way of securing their military communications. In 1943 they offered a contract to the local company Zellweger from Uster to design and build their own ciphering machine to replace the Enigma. The machine was designed by a team consisting of Hugo Hadwiger, professor of mathematics at the University of Bern, engineer Heinrich Weber and Hadwiger's student Paul Glur. The device, baptized simply Nema (from *neue Maschine*—the new machine) only came into use after the war, in 1947, and was used by the Swiss army until the late 1970s.

A little later, Polish cryptologists achieved yet another success in their struggle with the ciphers of the rotor machines, although it was a success of a somewhat paradoxical nature. Communications between Ekspozytura 300 and London were enciphered using the Polish cipher machine LaCiDa, which finally saw operational use. Ten years passed between the machine's design and its launch. During this period, the theory and practice of attacks on machine ciphers was revolutionized, mainly by the hands of Polish cryptologists. It was somewhat surprising that, having a team of the world's best specialists in attacks on machine ciphers, Polish Cipher Bureau did not commission them to examine the security of its own machine. Under these conditions, it was understandable that the operational launch of the device was not without its problems. In the first days of July 1941, Rejewski, having heard that LaCiDa was used to encrypt exchange with London, asked Langer, out of professional curiosity, to provide him with an example message. Having received one of the current dispatches, he sat down to work on it, accompanied by Zygalski, and after a few hours' work presented the open text of the message to the surprised commander. LaCiDa's design is known only in outline. The information available suggests that it represented a level of security situating it between the commercial and military models of Enigma. Nor do we know the nature of the two codebreakers' attack on its cipher. In his memories Rejewski admits that during his conversations with Antoni Palluth, he obtained shreds of information about some features of the machine's construction. In particular, he was aware of the different number of contacts in successive rotors and identified this feature of the cipher as the basis of his attack. However, breaking the cipher in a matter of hours, based on the single ciphertext, was a spectacular demonstration of their competence and experience. Langer and Ciężki must have been shocked. The dispatches exchanged with London contained extremely sensitive information. Their breach by an adversary could in particular have revealed the success of the Poles with Enigma. However, Ekspozytura 300 had no other encryption method than LaCiDa. After a much-delayed consultation with codebreakers, it was agreed with London that the messages would be first enciphered using hand cipher, and then re-enciphered with LaCiDa. The system proved successful; neither in the captured German archives nor during the interrogation of German signals staff were there any traces of a breach of the P.C. Cadix traffic.

To conclude our review of Enigma's relatives and the role they played in the wartime struggle, we must return to BP and Knox. When the Italian Enigma messages disappeared from the ether in summer 1941, another challenge was already waiting for him. For some time now, British listeners have been intercepting cryptograms of a heterogeneous nature and broadcast from various locations in Europe. The structure of some suggested the use of hand ciphers, while others appeared to be machine-encrypted. The location of the stations and the structure of the information exchange indicated that particular radiograms came from German intelligence service, the Abwehr. The network's main outposts used machine ciphers, while agents working underground used hand ciphers. The two types of ciphers required the use of completely different methods of attack, so two sections were created in BP and entrusted with breaking Abwehr's dispatches. Hand ciphers became the domain of a section headed by Oliver Strachey and named after him—ISOS (*Intelligence Services Oliver Strachey*, although the abbreviation has also sometimes been developed as *illicit services*). Dilly Knox headed the section attacking the machine cipher, so it became known simply as ISK (*Intelligence Services Knox*). Abwehr cryptologists remained faithful to traditions dating back to the First World War and preferred transposition ciphers. Breaking them was labor- and time-consuming, but they did not pose a significant challenge to British codebreakers.

Looking slightly ahead, let us note that the Abwehr used a rather unusual model of Enigma. Of course, the wiring of rotors in the Abwehr machine was different from other Enigma models. It lacked a plugboard, but had a completely different mechanism for advancing the rotors than the military model. Unlike the military model, the Abwehr's Enigma rotors moved very erratically during encryption. Each rotor had a gear wheel with a different number of teeth for each rotor. The three rotors contained 11, 15 and 17 teeth respectively (which gave rise to one of its names; *Enigma 11-15-17*). The machine's set of gears resembled a clockwork in appearance, so the model of the machine became also known by the common name *Zählwerks-Enigma* (clockwork Enigma). In the opinion of most cryptologists, the Abwehr's Enigma offered a lower level of cipher security than the military model. In this context, its use by the intelligence service, whose outposts were deployed in neutral countries, among others, may have seemed a controversial decision. However, there is some justification for the Abwehr's choice. The same model of machine was used not only by intelligence, but also (until 1943) by German military attachés throughout the world. The *Zählwerks-Enigma* appears to have been designed specifically for service abroad. It was being used in conditions considered safe enough to allow use of the ciphering machine. At the same time, the very location of these outposts offered adversaries the chance to pay a discreet visit to a German facility and learn about the machine's construction. The use of a standard military model at the attaché's premises would in that case lead to the disclosure of its design and use. Using a machine with a different design allowed the secret to be protected even in such a case.

When Knox launched his attack on the cipher in August 1941, he had no indication other than the structure of the ciphertexts. The messages were broadcast in groups of 5 letters, but at the beginning of each message there was a group of 8 characters. Previous contacts with Enigma ciphers allowed the assumption that this was a four-letter message key, encrypted twice. The structure of the key led to the assumption that, in addition to three moving rotors, the machine also had a moving reflector. Dilly was determined not to stumble this time at the first hurdle—the wiring of the entry drum. The construction of machines broken so far dictated two variants; alphabetical and based on the QWERTZ keyboard layout; Dilly bet on the latter and won. Knox's additional asset was the agents' dispatches broken by ISOS. Experience permitted to assume that some agents' reports originally encrypted with a hand cipher would be re-encrypted with Enigma at one of the regional Abwehr centers and sent to headquarters in that form. Identification of such a pair of messages provided a 100% effective crib. When, armed with this knowledge, Knox applied the methods that had proved successful against the commercial Enigma, he was disappointed. Nowadays, we know that the obstacle proved to be the irregular movement of the rotors, something Knox was unaware of at this stage of his work. Although, had GC&CS paid the necessary attention to the evolution of German cryptographic systems between the wars, Knox should have known the source of his problem. The Scherbius's company was offering the *Zählwerks-Enigma* machine to the British market already in 1926. The commercial offers from this period contained enough information to reconstruct the functioning of the machine.

When the buttoning up didn't work, Knox decided to follow in Rejewski's footsteps and launch an attack on the double-encrypted key. However, the cycles, laboriously reconstructed from the letters of the indicators, did not bring progress. At this stage Dilly was already aware that the key to success is to take into account the irregular movement of the rotors. He identified a situation where all four rotors moved simultaneously to the next position, then repeated this movement four characters later. He described such a situation as *crab*, presumably by analogy with the movement of the crustacean's legs. However, on reflection, he concluded that a less complex scenario—the simultaneous movement of just four rotors, which he termed a *lobster*—would be sufficient to break the cipher. ISK staff started a big *lobster* hunt.

The essence of Knox's idea was to combine his own methods of Enigma analysis (*rodding* and *buttoning-up*) with Rejewski's approach based on a cyclic cipher structure. The four-letter message key invited cipher clerks to use predictable phrases, from repeates **AAAA** characters, through sequences of consecutive keys (**QWER**), to four-letter German words—names and their diminutives, curses, etc. The *buttoning up* required that during the encryption of the probable word, no movement of the central and right rotor of the machine took place. This assumption was met in 21 out of 26 cases in the services Enigma, but the behavior of the Abwehr machine was unpredictable. The cycles borrowed from Rejewski made it possible to confirm whether a *lobster* occurred in a given case, and its occurrence meant that between two machine positions the relative position of all the rotors remained unchanged and buttoning-up could be used. Further work on Abwehr Enigma required time and effort, but was rather routine in nature. One of Abwehr's veterans later admitted that communication discipline in his service was rather weak and that there was no section responsible for communication security. The mistakes of the German cipher clerks, Knox's invention and the patience of his female staff permitted Mavis Lever and Margareth Rock to break a first message encrypted with Abwehr Enigma on 8 December 1941.

Comments of several codebreakers who had the opportunity to cooperate with Knox suggest that something in Dilly's personality made it difficult for him to cooperate with people he perceived as equal in experience or knowledge. The mathematicians and physicists who came under his wing in the early days of the war quickly emancipated themselves and took on independent roles. Knox remained at the head of a purely female section whose pillars were Mavis Lever and Margaret Rock. Dilly was jokingly referring to their role in his team by saying: "Give me the Rock and the Lever and I will move the Earth.[2]" Amazingly, but among his *girls*, he found not only a balance of spirit, but also character traits that determined his past successes—patience, perseverance and perceptiveness. The adventure with Abwehr Enigma was Knox's last venture, not only as a codebreaker, but in general. The cancer was consuming his strength on a par with the stress of the cipher challenge.

---

[2] A travesty of Archimedes' saying, '*Give me a point of support and a lever and I'll move the Earth.*'

After yet another surgery, he spent most of his time at home, continuing to work on his last challenge. Together with Tiltman and Foss, they created an elite caste of old-fashioned codebreakers who, without any technical or mathematical background, managed to find themselves in a new era of machine cryptography. Knox passed away on 27 February 1943.

Chapter 16
Ocean Hunting

When just before the outbreak of World War I, the distinguished reformer of the Royal Navy, Admiral John Fisher, asked Churchill, then acting as the First Lord of Admiralty, about the consequences of Germany's unrestricted use of submarines, Churchill replied that a civilized country would not dare such an action. Churchill should have probably recalled the answer to the same question given already in 1882 by the then French Navy Minister, Admiral Hyacinthe Théophile Aube. He refused to consider any legal restrictions if they did not correspond to the nature of weapons he considered useful: "War represents a denial of law… Everything is allowed and therefore legal, if it is directed against the enemy". As a result of Gallipoli fiasco Churchill had no direct contact with the Navy when in February 1917 German U-Boots started an unrestricted submarine campaign, strictly according to the French admiral's prescription. However, in the beginning of another world conflict, both Churchill and his earlier enemy, the German U-Boot, returned to their former positions.

It took a German U-Boot navy a long time before it was able to implement in full scale the wolf pack tactics developed by Admiral Dönitz. The premiere was not spectacular—the operation of six U-boats between 10 and 19 October 1939 ended with the loss of three submarines. It resembled its animal original in that its operations were commanded by the captain of one of the participating ships, Werner Hartmann. But Hartmann, having chased two neutral ships (which he sank), lost contact with the rest of the pack. The high losses during this operation confirmed Dönitz's intuition that the only practical way to control a submarine pack was to command them directly from the headquarters, from where the course of the operation can be seen better than from a submarine conning tower flooded by the ocean waves. The winter storms, then the mobilization of the Kriegsmarine for Norwegian operation, and the subsequent relocation of ships to the French ports have delayed the next act until the late summer of 1940. But between 30 August and 9 September five U-boats commanded from the land attacked the SC2 convoy, sinking five ships without any loss of their own. Having confirmed the effectiveness of the central

command, Dönitz had to solve yet another problem, resulting from a change of
U-Boot operational area. So far, paralyzing the Allied communications between the
British Isles and the continent had been the main objective of the U-Boot fleet.
Victory in France and the lost Battle of Britain had changed the objectives of the
submarine war. U-boots would take on an independent strategic role—bringing the
British Empire to its knees by cutting off supplies from the dominions and the
United States.

Their new role required the expansion of the operational area. Until now, the
Atlantic voyage required long and difficult navigation through the stormy waters of
the North Sea, full of minefields and Royal Navy ships. From the ports of Brest, St
Nazaire, Lorient, La Pallice and Bordeaux convenient trails led to the convoy routes
in Atlantic. The first U-Boot entered Lorient on 6 July, the Befehlshaber der U-Boote
(BdU) installed himself initially in Paris, moving his headquarters in November to
Château Kerillon in Kernéval, near the Lorient submarine base. Bringing the staff
closer to the U-Boot bases improved communications and crew morale, but did not
contribute to solving the problem that Dönitz considered to be fundamental—intelligence regarding Allied convoys. In narrow waters where his ships have operated
so far, spotting of a potential target by a patrolling submarine was the rule rather
than a surprise. The situation was changing in the open ocean. In good weather, the
visibility from the submarine conning tower reached ten miles. A minor difference
of the U-Boot's and convoy's courses was enough for them to pass each other unnoticed. Air reconnaissance, the most logical solution to Dönitz's problem, was hardly
available. An observer in an aircraft flying kilometers above the ocean could spot
the convoy from the distance of several tens of miles, and aircraft's range permitted
to explore an area of tens of thousands of square miles during one patrol. However,
the German naval aviation was in its infancy, and the turf wars between the services
hindered its development. Göring considered air to be his exclusive domain, and
taking advantage of his position at Hitler's side, he prevented in mid-1930s the creation of the Kriegsmarine's air arm arguing that the Luftwaffe would take also care
of the sea reconnaissance. In the reality only 15 of the planned 62 squadrons of
naval aviation were ready at the outbreak of the war, and their aircraft were mostly
unfit for the long-range operations.

Unable to count on the Luftwaffe, Dönitz returned to the ideas he was developing
in early 1930s, when he was acting as the U-Boot fleet commander without the
U-Boots. He was simulating the operations of the submarine wolf pack substituting
the missing U-Boots with the torpedo boats. His ships were forming the patrol lines,
capable of watching a wide stretch of the ocean in search for the enemy convoys.
The first dress rehearsal of this tactic during the war took place rather late and was
not very successful. When in early March 1941 the German airplane spotted in the
Atlantic the OB 292 convoy, Dönitz assembled across its predicted route a patrol
line consisting of six German and two Italian submarines, but none of them managed to spot the convoy. Later into the war patrol line tactics was gradually perfected and proved to be very efficient. Spotting the convoy by one of the patrolling
U-Boots was initiating a flurry of the wireless traffic. The task of the U-Boot having
spotted the convoy was providing the distant HQ with information as to the place of

meeting, convoy's composition, course and speed. On this basis, the HQ was issuing a series of orders, instructing the remaining ships of the wolf pack to concentrate for attack. The first spotter was supposed to keep in touch with the convoy, reporting by radio possible changes of course taken by the target.

This tactic had a few weaknesses, too. The Allied direction-finding stations were usually able to intercept the initial signal and calculate the approximate sender's position. Comparing it with the known convoys' routes the Admiralty could identify the spotted convoy ordering it to change its course trying to break away from the enemy. The escort ships, notified about the company, could start a counterattack forcing the German ship to go deep and causing her to lose contact with the convoy. Frequently, the loss of contact was simply due to bad weather, and limited visibility. However, if the convoy was discovered by the patrol line, did not change course or the spotter managed to keep its track, only a good deal of luck could save it from losses. Little could be done to eliminate the greatest weakness of the wolf pack tactics—the lively exchange of wireless messages. Transmission from the HQ to the participating U-Boots represented a lesser evil, at least assuming the security of the Enigma ciphers. Unable to decipher the messages the enemy could not draw useful conclusions from them. Since the beginning of the war Kriegsmarine messages did not include any call signs: all the messages were broadcast on the common frequency marking the addressee only in the message text. It allowed U-Boot commanders at sea to monitor the activities of other ships, strengthening the sense of community and making the traffic analysis almost useless. Increased radio activity during the wolf pack's organization and attack was easily disguised by the practice of broadcasting dummy messages during the periods of lull.

Messages sent by the U-Boots to their HQ seemed to represent the real danger. The Kriegsmarine knew that the British direction-finding stations are covering the whole northern Atlantic. Each message transmitted by the U-Boot could and usually was registered by several stations. The intersection of bearings they were taking indicated the approximate U-Boot position. Assessing the British capabilities by their own standards German submariners were assuming that their position could be fixed after any transmission lasting longer than about a minute, necessary for the DF station to take the bearing. Constant fear of over-transmission (*zu viel Funken*) was a frequent subject in their conversations. One of the aces of the U-Boot fleet, Otto Kretschmer, has earned the nickname of "Silent Otto" due to his sparing and reluctant use of radio. Kriegsmarine's security officers shared some of their concerns and took action to mitigate the risk. The best countermeasure was to reduce transmission time below the estimated opponent's reaction time; reducing it to just a few seconds seemed sufficient to ensure security. The easiest way to achieve this goal was to assign short code equivalents to the most common report types. Most of them were included in two documents; the short signal code (*Kurzsignalheft*) and the weather code (*Wetterkurzschlüssel*). The first one appeared during the war in several editions. Diverging for a moment from the real chronology, let's explain its functioning on the example using the version that came into force on 1 January 1942.

Let's assume that the U-276 spotted a convoy heading south in the square BE4131 of the Atlantic map used by the Kriegsmarine. When preparing the report, the officer selected the following four-letter code equivalents for each of its elements:

CKSA	-	convoy	(*Geleitzug*)
RBXO	-	BE41	(map square designation)
MBGV	-	31	(designation of the square sub quadrant)
QQYY	-		course 180 degrees

The report ended with a two-letter ship signature—YN (U-276). In its natural form, the short signal code would not pose a challenge to enemy codebreakers. The messages were extremely stereotypical in their nature, most of them could be linked to specific events, taking into account the time and place of sending. The reports in the short signal code had to be additionally protected, and the logical solution to the problem was Enigma. After encoding the report in the short signal code, the cipher clerk was selecting from an additional table a three-letter message indicator (in our example OOD, corresponding to the true rotor start position VEF) and, using the machine settings obtained in this way, enciphered the following open text

CKSARBXOMBGVQQYYYN,

receiving a cipher text in the form of

QLTISCYKAZEETNFXOJ.

The enciphered message was preceded by the prefix ββ indicating the use of the short signal code and the OOD indicator, repeated at the message end:

ββOODQLTISCYKAZEETNFXOJOOD

The whole resulting report was only 26 characters long, which allowed to transmit it in just few tens of seconds. The short signals code included messages corresponding to most standard reports that U-Boats were expected to send; position reports, fuel and torpedo supplies, convoy spotting, combat reports, meeting points with supply ships, markings of place, day and time, speed and course of the ship, etc. (Fig. 16.1).

The weather code was constructed on a similar basis. The need to use it resulted, among other factors, from the action taken by British troops on the same day that Germany attacked Belgium and the Netherlands. On 10 May 1940 British ships landed an infantry brigade in Reykjavik. The local government protested against British action, a bit *pro forma*, given that it later used the term "friendly troops" to refer to British and American units. The Allies have also occupied Greenland. In this case, the status of the Allied forces was more complicated. After the occupation of Denmark by German troops, the Danish governors of Greenland, Eske Brun and Aksel Svane, asked the Danish envoy in Washington to negotiate with the US

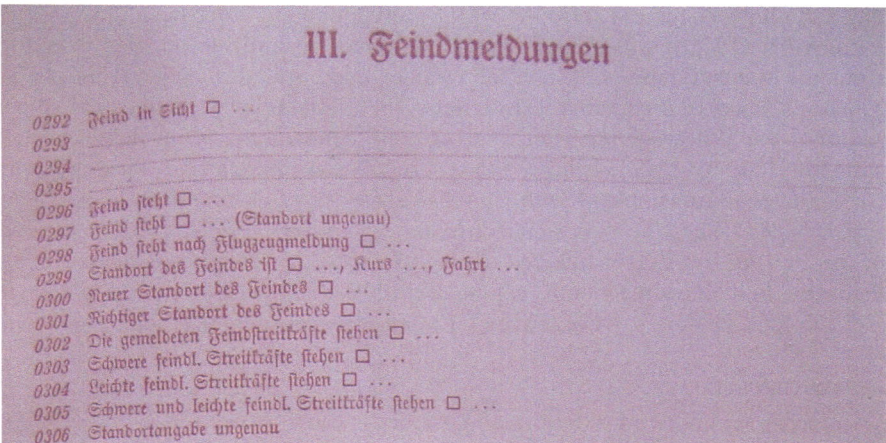

Fig. 16.1 Fragment of Kurzsignalheft (version introduced in 1942). (cryptomuseum.com)

government an agreement authorizing the United States to defend Greenland from German aggression. Kaufmann's proposal must have been prepared among a great deal of confusion, because it suggested their presence not for the period of war, but indefinitely. Worse still, Henrik Kaufmann was not authorized to conclude an international agreement on behalf of Denmark. His action had cost him the charge of high treason and death sentence, which was not repealed until after the Allied victory.

Acceptance of Kaufmann's proposal put the United States dangerously close to participation in the war. The location of Greenland at the gateway of the American continent and the threat of the emergence of Nazi air and sea bases on its coasts prevailed over the diplomats' and lawyers' doubts; on 7th June 1941 president Roosevelt signed the "Treaty of Defense of Greenland". Soon, engineering teams building new airstrips appeared in Iceland and Greenland. Both islands have become unsinkable aircraft carriers serving the aircraft covering the northern convoy route, but also very important weather stations. The patterns of sea and air currents over the northern hemisphere cause the weather for Europe being shaped mostly in the Arctic areas, where the warm air carried from the south by the Gulf stream collides with the masses of cool polar air. The observation of the Arctic weather allowed for the development of good quality weather forecasts for Atlantic and Europe, which was of paramount importance for both sides. Germany needed precise weather forecasts for the planned invasion of Britain and the preceding air offensive. Good forecasts were important for the Royal Navy and the Allied air offensive over occupied Europe. The occupation of the polar islands by the Allies and the removal of the German meteorological facilities operating there forced the Germans to use somewhat unconventional methods of weather observation.

Despite Dönitz's objections U-Boots were included into the weather reporting service. Every ship was obliged to submit a weather report a couple of times every day and on every request from the HQ. By accepting the order leading to *zu viel Funken* the Kriegsmarine signal service attempted to mitigate the risk resulting from the weather reports. A special weather code was designed permitting to squeeze a standard report into 9–13 letters. The first two letters defined the ship's position, the following reporting pressure, air temperature, wind direction and strength, cloud cover and visibility. For security reasons, the encoded weather report was additionally enciphered with Enigma. The encryption system was even less sophisticated than in the short signal code; the cipher clerk selected from a table one of the 26 rotor settings, enciphered the report appending a signature and prefixing the result with letters `WW` identifying the use of the weather code and a single letter indicating the key used. An example of a weather report could take the form:

`WWJKZNISFQEFLJ`,

where `WW` represents the weather code marker, `J` identifies the cipher key, `KZN` defines the sender's position (61°N, 17°W), `I` stands for the atmospheric pressure (1034 or 982 mbar; the actual value being selected by the recipient basing on his general knowledge of weather conditions), `S`—air temperature (+7 °C or −19 °C), `F`—cloud cover (6/10 or 9/10), `Q`—visibility (10 nautical miles), `E`—wind strength and direction (SW, 1–2° Beaufort), `F`—sea state (small waves from the SW direction) and finally `LJ`—sender's signature (U-530).[1] Such a report could be transmitted within just few seconds (Fig. 16.2).

— 14 —

Tafel 11.

T T = Lufttemperatur in ganzen Celsius-Graden.

+ 28°¹) C = a	+ 15° C = n	+ 3° C = a	— 10° C = n
+ 27° = b	+ 14° = o	+ 2° = b	— 11° = o
+ 26° = c	+ 13° = p	+ 1° = c	— 12° = p
+ 25° = d	+ 12° = q	0° = d	— 13° = q
+ 24° = e	+ 11° = r	— 1° = e	— 14° = r
+ 23° = f	+ 10° = s	— 2° = f	— 15° = s
+ 22° = g	+ 9° = t	— 3° = g	— 16° = t
+ 21° = h	+ 8° = u	— 4° = h	— 17° = u
+ 20° = i	+ 7° = v	— 5° = i	— 18° = v
+ 19° = j	+ 6° = w	— 6° = j	— 19° = w
+ 18° = k	+ 5° = y	— 7° = k	— 20° = y
+ 17° = l	+ 4° = z	— 8° = l	— 21°²) = z
+ 16° = m		— 9° = m	

Temperatur wegen Schadens am Meßgerät nicht meßbar: x

¹) oder mehr. ²) oder weniger.

Anmerkung: Bis 0,4° nach unten, von 0,5° an nach oben abrunden.

Beispiele: Abgelesene Temperatur + 7,4° = + 7° = v

" " + 7,5° = + 8° = u

Fig. 16.2 Air temperature encoding using Wetterkurzschlussel. (cryptomuseum.com)

[1] For Ralph Erskine, *Kriegsmarine Signal Indicators*, Cryptology, Vol. 20 No. 4, 1996. A telegram from U-530 on 03.03.1943, the then valid version of the weather code.

Vitata Charybdi in Scyllam incidi—that's how one could summarize the effects of the Kriegsmarine's weather and short signal codes. The designers of the Kriegsmarine communication system had put a lot of effort into protecting it from operators' mistakes. They had introduced rules preventing the use of stereotyped formulas in predictable places. They had eliminated operator's ability to select the predictable message keys. And finally, they had negated the results of the previous steps by introducing two code systems resulting in messages with perfectly repeatable and predictable content. Both codes have been specially designed to minimize the report length. The text of the standard Enigma messages could be modified to avoid repeatable content. One could swap message parts, paraphrase its content, use synonyms or abbreviations. Both codes forced the operators to encrypt perfectly repeatable and predictable texts. Repetitive structure and predictable content were not the only problem with the weather code. The messages sent by the ships were transmitted to the Kriegsmarine meteorological center, from where they were sent to powerful radio station at Norddeich, using the call sign **DAN**, which retransmitted them in a cipher, whose key was shared among all Kriegsmarine ships and units. The **DAN** meteorological report was broadcast in a constant and commonly known format; international weather code. Meteo data are not secret by their very nature, so the method of message encryption (substitution of trigrams based on tables changing six times a day known as *Germet-3* cipher[2]) was not particularly sophisticated. In the early period of the war a section of meteorological codes was organized in BP, consisting of about 10 people under the direction of Dr. George C. McVittie, a mathematician and meteorologist, seconded to Bletchley by the Meteorological Office.[3] During the war the section was breaking the weather codes and ciphers of many countries. The attack on the *Germet-3* cipher brought the first results on 8 February, 1940. Over time, the methodology of attack was improved by using IBM sorters digging through the intercepted messages in search of trigram repeats. Accelerating the breaking of the cipher was important, as each table of trigrams was valid for five days only. If the codebreakers wanted to obtain from meteo messages the cribs permitting to break other ciphers, they had to break the *Germet-3* in less time than its validity period. Fortunately, the authors of the cipher also came here with some help. George McVittie's section soon noticed that after the five-day key period, the same set of tables is reused again a month later. Thanks to this observation, the cipher was broken regularly and effortlessly, providing excellent cribs for the bombes.

The vulnerability of the Kriegsmarine weather and short signal codes must have been obvious for their developers, but the risks involved were considered acceptable for a number of reasons. First of all, their weakness depended on enemy's ability to locate the transmitting U-Boot. Such an assumption seemed to be false by definition; both codes were designed precisely to reduce the transmission time to absolute

---

[2] A radio station in Berlin with a DDX or POG call signs retransmitted similar messages for the *Luftwaffe* in *Germet-4* code.

[3] The section represented originally a section in *Hut 8*, later was reorganized as *Hut 10*.

minimum, preventing the opponent from locating the sender. Secondly, the adversary could only use a fixed message structure if he knew it. Both codebooks were printed with water-soluble ink and classified at the same level of security as Enigma keys. It seemed that their security was assured just in the same grade as Enigma cipher itself. Thirdly, the adversary he could only get an insight into the message structure by breaking both the code and the Enigma cipher, providing additional level of protection. And yet, the latter was regarded by the German cryptologists as unconditionally safe.

At the time when the U-Boots' communication system was being designed, only the first assumption was false. Thanks to the pioneering works by Robert Watson Watt the British methods of direction finding were far more advanced than their German adversaries were willing to accept. During the war, the Germans still relied on the traditional direction-finding methods, which required rotating the antenna to determine the direction assuring the strongest signal. British scientists developed a design permitting to take the bearing without moving the antenna, and supplemented it with a device registering the direction at the oscilloscope screen, even if the transmission was extremely short. A significant part of this chapter will cover the events leading to the collapse of the remaining two assumptions.

We parted with *Hut 8* codebreakers after Foss's success confirmed the theoretical effectiveness of *banburismus*, but did not lead to a breakthrough in attack on the Kriegsmarine ciphers. During the long months of autumn and winter 1940/1941, the codebreakers of *Hut 8* had few reasons to celebrate. Part of their frustration was due to scarcity of equipment. At that time only two bombes were available at the BP, having been used almost entirely to break the Luftwaffe keys. When in February 1941 the bombes got available for a brief moment, they demonstrated their effectiveness relative to the Kriegsmarine ciphers. For the first time the menu for all 336 rotor settings was tested, successfully! However, this success had only theoretical significance. The bombe managed to break the key for 28 April last year, of no operational value. Besides, without the bigram tables, the recovery of the machine settings did not permit the routine reading of the German navy's messages. This event offered the codebreakers only a bit of perverse satisfaction; the menu used was constructed basing on a crib previously rejected by intelligence team of *Hut 4*. Thanks to the earlier Foss's achievement and the recent bombe success, *Hut 8* codebreakers realized that their methods of attack could probably permit to crack the Enigma cipher as soon as they would manage to find an entry point into the Kriegsmarine system. Neither the bombe, nor the *banburismus* were enough to make the first breakthrough. Fortunately, in the spring of 1941, help came from the outside.

The sources of the operation, which opened the gateway to the Kriegsmarine ciphers, were to be found on the beaches of Dunkirk, among the rubbish left by the evacuated British Expeditionary Force. After Dunkirk Churchill was perfectly aware it would take a long time before British troops could return to the continent and face the Wehrmacht in an open combat. He planned, however, to keep the enemy alert and ordered the creation of elite units, whose spirit, training and equipment would allow them to strike at critical points of German defenses in the "hit and run" operations. The concept had its origins in PM's own recollections of the Boer war, when the Boers reached for the tactics of the guerrilla warfare; therefore the troops being created were named after their Boer original—*commando*. Selection of volunteers and training of the first commando units began soon in Scotland. Since their enemy was waiting behind the seas, the training naturally focused on amphibious operations. The first dress rehearsal of the landing operation was not a success. The officer commanding the unit woke up too late and managed to assemble only a handful of troopers before the landing craft bounced off the side of the transport ship.

Perhaps as a result of this failure, Admiral Roger Keyes came to conclusion that an operation in the face of a real enemy would mobilize his soldiers better, and at the same time would provide a more authoritative test of their combat readiness. On the other hand, setting too ambitious goals could result in another failure, so a reasonable compromise was found. Two commando units were to seize the fishing ports in the Norwegian Lofoten islands, burn the cod-liver oil refineries, sink the ships found in the harbor, and on their way back to Britain embark the Norwegian volunteers for service in the Norwegian army in exile. A group of Norwegian soldiers under the command of Captain Martin Linge joined the commandos, partly as interpreters, partly as recruitment agents. The operation "Claymore" carried out on 3–4 March 1941, was a complete success. The 3rd and 4th Commando landed without any obstacles in the main towns of the archipelago, blew up cod-liver oil plants, sunk seven German ships moored in the ports, captured more than two hundred German soldiers and a dozen or so Norwegian collaborators, and assembled a group of more than three hundred volunteers. The operation went smoothly, and the only resistance the forces encountered was an armed trawler under the Kriegsmarine flag, NN04/*Krebs*. Around 6.20 AM the ship was spotted from the HMS *Somali*, when trying to slip out of the port of Svolvær. The usual *Krebs's* task was supervising the nearby fisheries and ensuring that Norwegian fishermen did not flee to the UK too massively. In the face of the enemy, *Krebs's* commander, Lieutenant Hans Kapfinger, sailed out to meet his destiny, saving the honor of Kriegsmarine. To the admiration of the *Somali* crew, the small ship answered the fire, punching the ensign waving from the destroyer's mast. This angered Captain Clifford Caslon, who ordered to fire three more rounds, which ended the duel.

The crew of burning *Krebs* ran the ship aground on the nearby island of Flesa. *Somali* could return to its role of a flagship and make sure that the landings were running smoothly. As the ship was moving away from the scene of the skirmish, one of the surviving crew members appeared on board of the German ship and started destroying documents. When two hours later *Somali* returned to Svolvær, her crew noticed that *Krebs* had slipped back into the sea and was drifting in the fjord with a group of sailors flying a white flag on board. Caslon was afraid of a German air attack and did not want to stop his ship by the side of the wreck. But his signals officer, Lieutenant Marshall Warmington, who had been trained in codes and ciphers, hoped that the panic-stricken Germans might have overlooked some valuable documents aboard. He convinced the captain that using one of the Norwegian fishing boats to approach *Krebs*, he would not procure any danger for *Somali*. Caslon allowed Warmington to investigate the wreck in the company of two sailors.

All three boarded *Krebs* at 9.10 AM, almost three hours after the battle. On the bridge they found the corpse of the ship's commander, but no secret equipment or documents. It was only when searching through the captain's cabin that Warmington found the Kriegsmarine map and a pile of papers, some of which appeared to be secret. Noticing that one of the drawers in the captain's desk was locked, he smashed the lock with a gunshot and found a wooden box inside, containing two metal discs with contacts. As a signals officer, he realized they represent the parts of the ciphering device and packed the box to his bag.

In the early afternoon the ships set off on their way back, leaving behind high columns of smoke from the burning oil plants. After the team reached Scapa, the booty from *Krebs* was dispatched to BP on 12 March. According to Alexander's notes, seizing documents related to the Kriegsmarine ciphers represented one of the goals of Operation Claymore, but he seems to anticipate the future events. The primary objective of the operation was checking in practice the scenario of amphibious operations and the commando's battle readiness. According to John Durnford-Slater, the founder and commander of 3. Commando, the objectives "indicated by the Ministry of Economic Warfare" were as follows: destruction of the oil refineries and German ships encountered in the harbors, collecting and transporting volunteers for the Norwegian army in exile. If the search for cryptographic material was indeed one of the objectives of the operation, it should be considered only partially successful. Between disabling and searching the enemy ship, *Somali* took a three-hour roundtrip through the fjords, which, according to Lieutenant Warmington's report, allowed the Germans to destroy secret equipment and most of the documents. Luckily enough, what fell into the hands of the British sailors, allowed *Hut 8 to* break the deadlock around the Kriegsmarine ciphers anyway. The Enigma rotors and the secret map found by Warmington were already known, but among the seized documents was a sheet containing the *Dolphin* network keys the for the whole February (being outdated it was closed in the captain's desk drawer, avoiding being burned). All the material went down in BP's history as the *"Lofoten pinch"*. The *Lofoten pinch* did not include neither the bigram tables, nor the *K-book*. They had to be recovered step by step, but the scope of the required work was truly daunting. The first step was to find the individual message keys using the *Einsing* procedure. On the same day that the *Krebs* documents reached BP, the employees of *Hut 8* transferred to OIC 10 broken messages, and the next day another 34. The breaking of individual message keys, however, was a tedious process; the reading of the February messages was delayed until May, which probably limited their practical usefulness for OIC. However, reading all or even a significant part of the traffic in the Kriegsmarine network during a full month gave a starting point for the next breakthrough—the reconstruction of the bigram tables and the systematic use of *banburismus*.

With the *Lofoten pinch*, the war suddenly broke into *Hut 8*. Until now, its operation has resembled a university faculty rather than a business organization, not to mention a military unit. The work, though intense, had a somewhat abstract and detached character. A theoretical problem was attacked rather than the adversary behind it. Certainly, the team members were living in the hope that one day the methods they had developed would prove their effectiveness, but when that day finally came, they felt a little surprised and not fully prepared to face new challenges. Codebreakers owed their success to the gifts that came from outside: this success had to be consolidated to make future work independent of the smiles of fortune. Between the theoretical knowledge of how to reconstruct the missing elements of cipher and the ability to read the messages when the Enigma keys remained unknown, a sea of work stretched out. Success was not an end in itself. The recipients of BP's decrypts expected that codebreakers would not only be able to break the enemy's ciphers, but above all that they would be able to read them at a time when the information would allow operational use. Neither the structure of *Hut 8* nor the existing staff were able to cope with both challenges. The solution to the structure's problem was relatively simple: at *Hut 6,* an organization designed some time ago by Welchman had been functioning successfully for months. Following its example was straightforward; in March 1941 *Hut 8* moved on to a three-shift system, which was valid until the end of the war. In April, the equivalents of basic sections in the Welchman system were organized: the RR (*Registration Room*), MR (*Machine Room*) and BR (*Big Room*). The labor-intensive but repetitive nature of their work made it possible to fill the newly created sections with almost exclusively female staff, more easily available in wartime. A more complex problem turned out to be strengthening the intellectual elite of *Hut 8*—the team of banburists. By the end of March, enough messages had been broken to be able to recover the bigram tables. This, in turn, opened the way for the application of *banburismus*, but early attempts brought disappointment. It turned out that the main reason for the failures were the distortions of the underlying statistics caused by the inclusion of dummy messages. After correcting this problem, *banburismus* turned out to be an extremely labor- and time-consuming task, and its final, decisive stage and its final phase was more like an art than a craft. *Hut 8* had to be urgently reinforced by several codebreakers. In December 1940, Turing invited Shawn Wylie, whom he met on a scholarship in Princeton. Wylie accepted the invitation without hesitation and appeared in BP in February of the following year, just in time to support the first successful bombe attack on German naval ciphers. The success achieved on the basis of the menu developed by mathematicians encouraged *Hut 8* to create its own crib section, which Wylie had just headed. In March Alexander was transferred from *Hut 6* to *Hut 8*. Probably his new assignment wasn't directly due to the desire to strengthen the team of banburists. BP's heads had to be concerned about whether Turing would be able to successfully steer *Hut 8* from the research to the production phase. They had to recognize that Alexander would be able to instill in his new environment the principles of work tested at *Hut 6,* and his mathematical skills would allow him to communicate with Turing. Alexander met expectations in both spheres, and by the way he became an unrivalled champion of banburismus. In April Welchman offered a job at BP to a Cambridge colleague, Leslie Yoxall, who had just defended his PhD and was preparing to teach mathematics at Manchester Grammar School. When he came to Bletchley in May, Yoxall probably didn't realize that he had just chosen a career for the rest of his life. The same month Arthur Chamberlain appeared at *Hut 8,* and in June they were joined by Michael Ashcroft, Geoffrey Charlesworth, Jack

Good and Rolf Noskwith. Good also just defended his doctorate, under Hardy's supervision, and specialization in Bayesian statistics made him a dream candidate for a banburist. As a result of the second wave of recruitment in the early autumn of 1941, the number of codebreakers at *Hut 8* reached its maximum during the entire period of war—16. In Alexander's opinion, the team formed in August 1941 was ready to take on all the challenges brought about by developments in the seas and oceans. However, before the decrypts went to OIC, *Hut 8* was to experience the mood swing again, partly due to changes in the Kriegsmarine ciphers, partly due to operations carried out by the Royal Navy, although planned inside BP.

Hut 8 owed its breakthrough in breaking the *Dolphin* network to the *Lofoten pinch*. A source of satisfaction for the codebreakers was the fact that the methods they had developed proved successful once the captured documents had opened a narrow gateway into a cipher. However, their anxiety persisted—the breaking of the Enigma's naval ciphers hung by a thread. They were able to recover the internal settings of the machine, provided they has a good crib. *Einsing* allowed to recover the individual message keys, assuming the knowledge of the Enigma base settings. Breaking the machine's base settings and individual keys of messages received within a dozen or so days allowed for the reconstruction of the bigram tables. The methods developed so far still did not allow to defeat the Kriegsmarine ciphers when none of the three main elements were known. The element of critical importance at that time was a bigram table. *Hut 8* was aware that each day approximated a moment of change in the bigram tables, reconstructed with great effort, thanks to documents acquired on board of *Krebs*. Meanwhile, a "clean" break into the Kriegsmarine ciphers, solely on the basis of theory and without the help of outside materials, was still beyond the reach of *Hut 8*.

However, events unfolded in such a way that when the Germans finally changed the bigram tables on 15 June, this did not arouse any significant excitement in *Hut 8*. One factor that enabled the codebreakers to continue their work uninterrupted was the enemy's problems with weather forecasts. The occupation of Iceland and Greenland by the Allies prevented the Germans from locating meteorological bases on both islands. Weather reports sent by U-Boots were not very useful for meteorologists. The Dönitz ships operated in packs, so the abundance of reports coming from the convoy routes was accompanied by a complete lack from areas of the ocean beyond them. From the point of view of meteorologists, the most acute disadvantage of using submarines to collect meteorological data was the lack of data from the polar regions. The U-boats did not venture into the far north for at least two reasons. Hunters always follow the game, and the frequent fogs, floating icebergs and ice packs under these latitudes meant that the Allied convoys did not take the routes leading north of Iceland. The polar summer, which lasted six months, meant that the sun hardly set during this period, making it impossible for U-Boats to recharge their batteries on the surface. Reconnaissance flights, undertaken by aircraft taking off from Banak base in northern Norway, were a substitute for permanent weather stations in the far North. Long-distance *Heinkel He-111* could collect weather information in specific positions and at repeatable intervals. However, neither the range of aircraft nor the regularity and frequency of the measurements were sufficient for navy and air force meteorologists. Precise weather forecasts were so badly needed that, starting with the Norwegian campaign, both sides of the conflict opened its new theatre, which went down in history under the name of "Meteo War". If the British occupation of Iceland was its prologue, in its first act the Germans sent an ex-Norwegian whaling ship, *Furenak,* towards East Greenland, where it landed a four-man meteorological mission. The German team was captured shortly after landing, and the British responded by attempting to install their own weather station on Jan Mayen island twice. The sinking of the ship with the supplies cancelled out the first approach, the second was prevented by storms. However, the island had the power of attraction; in the autumn of 1940 the Germans also made an attempt to install a permanent weather station there. The mission's codename, 'Graf Finckenstein', was chosen in a not-so-conspiratorial way given that the real Graf Finckenstein participated. In mid-November a team of meteorologists was delivered to the coast of the island, where the British cruiser *Naiad* acted *as a* welcome com- mittee. The German captain decided that the only way out of the trouble is to crash the ship on the rocks, which two crew members paid with their lives. The rest of the team, including Graf Finckenstein, spent the next few years in PoW camps. After these failures, the winter storm season put a temporary end to the attempts to orga- nize permanent weather stations.

Regardless of the operations described throughout 1940 the Kriegsmarine sur- face ships tried to compete with the U-Boots in their actions against Allied shipping. With the Royal Navy dominating the waters directly adjacent to the British Isles, the only safe road to the Atlantic led through the Danish Strait, between Greenland and Iceland. Passing through the northern waters avoided premature contact with the enemy, but was associated with the risk of encountering storms, fogs and icebergs. In order to avoid meteorological surprises, ships were sent in front of the main mis- sion forces to provide weather information. Due to the specificity of the area, small vessels were usually assigned to this role, most often whaling ships, whose civilian crews were supplemented with a handful of trained radio operators and meteorolo- gists. Despite their peaceful appearance, the ships were gaining the status of German navy support ships, and their pre-war names were replaced by *Wetterbeobachtungsschiff* (WBS). Ether over polar waters filled with radio dispatches, since each ship reported on the weather several times a day. The messages were also received by the British listening stations, but until March 1941 no one saw any reason to try to break them. Instead, weather reports stimulated Harry Hinsley's imagination. When his col- leagues at *Hut 8* enjoyed the taste of their first successes, Hinsley was worried about the future. As a historian by education and passion, he did not get into the subtlety of bigram tables, day keys, cribs, cycles and bombe settings. However, he has heard enough about the naval Enigma to know that an unexpected success can last only a while. Anyway, did the state of affairs shaped by the *Krebs* prey really deserve to be called success? Because of his role in BP, Hinsley was more likely to take the point of view of end users. From this perspective, *Hut 8's* achievements looked less spec- tacular. The last messages from February were read in May. They allowed to break the Enigma's keys in the following months as well, but the interval between intercepting the message to breaking it ranged from several days to several weeks. Only in rare cases did the messages have an actual operational value.

The method of a historian's work is to search and analyze sources. Knowing that the imminent change in the Kriegsmarine cipher system will blind BP again, Hinsley wondered where would it be possible to find the same kind of a hint as the one found aboard the *Krebs*? The answer was largely contained in the question itself; if a small coast guard ship was equipped with the Enigma, it could be assumed that machine could also be found on board of the weather ships. The logic of the argument was a little disturbing to the author himself: it resulted in a solution as unexpected as controversial. Small and essentially defenseless ships, sent for weeks on to waters where they were on their own, announcing their presence and position through regular wireless transmission, would carry on board one of the greatest secrets of the German navy—the Enigma and accompanying documents. But in science, the logical and simple solution is usually true at the same time: when preparing to talk to Navy intelligence officers, Hinsley knew he had good arguments. The presence of the Enigma on board the weather ships was at least likely. The British had to receive signals from the ship *Sachsen* during its 86-day voyage last autumn. Hinsley concluded from this that, when spending a long time at sea, weather ships must be equipped with a set of keys covering more than one month. Finding a ship should not be a problem—frequent radio transmissions made it possible to determine its position. But finding a ship was only the beginning of the problem, not the solution. It could have been predicted that when the enemy was spotted, the crew would take care to throw overboard both the machine itself and the secret documents. Here, Hinsley followed Herivel's footsteps, putting himself in the position of a German radio operator who, in the face of imminent danger, has to choose between his own safety and the secret entrusted to him. If the Royal Navy ships arrange a spectacle suggestive enough to induce the WBS crew to leave the deck in panic and haste, the Germans will probably get rid of the machine and the current keys. However, it is likely that they will repeat the mistake of the *Krebs* crew, forgetting the keys for the previous or next month. After last year's round Hinsley enjoyed considerable authority in Admiralty, so it wasn't difficult to approach the Navy intelligence officers and present his concepts. It is possible that during the discussion his interlocutors may have exchanged glances, the significance of which escaped Hinsley. In 1940, the British intercepted three German weather ships which, if Hinsley was right, had to carry the keys to the Kriegsmarine code. First *Furenak*, caught off the coast of Greenland. Then *Adolf Winnen*, sunk in October by *Somali*, which seemed to follow the principle "first shoot, then ask questions". Finally, *Hinrich Freese*, crashed on the rocks of Jan Mayen Island. Even in this case, the British sailors were more preoccupied with rescuing the ship's crew than with professionally searching it.

The plan to intercept one of the German weather ships was at least risky. There may not have been an Enigma on board. The crew could have effectively disposed of the documents. It could also have radioed a distress signal resulting in the keys being changed. On the other hand, if the Germans did not decide to change the system after losing three ships in the previous year, there was a chance that the next case would not cause an alarm either. Aware of previous negligence, Hinsley's interlocutors were strongly motivated to bring this action to fruition while avoiding the mistakes made earlier. Its potential executors also approached the proposal with enthusiasm. When the escort ships fought a constant war in the Atlantic, the destroyers and light cruisers assigned to Home Fleet spent most of their time at the grim anchorages of Scapa Flow. The Royal Navy base offered so much entertainment that the most popular way of spending free time was to stay on board the ship. The officers bid on each other, for how long they hadn't set foot on shore, so the prospect of combat action was received with admiration. Hinsley first shared his thoughts with Clive Loehnis, OIC liaison officer at Bletchley. Together, they convinced Captain Haines, the OIC's Deputy Head of signals intelligence, who accepted the operation and drew up its operational plan.

The Hinsley Memorandum, which was the basis for the intended operation, was dated 26 April. When OIC officers were planning their operation, the WBS5 *Ostmark* had been circulating for almost a month in an area known to German meteorologists as *Operationsgebiet* 2 (OG2), located to the northwest of Iceland. In the first days of May, the British received radio signals confirming that her replacement, WBS6 *München*, has just left Trondheim and was heading north. The *München* sailors were pleased to see that their current mission is promising better than the recent one, when the February storm hit everyone. They didn't know that the Royal Navy appointed a strong squadron of ships, consisting of three cruisers and four destroyers, to hunt them down. Squadron was led from the deck of the cruiser *Edinburgh* by Vice-Admiral Lancelot E. Holland, with BP representative in the operation, Captain Jasper Haines, at his side. The main role in the operation again fell to an old friend from the Lofoten, the *Somali*. But this time on her bridge stood former Clifford Caslon' deputy, Henry Stuart-Menteth. Lieutenant Warmington, promoted to the rank of First Officer, could have hoped that the capitain's impetuosity would not hinder the operation.

The route of the squadron led through the waters patrolled twice a day by German aircraft. If the presence of British ships was to be a surprise to *München*, the Luftwaffe scouts had to be persuaded that the purpose of the mission had nothing to do with the presence of a weather ship. To this end, the ships carried out a series of maneuvers suggesting exercises in operations against German pocket battleships, followed by a cover for minesweeping operations. As a result, the ships reached the area where *München* was expected to be found on 7 May in the early afternoon. The nice weather, which the German sailors enjoyed, was an ally of the British. Around 5 PM, almost simultaneously from the deck of *Edinburgh* and the *Somali*, a streak of smoke was seen on the horizon. *München* could not avoid her fate. British ships were approaching at a speed of over 30 knots; the Germans spotted their masts shortly after they were detected. A nervous buzz has begun aboard the trawler. Fritz Rebelein, one of the weather observers, ran to the radio cabin where, with the help of radio operator Heinrich Wiggeshof, he placed the Enigma itself and secret documents in a lead-laden bag. At that time, the ship's captain was trying to buy the time necessary for his subordinates. He could not scuttle his vessel; when she was converted for operations in polar waters, sand ballast was replaced by poured concrete, less susceptible to shifting in case of sudden heel. This modification cut off access to the kingstones allowing the ship to be scuttled in an emergency. He therefore hid behind a smoke screen and attempted to maneuver to delay contact with the enemy. When Rebelein threw a bag of secret material overboard, Wiggeshof transmitted an open text message that the ship was being pursued by the enemy. This was the last signal from *München to* reach the German stations. Shortly afterwards, *Somali* and *Eskimo* were at a distance of effective fire, and the rounds from their guns and columns of water rising closer and closer forced the crew to abandon ship.

When *Somali* moored at *München*'s side with an elegant maneuver, the captain and the first officer paid tribute to the courage of the small ship's crew, personally leading an prize team. If Warmington hoped to repeat his success with *Krebs*, he was quickly disappointed. One glance at the radio cabin was enough to see that Wiggeshof and Rebelein did their job well, leaving not even a scrap of paper. Stuart-Menteth and Warmington withdrew from the deck as soon as Jasper Haines and his team arrived. Haines didn't even try to search the radio cabin—he headed straight for the officers' cabins. It also took him little time to confirm the validity of Hinsley's hypotheses; after a dozen or so minutes, he appeared back on board, carrying a handbag of papers. The whole operation took just over an hour. The crew of the *München* was fished out of the sea and brought below deck of one of the ships so that they could not watch the search. Haines moved with the documents aboard the destroyer *Nestor*, which took a course directly to Scapa Flow. British stations reported that Royal Navy intercepted a German vessel in Arctic waters, whose crew scuttled its own ship. The only decision that could have compromised the success of the operation was made by the force commander allowing *Somali* to tow the captured unit to one of the British-controlled ports. The return route led through the waters patrolled by Luftwaffe. Spotting of a ship, which according to the reports should rest on the bottom of the ocean, would have to arouse Germans' suspicion. Luckily for BP, both ships managed to slip through unnoticed, which opened the door to another risky decision. Two years later *München* was sold to fishermen from the Faroes and, under the name *Froyen,* sailed in waters patrolled by the German air force and U-Boots, remaining unrecognized by the former owners.

When Haines reached BP on 10 May and started unpacking his loot box, its contents confirmed the accuracy of Hinsley's intuition: among the documents found in the officers' quarters he has found the weather code and the Enigma settings for the whole of June. Precious finds were temporarily shelved—May was full of events, which forced codebreakers to focus on current affairs. Barely had Hines had time to return to BP when the center's representatives had to return to Scapa. A bit of luck and the initiative and courage of the sailors placed in British hands a spoil whose importance surpassed that of the *München* finds. The positive heroes of the story this time were the officers and sailors of the 3rd Escort Group, accompanying convoy OB318 in the waters south of Iceland. The victim was a U-110, commanded by Captain Julius Lemp. On 8 May 1941 at 17:00 Lemp spotted a convoy heading west. He immediately reported to the base, which allowed to direct the U-201 under the command of Captain Adalbert Schnee to participate in the attack. An almost cloudless sky, a calm sea and a brightly shining moon made it impossible to perform the U-Boots' favorite maneuver—a night surface attack, so Lemp shadowed the convoy, waiting for the arrival of the U-201. U-201 finally appeared on the morning of 9 May; the two ships sailed in parallel on the surface for some time, at a distance that allowed the captains to agree on a plan of attack. They decided that U-110 would attack first and U-201 would join half an hour later. When the first three torpedoes left the U-110 launchers at 11.58, they were detected by the corvette *Aubrietia's* sonar even before they hit their target. When the U-110 was turning underwater preparing for the salvo from the stern launchers, *Aubrietia* was already charging the detected enemy. The first series of depth charges missed the target, but the second and third caused damage that left Lemp no choice: in order to save the crew, he had to surface the ship.

On the surface, Lemp gave orders to open the kingstones and the ship's crew to leave. However, the evacuation proved to be risky. After *Aubrietia's* report accompanying destroyers—*Bulldog* (under Joe Baker-Cresswell, also commander of the Escort Group) and *Broadway*—followed with help. Both fired on the surfaced enemy ship, forcing the crew to abandon the deck. Silence fell over the battlefield, interrupted only by the distant sounds of depth charges with which other ships of the group attacked the U-201. *The Aubrietia* was busy picking out of the water the U-Boot crew members and the sailors from the two merchantmen sunk by torpedo salvo. In the meantime, Baker-Cresswell has come to the conclusion that he has a unique opportunity to search the submarine and maybe even get it into British ports. He sent a nine-person prize group under the command of Lieutenant David Balme to the U-110. As they paddled towards their prey, the ship, which the crew considered to be sinking, remained surprisingly stable on the surface. Balme discovered the possible cause when the whaler hit the U-110, and he climbed the ship's conning tower. He saw closed and screwed hatch. The discovery filled him with anxiety: any commander leaving the ship would leave the hatch open to allow the ship to vent and accelerate its sinking. Is it possible that one of the crew members is staying inside and is just activating the charge fuses to sink the ship? Despite the uncertainty, Balme opened the hatch, and went down. Inside the ship he was struck by a feeling of peace and normality. The lights were on, there was no sign of major damage. The ship's rooms looked as if they had just been abandoned by their tenants in a hurry, but without panic. Balme summoned the other members of the team who formed a human chain leading to the hatch and began to pass on to the outside the documents and devices of some interest. When Balme focused his attention on what he knew best, namely sea charts and navigational instruments, the group's radio operator, Alan Long, went to the cabin of his German counterpart. He noted down the frequency settings in both of the transmitters he found, grabbed the documents in his bag, and then became interested in a special typewriter, screwed to the desk. He was puzzled by the fact that pressing any key lit a lamp marked with a different letter than the key. The peculiar behavior of the machine made him unscrew the device and attached it to the collection of items sent to the mother ship. At this time, the *Bulldog*'s crew flipped a towline aboard U-110 and attempted to bring the prey to Iceland, but on 10 May, around 11:00 AM, U-110 leaned over the stern and sank into the waves. The equipment and documents captured aboard the U-Boat remained on board the *Bulldog* as a consolation prize.

On 13 January *Bulldog* entered the base at Scapa Flow, where BP representative, Lieutenant Allon Bacon, was already waiting. He wasn't prepared for the quantity and quality of the documents. The quantity problem was solved on the spot—additional crates were ordered and all the documents were photographed in case any misfortune befell the originals on their way to Bletchley. Inventory and evaluation of the documents were carried out by codebreakers in the following weeks. The least important element of the prey was the Enigma with a set of documentation. However, every subsequent document extracted from the crates must have been applauded. The first was a copy of the short signal code (*Kurzschlüsselheft*). It was a natural complement to the weather code acquired aboard *München*. The system designed by Kriegsmarine cryptologists seemed safe and was safe as long as none of its elements fell into the hands of the enemy. Documents captured on board of *München* and U-110 worked like the first knocked over domino cube. The knowledge of both code books allowed the British to deduce the content of a message from the circumstances of its transmission. Such a probable content of the message, encoded using one of the code books, became a perfect bombe crib. Among the documents found on board the Lemp's ship was a copy of a reserve hand cipher (*Reserve Handverfahren, RHV*).[4] The RHV was on board every ship using Enigma. It was used in case of machine failure. The GC&CS section analyzing hand ciphers of the German navy (RHV was not the only one in the family) could count on an average of a dozen or so encrypted messages daily. Under normal circumstances, this would not be enough to launch an effective attack on the cipher, however, the capture from U-110 facilitated the breakthrough.

As BP became familiar with the different encryption systems of its opponent, it became gradually easier to spot the situations when identical or similar open text was encrypted in different systems. Such situations were described in BP as *re-encodements* and more familiarly as *kisses*. *Kisses* are irrelevant to the codebreaker if he cannot break any of the systems in which they were enciphered. However, if he had managed to break one of them, he gets a perfect crib, allowing him to effectively attack the other systems involved. An example of such a compromise of the cipher was a message sent in the beginning of 1942 by Admiral Dönitz to all units of the German navy, informing about his taking command of the *Kriegsmarine*. The message of extraordinary length, was retransmitted in virtually all communication networks and the code and cipher systems. We do not know whether the German sailors greeted this dispatch with particular enthusiasm. However, it has certainly been warmly welcomed at BP, where it has helped solve some serious problems.

BP did not always have to wait passively for opportunities created by the enemy. One could just as well, by one's own actions, have provoked the enemy into sending a message with a predictable content. The favorite BP method used for this purpose was innocently referred to as *gardening*. The British captured German nautical

---

[4] The timing and circumstances of obtaining a copy of the *R.H.V.* code are subject to some controversy in the sources. Most of the sources link the prey to the U-110. However, Mahon's memoirs mention that BP acquired a copy of the *R.H.V.* in December 1941, indicating that it was captured during the second commando raid on the Lofoten.

charts in the early stages of the war, including minefield passage identifiers. *Gardening* consisted, for example, of sending a plane over the enemy coast on a clear moonlit night, ostentatiously dropping mines precisely over the minefield crossings. Knowing that the crossing was marked on enemy maps with, for example, the number 7, one could have expected the nearby harbor command to send a message saying **WEGSIEBENGESCHLOSSEN**.[5] The predictable content of the message itself gave a perfect crib. But the effects of the provocation usually went a step further. The harbor captain had to give a warning to all ships that were expected to enter the port in the near future until the route was cleared. Different classes of German ships used different ciphers. Most major ships encrypted their reports in the *Dolphin* network. However, for a large group of smaller units, especially those operating off the coast of Norway, the main system was the *RHV* code. Vessels sailing in internal waters and on the Baltic used an even simpler hand code called *Werftschlüssel* (shipyard cipher). If the British chose for *gardening* a harbor frequented by ships of all three categories, they could count on the harbor station to report in all three cipher systems, and quite often the sender would make the warning sound identical in all versions.

The *RHV* system represented a well-designed code. Messages encrypted with it looked outwardly identical to their Enigma-encrypted counterparts, making it difficult for the adversary to identify those more vulnerable to attack. The clever design of the cipher and the modest number of available ciphertexts prevented BP from reading the cipher until the key was captured.[6] The shipyard cipher was far less complicated, representing a simple bigram substitution. However, the biggest mistake of the German cryptographers and cipher clerks was not in the strength or weakness of any single cipher, but in encrypting identical texts using different cipher systems. As a result, the whole system of their communication became as strong as its weakest link was. The breaking of the unsophisticated *Werftschlüssel*[7] made it possible to find a reliable crib to break the *Dolphin* network. Over the next few years *kisses* have become one of the basic tools in BP's toolbox.

The keys to the RHV cipher and the weather code do not finish the list of U-110 prey. Yet Polish codebreakers have established that Kriegsmarine ciphers come in three variants: ordinary, officer and staff. The purpose of implementing the three cipher levels was to provide an additional level of security for certain categories of messages. The encryption in officer and staff variants was two-stage. In the first stage, the officer selected one of the 26 officer/staff cipher keys provided for the day,

---

[5] *Track 7 closed.*

[6] The British captured several versions of the R.H.V. key. On board the U-110 they have obtained the version marked by *Kriegsmarine* as *Bach*, then on 11.06.1941 on board *Gedania* they have obtained the version marked as *Teich*, finally on 01.01.1942 on board the VP5904 they have obtained *Ufer* and *Strom*.

[7] It seems that the original break into *Werftschlüssel* was due to the captured key. Sources suggest that the key to the cipher could have been recovered from the wreck of U-13. Although U-13 was sunk before the date of entry into force of the first edition of the shipyard cipher (15 May 1940), the distribution of the key had to precede its first date of use.

set the Enigma rotors according to the selected key and encrypted the open text of the message. He would then add in front of the encrypted text the identifier of the cipher variant and the key used (e.g. **OFFIZIERGUSTAV**, where **OFFIZIER** indicates the use of the officer variant and **GUSTAV** is an extension of the letter **G**, identifying the key used). He would pass the message in this form into the hands of a cipher clerk, who would cipher the whole thing in the usual manner of the regular cipher variant. When the cipher clerk on the receiving side decrypted the message, he received mostly unreadable text, but the **OFFIZIER** header at its start instructed him to put the result in the hands of an officer, who would set the Enigma in a position corresponding to the **G(USTAV)** key and decipher the open text. In the officer variant, the first level of encryption was performed with the same internal settings of the machine as in the regular one; the difference concerned different settings of the plugboard and the starting rotor position.[8] In the staff variant, the first level of encryption was realized with machine settings completely different from the regular cipher (the selection and order of the rotors and setting of the rings on the rotors were also different). The Poles had already broken the regular and officer variants of the cipher before the war; Rejewski's memoirs show that they had too modest a number of dispatches at their disposal to launch an attack on the staff variant as well. As they had communicated their findings to the Allies, the use and nature of the officer and staff variants was not a surprise to *Hut 8*. However, the documents from U-110 also included the keys to the *Oyster* network, an officer variant of the *Dolphin* cipher for June 1941.[9] The possibility of even a temporary insight into the content of messages encrypted with the officer variant was a gift of inestimable importance to BP. The effectiveness of the vast majority of cipher attack methods used at BP depended on the availability of a good-quality cribs. The ability to read *Oyster's* messages throughout the month has become a source of knowledge about their content, allowing to identify practical cribs for the future. Alexander describes two types of cribs routinely checked when trying to break the *Oyster* code. The Kriegsmarine central station regularly sent out to ships at sea a list of identification signals valid for the following month in an officer cipher variant. When the *Oyster* key captured on the U-110 was due to expire, this type of message was used as a crib, with success. The format of the list had to remain constant over a longer period of time, as the bombe menu based on it was given its own name: *ES Programme*.[10] The second method of attacking the officer variant derived directly from Polish

---

[8] Later in the war, the cipher systems in force in different geographical regions began to differentiate. In the result, simplified versions of the *Offizier* cipher appeared, which differed from the general variant only in a different starting position of the rotors (maintaining identical switch settings).

[9] There is some confusion in the sources about the range of prey on board the U-110. C.H.O'D. Alexander states in his memoirs that the key to the officer's variant was obtained '*for June but not July of Offizier having been pinched*'. In his dissertation, submitted in 2005 at the University of Braunschweig, Heinz Ulbricht states without any indication of source that on board the U-110, the British acquired the '*key to the officer variant for April 1941*' ('*die Unterlagen für das Chiffrierverfahren 'Offizier' für April 1941*'). The discrepancies do not significantly affect the reconstruction of further events concerning the officer's cipher.

[10] Probably short for *Erkennungssignale*.

pre-war successes. The small number of messages in the *Oyster* network made it easy to identify the multi-part ones. Knowing the time of transmission of the previous part (for instance 23.30), it was possible to attack the next part, assuming the crib in form **FORTZWODRE IDRE INUL**. This method was initially disappointing, but over time codebreakers noticed that cipher clerks place the continuation indicator not at the beginning of the message, but anywhere in the content. Given that the typical length of the *Oyster's* message did not exceed 100 characters, this meant that BP had to check a maximum of 100 bombe runs for a menu corresponding to all possible crib positions in the message. Moreover, the menu was only tested for a specific rotor order, known from the general variant of the cipher.

Independent of the attack methods based on cribs, two methods based on other properties of the cipher were developed in Hut 8; the so-called *dottery* and the *E-Rack*. We do not know the author of the first method; the creator of the second one was Turing, who proposed it as early as 1939, without indicating its potential application. When Leslie Yoxall analyzed the possibility of attacking *the Oyster* two years later, he noticed that the *E-Rack* could be helpful. Thanks to his observation, *E-Rack* applied to the officer variant went down in the history of BP as *yoxallismus*. The method's first success was remembered in *Hut 8* thanks to the style in which Yoxall defeated the cipher. Only the day before Turing, the author of the *E-rack*, expressed the opinion that the method could prove effective for messages of a minimum length of 200 characters. It was a frustrating axiom, as the length of the *Oyster* messages rarely exceeded half of that number. Yoxall managed to break the key, however, using a message only 80 characters long. Despite his success, Turing's authority did not suffer on this occasion. In the hands of other codebreakers, the method consistently failed for messages under 100 characters. Cryptologists concluded that Yoxall's success was due to an exceptional surge of inspiration.

Using all the methods together, the keys of officer variants of most Kriegsmarine networks were broken in BP from summer 1941 onwards. This was often done with considerable effort, not always justified by the results obtained. One of the *Oyster* messages gained particular fame. In it, the U-boat officer demanded that the furniture be moved from his current quarters ashore to his new mistress's apartment. The cruise must have been long and difficult, and introducing a cipher clerk into personal affairs was not the right thing to do, therefore the use of the officer's variant was most appropriate. While the officer versions of the Kriegsmarine ciphers were being broken more or less regularly, there was no determined attack on the staff version. It was only in August 1944 that the staff variant of the *Dolphin* network was incidentally broken, using long *kiss* from another network. The nature of the staff messages, which was probably recognized on this occasion, and the scarcity of source material discouraged BP from further attempts to penetrate the secrets of the Kriegsmarine staff messages.

The extensive list of documents gained aboard U-110 and the successes achieved with them somewhat obscured an obvious gap in the inventory: the key to the *Dolphin* network for the current month. Had one of U-110's crew members not taken care to destroy it, operation *Rheinübung*, which began less than 10 days after the ship was sunk, might have taken a different course. Analysing the course of the

Battle of the Atlantic years later, we are inclined to equate it with the struggle against the wolf packs. However, an observer of the early phase of the battle, covering the years 1940–41, would have trouble identifying the source of the greatest threat to Allied shipping. Almost on a par with U-Boots, he would probably put, besides mines and aviation, also surface raiders. Since the first days of the war, Kriegsmarine has been sending to the Allied shipping lanes the surface ships. Sometimes they represented the elite of the German navy: battleships, pocket battleships, and cruisers. Alongside them, the oceans were traversed by auxiliary cruisers: large and fast merchant ships, adapted to the new role by adding strong, although discreetly hidden weapons. Exactly on the eve of the sinking of the U-110, in distant waters off the Seychelles, the cruiser *Cornwall* sank one of German raiders, *Pinguin*, whose career is a good illustration of the danger to the Allied shipping. In the course of its 10-month long raid *Pinguin* has captured or sunk 32 Allied ships with a total capacity of 154,619 GRT, thus single-handedly equaling the result achieved by the U-boats combined during the same period. While *Pinguin* and its counterparts represented a deadly threat to unarmed merchant ships, they were almost defenseless against the Royal Navy ships. However, there were also vessels that posed a more serious threat. On 5 November, 1940, the German pocket battleship *Admiral von Scheer* came across a convoy HX84, escorted by the lone auxiliary cruiser *Jervis Bay*. Commander of *Jervis Bay*, Captain Edward S. Fogarty Fegen, sacrificed himself and the ship in a hopeless battle with a stronger opponent, giving the convoy time to disperse; thanks to the heroism of the *Jervis Bay* crew, the convoy lost only five ships. The Royal Navy was able to effectively deter German pocket battleships, but in the harbors of the North and the Baltic Sea, on the Atlantic coast of France and in the Norwegian fjords, there was a constant danger lurking in the eyes of the commanders of the British Home Fleet: German battleships and heavy cruisers. Ships so strong and fast that even the newest British battleships would hardly be able to face them alone. *Scharnhorst* and *Gneisenau* left the base in Kiel at the end of January 1941. When they entered Brest on 22 March, they had 60 days and 18,000 navigation miles behind them and 22 Allied ships captured or sunk. When the British attention was focused on both battleships, on 1 February heavy cruiser *Admiral Hipper* left Brest. She stayed in the sea for only 2 weeks, but managed to sink 7 of the 19 ships of the unescorted convoy SLS64. Both raids have become a kind of dress rehearsal for operation *Rheinübung*. On 18 May, 1941, two new Kriegsmarine ships left Gdynia on their first combat cruise: the battleship *Bismarck* and the heavy cruiser *Prinz Eugen*. The first one, in particular, posed a threat to the Royal Navy of an unknown and thus disturbing scale. *Bismarck* was considered to be the most powerful battleship in the world at the time. It outranked all British ships capable of matching his speed. Thanks to its construction and armor, she could endure blows that would be deadly for other ships. Its speed allowed her to break through protective barriers and get to the convoy routes, where merchant ships and escorts would be as vulnerable to her power as paddle boats.

The British were aware of the imminent danger. In the period preceding *Bismarck's* leaving her base, they received and decrypted messages ordering the Luftwaffe to perform special meteorological reconnaissance over the waters of the far north. On 21 May BP broke a message ordering the supply of Atlantic charts to the ship and the embarkation of a prize crew, which clearly indicated a corsair cruise. A day after leaving Gdynia, a German squadron was spotted during the crossing of the Great Belt. The next day the reconnaissance *Spitfire* photographed both ships in a fjord near Bergen. On the evening of 21 May, both ships slipped under the cover of darkness from the fjords and sailed along the north and north-western course towards the ice pack border. On 23 May both ships were detected by the cruiser *Suffolk*, patrolling with her twin ship *Norfolk* the passage between Iceland and Greenland. The cruisers' task was merely to shadow the enemy ships. Facing the danger was the task of the team coming from the south: the battleship *Prince of Wales* and the heavy cruiser *Hood*. *Hood*, despite its considerable age, was the pride of the Royal Navy. Heavily armed and fast, it was until recently the most powerful warship in the world. Its final dethroning from this position took *an* exceptionally dramatic course. Both squadrons noticed each other on 24 May around 5.45 AM. At 5.52 AM *Prince of Wales* fired the first round in the battle, whose fate was decided after just nine minutes. At 6.01 AM the round of *Bismarck's* main guns caused the explosion of *Hood's* ammunition chamber, the ship broke in half and sank in seconds. Admiral Holland (who just two weeks earlier had led a raid against *München*) died aboard *Hood*, along with 95 officers and 1324 sailors.

Remaining British ships have been busy trying to rescue the surviving crew members, have lost contact with the enemy. Admiralty did not know the where-abouts of German ships for more than 31 hours, but assumed that they would continue their Atlantic cruise. During the battle, a shell fired by the *Prince of Wales* damaged one of *Bismarck's* fuel tanks. The loss of some fuel meant that, sailing at top speed, the ship would not be able to reach the continental bases. The commander of the German squadron decided to abort the mission and took a course to Brest. And perhaps *Bismarck* would have managed to escape the pursuit if it had not been for Admiral Lütjens' fatal decision to break radio silence. Until *Hood's* sinking, the signals intelligence wasn't involved in events. However, after the battle Admiral Lütjens decided that by sinking Britain's largest warship, he announced his presence and position in a way that made the radio silence pointless. Beginning from the morning of 25 May, Bismarck exchanged 22 messages with her land HQ, which were received by British listening stations.[11] BP, however, has failed to break any of them. *Bismarck* was using a network described by Kriegsmarine as the *Kernflotte* key (flagship key), known in BP under the name *Barracuda*. Throughout the war, *Barracuda* has never been broken, confirming opinions about the security of the Enigma cipher when properly used. In *Bismarck's* case, it was a mistake to use the radio itself.

During the 25 May successive messages allowed to determine her approximate position. The ship's course plotted on their basis indicated that the *Bismarck* was

---

[11] On May 25th, at 8.52 a.m., a radiogram began to be transmitted from the *Bismarck* deck, which lasted for another 36 minutes! Further exchange of messages included birthday wishes from Adm. Reader to Adm. Lütjens, an application for the award of the Knight's Cross to the commander of the ship's artillery for sinking *Hood* and information about the awarding of the medal.

heading for ports in France. Shortly afterwards, the traffic analysis added a new argument: the British noticed that *Bismarck* stopped exchanging messages with Wilhelmshaven, switching to Paris. And it was at this very moment, literally minutes after the ship's new course was revealed, that BP was able to join the game. Paradoxically, this was not due to *Hut 8*, but rather *Hut 6* responsible for decrypting Luftwaffe ciphers. *Hut 6* deciphered the message in which a Luftwaffe officer stationed in Athens asked where he should wait for his son serving aboard the *Bismarck* to return from the sea. In response he was directed to Brest. British battleships and cruisers appeared on the field of the last battle of *Bismarck* on 27 May in the morning. On a battleship encircled from all sides, the towers of the main artillery went silent one by one, and finally at 10.39 the ship sank.[12] *Hood* was avenged.

The breaking of Luftwaffe dispatch was seen by the BP team as confirmation of the codebreakers' decisive role in the final triumph over Bismarck. In fact, it played a more modest role, confirming only decisions previously taken on other grounds. Although BP played an incidental role in this victory, it was *Hut 8* that wrote the spectacular epilogue for operation *Rheinübung*. The cruise of two heavy ships into the Atlantic was a complicated undertaking, requiring the participation of numerous auxiliary units. From the far North, weather ships were sending weather reports. Tankers were waiting in pre-agreed sectors of the ocean. The Atlantic was covered by vessels, prepared to take over from the *Bismarck* prisoners from captured or sunken ships. Although the two ships were exchanging messages in the impregnable *Barracuda* network, their auxiliary units were communicating with each other and with their land HQ using the *Dolphin* key. Meanwhile, the next day after *Bismarck* was sunk, as if for dessert, *Hut 8* finally managed to break *Dolphin* key, and decipher a package of messages detailing the positions of eight ships supporting operation *Rheinübung*. The Royal Navy faced temptation combined with danger. The memory of *Hood's* fate prompted to seize the opportunity to sink or capture vulnerable vessels. However, potential targets maintained radio silence, eliminating routine camouflage for broken Enigma messages. Their waiting areas were designated away from shipping lanes—accidental spotting by a patrolling ship would be uncredible. Under these circumstances, the finding and sinking of all or even most of the vessels must have raised concerns on the German side about the security of communications. Despite doubts, Royal Navy decided to take advantage of the opportunity offered by the success of *Hut 8*. The decisive argument was probably the dual role of the ships. On the one hand, they were to support the operation *Rheinübung*. On the day-to-day, however, they were supplying the U-Boats with fuel, torpedoes and food. The elimination of supply points forced the U-Boats to return to bases, reducing their operational time.

The Admiralty decided to confuse the enemy at least a bit, allowing two of the vessels to escape unscathed. This plan had only one disadvantage: it did not take

---

[12] Only years later the discovery of the *Bismarck's* wreck settled the controversy whether the direct cause of sinking were torpedoes fired by the cruiser *Dorsetshire* or charges detonated by the ship's crew. Bismarck's construction allowed him to withstand all the blows inflicted by British ships. It ended as a result of self-scuttling by its own crew.

into account the role of chance. The Royal Navy ships threw themselves on the easy prey. The first of the ships was spotted by the British even before orders were given to sweep the ocean. On 29 May, the auxiliary cruiser *Malvernian* spotted on the waters between Iceland and the Faroe Islands the fishing trawler *August Wriedt*, which at that time acted as a weather ship WBS8. The crew surrendered their ship, which was manned with a prize crew, brought to British ports and returned to peaceful use as the trawler *Maria*. On 3 June the tanker *Belchen* was surprised in waters around Greenland supplying U-93 and consequently sunk by cruisers *Aurora* and *Kenya*. The next day in the middle of the Atlantic, the captain of the tanker *Hamburg*, having spotted the approaching heavy cruiser *London* and destroyer *Brilliant*, scuttled his ship. The same British squadron forced the tanker *Egerland* to scuttle on 5 June. On 12 June, the cruiser *Sheffield* did the same with the tanker *Friedrich Breme*. On 15 June, north of Cape Verde the captain of the tanker *Lothringen* surrendered his ship to cruiser *Dunedin* (*Lothringen* later served the Allied navy as *Empire Salvage*). The eighth of the support ships, the reconnaissance ship *Gonzenheim*, was one of two ships to be spared by the British. However, it was haunted by bad luck all the way. First, she encountered the auxiliary cruiser *Esperance Bay*, managing to escape using her speed. She was sailing quietly for a few days, but before reaching safe waters she was detected by a reconnaissance aircraft from the aircraft carrier *Victorious*. Its radio report brought to *Gonzenheim* disaster in the form of cruisers *Renown* and *Neptune*. Although the British squadron did not participate in the hunt, its commander was not clearly ordered to ignore the presence of a German ship. Under fire from the *Neptune* guns, *Gonzenheim's* captain decided to evacuate the crew and scuttle the ship. The other ship that was supposed to survive the slaughter was *Gedania*. However, by chance she came across the auxiliary cruiser *Malvernian*, which forced the ship's captain, Heinrich Paradeis, to surrender the ship, which was brought to Greenock. It was a risky decision, but we will see that, by an unbelievable coincidence, it resulted in the Germans acquitting Enigma of the tankers' slaughter and placing all the blame on the captain of the *Gedania*.

Only three days after the sinking of the *Bismarck,* the *Dolphin* keys captured at *München* and U-110 came into force. They brought about a rapid change in the way of *Hut 8* functioning. Even in the last days of the previous month, it took from a few to a dozen or so days to break the messages that gave rise to hunt for auxiliary vessels. On the first of June, 18 minutes after midnight, the first message enciphered with a new key was received. Breaking its individual key, deciphering, translating and editing took less than four hours: at 4.58 AM, the result was teletyped to the OIC. In the following days, the average time between the interception of the message and the sending of a report based on its contents to the Admiralty stabilized at less than six hours. In this early period of operational use, the Enigma decrypts did not immediately meet the hopes of codebreakers and intelligence officers, and did not become the magic weapon in the fight against U-Boots. For the first time during the war, the OIC was able to anticipate an impending attack by wolf packs and steer convoys on a course that avoided their patrol lines. However, the beautiful summer weather and the growing number of submarines offset the effects of defensive measures and caused the tonnage of sunken ships to fall only slightly in June compared to May, still remaining above 300,000 GRT. The deciphered Enigma messages helped the Royal Navy to avoid surprises also from the surface ships. When on 12 June the heavy cruiser *Lützow* sailed out of its hiding place in Norwegian Lindesnes in the company of five destroyers, intending to get to the Atlantic and carry a cruiser war there, British reconnaissance planes were already waiting for it. Only two hours later, the squadron was attacked by *Bristol Beaufighter* torpedo craft from the 42nd Squadron. One of the pilots used an overheard Luftwaffe identification signal to mix into the official air escort of the cruiser, and dropped his torpedo from the near distance. The hit damaged the ship's power generators and forced it to return immediately to Kiel. *Lützow* was lucky in his misfortune: a torpedo hit also triggered a smoke generator, which quickly covered the ship and its surroundings with a dense curtain, protecting the cruiser from attack by other aircraft.

In the new reality the Admiralty had to master the new art of disguising the actions taken on the basis of broken Enigma messages. The principle dictating that no action should be taken solely on the basis of a broken message forced, in many cases, the surprise of the enemy to be abandoned. After the rout of German tankers in the Atlantic, Enigma brought further details of the functioning of the U-boat supply system. The Admiralty resisted the temptation to use this information directly, but organized a system of patrolling the seas off the busy shipping lanes, where German support vessels were usually hiding. The system quickly produced results. On 6 June, the blockade runner *Elbe* was sunk near the Azores by aircraft from the aircraft carrier *Eagle* (Germany lost not only the ship, but also a valuable load of rubber). On 20 June, the captain of the blockade runner *Babitonga*, temporarily delegated to the role of supply vessel for the auxiliary cruiser *Atlantis*, had to scuttle his ship under threat from the cruiser *London*. Three days later, the supply ship *Alstertor* survived the air attack, but surrounded by British destroyers, it also chose scuttling. A total of seven supply and weather ships were sunk during the month.

When it seemed that seizing of *München*, U-110 and other prizes brought about a new era in the history of *Hut 8*, an unexpected blow fell on the codebreakers. In the middle of the month, the Germans changed the bigram tables of the *Dolphin* network. The sudden change meant at least two types of serious problems for BP. Most pertinent was the thesis that the reason for the sudden change was the enemy's fear for the security of its own ciphers. The extraordinary, emergency character of the change was indicated by the moment of its introduction in the middle of the validity period of other cipher elements. If the change of the bigram tables were to be a sign that the Germans considered their ciphers to be compromised, an avalanche of further changes had to be reckoned with, which would destroy *Hut 8's* ability to read *Dolphin*. The problem of a smaller scale, though acutely felt on an *ad hoc* basis, was the need to reconstruct new bigram tables. In the spring of 1941, it was possible to recover them thanks to the "Lofoten pinch". At that time, however, codebreakers had a key to the cipher that allowed them to break all *Dolphin* traffic for a full month. The worrying and unusual moment of the recent change meant that *Hut 8* had at best a key for just two weeks—not enough to reconstruct the bigram tables.

Participants in the events left no testimony to the discussions taking place in *Hut 8* at the time. They must have been heated and, judging by the results, did not bring many positive proposals. As a result, on 19 June Hinsley sat down with a heavy heart to write a report postulating the repetition of the operation against one of the weather ships. The action against *München was* risky, an attempt to repeat it bordered on irresponsible bravado or desperation. If the adversary's vulnerability to threats to the security of its ciphers was increased by the loss of a dozen ships in less than a month, the loss of another one could have set off alarm bells resulting in a complete change of cipher systems. On the other hand, if a change was really coming, it was worth taking the risk of prolonging the period of reading Kriegsmarine's messages—knowledge of their structure and specific features could facilitate the attack on the *Dolphin's* successor. The Admiralty accepted a proposal equivalent to putting all the money back into the same lottery ticket. This time the target was WBS3 *Lauenburg*, which went out to sea on 25 May and circulated in an area northeast of the island of Jan Mayen, described by Kriegsmarine as *Operationsgebiet 3*. This time *Somali* did not participate in the operation. British squadron consisted of her twin ship *Bedouin* (veteran of operation against WBS6 *München*), *Tartar* (participant in the Lofoten raid) and *Jupiter*, until recently having operated in the Mediterranean. Squadron Commander, Vice Admiral Harold Burrough, boarded the cruiser *Nigeria*, and BP's representative, Allon Bacon, sailed on *Jupiter*. Fixes of *Lauenburg's* transmitter indicated that British ships would have to reach further north than during the hunt for *München*. The imminent ice pack, the danger of drifting icebergs and the frequent at these latitudes fog were more dangerous for them than the three guns installed on *Lauenburg's* board.

On 28 June around noon British ships reached the operational area and started to sweep the sea. The long polar day was facilitating the search. Shortly before 7 PM, an observer on the *Tartar* noticed in the rays of the low sun just above the horizon the shape of a ship hiding behind an iceberg. The rest of the operation followed a scenario developed by Hinsley and practically tested during the *München* adventure. Several rounds from the guns of approaching ships made almost the entire crew of the trawler to change into the boat. The two seamen remaining on board had just thrown the Enigma overboard and finished burning the secret documents when *Tartar* nailed on the side of their ship. Shortly afterwards, a boat from *Jupiter* also appeared, delivering Bacon, delighted to see a pile of papers rolling on the floor of the bridge and the navigation cabin. While the *Tartar* prize crew was busy pushing the finds into thirteen bags, *Bedouin's* crew took care of *Lauenburg's* sailors by bringing them underboard, blindfolded. The experience of the previous mission allowed to complete the search within an hour. Admiral Burroughs resisted the temptation to bring the prize to British ports and gave the order to sink the trawler. On his way back to Scapa Flow, Bacon switched to *Tartar* and locked up in the captain's cabin with the documents. Hinsley's intuition was confirmed once again: among the papers he found the *Dolphin* network key for July. The documents and their guardian reached BP almost in time—2 July. The average breaking time has again dropped below three hours. However, a much more important result of the *Lauenburg* raid was reading of all the network's messages within a month, which allowed for a full reconstruction of the bigram tables changed in June—*Hut 8* returned to the game. The Admiralty has apparently managed to develop methods to make effective use of the information coming from BP. In July and August, shipping losses remained stable at a level slightly above 100 thousand BRT per month, amounting to only one third of the losses in the preceding months.

It didn't take too long, however, for a signal that put a question mark next to all of *Hut 8*'s achievements so far. On 27 August, 1941, the *Hudson* from the 269th Coast Guard Squadron under the command of Squadron Leader Thompson caught the U-570 at the surface and dropped depth charges right next to the ship. One can imagine the pilot's astonishment when he saw the crew waving a white flag improvised from the captain's shirt. The captain of U-570, Hans-Joachim Rahmlow, was certainly not one of the aces of the submarine fleet. This was the first cruise of the freshly trained crew, and during the passage from the base to the Atlantic, the ship took lessons of real war, depth charges and damage, which undermined crew's morale. As a result of the last attack, the ship filled with toxic chlorine from damaged batteries, which prevented it from diving. In the situation described most U-Boots chose scuttling, however, Rahmlow decided to surrender the ship. The pilot didn't really know how to accept this surrender, so he was circling over the U-570 trying to get the Royal Navy ships to the action site. The first aid to arrive was a *Catalina* flying boat, followed shortly after by the trawler *Northern Chief* and destroyers, British *Burwell* and Canadian *Niagara*. The ships evacuated the crew, after which the trawler *Kingston Gate* took the U-570 in tow and led it to Thorlákshöfn Bay, Iceland, where it was deposited in the shallows for examination and provisional repairs. The ship later underwent a complete overhaul. In October 1941, the Royal Navy flag was raised on her, where she was known as HMS *Graph*. The fate of the original captain of the ship and his crew was complicated. In the PoW camp, only the armed intervention of the guards saved Rahmlow from his fellow inmates. First Officer Berndt was less lucky. He was sentenced by an officer's "court of honour" under Kretschmer to try to break free, reach a ship and sink it. He paid the attempt to execute this "sentence" with his life.

The survey of the ship has brought a lot of interesting information. First of all, the British realized that the construction of the U-Boots allows them to dive deeper than previously assumed. As a result, the charges used by the British had to be equipped with new fuses, effective up to a depth of 300 meters. The electrically propelled torpedoes captured on board were deemed worthy of copying by the Allies. The *Hudson*, circling over the ship, couldn't prevent the crew from getting rid of Enigma and all secret documents, so the intelligence officer delegated to investigate the ship had to fly to Iceland without too much hope. And yet, as Edward Thomas described in his memories,[13] he experienced on board the U-570 a moment of emotion caused by an ordinary piece of metal, in which he recognized the Enigma cover. The cover had four rotor cut-outs instead of three! Thus, BP received its first warning of a new challenge that was to materialize early next year.

---

[13] F.H. Hinsley, A. Stripp (editors), *Codebreakers. The Inside Story of Bletchley Park*, Oxford University Press, Oxford, 1993, str. 41.

Chapter 17
Mediterranean Interlude

Before the outbreak of war, deep pessimism prevailed among British cryptologists about the prospects of breaking the military Enigma. In view of the slim chances of breaking the cipher, there was no point in building a structure for distributing information from broken enemy messages. When an unexpected gift from Poland changed the mood, the missing signals intelligence structures had to be built from scratch. There was not enough time for this during the short campaigns in Norway and France, so the breaking of Enigma messages did not have a significant impact on them. The war in North Africa was to be the first stage in a conflict that lasted long enough for all the parties to treat it as a training ground for modern war. The special nature of the conflict in Africa also opened wide perspectives for codebreakers, who for the first time were able to provide substantial support for the troops fighting in the desert, as well as for sailors and airmen supporting and supplying them. During the African campaign the British learned not only to break the enemy's ciphers, but above all to use the information gained in this way effectively. Learning has not been easy and painless; part of it were spectacular disasters suffered despite having precise information about the opponent's intentions, coming from his broken ciphers. But at the end of the Mediterranean interlude, the whole of Allied signals intelligence represented a mature system, ready to play a decisive role in operations leading to victory.

The conflict in the Mediterranean was obviously asymmetrical. For Hitler, Africa was a peripheral theatre of conflict, where the war could not be won. He treated the African campaign as an onerous duty to help his battered Italian ally. From the British perspective, the same campaign looked completely different. A victory over Italian-German forces in North Africa did not allow the British to achieve any strategic goal. It would perhaps free up a few divisions, which might prolong British resistance to the Japanese offensive in Malaya and Burma. But the defeat in Africa would have fatal consequences for the empire, including losing the war. This would result in a disruption of communication between the UK and part of the dominions. The Iraqi oil fields, the main source of fuel for the British army and navy, would fall

M. Grajek, *Enigma Myth Deciphered*, History of Information Security,
https://doi.org/10.1007/978-3-031-65475-6_17

into German hands; given the role of the fuel crisis in Germany's defeat in 1944/45, one can imagine the consequences of reversing the situation. The presence of Rommel's victorious divisions in Syria would not leave the choice for Turkey. The conquest of the Middle East would open the way both to the east—to meet the divisions of the Imperial Japanese Army, and to the north—towards the oil fields of Baku. Great Britain could not win the war in North Africa, but it could certainly lose it there.

The war in the African desert was in many ways unique among the World War II campaigns. Events were unveiling in an area virtually devoid of natural obstacles, where *Blitzkrieg* principles could be implemented under almost laboratory conditions. However, it would be a mistake to limit the importance of the campaign in Africa to a few hundred miles of sand and rock along the northern coast. The armored clashes in the desert were only the tip of the iceberg, representing the entirety of the struggle. On the spot, both armies found enough sand and stones to fill sacks with which to strengthen their positions in the rocky desert, but even supplying the troops with drinking water often required transporting it over many miles. All the rest of the supplies had to be transported by sea. In the case of the Italo-German army, supplies had to be transported across the Mediterranean and then delivered overland, to a front about a thousand miles away. In the case of the British, supplies travelled around the whole of Africa, a distance of more than 15,000 miles. The outcome of clashes in the desert was determined by struggles over supply routes. This nature of war made signals intelligence one of the key weapons of war. Breaking a dispatch reporting the voyage of even a single tanker could have immobilize entire armored divisions or keep the entire air force on the ground.

We bid farewell to the British forces in the Mediterranean celebrating victory in the naval battle of Matapan. Combined with the effects of the air attack on the Italian naval base at Taranto, it marked the end of the Italian naval threat. At this stage of the war, the British army was also doing quite well. The victories achieved in Abyssinia, Egypt and Libya could and should have completely cleared the North African coast of Italian troops. Immediately after the victory at the Battle of Beda Fomm, General Richard O'Connor sent an officer of his staff, Brigadier Eric Dorman-Smith, to Cairo to plan the operations necessary to capture Tripoli. In Cairo, however, Dorman-Smith found the walls of the headquarters hung with maps of Greece and learned that the victorious army was to transfer its best divisions for the planned operation in the Balkans, going on the defensive itself. The first British troops landed in Piraeus on 7 March, 1941. The Greeks and British did not have much time to establish a joint strategy—on 6 April the Wehrmacht attacked Yugoslavia and Greece.

The Greek campaign appears to have been the first land-based clash in which Enigma decryption played a role. This role was largely a result of the reorganization of British signals intelligence carried out in November 1940. Local listening stations in the Middle East were already receiving a fair number of dispatches that were not heard in the UK. Transmitting them to BP, breaking the cipher, and transmitting the contents of the broken messages back to Cairo took too much time for the information still to be useful. As a result, a local BP branch, the Combined

Bureau Middle East (CBME), began operating in the requisitioned King Farouk Museum building in Heliopolis. The CBME's job was to read the messages on the spot using keys cracked by BP and relay the decrypts to local British HQs. In mid-March 1941 a direct radio link with BP was established at CBME, and on 27 March a link with General Maitland Wilson's headquarters in Greece became operational. Just in time to prevent a disaster for British troops in Greece. As early as 8 April, a broken dispatch reporting the progress of the German XL Corps at Monastir allowed British troops to be withdrawn in time to the defensive line at the foot of Mount Olympus. A week later, another dispatch regarding German penetration to the rear of the British position permitted to avoid their encirclement; Wilson moved his troops into position at Thermopylae in time. The very name of the new position raised hopes of a heroic defense in a British Prime Minister sensitive to historical parallels. However, on 21 April, another dispatch from BP not only confirmed the Greek surrender at Epirus, but also the progress of German tanks along roads parallel to the coast to the British rear. In the new situation, the British command took the decision to evacuate its troops from Greece. Operation *Demon* became the equivalent of Dunkirk, allowing 50 of the initial 57 thousand British troops to be evacuated, at the cost of the loss of most equipment and heavy Royal Navy losses. General Wilson himself accepted the end of the fighting in Greece with relief, summarizing the campaign in a letter to his wife in a tone of calm resignation: "This is how the military adventure ended, which I hope I will never take part in again. Political considerations have prevailed over military ones, which has led us into a risky game (…)".[1] If BP codebreakers watched the effects of their work, they had reason to share Wilson's mood.

History has written an epilogue for the Greek campaign, in which the Enigma decrypts may have played a key role. The Royal Navy base at Suda Bay in Crete became a transfer point for a large part of the troops evacuated from Greece. At the end of April 1941, nearly 30,000 soldiers from various units found themselves on the island, practically without heavy equipment, disorganized and in poor condition after the defeat suffered in Greece. On 30 April, General Bernard Freyberg was appointed commander of all forces on the island. His first task was to transform the mass of refugees from the continent into an effective fighting force. Commanding the New Zealand Expeditionary Corps Freyberg was a living legend in the imperial army. He was respected by Prime Minister Churchill, whom he met in the very first days of World War I and to whom he owed his first commission in the British army. He was to face a difficult task in Crete. In addition to his own New Zealand division, he had to integrate a mixture of the troops of the British Empire, consisting of the parts of the British Armored Brigade and the Australian Division, and ensure the co-operation of the local Greeks, definitely disappointed by the loss of the valiant Cretan division during the fighting in Greece. Crete has so far occupied a distant place in the British strategy for the Mediterranean. Its measure was the change of

---

[1] Archive of the Wilson family, cited after *Churchill's Generals*, edited by John Keegan, Zysk i S-ka, Poznań 1999.

six commanders of the forces on the island in the preceding six months. Freiberg's predecessors held their position for a month on average; his enemy gave him even less time.

The German side also lacked a clear strategy for the Mediterranean. Some commanders, especially in the ranks of the Kriegsmarine, postulated the destruction of the British empire's main communication and logistical artery through the capture of Gibraltar and Malta. Hitler seemed to favor an attack on Gibraltar, but he failed to win over a natural ally—General Franco—to the idea. Malta was a thorn in the side of the communication lines linking Italy with North Africa. Admiral Erich Raeder, who commanded the Kriegsmarine, argued that it was necessary to conquer the island before any involvement of the Third Reich in Africa. The unexpected winner in the debate, however, was Göring, who favored an attack on Crete. He was in command of the Luftwaffe, whose airmen and paratroopers had just performed brilliantly during the Greek campaign. His forces were still concentrated on airfields in Greece and ready for another show of Luftwaffe's power. And furthermore, among all those involved in the discussion, Göring stood highest in the Nazi hierarchy of power and was closest to Hitler. Thus was born the airborne landing operation on Crete, code-named *Merkur.*

Entrusting the operation to the Luftwaffe was a fortuitous circumstance for BP. German Air Force cipher clerks had a reputation for being the least concerned about communications security. As a result, the keys to Luftwaffe ciphers were broken at BP with a regularity that was never matched by the ciphers of other types of German services. In his memoirs Churchill recorded that "after the occupation of Athens, the German headquarters were less cautious than usual, and our agents acted with great courage. In the last week of April, we received information from some sources about the next German venture".[2] The Prime Minister's memo is a classic example of the disinformation that surrounded the breaking of the Enigma ciphers until the 1980s. The source texts of the decrypted dispatches, which are now available in British archives, are striking in the detail of the data subsequently passed on to military commanders. The first indications of paratroopers' preparations reached headquarters as early as 25 March. They were not yet concerned with Crete, but with landing in the area of the Corinth Canal. On 30 March, decrypts reported the arrival of the 7th Airborne Division headquarters in Bulgaria. Even earlier, on 24 March, an order coming directly from Göring had reserved a significant number of transport aircraft for a 'special operation'. The landing of paratroopers in the area of the Corinth Canal on 26 April may have disconcerted British staffs as to the purpose of this operation. However, on the very same day, the dispatches exchanged between the XI Air Corps and the 4th Air Fleet regarding the selection of airports for the 'Crete operation' were broken. General Wavell, troubled by numerous crises within his command during this period, was inclined to treat the news as German disinformation. He was considering at the same time the possibility of an

---

[2] Winston S. Churchill, *Second World War*, Polish edition Phantom Press International/Refren, Gdańsk 1995, volume III/book 1, p. 269.

attack on Cyprus, German support for the anti-British insurgency in Iraq, and intervention in Syria, alongside Vichy forces. However, when he visited Crete to put General Freyberg in charge there, he drew his attention to the threat of an airborne invasion.

From then on, a stream of data on German plans began to flow to Freyberg's staff. On 1 May the VIII Air Corps announced a ban on bombing airfields on Crete. On 4 and 5 May, the same unit requested urgent photographic reconnaissance over the island. Finally, on 6 May, BP delivered the text of a lengthy dispatch containing a complete plan for the German operation, detailing its objectives, the forces involved and an instruction to complete preparations by 17 May.

Freyberg was one of the best-informed commanders in military history on the enemy's plans, but he clearly experienced changing moods. Immediately after taking up his position, he was quite pessimistic about the prospects: "With the forces at my disposal, I am completely unable to face the expected attack. (...) If, for one reason or another, immediate assistance is not possible, I believe that the question of maintaining Crete should be reconsidered".[3] Perhaps the news brought by the broken Enigma messages contributed to the rapid change in his mood: "I don't understand all this nervousness; I'm not afraid of an airborne landing at all. (...) Only the combination of airborne and sea landings changes the situation completely".[4] Freyberg's optimism was shared by Wavell, who reported back to London: "Colorado perfectly prepared to fight the plague of cockchafers".[5] The optimism of both generals was not without foundation. Freyberg knew the moment of the attack, its goals and the forces he would face. He had troops somewhat battered in the Greek campaign, but brave and stronger than anything the Germans could throw at the island in the first phase of the landing. The failure therefore raised natural questions about its cause. Historians have identified the main source of defeat as the very element that could and should have been the foundation of victory: the Enigma decrypts. The paratroopers, whose drop was to open the battle, were too few in number to ensure victory. Therefore, a key element of the German plan was to capture at least one of the airfields on the island where the transport planes with reinforcements were to land. Freyberg was perfectly aware of this plan and deployed his forces so that they could effectively defend the airfields. His plan worked in the case of Heraklion and Rethimno and failed in the case of Maleme. Churchill recorded in his memoirs: "our defense forces were deployed to protect the airfields". Apparently, immediately after taking command, Freyberg noticed a weakness in the grouping of his troops near Maleme. They were concentrated exclusively on the east bank of the Tavronitis River, leaving the vast plain on its western side undefended. However, when he attempted to redeploy part of his own 5th Brigade, adding a Greek regiment, he was reportedly met with a ban from Wavell. From the account given by the general's son, we know that "My father told me that

---

[3] Freiberg's letter to Wavell of 1 May 1941.

[4] Freiberg's letter to Churchill of 5 May 1941.

[5] Wavell's letter to Churchill dated May 15, 1941. "*Colorado*"—a code name for Crete.

Wavell sent him a telegram: the authorities in London would rather lose Crete than risk exposing our source". The reason for the loss of Crete was therefore supposed to be the intention to protect the Enigma secret. This is a rather risky concept. The deployment of the defenders' main forces in the area of the airfields, rather than at key points along the coast, clearly confirmed their knowledge of the nature of the impending attack. A key reason for the defeat appears to have been the inexperience of British commanders at all levels in directing modern operations, taking place in three rather than just two dimensions. The experience of the First World War taught Freyberg to plan and analyze operations in terms of holding and conquering ground. He did not take into account the changes resulting from the massive use of aviation and air transport. As a result, he failed to convert knowledge of the enemy's plans into victory.

Enigma decrypts consistently yielded knowledge of enemy activities. On 15 May, Freyberg learned that the Luftwaffe had requested a 48-hour postponement of the readiness originally set for 17 May. On 19 May, he received a dispatch setting the attack for the following day. The next morning Freyberg was just drinking tea on the terrace of his quarters when the first parachutes appeared over the island. His adjutant must have been astonished when Freyberg greeted their appearance with the comment "They are deadly punctual". Enigma also brought key information during the battle. With it, the Royal Navy was able to disperse two convoys of replenishments transported by sea. Nevertheless, on 26 May Fryberg had to recognize his defeat and notified Cairo to evacuate Crete. In the course of evacuation, an incident occurred which could actually reveal the secret of Enigma. German paratroopers searching the buildings recently occupied by Freyberg's staff found a document evidently containing information from the Enigma decryption. They attached it to the official report of the battle, but apparently in the euphoria of victory, the inconspicuous piece of paper attracted no one's attention. The German intelligence officers probably thought that if the enemy had known their plan, the fate and outcome of the battle would have been different.

When the best divisions of Britain and the Dominions were engaged in the ill-fated Greek campaign, Enigma brought a warning of a new threat to the troops in Cyrenaica. Hitler recognized that a total Italian defeat in Africa could deprive him of one of his few allies. He therefore decided to prevent its defeat rather than participate in the victory. In late 1940 and early 1941, X Air Corps was redeployed to airfields in southern Italy. Numbering over 300 aircraft, the unit specialized in attacks on British shipping in the Channel. When the season of bad storms deprived it of targets, it was thought it would find them in the Mediterranean. BP gave first warning as early as 27 December 1940, supplementing it on 4 January with information about the installation of German bombers in Sicily. At the time of sharing this news with Kair there were only seven of them, but the Luftwaffe forces were to grow rapidly. On 10th January British ships participating in a convoy codenamed "Excess" were surprised by the attack of two *Ju-87* squadrons from Trapani base. The air carrier *Illustrious* has been badly damaged having to seek refuge in Malta, where the ship was patched enough to survive the voyage to the US shipyard. The next day the same squadrons attacked British cruisers; *Gloucester* was damaged, but the fire of the engine room in *Southampton* couldn't be controlled—the burning wreck was sunk by the destroyer *Diamond*. On 21 January BP received information about the deployment of German bomber groups at airports near Benghazi and in Rhodes. Less than a week later, another message informed that aircraft adapted for sea mining operations are being deployed to Rhodes. The location of the airfield and the type of aircraft left little doubt as to their purpose. A few days later, the planes began operations to mine the Suez Canal.

However, the air operations were only a prologue to the German engagement in Africa. During a meeting on 19 and 20 January 1941, in Salzburg, Hitler convinced Mussolini to also accept the support of ground troops. On 4 February, the *Oberkommando der Wehrmacht* began preparations for Operation Sonnenblume, the redeployment of a German armored group to Africa. Hitler placed General Erwin Rommel at the head of the German forces. It was a controversial choice. Rommel had become famous fighting against the Italians in 1917. His raid during the German-Austrian offensive at Caporetto earned him the 'Blue Max', the highest decoration of the German Imperial Army, and a decidedly low opinion of the military talents of his then opponents and current allies, the Italians. The British did not know the outcome of the Salzburg talks, but they did not have to wait long for the results. The Polish intelligence network operating in Italy reported rail transports of supplies useful in desert conditions heading for Naples. Air reconnaissance missions commissioned on this basis allowed the stockpiles accumulated on the waterfront of the port of Naples to be estimated at half a million tons on 3 February 1941. On 9 February, an Italian Air Force dispatch confirming the Luftwaffe's involvement in covering convoys bound for Africa was broken. RAF intelligence drew the conclusion that German troops were being transported to Africa, but failed to convince their superiors of its veracity. The transmission of the information to Wavell's headquarters in Cairo was blocked for 10 days. When, on 15 February, further decrypts brought news of numerous flights of German aircraft between Sicily and Tripoli, analysts were not taking any chances again. As expected by their superiors, they interpreted the flights as German assistance in the evacuation of Italian troops from Africa. In fact, on board one of the 'evacuation' flights, General Rommel arrived in Tripoli on 11 February.

Only four days earlier the Italians had suffered a humiliating defeat at Beda Fomm. The surviving Italian forces were unable to hold the defensive line in Sirte and prevent the loss of Tripoli. It seemed that Rommel was late to the show. However, as we know, General O'Connor's troops stopped after the victory and then transferred the best divisions to Greece. The decision to suspend the offensive towards Tripoli was taken largely on the basis of the Enigma decrypts. In the preceding period, they brought a wealth of information on German plans in the Balkans. At the same time, they were silent on operations in North Africa, as German intentions in this area had not yet been defined. A natural property of the human mind operating in a state of information deficit is to pay more attention to the issue about which we know more. With more information on the Balkan campaign, British political leaders attributed more importance to it than to the war in Africa. Back in January 1940, Wavell questioned instructions sent from London to give higher priority to the upcoming campaign in Greece than O'Connor's offensive in Africa. But Churchill had an impressive collection of broken Enigma messages to dismiss the doubts of his commander. Prime Minister's instructions were clear: "We expect and demand the rapid and effective execution of our decisions, for which we are fully responsible". Churchill did not take into account that even the most firm orders reflect only the intentions of the opponent, who has to face numerous difficulties; the state of the roads, the weather, the supply of troops, political arrangements, etc. The messages Churchill received in the first decade of January suggested that the Germans should launch an offensive around the twentieth day of the same month, so he considered haste necessary if British troops were to arrive in time. Soon, however, other broken messages revealed the state of negotiations with the Romanian and Bulgarian authorities on the transit of German troops. It became clear that the start of the campaign in the Balkans would be delayed by several weeks. As a result, on 11 February the head of British Military Intelligence presented an assessment of the situation in Greece, according to which German divisions were to occupy Thessaloniki after the first week and Athens after three weeks of operation. Nevertheless, Churchill pushed through the decision of the Defense Committee to continue preparations for sending troops to Greece, sweeping aside the opinion of General John Dill, the then acting *Chief of Imperial General Staff* (CIGS), on that occasion. The Prime Minister's opinion was not even changed by the first clash between British and German troops in Africa, on 24 February 1941. Three days later the British Cabinet upheld its decision to send troops to Greece, and on March 4 the first convoys set off. In the meantime, even Wavell has accepted the superiors' dispositions. Broken Italian dispatches concerning convoys transporting German troops to Africa may have contributed to this. Wavell deduced from these that the new troops would not reach readiness before the summer.

Meanwhile, Rommel made an aerial reconnaissance over the front line the same day he landed in Tripoli. Three days later, the first convoy arrived in Tripoli with German troops on board. In order to speed up its unloading, Rommel ordered it to continue also at night, under lamplight. The following day, a parade of the first German troops leaving for the front took place in front of the local Italian headquarters. To amplify its moral effect, the same vehicles paraded several times. British intelligence in Tripoli appears to have been virtually non-existent at the time, as it overlooked both events.

At the time, the British front line was held by the 2nd Armored Division (less one brigade sent to Greece) and the Australian 9th Division. The troops sent to Greece took with them most of the vehicles available in Africa; for lack of transport, the Australians had to leave a part of troops in distant Tobruk. None of the units had combat experience. Meanwhile, Rommel was quick to understand the fundamental laws governing desert warfare upon his arrival in Africa: "In the African desert, non-motorized troops have no value in a clash with a motorized opponent. This is because the opponent can easily bypass any fortified position. (…) Even the best soldier is useless without vehicles, tanks and weapons. If you destroy the enemy's tanks, you take away his ability to fight without great casualties".

Wavell was aware of the weakness of his own troops in the desert. So, he visited General Neame in command at the front with a reassuring message about the low likelihood of a German attack in the coming weeks. In the event of a surprise, he gave him permission to withdraw towards Benghazi, and even surrender the port and all of Cyrenaica. The news of no imminent threat may have come from BP's first success in the new campaign. Even before the first German units landed in Africa, a new network of German liaison, referred to in Welchman's color code as *Light Blue*, was on the air. This was the network of the X Air Corps, relocated to airfields in Sicily and North Africa. During the early period of operations in Africa, Rommel had no communication of his own with Berlin and had to use the Luftwaffe network. BP broke the network key for the first time on 28 February and from then on broke it systematically with a delay rarely exceeding 24 hours. From the broken dispatches, the British learned the name of the new adversary and his position as Afrika Korps commander, which suggested the future size of his force. They also learned that the 15th Armored Division, which was to form its nucleus, was not due to arrive in Africa until early summer. The opinion of the staff in Cairo was not altered by the orders deciphered on 10 March to carry out an aerial reconnaissance over Agedabia, the confirmation of the arrival in Tripoli of convoys number six, seven and eight, or even the order, broken on 19 March, cancelling all leave and passes in connection with the planned start of the offensive. Indeed, Rommel made a brief visit to Berlin, where he tried to bargain for a strengthening of his forces. Before departing, he ordered readiness to attack for 24 March. He returned to Africa irritated by the lack of understanding from the OKW and immediately threw his troops into the attack.

The speed of Rommel's troops made the Enigma decrypts temporarily irrelevant. The time required to break a new key, decipher the messages and transmit their contents to the field headquarters meant that the information thus obtained referred to past events. Even when the dispatches contained useful information, they were sometimes misinterpreted. The core of Rommel's force at the time was the 5th Light Division. However, from dispatches concerning it, British intelligence deduced that the 5th Armored Division was also present in Libya. When its troops were not found to be present at the front during the fighting, the British concluded that the enemy retained a powerful armored reserve. In addition, on the night of 6/7 April, a German patrol took captive Generals Neame and O'Connor (who had interrupted his convalescence in Cairo hospital and was attempting to return to his troops). Disorganized and lacking effective command, the British troops on 11 April found themselves back on the Egyptian border, from where, on 9 December of the previous year, they had embarked on their victorious offensive.

The crisis in the Libyan desert was not the only problem General Wavell had to deal with. During the same period, the threat was maturing in other hotspots of his area of command: in Syria and Iraq. On 1 April, an anti-British uprising broke out in Iraq. As early as 10 April, Hitler gave the order to support the rebels. Without much effort, he also got the Vichy authorities to provide airfields in Syria to serve as bases for German aircraft. Aircrews and ground crews were recruited from among volunteers and the aircraft bore the markings of the Iraqi Air Force. However, the

Luftwaffe provided the logistics and communication for the operation. This was a fortuitous circumstance for the British, due to the regular breaking of *Red* network keys. This proved particularly useful when, in early June, Wavell decided to end Franco-German cooperation and the 7th Australian Division entered Syria from Palestine. In the course of the fighting, the French attempted to transfer supplies and reinforcements from the metropolis. In order to avoid the bombing of French ships and merchantmen by the allied Luftwaffe, they passed on information about their voyages to their then allies. The same information was repeated in *Red* network, providing the British with data to systematically attack and sink French ships. There were Abwehr agents operating in the country, conducting political and military intelligence. However, they did not have their own radio network to transmit reports to Berlin, so they used the kind of the Luftwaffe. Breaking the keys of the *Red* network therefore also allowed a peek behind the political scenes of the campaign. The British learned from this source that Vichy was planning to send a strong squadron of ships towards Syria: a battleship and four cruisers escorted by destroyers with significant reinforcements on board. Their cruise was to take place under Luftwaffe air cover, which would have been tantamount to the Third Reich taking an active role in the Syrian conflict. Vichy France would thus have already fully and officially moved into the camp of the Axis powers. A state of uncertainty continued for several days, after which another broken Enigma message brought the news that the operation had been cancelled. The operation soon lost its raison d'être—Vichy troops in Syria capitulated on 11 July.

Disorganized after the defeat at Cyrenaica in April, British troops were recovering in Egypt. Time to reorganize their troops was also being bought by the Australians of the 9th Division, who had garrisoned the fortress at Tobruk during the retreat. Rommel himself claimed that in the desert any defensive position meant anything only within range of its guns and could simply be bypassed across the desert. His problem with Tobruk was that the only supply route for his troops on the Egyptian border was within range of Tobruk's guns, and General Leslie Morsehead commanding Australians was an experienced gunner. Rommel could not afford to attack in Egypt before capturing Tobruk. At the same time, after the raid through the Cyrenaica, his own forces were insufficient for a frontal attack on the fortress. Reassuring news also came from BP. One of the broken dispatches brought the news that, by continuing the advance as far as the borders of Egypt, Rommel had flagrantly transgressed the orders of his nominal superior, Marshal Gariboldi, commanding the entire Italian-German army in North Africa, provoking his fury. On 14 April, the broken dispatch brought news that the German-Italian army was going on the defensive after reaching Sollum on the Libyan-Egyptian border. On 26 April, BP broke the another dispatch addressed to the Luftwaffe commander in Africa; the German air force was to shift the focus from supporting ground troops to fighting British shipping in the area of besieged Tobruk. Egypt was temporarily safe. The race began as to who would be the first to bring replenishments and supplies to the front to enable the offensive to be resumed. In early April, information reached Wavell's staff that the German 15th Panzer Division, which was to become the main striking force of the Afrika Korps, was already in Italy, was being redeployed gradually to Palermo and was expected to arrive at Tobruk in early May. Subsequent decrypts have gradually postponed its arrival date until the middle of the month.

Understanding the significance of its appearance on the battlefield, Churchill decided to play *va banque*. He ordered the entire current production of tanks and 50 Hurricane fighters to be loaded onto five fast transport ships and sent them to Alexandria via the Mediterranean. Had the convoy, code-named *Tiger*, sailed along a safe route around Africa, it would have reached Egypt too late to counter the enemy reinforcement. Rommel was also in a hurry. Berlin was concerned enough about his commander's excessive initiative send in a controller, General Friedrich Paulus, to Africa. Paulus arrived at Afrika Korps headquarters on 27 April, just in time to express fundamental reservations about the assault on Tobruk scheduled for 30 April. He was watching the attack, which made a breach in the defensive line but did not lead to the capture of the fortress. He summarized his impressions in an instruction to Rommel, copies of which he also forwarded to Berlin, via *Red* network, of course. An extract from a dispatch from Churchill to Wavell confirms that a copy of the instructions was also on the Prime Minister's desk the following day. In his diaries Churchill traditionally attributed knowledge of the enemy's intentions to the action of a mysterious agent: "We had a spy in close contact with Rommel's headquarters at the time. This man gave us the exact information about the extraordinary difficulties of Rommel". General Paulus, who would have been surprised to learn that he was involuntarily playing the role of a British spy, advised Rommel to take a defensive attitude; to protect Cyrenaica, to hold the Sollum line and the Halfaya Pass, to improve supplies and create mobile reserves. After the experience of the attack on Tobruk, he forbade any future action against the fortress. He also forbade the crossing of the Sollum line without a clear order from OKW before the arrival of the entire force of the 15th Armored Division at the front.

On this basis, Churchill recognized that the situation called for decisive action and again imposed his views on his tormented C-in-C in the Middle East. Wavell gave a series of lectures in Cambridge in 1939, dealing with relations between politicians and the military. In the third lecture, he stressed that "soldiers should remain submissive to politicians". He did not always follow his own recommendation, but in this case, he succumbed to pressure from the Prime Minister and launched an attack on Sollum and Halfaya Pass even before the *Tiger* convoy arrived. On 15th May the British forces commanded by General Gott set off for Operation *Brevity,* broke through the defenses at the pass, captured Sollum and Fort Capuzzo. Later, however, Rommel easily outmaneuvered the attacking troops and inflicted losses on them by attacking from different directions. As a result, the time of success of Operation *Brevity* was adequate to its code name. The same afternoon Rommel's counter-attack forced the British to surrender Fort Capuzzo, and the next day also Sollum.

The British did not learn one of the key reasons for the defeat until many months later. They owed their failure to the activities ofthe 3/56th German Signals Intelligence Company, which had disembarked in Tripoli just three weeks before the start of Operation *Brevity*. Commanded by Lieutenant Albert Seebohm, the unit proved to be an extremely effective weapon for German intelligence, especially given the poor discipline of communication within the British units. Recognizing British intentions did not require breaking ciphers. British commanders were in the habit of talking quite openly over the radio about planned and ongoing operations. Sometimes they masked the subject of discussion by using vocabulary referring to polo—a favorite pastime of Indian army officers. The only effective obstacle for German listeners was the occasional use of Urdu, *the lingua franca* of the Indian army, in radio conversations. In planning Operation Brevity, Wavell and Gott were convinced that Rommel's armored reserves were grouped in the Tobruk area, preparing for another assault. Meanwhile, British radio indiscretions caused Rommel to pull them up towards the front near Sollum, where they waited for the British. The use of signals intelligence was clearly not the monopoly of the British side.

Wavell's only consolation was the safe arrival in Alexandria of the Tiger convoy, which lost only one ship along the way. Having reinforced his armored troops with the arriving tanks, Wavell was able to plan another attempt to unlock Tobruk. This time a more heroic codename was chosen for the operation than for its predecessor—*Battleaxe*. In other respects, however, the operation proved to be a repeat of *Brevity*. The field commanders again discussed the plans so openly that Seebohm's radio operators worked them out in detail. When British tanks attacked Halfaya Pass on 15 June, they were met with an artillery trap that quickly had them renaming the pass as Hellfire. On the morning of the first day of the operation, a complete plan of the operation fell into the German hands, including the call signs of all participating units. It was immediately placed in Seebohm's hands, who could henceforth closely monitor the course of the battle and relay key information to Rommel. By the afternoon of the first day of the operation, the British had lost more than half their tanks, which determined the fate of the battle. As a result of the defeat, on 22 June, an order arrived at headquarters in Cairo dismissing Wavell from his C-in-C position in the Middle East and transferring him to India.

After the operations described, the guns in the desert temporarily fell silent and events in the Mediterranean centered around Malta. The small island has already once played a significant role in the history of Europe—in 1565, when under the leadership of the Knights of Malta, it faced a great siege by Turkish troops. The events around the island starting in 1940 have gone down in history as the second Great Siege. This second siege was fundamentally different from the first one. Not a single enemy soldier set foot on the island. No enemy fleet operated in its waters. No storming of the island's fortresses took place. And yet the island's inhabitants and garrison experienced all the horrors of modern warfare. Bombs rained down on homes and barracks almost every day. The defenders were haunted by a deep feeling of loneliness. Everything was in short supply: from water and food to fuel and ammunition. And all this lasted many times longer than during the siege by the Turks. A glance at the map makes clear the importance of Malta to the Mediterranean campaign. In the central Mediterranean, the distance between Sicily and Cape Tib in Tunisia is less than 60 nautical miles. Virtually every ship sailing from any Italian port towards Tripoli must pass through the waters surrounding the island. Photographs taken by reconnaissance aircraft stationed on the island could be used for accurate accounting of supplies to Africa. Aircraft and ships operating from Malta could have decimated African convoys. Particularly given that the British generally knew the place and time of their departure from port and the course leading towards the African coast. At one time, the decryption of a commercial model of Enigma led to a British victory at the Battle of Matapan. Shortly afterwards, the Italians stopped using the Enigma, replacing it with the C-38m cipher machine. On 23 June, the first broken ciphertext of the new network arrived in Alexandria from BP. Starting on 10 July, BP systematically broke the keys of this network. As a result, a group of BP-trained codebreakers was installed in Malta, tasked with cracking the intercepted dispatches on the spot and relaying the results directly to the airmen and sailors stationed on the island.

The results of the codebreakers' work were passed on to airmen whose patrols provided an alibi for breaking ciphers. Confirmations of spotting a convoy at the expected place and time went to the captains of submarines stationed in Malta, which suddenly and unexpectedly began to have great successes. The Italians converted the pride of their shipbuilding industry and pre-war passenger fleet into troop carriers; the liners *Neptunia*, *Oceania*, *Vulcania*, *Marco Polo* and *Esperia*. *Esperia* sank on 20 August, torpedoed by the HMS *Unique*. When on 18 September HMS *Upholder* fired a torpedo volley which sunk *Neptunia* and *Oceania*, her twin ships *Unbeaten* and *Ursula* were waiting nearby ready to finish the job. One of the creators of the Ultra system, Frederick W. Winterbotham, described with amusement an episode in which he participated during his visit to Malta. Taking part in the reception issued by the governor, he noticed a young Royal Navy officer, whose uniform was decorated with a large gallery of medals, and congratulated him on his achievements. In response, the sailor told him of his "extraordinary fortune; whenever he was ordered to patrol in a specific sector, a convoy sailed right in front of him with supplies for Rommel". The catastrophic losses of convoys gave rise to suspicions of information leakage. Fortunately, the Germans had a ready explanation for the problem at hand—betrayal within the ranks of the Italian allies. Their suspicions reached the highest echelons of the Italian navy, up to the level of its intelligence chief, Admiral Franco Maugera.

In mid-August 1941, the name of a new unit, the 361st Infantry Regiment, began to appear in broken Luftwaffe dispatches. This was the first sign that Rommel's efforts in Berlin to strengthen his forces were bearing at least partial fruit. The new regiment became the start of a somewhat improvised unit; the Division zbV Afrika (special purpose division Africa), later renamed the 90th Light Division. The same dispatches brought the information that, in view of the losses of convoys between Sicily and Tripoli, Rommel's corps will be supplied by convoys circulating between the ports of Greece and Benghazi, out of reach of the aircraft and ships from Malta. Moreover, in September, decrypts also reported that part of the X Air Corps had been reassigned from Greece to Sicily. This could only have meant the intention of an air offensive against Malta. The island was preparing for a long and difficult winter.

On the Libyan-Egyptian border, both sides used the pause in operations to restructure their troops. The British converted the former Western Desert Force into the 8th Army. Wavell's successor as C-in-C, General Claude Auchinleck, placed it under the command of General Alan Cunningham, younger brother of Royal Navy commander in the Mediterranean, Admiral Andrew Cunningham. Given that the RAF commander in the Middle East was Deputy Marshal Arthur Conningham, German intelligence must have encountered considerable difficulty in distinguishing between key commanders. On the opposite side of the front, Italian-German troops were renamed Panzer Group Africa on 1 September. Its nucleus was the Deutsches Afrika Korps, consisting of the German 15th and 21st Panzer Divisions, the Italian Savona Division and the Division zbV Afrika in the process of being organized. The Italians had the XX motorized corps (Ariete and Trieste divisions) and the XXI infantry corps (4 divisions). The reorganization also included signal intelligence structures. Shortly after Auchinleck took command, Kenneth McFarlan, whom we had previously bid farewell when he departed Cazaux airfield after the defeat of France, landed in Cairo. When he landed in Egypt, the local BP outpost was still called the Special Signals Unit. The first letters of this name evoked too many associations with the Secret Service, creating undue interest. Under McFarlan's command, the unit was rebranded as No. 5 SCU/SLU. It was not until late 1941 that the Ultra system began to take on the mature structure. Just in time, for during the same period, on the one hand, the field commanders' need for information from BP increased, while on the other, its codebreakers began to succeed in new areas. By mid-1941, *Hut 6* was regularly breaking the keys of Luftwaffe networks, but had no significant success in wrestling with the ground forces' communications. Wehrmacht signals officers remained faithful to General Fellgiebel's doctrine, reiterating that the "das Funken ist Landesverrat" (use of radio is treason against the country). In Germany and occupied Europe, German troops mainly used wired communications, inaccessible to British listening stations. The situation changed in June 1941, with the German attack on the Soviet Union. The ether simply exploded with radio dispatches exchanged by the three German Army Groups fighting in the east. Back in June, *Hut 6* codebreakers managed to crack one of the Wehrmacht keys used in Russia—*Vulture*. Thanks to the messages they read, they learnt the typical structure of orders and reports exchanged by the Wehrmacht, which provided an abundance of good cribs for bombes. This, in turn, enabled the key of one of Rommel's army networks, *Chaffinch*, to be broken in September. However, this key proved to be a difficult opponent. Breaking it required testing cribs for all possible rotor orders. Success, if it came, required about a week of bombe work. Despite all the problems, *Chaffinch* proved to be an interesting target for codebreakers. It was, in fact, a conglomerate of three communications networks, each of which transmitted extremely valuable information. Two networks linked Rommel's headquarters to the logistical bases of his troops in Italy and Greece. Precise reports on the condition of the troops, their stocks of ammunition, fuel, rations and water flowed from Africa to the bases. In the opposite direction, the voyages of supply ships were reported, giving their ports of departure and destination, dates of departure, cargo and route. A third network linked Rommel's headquarters with commands in Rome and Berlin. Through it, precise information on the numbers and condition of his own troops, assessments of the position of his own and the enemy's troops and information on his own intentions were transmitted daily.

A synthesis of the information exchanged in the Luftwaffe and *Chaffinch* networks fully revealed Rommel's intentions. In mid-July, *Hut 6* learned of ongoing negotiations with Vichy over the sale to Germany of heavy siege guns from French depots in Tunisia. On 12 September, a decrypt of the *Light Blue* network revealed that the Germans had ordered from the Italians detailed plans for the fortifications of the Tobruk fortress (previously constructed by Italian troops). At the same time, Rommel's logisticians ordered a shipment of flamethrowers, in desert conditions useful only for assaulting fortified positions. A series of German-Italian staff conferences, of which Enigma informed systematically and punctually, the relocation of troops and the continued preparation of Luftwaffe units stationed in Africa for action against Tobruk reinforced the British side's view that an attack on Tobruk was imminent. The only unknown issue was the timing of the planned operation. Dates appearing in enemy dispatches were postponed many times. On 3 October, a broken dispatch from the *Chaffinch* network brought the information that problems with transport across the Mediterranean had delayed the return of soldiers from leave by about 2–3 weeks. On the same day, Auchinleck approved the 8th Army commander's plan for the offensive that would unlock Tobruk—Operation *Crusader*.

The race against time has begun. Rommel ordered his troops on 26 October to reach readiness to storm Tobruk between 15 and 20 November, then set off for Rome and Berlin in an attempt to overcome OKW's objections to his plans and to provide his troops with better supplies. Launching Operation *Crusader* in a situation where the enemy would be prepared or even committed to storming Tobruk increased the chances of success. On the other hand, if the storming of the fortress occurred or even succeeded before the Eighth Army could intervene, its plans would lose their basis. In mid-October, the 8th Army headquarters commissioned BP to assess the enemy's intentions from decrypts. It concluded that Rommel "expects a British offensive (… and) is far from undertaking his own offensive operations". However, the breaking of the *Chaffinch* key on 2 November completely changed the assessment; the Germans not only ordered accurate maps of the fortress, but also significant quantities of assault equipment. On 10 November, the broken *Chaffinch* message informed not only that Panzergruppe Afrika's leaves had been completely halted, but also that the plans for the attack on Tobruk had been supplemented by a naval landing. When the 15th Armored Division urgently ordered a delivery of 800 dog-tags on 12 November, and another dispatch on 16 November announced the arrival of Italian Marines battalion *San Marco*, the staff in Cairo was strengthened in its opinion of an imminent attack on Tobruk. On 16 November, a dispatch from the OKW to Panzergruppe Afrika was read, according to which "the imminent attack on Tobruk, planned by General Rommel in consultation with Duce, is desirable". One element was missing from this avalanche of confirmations, but it was the most important element—the date of the attack. In his memoirs Churchill wrote that "the liquidation of Tobruk was the main objective of the German command, which planned to attack this point on 23 November". It seems that such a precise timing of the German operation was the result of an *ex post* analysis. General Auchinleck mentioned that "with all plans during the war, there is always some 'but', and in the case of Tobruk, this 'but' came down to the fact that we did not know the date Rommel had chosen for his attack".

An analysis of the first phase of Operation Crusader leaves no doubt as to its nature. It was a plan to strike at the rear of the enemy engaged in an attack on the fortress of Tobruk. The XIII Corps, consisting mainly of infantry, was to tie down the garrisons of Sollum, Halfaya Pass and Bardia, outflanking them from the south, and after neutralizing them, move towards Tobruk. At the same time, the XXX corps, consisting of armored and motorized units, was to bypass the enemy's right flank and proceed to raid its rear. It was an unfortunate attempt to mimic the opponent's earlier tactics. Rommel was in the habit of tying up the enemy frontally using infantry units, especially the Italian infantry, while sending his own armored units to raid the enemy's undefended hinterland, disorganizing his command, communications and depots. The actions of the XIII Corps corresponded to the first part of this scheme. However, the planned actions of XXX Corps brought its units right into the middle of an enemy grouping preparing to attack Tobruk. If these units were not engaged in combat with the fortress garrison, their mobility and flexibility allowed them to turn around and face the new threat with relative ease. The corps' task was defined in Churchill's memoirs as follows: "XXX Corps was to disperse for flanking operations, finding and fighting Rommel's armored troops, or at least tying them up by fighting to protect the XIII Corps". The key word in this sentence is 'to disperse'. Rommel strictly respected one of the principles of armored combat formulated by Heinz Guderian: "Klotzen, nicht kleckern" (splash, don't spill), keeping his armored formations focused. The planned dispersal of the XXX Corps units meant that they were to enter the enemy's concentration area one at a time, with no communication and no possibility of mutual support. An analysis of the plan and the actual course of the battle may indicate that Enigma decrypts played a not very fortuitous role in it. Despite the above-mentioned flaws, Auchinleck and Cunningham constructed a reasonable plan that could have been fully successful, provided the operations were precisely coordinated. The date given by Churchill for the German attack on Tobruk, 23 November, was correct. Had the Eighth Army attacked on that or one of the following days, its operation the operation could have been a complete victory. However, the British had already attacked on 18 November, just five days too early for Operation *Crusader* to be carried out under the conditions for which it was planned. Perhaps the British commanders feared detection of a concentration of XXX Corps troops on the far flank of the enemy position. The second possible reason is Churchill's impatience. The Prime Minister was annoyed by further delays in the offensive of his troops and expressed his moods in correspondence with their commanders. It seems that under the combined influence of both factors Auchinleck decided to launch his own attack, even though Rommel delayed the attack on Tobruk.

The fate of the battle was complicated and was in the balance for quite some time. At the time of the battle's crisis on the British side on 25 November, Churchill ordered Auchinleck to "burn all messages and special material", clearly fearing for the safety of the Enigma decrypts. Cunnigham, seeing the wrecks of his own tanks burning up in the desert, was ready to acknowledge defeat. At this critical moment, the situation was saved by Auchinleck and the codebreakers. Auchinleck observed the course of the battle at 8th Army headquarters. Perceiving in Cunningham a decline in morale, he relieved him of his position as army commander and replaced him with General Ritchie. From the dispatches broken on 23 November, Auchinleck knew that the enemy no longer had reserves. The day before, Enigma brought news that the Royal Navy had forced an Italian convoy carrying new tanks for Rommel's troops to turn back. Two days later, a message intercepted earlier was read out at BP informing of the critical state of fuel stocks in Panzergruppe Afrika. As a consequence, Rommel had to switch to strictly static defensive operations on 22 November. The crisis on the British side was overcome, while the fighting crisis on the German-Italian side just began. On 27 November, general Freyberg's New Zealand Division merged with the Tobruk garrison in El-Duda. The next day, the British captured the staff car of the 15th Armored Division, where they found keys for two German communication networks—*Chaffinch* and *Phoenix*. The first one was broken by BP earlier, but usually with a long delay. The keys transmitted to Bletchley allowed Hut 6 to read all of the outstanding November messages without delay until the end of the month. The documents included *Phoenix* network keys for seven days only, but the coming week was to be decisive for the fate of the battle. Fighting in the corridor connecting the 8th Army to Tobruk raged for the next two days and nights. Rommel gradually saw that his exhausted troops were facing annihilation. In addition, a dispatch from Rome arrived on 5 December informing him that he could not count on new reinforcements until the Luftwaffe had crushed Malta's resistance. The commander of Panzergruppe Afrika had to make a decision he had long tried to avoid: to save his troops he had to sacrifice the garrisons fighting on the Egyptian border, give up hope of maintaining the Tobruk blockade and withdraw the rest of his troops to the east. In a fighting retreat the Panzergruppe Afrika reached on 6 January Agedabia, the starting point of Rommel's offensive 8 months ago.

This is how the battle described as the fourth Libyan campaign ended. The next round was to take place in a significantly changed global situation. On 7th December 1941 the Japanese attack on Pearl Harbour opened a new theatre of war and confirmed its truly global character. Soon a series of British defeats in Asia transformed the 8th Army's situation. India, Australia and New Zealand, preoccupied with their own defense, ceased to be a reservoir of reinforcements. Some of the dominion divisions came to be sent back to Asia. Malta, which made a great contribution to the victory of Operation Crusader, was also affected. Between June and October, Italians lost an average of 16% of the tonnage of convoys sent. However, starting in November, the increasingly better cooperation of codebreakers, reconnaissance planes and the cruiser squadron based on Malta increased the level of convoy losses to 62%. However, this was the height of the island's efforts, for which a difficult time was looming. On 18 December, a cruiser squadron based on Mata lost three ships in a minefield. On the same day, Italian live torpedoes penetrated the port of

Alexandria and damaged two Royal Navy battleships, the *Queen Elisabeth* and the *Valiant*, immobilizing them for weeks. In September BP picked up and broke a message indicating that a group of German U-Boots were moving to the Mediterranean. The transfer to Sicily's airports of the 2nd German Air Corps has reduced the potential of Malta's "unsinkable carrier" almost to zero. During the first 6 months of 1942 there was only one day, during which Malta was not bombed. The only surviving cruiser, HMS *Penelope*, was so punched up with shrapnel that her own crew renamed her HMS *Pepperpot*. The pessimistic mood of the island's defenders was not improved by the appointment of its new governor. Lord Gort was respected in political and military circles, although at the same time he was regarded somewhat as a specialist in honorable defeats (he commanded the forces evacuated from Dunkirk). Accompanying him as the island's new air force commander was Battle of Britain's hero, Deputy Marshal Keith Park. In bidding him farewell before departure, Churchill remarked that he might have the unpleasant duty of lowering the British flag on the Governor's Palace. It was clear that Malta would not be supporting the British forces fighting in the African desert in the coming months.

Rommel retreated to Agedabia, but was far from declaring the battle over. His attitude was only partly due to his offensive spirit. In fact, he had almost as good sources of information about the condition of the British troops as they had about him. His on-call source of information on the British, Lieutenant Seebohm's signals intelligence company, was effective during the fighting. When troops rested after combat or maintained radio silence in preparation for the next clash, its usefulness diminished. However, Rommel also had another source providing information of a strategic nature. His unwitting collaborator was Colonel Bonner Frank Fellers, the American military attaché in Cairo. The West Point academy graduate arrived in Africa in October 1940 and energetically set about gathering information of interest to US military intelligence. British sources maintain that neither Wavell nor Auchinleck passed sensitive information to Fellers. Judging, however, from the variety and abundance of information passed to Washington, these assurances did not always correspond to the truth or Fellers must have acquired other, alterative sources of information in the Cairo staffs. British officers hosted him at their own clubs, quite openly expressing their opinions on the ongoing operations. Fellers visited battlefields and camps of British divisions. He was a competent and intelligent officer, so was able to fill in the missing bits of information himself. When the US entered the war in December 1941, the hosts' openness towards the ally's representative further increased. Fellers was sending his reports, coded in *Brown* code, by telegraph to MILID WASH (Military Intelligence Washington). However, US military intelligence was not the only recipient of his dispatches. Beginning in August 1941, his dispatches were also read by German codebreakers, who forwarded the broken reports immediately to Rommel's staff. The story of how the *Brown* code was broken is convoluted and not entirely clear. Most authors claim that a copy of the code book was stolen from the American embassy in Rome by Loris Gherardi, who was employed there as a messenger. Others claim that it was stolen by the alluring daughter of one of the fascist dignitaries during an operation of a not-so-military nature. Whichever option corresponded to the truth, the head of Italian military intelligence, General Cesare Amè, forwarded copy of the codebook to his German counterpart, Admiral Canaris. This did not solve all the decryption problems; the Fellers' dispatches were additionally superenciphered. Another German ally came to the rescue; the Hungarians managed to photograph the additive tables and hand over copies to the Germans. Synthetic assessment of the importance of Fellers' dispatches to the German military machine was conveyed by Hitler himself, who, over lunch with Göring, expressed the hope that the "US Mission in Cairo will continue to keep us informed about British military plans, thanks to its poorly secured cryptograms".

Thanks to Fellers' dispatches, Rommel knew that the best British divisions had either been sent back to Asia or were resting in the rear. The front opposite his troops was held by divisions freshly arrived in Africa, partly undergoing acclimatization and training.

In a letter to his wife dated January 17, Rommel wrote: "The situation is changing for the better and my head is full of plans that I don't even mention to those around me. They'd think I was crazy". There was something, however, that did not give the German commander peace of mind. The course of Operation Crusader had made him suspicious that the British knew too much about the condition and intentions of his troops. He was determined that the planned attack would be a surprise to the enemy. To this end, he led a retreat further than tactical necessity dictated. His soldiers were told of the need to retreat even further. He forbade aerial reconnaissance over enemy posittions. He did not even notify the commands in Rome and Berlin of his plans! He pulled up reserves at the front in stages, exclusively at night. Finally—he abandoned radio communication altogether, hanging orders on telegraph poles along the coastal road. BP gave no warnings of an imminent attack. Only one dispatch from the Light Blue network may have indicated offensive intentions (the Luftwaffe commander in Africa complained of an insufficient supply of bombs "in view of operations planned in the coming days"), but it was too vague to alert British headquarters. As a result, Rommel's attack on the night of 21/22 January came as a complete surprise to the Eighth Army. Surprise, however, was not the only reason for the poor performance of some British troops. Numerous tankers of the newly arrived to Africa 1st Armoured Division recognized that there were faster means of evacuation than the tank. So, they left operational tanks on the evacuation routes, which were then used for escape by the infantry soldiers following them. By 6 February, the front line had moved back to Gazala.

Rommel again went on a visit to Hitler's headquarters in an attempt to bargain for the extra three divisions needed to attack Egypt. Hitler did not succumb to the vision of the Orient. His staffers stressed that two obstacles needed to be eliminated before crossing into Egypt: Malta and Tobruk. On 7 February 1942, broken Luftwaffe dispatches announced the redeployment of XI Air Corps units to the Reggio Calabria area. The British were well aware of this unit from Greece; its aircraft had transported paratroopers who had made landings in the areas of the Corinth Canal and Crete. Their presence in southern Italy could only mean one thing: the planned landing in Malta. After the costly victory on Crete, the paratroopers were not organized in a single tactical unit, but dispersed as elite infantry in different areas of Wehrmacht operations. In March, BP began to intercept information about moving them from their current garrisons to southern France and Italy and forming them into a brigade. The dispatches also brought news of Colonel Hermann-Bernhardt Ramcke taking command of the brigade, after whom the brigade was named. When it seemed that the invasion of the island was imminent, suddenly the intensity of its bombardment decreased and decrypts confirmed the withdrawal of the 2nd Air Fleet unit from Sicily. On 2 May, broken dispatch brought the news that the Ramcke Brigade, which was to be the nucleus of the force attacking Malta, would be moved to Africa as a reinforcement of Rommel's forces. Paradoxically, Malta may have owed its salvation to one of Fellers' dispatches. In one of his reports, he described the desperate state of the island's defenses; no food or ammunition, no operational aircraft or ships, and finally no fuel for those still serviceable. In such a state, the island could not contribute damage to the convoys, so it was a shame to waste energy conquering it. And Rommel enthusiastically welcomed the reinforcement of his forces with an elite brigade of paratroopers.

On the British side, the need to commit three experienced divisions to Asia limited the possibilities for offensive action. Auchinleck reasonably assumed that in the present situation the task of his army was not so much to win the campaign as not to lose it. Had the fate of the campaign been shaped in Cairo, the 8th Army would probably have remained on the defensive throughout 1942. However, Churchill, who was receiving news of defeats in Asia, had to endure Stalin's pressure regarding the second front and was increasingly facing opposition in the House of Commons, sought victory in the only theatre of operations where it seemed achievable—Africa. The only obstacle, except for the German-Italian army, was the fact that "General Auchinleck could not be convinced". In his memoirs, Churchill bluntly described "the growing pressure that we exerted on him, ending with categorical and formal orders to attack the enemy".

As Churchill and Auchinleck fought an internal battle over campaigning strategies, Rommel planned his own operations essentially ignoring Berlin. As early as April, he set the date of the attack at 26 May, and the messages broken by BP provided the British with increasingly precise information as to when it would strike. At the end of April, a *Chaffinch* message announced that in the end of May, a Ramcke brigade will be transferred to Africa. On April 30, *Panzerarmee Afrika*'s quartermaster's office placed an order for the most urgently needed supplies, stressing that they must be delivered before the end of May. Intercepted and deciphered Kesselring's message included an assessment that the attack planned by the British for June would be late. On 3 May, the Rommel's quartermaster informed about increasing the fuel consumption standard as of 1 June, which signaled the start of mobile warfare. On 9 May, the British Intelligence Committee assessed that Rommel's attack could take place at any time after the third week of the month. As Auchinleck refused Churchill's request to come to London to discuss the situation, he received a categorically worded dispatch on 17 May: "It is imperative that I receive a statement of your general intentions in response to our recent telegrams". Two days later, he bowed to the will of his political superior and announced the launch of an offensive correlated in time with the departure of convoys of supplies for Malta. Relieved by his general's concession, Churchill tried to express his confidence in Auchinleck by suggesting that he take direct command of the 8th Army and move the New Zealand division from Palestine to Egypt. After years Churchill noted sarcastically that "General Auchinleck did not accept the last two suggestions. We'll see how events will force the General to take both of these steps. But, unfortunately, too late!".

Based on their assessment of Rommel's offensive intentions, clearly signaled in the broken dispatches, Auchinleck and Ritchie had been planning a defensive battle for weeks. Pressure from the Prime Minister forced them to adopt a completely different concept of action. In accordance with the original battle plan, Ritchie grouped his army unequivocally defensively. Between Gazala on the coast and Bir Hakeim, he prepared a deep system of minefields and barbed-wire fences. Behind the protective line of minefields, every few miles, he put up reinforced points, which he described as "boxes". Each "box" was prepared for the circular defense, had sufficient supplies for a long period and was manned by a brigade group. Ritchie knew that the line could be bypassed from the south. So, in the back of the front, he placed additional "super boxes" and an armored mobile reserve. The two most important "super-boxes" were the *Knigthsbridge* located a dozen or so miles behind the first line and the Tobruk fortress, which was also the logistical depot of the front. Under the influence of the Prime Minister's expectations, he had to, as quickly as possible, move the troops from the defensive grouping to the starting offensive positions. JFC Fuller commented on the modified plans with some sarcasm: "Of course, the two tasks were contradictory, because the defensive and the offensive do not match each other".

Consistent with earlier assessments, the dispatches broken in the third week of May confirmed an imminent attack by Rommel's forces. On 22 May, Kesselring's order arrived in Cairo to complete preparations for the attack by 24 May. On the same day, a report was received that the DAK commander, who was on leave, was expected in Africa on 24 or 25 May. On the eve of the expected attack, as many as three dispatches confirmed that an attack was a matter of hours away. Finally, on 26 May, British stations picked up the code name *Venezia*, which they correctly interpreted as a signal of the start of the offensive. In the afternoon of that day, Rommel gathered his armored and motorized divisions in the central part of the front and demonstratively led them towards the sea. After dusk, however, the entire cavalcade of vehicles turned back and headed towards Bir Hakeim, which formed the southern end of the British defensive lines. The maneuver was quickly detected by British armored car patrols and confirmed at dawn by aviation, but there was little Ritchie could do to counter the bypassing of his defensive lines. According to the original defensive battle plan, his armored divisions were grouped in the northern section of the front and dispersed due to regrouping for the attack. In the morning, concentrated German armored divisions attacked scattered British brigades. Successively the 2nd, 22nd and 4th brigades were smashed. German tanks swept through the HQ of the 7th Armored Division, whose commander, General Messervy, and much of his staff were taken prisoner.[6] The only consolation for the British was that all the 'boxes' were successfully defended.

With all the mistakes made, fate and the tough defense of the Bir Hakeim box by the French handed the British the chance to not only defeat but even destroy the German armored divisions. The successful defense of all the boxes meant that no supplies reached the fighting Rommel's divisions. Their commander made a potentially risky decision. He moved the entire DAK to the west, where his wings and rear were protected by British minefields, shielded himself from the advancing British with a line of anti-tank guns and concentrated all his energy on breaking through the minefields that would allow his troops to be resupplied. Rommel's operations confirm that Auchinleck and Ritchie's original plan may have been successful. Despite losses in the first phase of the battle, the 8th Army still had 420 tanks at its disposal, twice as many as the enemy. Ritchie was able to divide his troops into two parts. One to bolt Rommel in the Cauldron, the other to use to break through the Italian troops and block Rommel's supply route from the west. He knew the gravity of Rommel's situation from the broken dispatches of the *Chaffinch* and *Phoenix* networks. Even before the battle, BP was breaking the keys of both networks, but with considerable delay. From the start of the battle, the number of intercepted dispatches increased so much that the keys of both networks were broken on the same day they went live. The British were aware of the lack of fuel, ammunition and, above all, water in Rommel's divisions encircled in the Cauldron. However, instead of simply blocking the enemy and calmly waiting for him to surrender, they launched several poorly planned and poorly executed attacks. Finally, box Bir Hakeim succumbed to the enemy's superiority, and Rommel's pioneers punctured the supply route for his divisions in the Cauldron. Rommel regained the initiative and his counter-attack resulted in a mass escape of 8th Army troops, remembered as the 'gallop from Gazala'. Pursuing the defeated British, Rommel could probably have entered Egypt unopposed, but he stopped his pursuit for a moment to pick a fruit from a tree that had escaped him in a previous hand of cards—Tobruk. The second siege of Tobruk lasted exactly one day. On 21 June, the garrison commander, General Klopper, capitulated. Churchill received the news of Tobruk's surrender during a meeting with President Roosevelt. The stress caused by the event determined the dismissal of Auchinleck, but the execution was suspended until the Prime Minister returned from across the Atlantic. Unaware of this fact, Auchinleck attempted to put his troops and frontline in order. On 25 June he removed Ritchie from command of the 8th Army and took command himself. As Ritchie had managed to evacuate Bardia, Sollum and Halfaya Pass early on, Auchinleck saw no point in defending Mersa Matruh and made a decision that saved the campaign for the British. He decided to withdraw the surviving troops to the El Alamein position, where the impassable Al-Qattar depression came within a few dozen miles of the

---

[6] In the course of the chaotic clashes that followed, General Messervy managed to escape and reach the British lines. On 12 Junehe again got into the middle of the German grouping and avoided captivity by hiding most of the day in a dried-up well. On June 19, he was dismissed from the post of commander of the 7th Division.

sea. For the first time during the African campaign, Rommel was to encounter a position that could not simply be bypassed across the desert.

Geography, however, was not the 8th Army's only ally. During the Battle of Gazala, there was a breakthrough in breaking the Enigma keys used in Africa. After an incidental break-in earlier in the year, Hut 6 broke the key again and broke it systematically from then on, generally in less than 24 hours. Over the course of several weeks of systematically breaking the dispatches, the codebreakers noticed that Panzerarmee Afrika was transmitting a daily report to Berlin every day, usually in the evening, covering the current state of the army, the losses suffered during the day and an outline of the intentions of operations for the following day. The timing of the report was a gift to British intelligence. A special procedure had been devised to fish out a long report from among others broadcast by the enemy. During the evening hours, the *Chaffinch* network key was usually already broken, so within hours the contents of the report would reach 8th Army headquarters.

Another novelty was the approach to the *Phoenix* network. This was a purely tactical network; its messages were transmitted at low transmitter power and were not received in the UK. Up to that point, it had taken too long to transmit the contents of the dispatch to BP, decipher it there and pass it on to Egypt for the information gained to still be useful. However, when the situation of the British troops in Egypt became dramatic, it turned out that this time could be shortened enough to learn the enemy's intentions in time. Even more significant was the breaking of another Enigma key, known in BP as *Thrush*. When Rommel's troops stood at El Alamein, it took days to transport supplies overland via the road leading from Benghazi. Consequently, the supply of key materials (ammunition, fuel) to the German troops was by air, from bases in Crete. The *Thrush* network allowed the Germans to coordinate these supplies and the British to send their own aircraft to the right place at the right time to cut off or at least limit supplies to Rommel by this route. Traditionally, *Hut 6* has been breaking the keys of numerous Luftwaffe nets, including *Locust*, *Gadfly* and *Primrose*. Of particular importance, however, was the breaking of the *Scorpion* network, used to co-ordinate ground troops and Luftwaffe. Luftwaffe liaison officers, *Flivos*, accompanied the leading German units, regularly informing their staffs of their location, direction of march and intentions for further action. Codebreakers noted that the key to this network was constructed in a particularly careless manner. Only the key for the first day of a new month was truly random; the keys for subsequent days were purely permutations of it. As a result, it became possible to transfer the decryption of the *Scorpion* network's messages entirely to the CBME, drastically reducing the time required to transmit their contents to the local staffs. As *Hut 6* struggled against the networks of land forces and the Luftwaffe, *Hut 8* broke in August 1942 the key of a network used by U-Boats operating in the Mediterranean. This meant that during the second half of 1942, British codebreakers systematically have been breaking all the German military networks used in the Mediterranean!

As the British gained insight into virtually all aspects of German military operations in the Mediterranean, Rommel lost both his sources of information. How the British found out about the problem of Fellers' reports remains an uncomfortable mystery. Presumably they broke the dispatches containing his reports themselves, then found obvious references to their contents in the broken Enigma dispatches. Churchill diplomatically drew President Roosevelt's attention to the weakness of the American code. He acknowledged that British codebreakers had broken the code, but stipulated that this had happened before the US entered the war. He also assured that no similar actions were currently being carried out against US communications. The message was clear; if our specialists had cracked the code, Axis specialists could have done so too. The allusion was only partially understood. In the spring of 1942, a two-person commission came to Cairo to investigate Fellers' security procedures. The commission confirmed the correctness of the attaché's actions and returned to the country. However, on 24 June the Germans made a nightmarish mistake. In one of dispatches broken in BP, they not only quoted Fellers' opinion that Rommel's seizure of Cairo and the Suez Canal was to be expected, but also identified the source of that opinion. In the following days, the same mistake was repeated several times. It turned out that some of Fellers' reports were equivalent to death sentences for the soldiers whose planned operations he was describing. On 11 June, for example, he sent dispatch number 11119 to Washington, in which he described a planned operation to facilitate the arrival of convoys carrying supplies to Malta. Fellers was optimistic about the operation's chances of success, writing: "the method of attack gives a great chance to destroy targets, and the risk is minimal in relation to the expected benefits". And so it would have been, but at 8 AM on 12 June the telegram was intercepted by German listening stations, deciphered before 10 AM, and at around 11.30 AM on the same day its translation reached Rommel's HQ. When Long Range Desert Group soldiers struck the German airfields the following day, their defenders were prepared and caused a bloodbath of the attackers. The action taken after the final confirmation of the source of the German information had to be radical. After 29 June 1942, neither the Italians nor the Germans were able to penetrate the reports of Fellers (who, by the way, was recalled from Cairo shortly afterwards) again.

Fellers' reports rarely referred to tactical problems. In this role, the signals intelligence company commanded by Lieutenant Seebohm performed far better. An alarmingly large proportion of the information it relayed to headquarters came simply from listening to officers' radio conversations in open text. According to the company's soldiers, the only problem they occasionally encountered was the use of Urdu, the official language of Indian troops, by some officers. Working for Rommel was not an easy piece of bread. Correct reception of enemy dispatches required that the receiving antennas be placed as close as possible to the transmitter, so Seebohm's soldiers were usually stationed close to the front line. This was the case in July, on the front line at El Alamein. The company rolled out its antennas at the immediate back of the front line, on Jesus Hill, Tell el-Eisa. The antennas must have been visible from a great distance, so they may have attracted the attention of the Australians occupying positions opposite. Or the attack on the morning of 10 July was simply dictated by the desire to gain a good observation post. In any case, at 3.40 AM the Australians of 9 Division wrapped their boots in rags and set off towards the positions occupied by the Italians of Sabratha Division. Most of the Italians had already woken up as PoWs, the only resistance was encountered in the posts manned by the elite *bersaglieri* unit. Their survivors swept through Seebohm's company positions in the morning. Its commander was a brave soldier, but the importance of his unit should have prompted him to evacuate immediately. By coincidence, he had recently done so in a similar situation, after which he was scolded by a senior officer for cowardice. This brawl determined the fate of the company and, to a large extent, the entire campaign. Seebohm organised his men for a circular defense, but in the face of the superiority of the Australians supported by the fire of powerful artillery, he had no chance of success. After a short battle, most of his men were dead, wounded or taken prisoner. Seebohm himself, severely wounded, was taken to hospital in Alexandria, where he soon died. The company's equipment and its almost complete archive remained on the battlefield. Reading the documents and interrogating Seebohm's unusually talkative deputy, Lieutenant Herz, the British learnt the full extent of their sins against the communication security. They found confirmation of the cracking of most of the 8th Army's field codes and the now somewhat superfluous evidence of the role played by the Fellers' reports. Most frustrating of all must have been the realisation of how much the lack of discipline in radio conversations contributed to the numerous defeats of the 8th Army about the last year and a half. In the coming months, the Germans managed to reconstitute the unit as the 621st Signals Intelligence Company. However, it never achieved successes comparable to its predecessor. Arguably, this was largely due to a radical increase in radio discipline on the British side. Rommel did not reveal to his subordinates the source of his knowledge of the enemy's actions and intentions. When asked about them, he always replied that he possessed a rare gift of feeling in his fingers—the *Fingerspitzengefühl*. On 10 July, around 9 AM, surprised by the lack of new information about the enemy, he asked the company liaison officer at his headquarters, Lieutenant Wischmann, about the reasons for the silence of his colleagues. When a perplexed Wischmann replied that the company's radio station had not responded to calls for three hours, Rommel asked where it was stationed. When Wischmann pointed to Tell el-Eisa on the map, Rommel's response was short: *Dann is sie futsch* (well, it's gone). In a crucial phase of the campaign, Rommel lost two sources of information in quick succession, and with them *Fingerspitzengefühl*.

At the same time, a steady stream of decrypts flowed to the Cairo headquarters. In the second decade of June, Auchinleck watched the aftermath of discussions between Hitler, Mussolini and their generals about the extent of the coming offensive. Rommel's original intention was to capture Tobruk and continue the offensive to the borders of Egypt. On 22 June, however, Rommel asked Hitler's permission to continue the offensive even after crossing the border. Mussolini was skeptical, but Hitler convinced him with the vision of a triumphal entry into Cairo ("The goddess of war visits the warrior only once. Whoever rejects her at such a time cannot hope for her next visit"). On 23 June, both leaders agreed to go on the offensive, accepting the price of these decisions—the postponement of the planned invasion of Malta.

Almost at the same time, other dispatches brought clarification of Rommel's detailed intentions. After handing over to Berlin an inventory of the material captured at Tobruk (which temporarily solved his army's logistical problems), Rommel demanded the delivery of detailed maps of Egypt and a ban on bombing the port at Sidi Barrani, which might have proved useful to his troops. The following day, air reconnaissance confirmed the march of DAK units towards Mersa Matruh. During the three-day battle for this village, the DAK requested from the X Air Corps in Crete detailed aerial photographs of the El Alamein position. Confirmation of the execution of this order arrived in Cairo the following day, followed by a copy of Rommel's order directing the DAK to march towards the last defensive line before Alexandria. During the decisive days between 1 and 3 July, decrypts reached Auchinleck's staff quickly enough to be used operationally. The general knew in time about enemy's orders to postpone the first attack on the El Alamein line from 20 June to 1 July. He also knew the positions and strike directions of the DAK units. As a result, he knew that DAK's march south represented Rommel's usual ruse. On the last day of this round of fighting, intercepted *Scorpion* network dispatch confirmed Rommel's intention to shift the attack to the coastal road and gave the attack axes of the 21st Armored Division and 90th Light Division. Finally, on 4 July, Cairo received copies of an order directing a halt to the offensive until replenishments could be pulled up.

Around the same time that Auchinleck was winning the clash that went down in history as the first battle of El Alamein, Churchill was wrestling in the House of Commons with a motion of no confidence in his government. After a two-day discussion, he won this clash on 2 July; only 25 MEPs supported the motion, with 475 votes against. The Prime Minister, in a combative mood after crushing internal opposition, insisted on going immediately on the offensive in Africa. His impatience was probably linked to the results of the just concluded British-American staff talks. In the course of these, it was agreed that a combined American-British force would make a landing in French North Africa in the autumn of the same year. An American landing in north-west Africa with the enemy knocking at the gates of Alexandria would have been a heavy blow to the honor of the British empire. Churchill therefore expected his troops to be able to wrest victory in Africa before the start of Operation *Torch*. Auchinleck was realistic in his assessment of the situation of his troops, believing that an offensive at this point would mean squandering the newly gained success and could end in disaster with unpredictable consequences. Churchill was on his way to Moscow for talks with Stalin, so he decided to stop in Cairo and settle the controversy on the spot. He arrived in Egypt on 6 August and immediately announced the decision to effectively remove Auchinleck from command in the Middle East. This was formally done by splitting his command into two independent areas: one covering Egypt, Palestine and Syria, the other Iran and Iraq. Auchinleck felt the Prime Minister's decision was not fair, so he declined the offer to take command in Iran and Iraq and decided to return to the Indian Army. His positions in Egypt were taken over by one of Churchill's favourites, General Harold Alexander. General Gott was to become the new commander of the 8th Army, but he was killed in a plane crash before his appointment was announced. In these circumstances, General Bernard Montgomery was pulled from the UK as a matter of urgency. At this stage he was not yet a Churchill's favorite. After a visit to the 8th Army staff on his way back from Moscow, the Prime Minister described the new army commander in not very encouraging terms: "I don't think any of you will like the guy". The Alexander-Montgomery duo was to go a long way together. However, they were embarking on it with considerable ballast from the past. When Alexander studied at the War Academy, Montgomery, who lectured there, referred to him as "an empty vessel". He also noted that academy staff had "concluded that Alexander lacks common sense, and they were right to do so".

The immediate reason for Auchilnleck's removal from command was his reluctance to take the offensive before September. Churchill, however, accepted without discussion his successor's proposal that any offensive should not begin before October. Prior to this, the 8th Army was to receive 300 new US *Shermann* tanks. Moreover, on the staff desks of his new HQ, Montgomery found a ready and complete plan of attack drawn up by his predecessor. For the Prime Minister's use, Montgomery exhibited another plan uncovered; the evacuation of troops up the Nile and into Palestine in the event of defeat at El Alamein. Its finding certainly helped to assuage any remorse the Prime Minister may having sacked the general planning defeats instead of victories. For those introduced to the context, however, the facts were clear: when Churchill visited the Eighth Army, its divisions were deployed strictly in accordance with Auchinleck's plan and prepared for the operations he had planned.

When Alexander and Montgomery took command in Egypt it was clear that the 8th Army would have to fight one more defensive battle before going on the offensive. On 18 July, Rommel wrote to his wife: "The last crucial days have been particularly bad for us. (…) In military terms, they were the worst days of my life". Only a week later, another letter testified to a change in his mood: "Our desert is filling up slowly. The worst is over". Divisions of Italian infantry, Ramcke's paratroopers and Folgore's Italian airborne division had arrived at the front. Hitler agreed to move 164. Division from Crete to Egypt. Decrypts flowing from BP allowed the Cairo headquarters to track the growth of enemy forces. They reported, for example, Hitler's decision to send 90 tanks from current production to Egypt (their number eventually increased to 110). The decrypts made it possible to sink one of the ships carrying them already in the port of Mersa Matruh. In the first half of August, the broken dispatches brought unspecific references to the planned offensive, then detailed that *Panzerarmee Afrika* troops were to reach offensive readiness on 15 August. This day passed without significant events, but it was then that a dispatch from Rommel's staff was intercepted and forwarded to Cairo two days later. In it, Rommel stressed that the pace of supply to both sides gave him a temporary advantage that would last until September. From then on, the balance would shift in favor of the 8th Army, which should be ready to strike in mid-September. The key to determining Rommel's intentions was the final passage of the dispatch, in which he advised that, as a result of Allied superiority in the air, his troops had to regroup for attack at night, under moonlight. The nearest full moon fell on 26 August, so the British commanders had a good approximation of the date of the attack. The only missing detail was its place, but the British did not leave Rommel much freedom of choice. One element of Auchinleck's plan was to provoke the enemy into repeating his favorite maneuver, that is, to bypass the defensive lines on their southern flank, across the desert. To this end, they have plotted a map of their lines, reflecting reality in a somewhat perverse way. In it, impassable desert sands were marked as roads and existing roads as minefields. The map was furnished with the necessary signatures, stamps, tea and blood stains, before being placed in a British staff car, which completely accidentally became entangled in a German minefield at night. When the British received an order to the Luftwaffe on 17 August forbidding air reconnaissance in the southern sector of the line, they have ascertained that their disinformation worked.

In the meantime, Rommel's superiors must have realized his enduring habit of presenting his own forces as weaker than they actually were. Early August saw the introduction of a new quartermaster report structure, which was greeted with unabashed delight in British headquarters. The report contained previously unavailable details: the number of officers, non-commissioned officers and privates, the number of tanks (including serviceable ones), guns, cars and other equipment. The real strength of the Panzerarmee Afrika, which until then had to be assessed on the basis of indirect indications, was presented with full precision, in a clear and synthetic form. On 21 August, the decryption brought news that Mussolini had approved the date of the attack for the 26th of that month. At this point, something jammed in the smoothly running machinery of German-Italian preparations. Most sensational of all was Rommel's personal report requesting that he be relieved of his position as commander of *Panzerarmee Afrika* for treatment in Germany. Two days later, the British learned that the fuel required for the offensive had not yet arrived at the front, and that the divisions' own vehicles sent to collect it would not return until the 28th. Probably due to the state of his own health and the supply situation, Rommel made a final decision to attack at 4 PM on 30 August, setting the start of the operation for the following night. The battle, which has gone down in history as the second Battle of El Alamein, was not interesting from a tactical point of view. Hardly had the DAK troops set off on their planned raid around the British positions, they got stuck in minefields. The time lost to overcome them meant that they had to turn north earlier than intended. In doing so, they drove straight into a fortified position in the Alam Halfa hills, which they were unable to force through. In the light of day, German columns stretched out in the desert at the foot of the hills and awaiting fuel deliveries were massacred by RAF bombing. In these circumstances, Rommel took the decision to retreat, being extremely surprised that the enemy had not cut off his path. Alam Halfa was a half victory for Montgomery. He defended his positions without much effort, fighting only at a distance and not allowing the enemy to get into a direct clash. But it was precisely for fear of a direct clash that he made no attempt to trap the DAK and destroy it in a convenient situation. Immediately after the battle, Rommel commented briefly on the enemy's actions: "If I were Montgomery, we wouldn't be here anymore". This did not change the truth, however, that for Rommel the failure of the offensive meant the end of his dreams of entering Cairo and the beginning of a series of defensive battles, fought with increasing disparity of forces.

Rejecting Rommel's attack meant, in JFC Fuller's view, that Montgomery "could quietly ignore him until the 8th Army was reinforced to the extent of crushing the enemy". The British were therefore able to concentrate on equipping the incoming divisions with new tanks, American *Shermanns* and *Crusaders* in a new version equipped with a six-pounder gun, and training them in desert warfare tactics. Rommel remained in Africa long enough after the Battle of Alam Halfa to prepare his troops to fight the inevitable defensive battle. Italian and German troops busied themselves with reinforcing the defensive position, the nucleus of which was a strip of minefields still laid by the 8th Army, captured by the DAK in the early stages of the operation. The developed system of minefields was transformed into what its authors termed the Devil's Gardens. Desert Fox did not accept defensive tactics voluntarily—he was forced into it by the Allied offensive against his lines of communication. During this offensive, Enigma decrypts played a key role. In mid-August they contributed to the sinking of the *Lerici* and *Pilo* freighters; on 21 August RAF aircraft sank the tanker *Pozarica*. On 26 August, the decrypts revealed details of the plan to send twenty ships to Africa via Greek ports: on the same day the RAF successfully bombed the Corinth Canal, through which the planned deliveries were to pass. On 28 August a message containing details of another plan to supply troops in Africa was broken in BP; it turned out that three of the eight participating ships had already been sunk. The next day, an Italian C-38m message reported the departure of a *San Andrea* tanker from Taranto, which went down on 30 August. As a result, Rommel started his attack on Alam Halfa, having fuel reserves for 4–5 days of battle. The slaughter of supply ships, directed by the Enigma, continued during the battle. On 2 September *Picci Fassio* was sunk and the *Abruzzi* was damaged; on 4 September *Bianchi* went to the bottom; on 5 September another three ships were sunk. Of the forty-eight Axis vessels sunk in the Mediterranean between 2 June and 6 November, only one vessel's voyage was not notified by decrypted messages and two others did not give precise indications on departure, cargo, port of destination and course. The remaining ships were victims of RAF aircraft or Royal Navy ships, but the primary reason for their mission's failure was the breaking of the enemy's messages. The source of the codebreakers' satisfaction was a number of dispatches describing the consequences of their actions. On the 28 August they broke a dispatch stating that a supply crisis at sea had forced the enemy to launch an emergency fuel transfer system by air from Crete. On 30 August, Rommel's despatch was read out, reporting that only a hundred of the announced 28,000 tonnes of fuel had reached his troops. On 1 October, Rommel's staff reported that the use of the troops' own vehicles to transport fuel had caused an acute ration crisis at the front. On 20 October, the staff reported that DAK had a supply of fuel for four and a half days of fighting, of which only a supply for three days was east of Tobruk. The DAK staff also signaled the disastrous condition of the soldiers, who, as a result of their long stay in desert conditions and insufficient food rations, suffered from mass illnesses. Rommel himself had to leave Africa on 23 September and go for medical treatment in Germany, having previously transferred command to General Georg Stumme, who was pulled from the eastern front.

There was, however, a shadow hanging over the perfect functioning of British signals intelligence, concerning the security of the secret of breaking the Enigma ciphers. Rommel's first alarm signal was the interrogation of a British

PoW, who confirmed that the British knew the date of the attack at Alam Halfa. The precision of the Allied attacks on Axis shipping must also have given cause for thought—not all of them could be explained by treachery within the ranks of the Italian ally. The real crisis, however, came when, in the first days of September, the press of the Allied countries brought speculations about Rommel's state of health. On 9 September, Churchill ordered an investigation into the leak, which soon produced further alarming results. There were further reports in the broken Enigma dispatches potentially concerning the security of German ciphers. Fortunately for the Allies, Rommel's thoughts were by this time revolving around a planned reunion with his family and treatment at the Semmering sanatorium. The crisis had temporarily passed.

The upcoming third battle of El Alamein has taken a position in world military history that does not correspond to its real significance. Churchill said of it: "Before Alamein we had never won. After Alamein we were never defeated". Neither of these statements corresponded to the facts. Montgomery's assessment of his (actually Auchinleck's) battle plan was extremely modest: "It was a master plan and only a master could have written it". In fact, the *Lightfoot* and *Supercharge* operations that made up the battle were a cross between the bloody operations of the First World War and trick and disinformation operations as impressive as they were meaningless. Montgomery began with disinformation designed to convince the enemy that the attack would come from the southern sector of his line. To this end, he began construction of a sham pipeline to supply his troops in the south with water. The pace of the work suggested its completion in November. The Germans did not swallow the bait; General Stumme warned his troops that an attack would most likely occur in the area of the Ruweisat hills, around 20 September. Even if the enemy had accepted the disinformation, its impact on the fate of the battle would have been negligible. On a front 30 miles wide, pulling up mobile reserves to any point on it was a matter of hours. Churchill, as usual, put pressure on his commanders to go on the offensive as early as possible. Knowing that the landing in French Africa would take place on 8 November, the Prime Minister expected victory in Egypt to create a counterweight to the expected American success. Enigma decrypts yielded information that the enemy had ammunition reserves in Egypt sufficient for about 16 days of fighting. Montgomery therefore set the date for the attack two weeks before the planned start of Operation *Torch* and announced that the fighting would last about twelve days.

Victory did not come easily for the British. Tell el-Eisa again became the scene of heavy fighting when dominion divisions managed to cut two narrow passages through minefields on 23 October. The attack caused chaos in the enemy ranks, as General Stumme died in its early hours of a heart attack. The corridors that had been won proved too narrow; the 1st and 7th Armored divisions became bogged down in the minefields. The Enigma decrypts, which had contributed so much to the planning of the battle, lost their significance once the battle began. The first useful piece of information received from BP during the battle was confirmation on 25 October that Rommel had interrupted his treatment, returned to Africa and reassumed command. Even if the British had not learned this from Enigma, they would have recognized the fact easily from the change in command style. For the first three days of the battle, the German armored reserves remained passive in the deployment areas. On Rommel's return, the 15th Panzer Division and the Italian *Littorio* Division launched a violent counter-attack on the New Zealanders. The following day Rommel mounted a two-pronged attack on the strategic point of the battle, known as Kidney Hill, repulsed by concentrated artillery fire by the British. The first stage of the battle was inconclusive. Montgomery failed to win a breakthrough; Rommel failed to close several gaps in his lines.

By this time, news had reached Churchill that Montgomery was withdrawing troops from the front line for rest, allowing them to swim in the sea and sunbathe. The Prime Minister, apparently not yet quite cured of his "guy-anyone-of-you-would-not-like" opinion, reportedly exclaimed "is it really impossible to find a general who can win battles?", after which he ordered Alxander to investigate situation directly at 8th Army headquarters. Alexander's report proved reassuring and the Prime Minister was able to focus on the political implications of Operation Torch, allowing Montgomery to plan the second phase of the assault. Planning the second stage of his offensive, Operation *Supercharge*, Montgomery initially intended to strike again in the north. Fortunately, there were several sober-minded officers at his headquarters, who from the beginning were skeptical about the idea of attacking the enemy at its strongest point. Air reconnaissance reports were received in time, indicating that as a result of the fighting around Tell el-Eisa and Kidney Hill almost all German forces were concentrated in the front sector adjacent to the sea. Convinced by his own staff, Montgomery changed the structure of the planned attack. In the northern sector, General Morsehead's Australians were to launch an attack towards the sea. Their assault should convince Rommel that the northern sector remains the main area for the developing battle, while the real focus of the *Supercharge* was to be moved further south, to the point of contact between German and Italian troops. The Australians did a great job; during the night of 30/31 October they reached the sea, cutting off several units of the 90th Light Division and engaging the Germans in attempts to break the encirclement. The actual operation *Supercharge* started before dawn on 2 November. Indomitable, though very exhausted, New Zealanders set off to attack behind the moving curtain of artillery fire and broke through the minefield belt. The 9th Armored Brigade attacked at dawn, as the silhouettes of the tanks drew in the rays of the rising sun and lost one hundred and two of their one hundred and twenty-eight tanks, failing to reach their obectives. However, in the midst of its struggle, the air raids and artillery fire gradually eliminated the line of German "ack-ack" guns and allowed to introduce into the resulting gap the 1st and 10th Armoured Divisions. On the 2 November, around 9 AM, the 8th Army's own signals intelligence recognized the outline of Rommel's counter-attack: the biggest and most decisive armored clash of the whole African campaign was approaching. For the next 24 hours, an armored battle raged around Tel el-Aqqaqir, in which both sides suffered heavy losses. However, when the battle expired in the morning of 3 November, the British still had about 650 tanks, while only thirty-five remained under Rommel's command. The Desert Fox had to accept the defeat and concentrate his efforts on saving the remnants of his army. The infantry divisions had already been ordered to march to the rear when Rommel's staff received a message from the Führer forbidding retreat and ordering them to fight to victory or an honorable end in battle. General Ritter von Thoma, commanding the DAK, commented on the Führer's order in a practical way. The next morning, he returned from headquarters to the front line, where on the hill of Tell el-Mampsra he tried to stop another British attack with the rest of his tanks (although Rommel himself and his chief of staff Beyerlein were convinced that he sought death). When the last battleworthy tank fired its last bullet, he stood amidst the burning wreckage of its tanks and corpses of soldiers, waiting motionless for the British to take him prisoner. It was 4 November—Montgomery achieved his 12-day victory scenario.

The third Battle of El Alamein was a great success for the nations of the British Commonwealth. However, the success had a high price and was not fully exploited. During the fighting, the losses of the British infantry divisions reached one quarter of their numbers. The armored divisions have lost almost half their tanks. During the twelve-day battle, the total losses of troops in the dead, wounded and missing were roughly equal to the number of soldiers of the Western Desert Forces, at the head of which at one time General O'Connor set out to claim victory at Beda Fomm. The success proved significant, but not conclusive. Rommel deftly slipped away from the procrastinating British pursuit. On 6 November, when a large part of German and Italian troops had already managed to avoid British maneuvers, heavy rains fell, which turned the desert into a sea of mud and prevented the pursuit from continuing. In the evening of that day, a dozen or so miles south of Mersa Matruh, the tankers of the 1st Armored Division looked helplessly from the turrets of their tanks, immobilized by the mud and fuel shortage, at the column of about a thousand vehicles, all that was left of the DAK. Two days later, the Allies landed in French North Africa. Hitler has so far skimped on his favorite general of soldiers and tanks that could have given him a victory. After the Allied landings, when the African campaign was lost to the Axis, German troops of about 150,000 soldiers were gradually transferred to Africa. For the next six months Rommel skillfully fought in retreat, biting back from time to time and considering about what he would have achieved in July 1942 with such an army.

Chapter 18
Yanks Are Coming

Before World War I, the United States did not have an effective cryptology service. During the conflict Herbert Osborn Yardley's self-educated cryptological talent emerged, around which a team known as the Black Cabinet was formed in the State Department. When Stimson ordered its dissolution in 1929, the United States returned for some time to a state of almost complete innocence in cryptology. The low status of the discipline was only partly due to the honorable attitude of American politicians of the time. The man with whom American cryptology was identified during and immediately after the First World War, Herbert Yardley, contributed to it in equal measure. When Stimson abolished his unit, Yardley couldn't get along with life, and with his back to the wall during the Great Depression, he described the history of his service in a book *The American Black Chamber*, which became as great a publishing success as the indiscretion and trouble for the author's former employers. Having been ostracized in the only environment that could provide him with a job as a cryptologist, he later took advantage of Chang Kai-Shek's proposal and went to China, where he was involved in breaking Japanese ciphers. His life-style at the time proved that Yardley was indeed no gentleman. However, after the end of his section, it turned out that Stimson allows for the breaking of someone else's messages on condition that the morally dubious procedure would not involve the State Department. The Black Cabinet's has been transferred to the Defense Department's Signal Corps.

It was managed by William Frederick Friedman, father of modern American cryptology. At a time when cryptology remained a domain reserved for diplomacy and the military, much of Friedman's output in this field was created during his work in a civilian research laboratory where he was employed as a geneticist. Friedman graduated in genetics from Cornell University, as no tuition fees were required for this course. Shortly after graduation, he found employment in the Riverbank laboratory organized by a textile merchant and a somewhat controversial amateur scientist, George Fabyan. Given that he was there to study the effect of the phases of the moon during sowing on wheat yields, one can understand that he

M. Grajek, *Enigma Myth Deciphered*, History of Information Security,
https://doi.org/10.1007/978-3-031-65475-6_18

quickly moved to the cryptology section of the same laboratory. It was rare to practice cryptology in a civilian institution at that time, but it was also a very special cryptology.

Elisabeth Wells Gallup claimed that in Shakespeare's works she found a hidden message in which Francis Bacon was not only the true author of the plays (wrongly attributed to Shakespeare), but also the natural son of Queen Elizabeth (wrongly referred to as the "virgin queen"). Fabyan became one of the followers of her theory; from 1912 Elisabeth Gallup worked at Riverbank Labs to tear out the dramas of their secrets. Luckily for Friedman, soon after he joined the team, the United States entered the war and the demand for true cryptology increased. The series of publications that Friedman developed for the cryptology courses organized by Fabyan's laboratory for the army has gone down in the history of cryptology. Later, Friedman spent some time in Europe, meeting with his British and French counterparts. After the war he returned to Riverbank Labs for a while, but in light of his new experience, research into Bacon's secret messages no longer met the criteria of science; he used his new relations to recruit himself as chief cryptologist in the Signals Corps.

When, on 10 May 1929, Friedman's section took over the documentation and tasks of the former Black Cabinet, he already had serious achievements in his new function. His success did not change the fact that the chief cryptologist of the American army was a commander without the troops; his only assistant was a clerk, an experienced boxer and less experienced typist. As soon as the responsibility of the unit was extended to cryptanalysis, it was renamed the *Signal Intelligence Service (SIS)*, and the first practical signal of the changes was an increase in staff. Friedman was looking for new associates around the same time as the course in cryptology was coming to an end in Poznań. Just like the Polish Cipher Bureau, he was looking for the candidate codebreakers among mathematics graduates, and decided to hire three candidates; Frank Rowlett, Solomon Kullback and Abraham Sinkov. All three had a great career ahead of them; Rowlett became a leader in the attacks at Japanese ciphering machines; Kullback developed his boss' theoretical ideas by publishing a description of his *phi* and *chi* tests in 1935; Sinkov was the main cryptanalyst at MacArthur's headquarters in the Pacific during the war. In September 1939 the SIS staff grew to 19 people, which put the US Army cryptology service in the position of Cinderella even compared to the Polish Cipher Bureau.

The office employees have been following the development of machine cryptography almost from the very beginning—Friedman's negative opinion of the Hebern's machine was the reason why the army rejected his offer. Friedman and Rowlett were involved in the construction of ciphering machines, and their patents in this field are dated 1933–1944. Even earlier, in 1927, Colonel Conger, American military attaché in Berlin, handed over an advertising brochure for a commercial model of the Enigma to Washington, and in May of the following year, the commercial Enigma, purchased for 144 dollars, sailed to the United States. The device arrived in the USA at an unfortunate moment, on the eve of the bureaucratic turmoil around the Black Cabinet and got stuck in the warehouse. However, the Germans have made an unprecedented attempt to attract the American military's attention to their ciphering machines. Twice, in 1930 and 1931, they invited Major Evans, American military attaché, to maneuvers. The invitation itself was not unusual, but the demonstration of the functioning of the ciphering machines arranged on that occasion was shocking. Evans described the device as "similar to a commercial machine and manufactured by the same company. (…) The only difference in the military model is that in the front of the machine (…) there are a number of sockets which can be wired together to change the current flow in the machine". What is even more unbelievable, the Germans shared with the interlocutor the information that the key stations in the land army network use a machine model with 10 rotors, while the navy signals center uses a machine with 20 rotors! Despite an invitation, American cryptologists did not pay attention to Enigma, probably focused on Japanese codes and ciphers.

When Yardley's revelations made the Japanese realize that the Americans had read their messages, they tried to seal their secrets by using machine ciphers. The first Japanese encryption device, known to Americans as *Orange*, was not a challenge for cryptologists, so it was soon followed by a new model, known to Americans as *Red* or *Type-A*. After its debut in 1932, it provided security for Japanese communications for a relatively short period of time, but two years later the principle of the machine was reconstructed and its ciphers were broken. The breakthrough was made by GC&CS codebreakers—Hugh Foss and Oliver Strachey. It was not until 1935 that Friedman's service launched an attack on the *Red* ciphers and succeeded after a year's work. The Japanese must have been aware of the weakness of the *Red* device, because as early as 1937 they started designing a new machine, which in their terminology gained the term "Type 97 printing device" or type B encryption machine (*Angooki Taipu-B*). It was not until the end of 1938 that *Red's* telegrams started to mention the new machine, which in the USA became known as *Purple*. The first message encrypted by the new machine came out on 20 March 1939 from the Japanese embassy in Warsaw.

In the Anglo-Saxons' reaction to the *Purple* there was a significant reversal of fortunes. During this time the British frantically tried to make up for the lag in the analysis of ciphers and codes of the Third Reich and did not take up the *Purple* challenge. On the other side of the Atlantic, Friedman's team took advantage of the fact that many Japanese messages were encrypted simultaneously in *Red* and *Purple* systems. Soon Americans also noticed that *Purple* was following in the footsteps of its predecessors, which separately encrypted vowels and consonants, making it easier for Genevieve Grotjan to make the first breakthrough. On September 20, 1940, she noticed a regularity in the messages that had so far escaped the attention of other team members. The throwing of all forces in the direction she indicated has brought rapid success; on September 27, 1940, the Americans have broken the first messages. Codebreakers must have a subconscious sense of historical moments. At one time, Rejewski had broken the Enigma ciphers less than a month before Hitler came to power. Friedman's team broke Purple's exactly on the day it was announced that Japan joined the Axis. In the autumn of 1940, the engineer Leo Rosen constructed a functional equivalent of Japanese machine to automate decryption.

Traditional rivalry of the US services meant that the Navy also had a *Code and Signal Section*. From 1 July, 1922, inside the naval headquarters, the section bore the designation Op-20-G, under which it was to go down in history. It wasn't until early 1924 that the head of Navy communications, Commander McClean, decided to set up a section within Op-20-G to conduct research on cryptological systems of other countries. At the very beginning, the section had only two staff members, but the founders' personalities balanced the modesty of their number. Lieutenant Laurance F. Safford became the head of the section. He had no cryptological experience, but he was a gifted organizer. His only employee was Agnes Meyer Driscoll, known among her colleagues as Miss Aggie. Miss Aggie already had a lot of experience in cryptology and in the early days of the research section she had to play a mentoring role towards her formal boss (this attitude became quite firmly established in her character). In 1920, Agnes Driscoll worked as a cryptographer in the Navy Signals Department, from where she was hired by Fabyan for a year and employed in the cryptology department of his Riverbank Labs. When she returned to the Navy, her first experience with machine ciphers was waiting. Edward Hebern has announced several texts encrypted with his new machine, describing them as unbreakable. Miss Aggie broke the machine cipher, which encouraged Hebern to hire her as a consultant. A year later, the company went bankrupt and Agnes Driscoll returned to the Navy, just in time to take up a position in the newly created section. In time, Joseph Rochefort joined the team, who would later give the navy the greatest victory in its history and experience the greatest humiliation on its part.

Until the outbreak of war, a state of friendly rivalry prevailed between the cryptologists of the two services, with rivalry clearly dominating friendship. It wasn't until the outbreak of war in Europe and a wave of changes to Japan's cryptographic systems that representatives of the two organizations began to talk. In mid-1940, the Army's chief of communications, General Mauborgne, proposed to coordinate the activities of the interceptions and codebreaking centers. However, Safford's obstructionism meant that the decision had to be made over the heads of the codebreakers. For the key problem, the breaking of Japanese diplomatic messages, a somewhat cursory compromise was agreed. The Army was to deal with the problem on even days and the Navy on odd ones. Or vice versa. This compromise placed the cooperation of the two sections within a clear organizational framework and at the same time granted a dispensation from the practice of mutual friendship on even days. And odd ones.

Faced with the possibility of dragging the US into war, Friedman concluded that it might be useful to exchange experience with counterparts across the Atlantic. In mid-1940 SIS successfully dealt with several diplomatic codes of Japan, but its achievements in other areas were not spectacular. Friedman's employees managed to crack four Mexican codes, two Italian and one German. Both Friedman and his military superior, Colonel Spencer Akin, were aware that the SIS achievements to date would not be an important support for the U.S. Army and diplomacy in the event of war, so they prepared a memorandum suggesting the exchange of "all material available to us on a fully reciprocal basis". At the time, a mission of US military was in London, tasked with identifying any areas where the US could come to Britain's aid without breaching the obligations of a neutral country. Friedman and Akin's reflections had to find their way overseas somehow, because on 31 August one of its members, General George Strong, made an offer that surprised the hosts as much as the other members of his team: the exchange of information in the sphere of cryptology. The British were surprised, for only a year earlier, a Royal Navy mission had attempted to interest its US Navy partners in exchanging information on the communications of the Imperial Japanese Navy; a proposal that was firmly rejected at the time. The British kindly accepted Strong's offer, which allowed him to send a message to Washington on 5 October with a question: "Are you ready for a full exchange of information on the codes of Germany, Italy and Japan? Are you ready for a continuous exchange of the most important intercepted communications-related messages from these countries? Please respond immediately." Although Strong received an answer to his question, it was not immediate.

The first obstacle was Safford, who questioned all forms of cooperation between his service and the outside world. Friedman, the alleged initiator of the proposal, was more open-minded, but there was also an obstacle on the part of the Army: in September 1940 SIS did not have much material to offer. It could be assumed that the British would be mainly interested in Third Reich communications. However, the Americans have so far only managed to crack one German diplomatic code, DESAB (*Deutsches Satzbuch*), in the simplest of its three variants. Friedman must have been embarrassed by the state of his assets when General George Marshall allowed further negotiations with the British on 11 October. Once again, the cryptologists were helped by their incredible sense of historical moments. Just over two weeks after Marshall's decision, Frank Rowlett's team broke the *Purple* code, providing Americans with a tradable commodity. Consistent obstruction from the Op-20-G delayed the start of the cryptological exchange, but in early December, combined pressure from President Roosevelt and the Defense and Navy Secretaries managed to break through Safford's and his superiors' doubts—it was decided to send a combined SIS and Op-20-G mission to the UK. Friedman's participation in the mission would have been a natural culmination of his efforts. However, the stress of working on *Purple* caused a breakdown in his health and forced him to spend some time recuperating. He was replaced by Abraham Sinkov, accompanied by Leo Rosen on behalf of SIS. The Op-20-G representation illustrates Safford's skeptical attitude to the venture; the Navy was represented by Lieutenant Robert Weeks, a Japanese-speaking linguist without cryptological experience, and Ensign Prescott Currier. On 24 January 1941 the battleship *King George V* with the new British ambassador Lord Halifax on board arrived in the port of Annapolis on its maiden transatlantic journey. Four members of the mission observed its entry into the harbor from the seafront, sitting in wait for embarkation on four crates containing secret material.

King George V reached base at Scapa Flow on 6 February. There were two flying boats waiting for the Americans to transport them to one of the ports near Bletchley, but it turned out that their hatches were too narrow to accommodate the crates brought from the States. Not wanting to part with the luggage Sinkov and his colleagues had to embark on the cruiser *Neptune*, sailing in a convoy to Sheerness. On its board they experienced a real war for the first time: German aircraft bombed the convoy, bombs fell not far from the ship, and the deck was strafed by machine guns—fortunately, the bullets did not damage the precious crates. The Americans were not aware that contrary to the positive answer to General Strong, the idea of cryptological exchange also aroused doubts in the UK. The British doubts had the face of Herbert Yardley. Indiscretions included in his book were detailed and damaging. Menzies, Denniston and their colleagues had to wonder when and in what form the secrets they could share with the Americans would come to the surface? Their fears were caused by Yardley's unexpected return to America. Yardley had resigned from his job for Chiang Kai-Shek and unexpectedly appeared in Washington, D.C., where, taking advantage of his acquaintance with General Mauborgne, he imposed himself as an SIS consultant on Japanese codes and ciphers. Discussions preceding the arrival of the Sinkov's mission concluded, that the scope of the exchange would not cover the main BP secret—the fact and techniques of breaking the Enigma ciphers. An internal instruction was issued that in the event of American visitors arriving at BP "steps will be taken to keep them at a distance from our most secret achievements".

Nonetheless, the Americans experienced a dignified reception, for a country at war. Four cryptologists have at their disposal the Shenley Park residence with numerous servants and a richly filled bar. During the official meetings at BP, the ancient sherry from Denniston's stores played the role of exchange stimulator. Sinkov and Rosen gave several lectures presenting the principles of construction and cryptanalysis of the *Purple* machine. During several trips, the guests were shown the most important elements of the British signals intelligence system: interception stations in Scarborough and Flowerdown, DF centers, radar and its underground command center in Dover, OIC in Admiralty and selected BP sections. It seems that the Americans were received roughly in the manner of an experienced guide guiding a group of young people through a museum; presenting spectacular exhibits to occupy the visitors' attention and facilitate discipline, while keeping them at a distance from the most valuable and sensitive objects. The British adopted a simple formula of conversation—they tried to answer guests' questions rather than offer them knowledge on their own initiative. The Americans, for their part, were too tactful (in the case of the SIS representatives) or too full of reserves (in the case of the Op-20-G) to ask questions provoking the hosts to open the curtain more widely. The next days of the Americans' stay in the UK have passed, while BP's main area of activity, Enigma, was still outside the field of discussion.

As the final instance, Churchill intervened. Observing the visit from a distance, he must have been full of concern that the excessive reserve of his subordinates might jeopardise the ongoing Lend-Lease Act negotiations. On 26 February, Menzies, reading his superior's intentions well, submitted a report indicating the preponderance of opinion in favour of revealing the Enigma secret to the Americans. The next day, the Prime Minister scribbled on the document "As proposed. WSC. 27.2". However, before the members of Sinkov's mission were introduced into the secret, the hosts took additional precautions. Delegation members had to sign commitment that they would relay the information obtained only to their immediate superiors, only orally, without any written notice. The Americans were initiated into the fact of breaking the Enigma ciphers, they were given the opportunity to meet and exchange opinions with a group of codebreakers, as well as demonstrated the work of the bombes, but without going into the details of their functioning. The guests visited *Hut 6* on 28 February, the day when the *Light Blue* key was first broken. Codebreakers celebrated the success with a dance of victory on the desks, which supposedly ruined the Americans' belief in British phlegm. At the end both teams were offered a paper Enigma model.

Sinkov's mission visit to BP a was accompanied by compromises on both sides. The composition of the team reflected the unequal interest of both American agencies in the mission. On the British part, there was a clear conflict of interests between the willingness to demonstrate BP's achievements, and the caution of its management in revealing the key secrets. On the SIS side there were officers with solid cryptological training. The Op-20-G team consisted of low-level staff members, with no cryptological experience. In the coming months, however, it was naval intelligence that was to grapple with German U-Boot dispatches, while the Army was to concentrate mainly on Japanese communications. The visit of US cryptologists to BP was arguably an unprecedented act in the history of the British secret service. However, the future has shown that it made a limited contribution to building mutual trust. Nor did it make a significant contribution to future, independent attacks by American codebreakers on the Enigma ciphers.

In fact, the Americans were not entirely honest with their hosts either. Not a word did they acknowledge that Op-20-G had been attempting to attack the U-Boats' ciphers for some time, and that Robert Weeks was one of the members of this team. Despite the neutrality of the US, war has been knocking on the gates of this country on a fairly regular basis. On 5 September, 1939 Admiral Stark formed the Neutrality Patrol—a naval force that was to make sure that the war did not get too close to the American shores Officers of this force identified the word 'war' in a particular way, associating it exclusively with German ships. They reported their observation to headquarters in such a way that the report would also be picked up by Royal Navy ships patrolling nearby. The flow of supplies from American ports to the UK also left no doubt that US neutrality has a pro-British color. From 1 September, 1941, the US Navy was to take over the escorting of Atlantic convoys to the shores of Iceland, and on 31 October of that year the US Navy lost the first ship torpedoed by the U-Boot—destroyer *Reuben James*. Anticipating the prospect of US Navy involvement in the Atlantic, Safford decided to reprioritize his service's work. He detached Miss Aggie from attacking Japanese codes, entrusting her with attacking U-boot ciphers. Originally, Driscoll had only former navy radio operator Milton Gaschk and two female aids at her disposal. In late 1940, four young US Navy officers, including Robert Weeks, joined the team. The shocking feature of this team was its complete lack of experience. However, Driscoll had already brought up so many generations of naval cryptologists that she did not see it as a problem. A much more serious inconvenience was the lack of cipher material; Driscoll gave the team only 131 dispatches. Despite all the limitations, the team got to work and within a few months had registered progress. Its members noted the distinct contribution of rotors and the plugboard to the structure of the cipher, and the specific role of the letters **Z** and **X** in the messages. The seriousness of these findings is somewhat diminished by the fact that they were mostly based on the previously cited report of an American attaché, whom the Germans had tried to interest in purchasing the machine.

After Sinkov's mission returned, the Americans complained about the censorship of information imposed by the British. They were somewhat right, the British certainly did not match the openness of the Poles from the Pyry meeting. But there were problems on both sides. Opponents of cooperation stressed that the exchange was not equal. The Americans handed over a reconstruction of the Japanese *Purple* machine, while the British did not reciprocate with an analogue reconstruction of the Enigma. This claim contained just enough truth to be able to manipulate the decision-makers, and it was probably formulated for this purpose. Clearly, the British did not provide the Americans with a copy of the device that could be used immediately to read German dispatches. However, the information contained in the paper model of the Enigma allowed its analogue to be constructed immediately. The Americans apparently did not know that, unlike the Poles, the British had never built an exact equivalent of the German machine. To decrypt its messages, they used a suitably modified copy of their own TypeX cipher machine. Handing over its copy potentially compromised the security of British own communications, and at this stage of the transatlantic relationship was not an option. After all, a little later, the Americans also refused to demonstrate their SIGABA machine to the other side.

The model of the machine allowed Driscoll's team to abandon the results of earlier work and begin a systematic attack on the Kriegsmarine's ciphers. Driscoll set out to design her own method of attack, which she declared was to be independent of mistakes made by enemy cipher clerks and economical. The first caveat signals that Sinkov's mission has learned about the "Herivel tip" and *cillies*. The word 'economical' probably meant that the method was to be independent of the design of the complicated machinery corresponding to the British bombes. Driscoll relied on the classic catalogue attack; its staff proceeded to laborious and time-consuming catalogue construction of all possible clear text bigrams and their encrypted equivalents, with all possible rotor settings. Usage of a catalogue attack was not a mistake in itself. After all, *Einsing* was used in BP as one of the auxiliary methods, which was also based on the catalogue. However, Driscoll ignored the problem of the Enigma plugboard and its impact on the generated cipher.

Miss Aggie's problem in responding to the Enigma's challenge was her unexpectedly conservative attitude, covering not only cryptological challenges. She was a supporter of Christian Science, whose followers put the healing power of prayer before the methods of conventional medicine. After a car accident in October 1937, she applied the principles of her faith in practice; when she returned to work a year later, the permanently deformed leg and the need to use a cane were merely outward manifestations of recent experiences. In her surroundings, Driscoll was perceived to have closed herself off. Perhaps it was during this period that her popular nickname 'Miss Aggie' was replaced by the more detached 'Madame X'. Driscoll, who raised generations of American cryptologists, reached for tried-and-tested solutions in tackling the Enigma cipher. Unfortunately, these proved inadequate, led to a dead end and resulted in the loss of about a year in the American codebreakers' struggle with Enigma.

Fortunately, not all this time was lost to the organization of cooperation between the cryptologists of the two countries. Denniston treated Sinkov's mission as reconnaissance and did not expect immediate results from it. While the Americans were touring the BP huts, he wrote to Menzies: "Now that we can cooperate fully and in every area, we are drawing up plans for continued cooperation for the period after their return to the US". The conduct of the visit quelled the voices of sceptics, despite a few childish gaffes by the Americans on their return trip to the country and immediately after their return. Hardly had the Royal Navy ship with all four on board set sail for Scapa Flow, from where they were to depart for the US, a disturbing intelligence report reached BP. One of the members of the Army delegation was overly enthusiastic and open about the cooperation of cryptologists. As the British swallowed its contents, a demand arrived from the US to hand over to them a 'cipher-solving device' sent in clear text. In 1941, Denniston decided that the situation was ripe for his visit to the United States.[1]

On his way across the Atlantic, he had to think about the goals he wanted to achieve and the available resources. The objectives, as far as we can reconstruct them, were obvious. Denniston intended to convince the two US agencies that at this stage they should not operationally concern themselves with breaking the Enigma ciphers. On behalf of BP, he was able to promise that the British would share any data that would allow the Americans to undertake research work on the cipher. The research work should prepare Americans to undertake practical decryption should the need arise in the future. For the time being, the Americans should focus their attention on Japanese codes and ciphers, which are of greater practical importance to them, while the British, engaged in a struggle with Enigma, had to relegate them to the background. Denniston's fear of an active role for his partners in the attacks on Enigma was due to two reasons. He feared indiscretions or leaks of information on their part. He knew how much time and effort was required to build a system for the secure transfer of information from decryption to the final recipients. Moreover, American commanders had a reputation with the British as impulsive cowboys, reaching for their weapons first, only later considering the consequences of such behavior. If decryption information were at the root of such an action, it could lead to the secret being betrayed. Denniston's second source of fear was the industrial power of the US. He was well aware that the main tool of his codebreakers, besides mathematics, were complex electromechanical devices, the number of which still remained modest in relation to needs. He could easily envisage a situation in which the Americans were mobilizing the power of their own

---

[1] On his way to Washington, he had to visit Canada and solve the Yardley problem, which in the meantime ended his cooperation with SIS, but cleverly took advantage of the resulting increase in his own credibility and started working for the Canadians, taking over the management of the newly established cryptology office under the name *Examination Unit*. After the strong intervention of the Americans and the British, who showed an astonishing consensus on the Yardley issue, the Canadians got rid of the cuckoo egg by the end of 1941.

industry, whose designs and products would move BP into the shadows. He was not the only one with concerns of this nature. In the instructions he received its author (judging by signature Menzies) advised against discussing the subject of a possible supply of tabulators to BP, as this might make the Americans realize the implications of their industrial advantage.[2]

Denniston visited the SIS in the first instance. Friedman had already returned to work and the two service chiefs easily worked out a middle ground. Friedman demonstrated to the Briton how the information brought back from BP allowed him to expand his service; new sections of the Vichy and Latin American codes were created, and progress was made on the Third Reich's diplomatic codes. Shortly before Denniston's visit, the team under Kullback's leadership switched to 24-hour operation, which allowed the use of tabulators and sorters of another staff section. Denniston was somewhat surprised by the modest degree of automation of the work at SIS. During his visit to BP, Rosen rolled out a vision of extensive use of the specialized devices. However, aside from analogs of Japanese *Red* and *Purple* machines, Denniston saw no other specialized equipment. He was impressed by the extensive use of sorters and tabulators, but probably not by the purposes for which they were used: "They use these machines on a much wider scale [than BP] to save personnel effort, but I don't think these mechanical devices are the way to success". The visit to SIS ended with the establishment of rules for future communication.

The meetings in Op-20-G had a somewhat surprising course for Denniston. After an introductory meeting with management, Driscoll took the baton. From the very first words she spoke, Denniston realized that Weeks and Currier did not respect the obligation to pass on their knowledge to Safford alone; Madame X made no secret of her knowledge of Enigma. Denniston's attention also did not escape the fact that during their visit to BP Weeks and Currier did not say a word about her attack on the Enigma cipher, which was already underway at the time. Denniston remained silent when Driscoll revealed her low assessment of BP's attack methods. In her view, they required long cribs, complex technical equipment and the availability of extensive cipher material. Driscoll briefly outlined her catalog attack, maintaining that it does not require the availability of many ciphertexts, the cribs are short and naturally structured, and the method is independent of mistakes made by German operators. Denniston was probably not prepared to comment on her speech, but he remembered that his subordinates had at one time considered the possibility of a catalog attack and concluded that it was impractical against Enigma. A little later, Turing calculated that breaking a single key using the Driscoll's method would require 72,000 man-hours. On the positive side, Driscoll not only did not demand the delivery of a copy of the bombe, but questioned its effectiveness, which was perversely in line with Denniston's priorities. The ensuing discussion must have been somewhat frustrating for both sides. Driscoll admitted that she did not understand some aspects of Enigma's operation, such as the double-stepping mechanism,

---

[2] '*I should feel inclined not to mention the Hollerith machinery, as this might be used as an argument for entrusting the American firm with the construction of the Bombe (...)*'. NA, HW 14/45.

despite the information brought by her subordinates from BP. Denniston promised to answer all the questions, but when he tried to invite her to visit the UK, where she could get a first-hand explanation of her doubts, he met with a firm refusal, motivated by her health condition. Finally, she promised to provide a more detailed explanation of her method of attack and the rationale for its superiority over BP's preferred methods. Since Driscoll's position found unequivocal approval from Safford, talks with Op-20-G hung in limbo.

Soon after Denniston's return to BP, it became apparent that the layers of mutual distrust had not fully surfaced during his visit to the US. The BP chief brought with him a list of questions from Driscoll's side. Some of these were understandable and concerned plugboard settings, bigram tables and the delivery of more intercepted ciphertexts. Some were somewhat confrontational: a commitment to deliver to Op-20-G copies of all intercepted ciphertexts, or a renewed request to provide a copy of the machine. The codebreakers at BP were most appalled by questions confirming her lack of knowledge of the basic principles of the machine's functioning. Turing and his colleagues asked the natural question: what value will the Driscoll catalogue represent if its authors do not know how the machine works? Despite their doubts, they prepared answers to all the questions and sent them to Washington on 2 September.

The problem for the Op-20-G team was the wrong direction of its efforts, something its bosses did not want to admit. Safford put his full trust in his mentor and tried to speed up the work by strengthening the team; by the end of 1941 it numbered 14. The nervous tension resulting from the lack of results contributed to increased stress for Driscoll, who vented it on those around her. Convinced of the lack of goodwill on BP's part, she saw it as the main cause of her problems. Apparently, she also convinced Safford of this thesis, who in turn mobilized his superior, Admiral Noyes, into action. Throughout November and December, BP was bombarded with letters from Washington formulated in a rather ultimative tone. The beginning of the exchange was the mysterious disappearance of the shipment of 1 October. It contained not only answers to Driscoll's earlier questions, but also Turing's memo, pointing out the weaknesses of her catalogue attack. An interesting aspect of the case was the disappearance of the memorandum, while Denniston's accompanying letter to Safford arrived reliably. Manipulated by his subordinates, Admiral Noyes did not entertain diplomacy. In a series of letters to Denniston, he accused BP of breaking the exchange arrangements, censoring information and sabotaging work carried out by the Op-20-G. Captain Hastings, a SIS liaison officer in Washington, informed Menzies at the end of November that "there is serious anxiety and disappointment surrounding the exchange of special intelligence. [Admiral Noyes says the Americans] are aware that you are blocking the handing over of some European code books and keys that should go to Washington by agreement".

Denniston felt that this was no time for disagreements, especially unfounded ones. Particularly given that the second half of 1941 was a difficult time, both for him personally and for the organization he led. Moreover, transatlantic cooperation

was in the interests of both partners: on 1 September, the US Navy took over responsibility for for escorting Atlantic convoys in the western part of their route. In this
situation, it was necessary to share knowledge of U-Boot operations with the
Americans or allow them to break the cipher themselves. Irritated by the accusations from Noyes, Denniston abandoned diplomacy for a moment: materials
requested by Noyes were sent in a package that was only partially confirmed; no
information regarding the Enigma is blocked by BP, which has complied with all of
Driscoll's requests except for the machine itself; Driscoll has so far failed to fulfil
her promise to provide information about her method of analysis. The truth had the
desired effect. Admiral Noyes must have recognized the manipulation of his subordinates and on 11 and 12 December sent two letters to Denniston retracting all
previous accusations and finding the explanations received convincing. On 13
December the British received additional confirmation of the source of the recent
problems. Someone from the Navy Department sent a telegram to BP, referring to
the biblical verse of the Luke Gospel: "And she found them. So, she called her
friends and neighbors and said, rejoice with me, for I have found the lost that was
strayed". The sobering up came just in time. For a week, the US had been a party to
the ongoing war.

Chapter 19
The Black Pit

The struggle that has gone down in history as the Battle of the Atlantic began modestly, with U-Boats operating in the shipping lanes leading to the main ports of Britain's west coast. When the Admiralty created a new naval command, Western Approaches, its activity made attacks in these areas too risky for submarines. When the British established an effective convoy system off the west coast, the *Befehlshaber der Unterseeboote* (BdU) responded by pushing back to the west zone where his *'grey wolves'* waited for the spoils. The Royal Navy was forced to escort ships between British ports and 20, 30, and finally 40° W. A convoy going out towards the ports of Nova Scotia and Newfoundland would follow in the company of an escort to a certain point in the western longitude, where the escort group would take charge of the convoy sailing in the opposite direction. The escort functioned similarly on the other side of the ocean, where the Canadian Navy would take care of a convoy leaving for the UK on a dangerous initial leg of the route, allowing it to sail unescorted in the open ocean until it encountered a British escort group. However, in May 1941, convoy HX126 was attacked at 41° W, as far west as any before it. The move of the U-boots into the open ocean forced the Royal and the Canadian Navies to eliminate the mid-ocean gap and escort ships all the way. From then on, ships of the Newfoundland Escort Force Command, mainly Royal Canadian Navy, accompanied the convoys heading east to a point known as MOMP (Mid-Ocean Meeting Point), located at 35° W, where escort groups based in Iceland took charge. These in turn escorted the convoy to EOMP (Eastern Ocean Meeting Point, 18° W), where Western Approaches took over. In addition to the surface escorts, the ocean was patrolled by aircraft of both parties. The German submariners were most likely to operate in the eastern Atlantic, where they could expect support from Condor aircraft taking off from bases in France. The U-boot captains claimed that fighting the convoy escorts and torpedoing the ships was the easier part of the job. The problem was finding a convoy in the vastness of the ocean—a report from a patrolling aircraft was the solution. However, where *Condors* reached, British aircraft also patrolled, and, taking off from bases in Newfoundland and Iceland, they covered most of the convoy route. The range of aircraft available to the Coastal Command at the time allowed continuous patrols to cover waters within about 400 miles of base and incidental patrols within 400–600 miles. In October 1941, a report was prepared for the Admiralty, according to which more than half of the ships lost in September fell victim to U-Boots outside the area of permanent air patrols, and the remainder in an area where aircraft appeared sporadically; no ship was lost in the area covered by continuous air cover. The facts dictated a solution—air cover had to be provided for the convoys along the entire route. The Allies did not have at the moment aircraft with sufficient range to provide air patrols over the central part of the convoy route. The area in the mid-Atlantic was referred to as the Black Pit, and its waters were gradually turning into the largest graveyard of ships in the world's seas.

The breaking into the *Dolphin* network thanks to documents found on board *München*, *Lauenburg* and U-110 brought temporary relief to the sailors of ships and escort vessels in the Atlantic. The great map of the ocean in the Submarine Tracking Room no longer gave only the current positions of convoys and the places where further U-boot casualties had been sunk; a large number of symbols appeared on it showing the positions and courses of enemy ships. *Hut 8* received complete keys covering June and July 1941, so there was no need to resort to the aid of the bombes that summer. Translated and commented enemy dispatches were landing on the desk of the duty officer in the OIC in less than two hours after transmission. If it was apparent from them that any of the convoys were sailing towards the U-boot patrol line, the Admiralty order directed ships to a new course, avoiding the position of the outermost ship in the patrol line from the open ocean....

The new tactic proved surprisingly successful; the U-boots waiting in patrol lines did not register contact with the convoy between 1 and 23 June. The unprecedentedly long period without contact with the enemy must have been of concern to Dönitz, especially as it followed immediately after the tankers' slaughter. Luckily for the British, the U-boot commander pieced together the facts and recognized that the enemy might have obtained the keys aboard the *Gedania*. Accordingly, on 20 June, he decided to disband wolf pack *West* and redeploy its ships, along with the new arrivals, into a long and fairly loose patrol line extending through areas of the mid and north Atlantic. But this was not his only response. Dönitz was aware of the crucial role that communications played in wolf pack tactics and was therefore extremely sensitive to the problem of its security. He was repeatedly persuaded by experts of the security of the Enigma cipher, so he directed suspicion in other directions. Dönitz was inclined to attribute possible knowledge of his plans to British and French agents operating in U-Boot bases or to treachery in Kriegsmarine headquarters other than his own. In many of his actions, the BdU demonstrated the attitude of a perfectionist convinced that only the areas of operation directly controlled by him can function reliably. Since he had no sympathy for his immediate superior, Admiral Erich Raeder, he did everything to consolidate his exclusive authority over the U-boot fleet. Autarkic tendencies also applied to the sphere of communications. From the beginning of the war, the U-Boots operated within a network common to all Kriegsmarine ships. Dönitz attempted to exclude his ships from the general network by transferring them to another over which he would exercise exclusive control. *Hut 8* first encountered the effects of this policy in April 1941, when the U-Boots' dispatches could not be read despite the *Dolphin* key being broken. Codebreakers were quick to explain that submarines used the same Enigma settings as the rest of the navy, but taking the rotor start positions in reverse order (if the key used by surface ships required a rotor setting GTK, U-boat ciphers set them in the position KTG).

When the tankers' slaughter and the loss of the *Gedania* made Dönitz aware of the possibility of compromising the keys for June, he was unprepared to take the next step and reached for an *ad hoc* solution. On 10 June, the *Gedania* was brought to Greenock; on the 16th of that month, BdU ordered a change in the way the ships' positions and other geographical coordinates were communicated. Until then, the positions were determined on the basis of a secret Kriegsmarine map, on which the Atlantic was divided into a grid of squares. Each square of about six nautical miles could be identified by its designation consisting of two letters and four numbers, e.g. the designation CA2745 indicated a point in the center of New York harbor. The methodical Germans were also prepared in case the map was captured by the enemy; on board each ship was a list containing the names of a number of points on the ocean and their geographical coordinates. From 16 June onwards, U-Boots ceased giving coordinates based on a grid of squares, replacing them with coordinates expressed in relation to notional points on the ocean, e.g. '30 nautical miles south-east of Fritz Point'. It was not a sophisticated system, but it passed the test as an emergency procedure. BP managed to ascertain the position of most of the points still during June, but the satisfaction was temporary. On 11 September, the old grid was restored to favor; however, the resulting square designations were further encoded before encryption by replacing a pair of letters based on the current bigram table. By making the change, Dönitz showed confidence in the security of the Enigma; information about the new procedure was passed to ships by encrypting it in a 504-group long message in the officers' *Dolphin* variant. Hardly had the described method of coordinate encoding fallen prey to BP, the Germans introduced another complication. Until now, a single table had been used to encode coordinates. From November 1941 onwards, each ship was equipped with several tables, and the choice of one was announced in a camouflaged manner; in the text of a dispatch from HQ there appeared an agreed name and address of the person who identified the table used to encode the coordinates and the value added to the square designation. U-boot command could, for example, issue an order containing the phrase 'Gerhard Buchholz, Gotteshilfestrasse 20'. The first letters of name and surname in the example identified the correct bigram table, and the house number indicated that the number 2000 should be added to the numerical part of the designation. As a result, the square CE2987 could, when encoded, take the form AD4987. The system resisted attack by codebreakers for a long time. Its complexity also meant that U-boot crews were sometimes asked to confirm whether they were actually transmitting from the center of the Sahara or a lovely valley in the Andes.

During June and July 1941, *Hut 8* deciphered all the dispatches of the *Dolphin* network. Apart from information of operational importance, the codebreakers gained a lot of information useful for their own workshop. They were able to clean up the probability tables used in the *banburismus* from the taint resulting from training and dummy dispatches. This allowed them to avoid a crisis when the validity of the keys acquired on U-110 expired at the end of July. The corrected tables made *banburismus* an effective tool that allowed the *Dolphin* network's messages to be broken within 1–2 days. In August, *Hut 8* continually delivered decrypts to the OIC; it only failed to break the keys for the 1–4, 24 and 25 of the month. After September, in which the key for all days of the month was broken, came October, in which the Enigma resisted the attack in four days. From 14 October onwards, the *Dolphin* key was broken every day until 7 March 1945, when the network disappeared from the air. But problems with the key in the first days of the month reminded codebreakers of the April surprise, when the U-Boots started using an inverted rotor setting. The mastery of other Kriegsmarine keys helped to identify the source of the problem. *Hut 4* was home to a team led by Dr. C.T. Carr, dedicated to breaking the manual ciphers of the German Navy, which at the apogee of its activities had mastered 20 of the 27 known German systems. Carr's codebreakers suffered from an unwarranted inferiority complex towards their colleagues from *Hut 8*, giving expression to it in a humorous way—have installed a sign on the door reading *Operation Cindrella*. But on 5 October they were the ones who experienced a moment of, enabling their colleagues to resume decryption. Among the broken *Werftschlüssel* messages they have found a dispatch that was a copy of another message, transmitted in the main U-Boot network. By using its fragment as a crib, *Hut 8* was able to break the key of the main network as well. It turned out that since 1 October the rotor starting position used by the U-Boots was not only different, but also unrelated to the *Dolphin* key; Dönitz had taken another step in his efforts to completely separate the communications of the submarine navy. Although a complete break between the communications networks of U-Boaots and surface ships was not to take place until the first months of 1942, a change in early October caused that the former be treated as a separate key. The communications network of U-Boots operating in Atlantic waters became known as *Shark* and was to become the main challenge to *Hut 8* for months to come.

November brought another surprise—at the end of the month, the Germans changed the bigram tables, temporarily rendering *banburismus* useless. Just a few months ago, Hinsley would have sat down at his desk with a heavy heart and started writing a request to the Admiralty to undertake a hunt for yet another weather ship. But thanks to the continuity achieved over the summer in breaking U-boot dispatches, this time the codebreakers have managed to overcome the crisis by their own efforts. After reading hundreds of dispatches, they were able to identify the habits of the captains and the standard formulas used in communication, so they had a set of pretty good cribs. Analyzing the structure of the keys in force on the following days, they noted that their structure was not entirely random and was subject to certain rules. The key had to include one of the navy-specific rotors; VI, VII or VIII. No rotor could remain in the same position for two consecutive days. The combined application of the known rotor selection rules meant that the number of rotor combinations was reduced from a theoretical maximum of 336 to 105. The final factor enabling a successful attack on the new bigram tables was an increase in the number of bombes. When *Hut 8* began its attack on the *Dolphin* network in August, BP had six bombes in operation. By the end of November there were fifteen, twelve of which had been placed at the exclusive disposal of *Hut 8* for a period of time. Turing and his colleagues knew good cribs, for each of them they only had to examine one-third of the theoretically possible rotor orders, and they had a sufficient number of bombes at their disposal for this purpose. Each broken day key opened the way to *Einsing* and the reconstruction of individual message keys. Each broken key allowed a section of the bigram table to be reconstructed. At the end of the year, *Hut 8* was feverishly working on reconstructing the bigram tables when an unexpected gift from outside made it redundant.

In December, the Royal Navy conducted two operations involving commando units, again targeting the Lofoten Islands. This time the possibility of searching encountered Kriegsmarine ships for secret documentation was included in the operation plan. A few days before the start of the operation, a boat driven by Jan Sigurdson, a fugitive from Norway, reached the Shetlands. Thanks to the information he provided, Allon Bacon pointed out to the departing vessels which of the ships they might encounter offered the prospect of seizing material of interest. As part of Operation *Anklet*, a squadron of ships under the command of Rear-Admiral Frederick Dalrymple-Hamilton entered Vestfiord on 26 December, where the sailors from the destroyer *Ashanti* searched the German coastguard ship *VP5904/Geier* before sinking it. Operation *Archery* continued in parallel, with Rear-Admiral Harold Burrough getting his squadron into Vaagsfiord on 27 December, where destroyers *Offa* and *Chiddingfold* repeated *Ashanti's* achievement by searching and sinking the coastguard ship *VP5102/Donner*, and *Onslow* did the same with the coastguard trawler *Föhn*. On board the German vessels, both the current bigram tables and the K-Book, replaced at the same time, were found. On New Year's Day both teams reached Scapa Flow, from where the captured documents were delivered to BP.

The possibility of abandoning the reconstruction of the bigram tables was good news for *Hut 8*. Unfortunately, it was accompanied by not one, but two pieces of bad news. German communications procedures allowed for the possibility of key compromise and included an emergency procedure. If the HQ had reason to suspect that the valid key had fallen into the enemy hands, it could issue a so-called *Stichwortbefehl*, or password, the contents of which influenced how to modify a basic cipher key. The coincidence of timing suggests that the British raids on the Lofoten Islands and the loss of several units prompted Dönitz to reach for this precaution. On 28 December, he issued the *Stichwortbefehl*, applicable from then on to the entire U-boot network. Mahon mentions that from 1943 onwards, the *Stichwortbefehl* was in force virtually continuously in the Atlantic U-boot communications network, with constantly changing passwords. The procedure for transforming the original key was also permanently complicated to the point of absurdity. German cipher clerks forced to perform mental gymnastics in connection with the

Stichwortbefehl did not learn that they were fighting a battle with the shadows—Allies only occasionally were breaking into the network on the basis of captured keys. The cipher-breaking methods they used on a daily basis did not distinguish between the original and modified versions. As a result, the confusion over *Hut 8*'s first-ever encounter with the *Stichwortbefehl* lasted only two days, after which the situation returned to normal.

Enigma Key Generation Based on Stichwortbefehl

In the captain's safe there were several envelopes marked with various passwords. Let us assume that the HQ issued an order reading `STICHWORTBEFEHL PERSEUS`. The captain would find the envelope marked `PERSEUS` and then take out a sheet of paper containing a different password, for instance `DANZIG`, the letters of which were used to modify the settings of the enigma machine. If the key for a given day stipulated that rotor numbered I, III and V should be installed in the machine, the officer would add the ordinal number of the first letter of the password (in the example, `D = 4`) to the number denoting the number of each rotor modulo 8. As a result of the modification, rotors V (resulting from the sum $1 + 4$), VII ($3 + 4$) and I ($5 + 4$ mod 8) were installed in the machine.

The next 3 letters of the password affected the ring setting of all rotors. Let us assume that the original key provided for a ring setting of `U, K, G`. The ring position of the first rotor must therefore be shifted by `A = 1` characters, the second by `N = 14` and the third by `Z = 26` characters. The new ring setting is `V, Z, G`.

The numerical value of the next character of the password (`I = 9`) had to be added to all the characters in the list plugboard settings. If, for example, the basic key provided for an `A-F` pair, it turned into `J-O`. If the password contained more than 5 characters, its remaining letters were meaningless.

The procedure was well designed because the password triggering its use (`PERSEUS`) was unrelated to the password used to modify the machine settings (`DANZIG`). The weakness of the procedure was the assumption that the envelopes do not fall into the enemy hands along with the compromised key (as happened on board U-110). From BP's point of view, the *Stichwortbefehl* procedure did not pose a threat, since under normal circumstances the keys were obtained by cryptanalysis, not by betrayal or capture.

The worse news was a document found on board one of the ships that confirmed earlier indications of the planned switch to a new Enigma version, equipped with four rotors. The first harbinger was the Enigma cover found on board U-570 with four windows. Soon references to the new version of the machine, referred to as the M4, began to appear in U-boot dispatches. When the bigram tables captured in the Lofoten Islands arrived at BP in early January, it seemed that 1942 was off to a promising start for *Hut 8*, but the good fortune did not last long. First, on 20 January the Kriegsmarine introduced a new weather code, depriving the British codebreak-ers of the cribs they were already starting to treat as good friends. Finally, on 1 February 1942, the predictions were confirmed: none of the cribs so reliable up to that point had had any effect; the M4 cipher machine came into use, and the *Shark* network finally and entirely separated from the *Dolphin* network (Fig. 19.1).

Thanks to broken messages, captured documents and the mistakes of German operators, Turing and his colleagues already knew quite a bit about the new Enigma model. Perhaps the most important piece of information was that calling the M4 a four-rotor machine was a bit of exaggeration; it was rather an adaptation of the three-rotor Enigma to work with four rotors. The adaptation was done by splitting the old reflector into a new, narrow one and inserting an adjacent narrow rotor. The new reflector and rotor were housed in an unchanged mechanism, which included elements of the rotor shift for only three rotors. As a result, the new rotor could be set to one of 26 possible positions prior to encryption, but remained stationary dur-ing operation. The ability to change its starting position meant that the cipher key covered four characters. The new reflector became known as the *narrow B* or *Bruno*, while the narrow rotor was referred to as the *beta*, giving rise to a series of rotors denoted by successive letters of the Greek alphabet and consequently called *Greek rotors* (*Griechenwalzen*) (Figs. 19.2 and 19.3).

Had the M4 Enigma been a fully four-rotor machine, *Hut 8* would have faced an almost impossible task. This would have meant that the choice of rotors could be made in one of $9 \cdot 8 \cdot 7 \cdot 6 = 3024$ ways instead of the existing 336. Checking all pos-sible combinations with bombes would have required time making decryption impractical. But the compromise way of introducing a fourth rotor made it easier to solve the problem. Not only did the new rotor remain stationary during operation, but it could not be placed in any other position but next to the reflector. This meant

Fig. 19.1 4-rotor U-Boot Enigma, M4. (cryptomuseum.com)

Fig. 19.2 M4 thin reflector. (cryptomuseum.com)

Fig. 19.3 M4 thin reflector beta, one of the Griechenwalzen. (cryptomuseum.com)

that codebreakers still had to deal with 336 combinations of the three old rotors, to which the beta rotor added 26 new ones.

Hut 8 had been warned of the new threat in advance and could prepare to face it. Mahon records that the first mention of the M4 arrived to BP in a document dated January 1941. The equipping of U-Boots with the M4 had to take some time, so it was necessary to allow for a situation where some ships used the M3 model and others the M4. To enable them to communicate with each other, the new model had a mode of operation in which it behaved identically to the M3. The instructions obtained informed that, with the ring of *beta* rotor in the position **Z** and the rotor itself in the position **A**, the set *beta* plus *Bruno* behaved identically to the reflector of the old model, enabling the two machines to work together. There were ships at sea with the M4 on board, using their machine in a mode that mimicked the behavior of the older model. Earlier experience suggested that one of the cipher clerks would make the mistake of transmitting the same text in the old and new variants of the cipher. Indeed, it happened on several occasions that a German radio operator would transmit a report encrypted at a *beta* rotor position different from **A**, after which he would re-transmit the same text encrypted correctly or inform the recipient in which rotor position the previous message should be read. Such situations, referred to in BP as *duds*, allowed *Hut 8* to determine the wiring of both the *beta* rotor and the *Bruno* reflector during the first half of 1941.

Once the mathematicians had done their work and reconstructed new Enigma elements, it seemed sufficient to entrust the engineers with the construction of a 4-rotor bombe, which would be the answer to the introduction of the M4 Enigma. However, it soon became apparent that the 4-rotor bombe was not merely a slightly more complicated version of its 3-rotor sister. If it was, the time to test a single menu for a given rotor sequence would have been 26 times longer than on the older model, making it impossible to find the key in time still to be useful. If the new version of the bombe was to serve a practical purpose, it had to run at least 26 times faster—its high-speed rotor had to rotate at around ten thousand revolutions per minute. At this speed, the bombe's logic had roughly one-thousandth of a second to test whether the current rotor position satisfied the condition corresponding to the menu. In the 3-rotor bombe, the circuitry to detect the stopping condition was relay-based, but the relay circuitry was not able to react within one-thousandth of a second. The premiere of the 4-rotor Enigma forced a technological leap from electromechanical technology to electronic circuits. In the latter part of 1941, Dr. C.E. Wynn-Williams, a member of staff at the Telecommunications Research Establishment in Malvern, set about designing a prototype of the 4-rotor bombe. TRE was the main center for radar work at the time, so its staff had a wealth of experience in electronics applications. Wynn-Williams sketched his device as an attachment to a standard 3-rotor bombe. The two parts were to be connected by a thick bundle of some 2000 wires, from which the device took its common name—*Cobra*. The Wynn-Williams version of *Cobra* proved to be a capricious animal and was far from tame when the *Shark* network appeared. Without the new version of the bombe, *Hut 8* could only sporadically read the dispatches of U-Boots operating in the Atlantic. The situation of the codebreakers came closer to that of the sailors and ships they worked for: they found themselves in their own equivalent of the Black Pit.

The crisis in breaking U-Boot ciphers was not the only one to affect BP in the second half of 1941. The center became a victim of its own unprecedented success. During the first twenty months of the war, it managed to cope with most of the challenges posed on numerous and varied fronts. It successfully broke German, Italian, Japanese and Soviet codes and ciphers, diplomatic and military, used on land, sea and air. It almost entirely replaced British human intelligence, which had been

344 19 The Black Pit

virtually supressed in continental Europe. Thanks to its successes, it grew at a pace that must have posed organizational difficulties under wartime conditions. The structures of control over BP's activities had not yet undergone any significant change from the state formed during the peaceful years. The increase in the scale of BP's operations was a severe test of the organizational capacity of its leadership, and the sudden usefulness of the center to the Army, Navy and Air Force must have rekindled ambitions and rivalries that had been swept under the carpet at the time of its creation. However, before the internal crisis manifested itself fully, Bletchley experienced a moment of glory; its importance to Britain's war effort was confirmed, expressed directly by the most enthusiastic recipient of the products—Churchill.

On Saturday, 6 September 1941, a cavalcade of government limousines drove through the main gate of Bletchley Park. The 'Room 40' veterans watching the parade of the Prime Minister and his entourage muttered something about the Prime Minister's understanding of secrecy, but for most BP staff Churchill's visit provided an unexpected and joyous interlude. A group of staff members gathered in the courtyard outside the manor building, and Churchill made a morale-boosting speech to them from the height of a felled tree trunk. His very first words evidenced some surprise at the presence of the audience; "You all look very… innocent". Afterwards, the Prime Minister recalled several situations in which BP information had proved crucial to his decisions, and referred to the changes in the way war had been waged since the days of Republican Rome, calling his cryptologists "geese that never gobbled, though they laid golden eggs". Churchill and his entourage then set off on a tour of the huts (which indeed resembled a goose farm). In *Hut 8*, Turing failed to overcome his intimidation and was replaced by one of his colleagues, who demonstrated to the Prime Minister how an Enigma message could be broken based on an re-encipherment from *Werftschlüssel*. The tasks of *Hut 6* were presented by Welchman, whose speech Travis attempted to interrupt after discussing two of the announced three points. The Prime Minister, however, rejected his objections by noting "I think there was supposed to be a third point, Welchman" and patiently listened to the codebreaker's speech to the end. News of the Prime Minister's visit quickly spread through BP, and the praise contained in his speech added motivation to people doing an onerous and monotonous job every day. The Prime Minister was also impressed, but it was fitting that he reserved his summary of the visit exclusively for Menzies' ears: "I know I told you to leave no stone unturned in search for the right people, but I did not think you would understand me so literally".

In the course Prime Minister's tour presumably no one dared to discuss the tension that had hampered BP for several months, although perhaps its awareness might be one of the reasons for the Prime Minister's visit. If that was the reason of Churchill's visit—it was unsuccessful. The problem facing BP's management lay at the very foundations of the GC&CS, shaped in the early inter-war years. After the victorious outcome of the First World War, the British Navy, Army and Air Force had to go on a peaceful diet. The perpetual paucity of resources then prompted them to relinquish the maintenance of their own signal intelligence services and to shift responsibility to the Foreign Office, whose budget had not been significantly trimmed. The structure set up seemed practical in peacetime, as most decrypts concerned diplomatic messages and the listening stations took over few enciphered military dispatches. However, the first two years of the war completely changed both the tasks of the organization and its environment. The vast majority of the dispatches going to BP were of a military nature. The addressees of the decrypts were also mostly officers of the three services. Although the most important functions in BP's sections were held by civilians remaining on the Foreign Office payroll, BP's analytical and intelligence sections were dominated by naval, army and air force officers. The day-to-day running of BP was increasingly dependent on the supply of personnel representing the army and its auxiliary formations—the multitude of WRENs, WAAFs and ATS members. The military accepted with less and less understanding a situation in which key positions remained under the control of Foreign Office officials. Traditionally the most influential Royal Navy had the least reason to participate in the competency tussle. Its interests at BP were represented by a separate division, consisting of *Hut 8* and *Hut 4*, whose staff communicated directly with the Admiralty. But the *Hut 6* and *Hut 3*, operating in parallel, were dividing their time and energy between the communication networks of the enemy's land army and air force. It was enough for the function of liaison between BP and Army and Air Force HQs to fall to ambitious officers to spark ruthless competition, which is what happened in the second half of 1941.

The competence dispute affected *Hut 6* relatively little: the military realized that the team of mathematicians was irreplaceable. In these circumstances, the conflict centered around that area of BP's operations in which the military felt both interested and competent: *Hut 3*. The bidding was started by the Army through the pen of its intelligence chief, who in early September issued a memorandum pointing out the inefficiency of the way BP's day-to-day operations were managed. While the center's administration and budget remained the domain of the Foreign Office, the setting of its operational priorities was handled by an interdepartmental body, the Signals Intelligence Board (Y Board), which included representatives from all the departments that used information flowing from BP: the three services, SIS and the Foreign Office. The head of military intelligence noted that the work of this body was dominated by Denniston, acting in his dual role as chairman of the Board as well as head of the center it supervised. According to the DMI, members of the Board were out of touch with the military planning sections, which did not allow the priorities of BP's work to be aligned with the war operations being planned. The cure for this weakness, in the DMI's view, was to be the shaping of signals intelligence strategy by the Joint Intelligence Committee, accountable to the Chiefs of Staff. The proposal fell through, but it sparked a debate in which the partisanship of the Army and Air Force played as important a role as the aspirations of the officers representing them. *Hut 3* was headed at the time by Malcolm Saunders, seconded from naval intelligence. This was a sensible arrangement, as it allowed him to play the role of arbiter in disputes between the Army and Air Force. But, at the same time, it was doomed to be contested by Army and Air Force officers, pointing out that the section's activities concerned the interests of the services they represented, so it was inappropriate to subject it to the supervision of a naval officer. Opposition to Saunders' leadership was the only issue on which officers reached consensus. The representative of RAF interests, Group Captain Humphreys, made no secret of his ambition to take control of *Hut 3*. Nigel De Grey describes the deft intrigues he undertook, and the equally aggressive actions of his Army counterpart, which forced Saunders to react decisively. Several attempts to impose compromise solutions, made in October and November first by Travis, then by Menzies, failed to deliver a solution, introducing an additional layer of bureaucracy into the operation of *Hut 3*.

Even if Denniston's name did not appear in the notes flying across the division line, it was incumbent on him to unravel the dispute. Denniston, however, remained indifferent to events, letting them run their own course. One of the reasons for his attitude was probably his health problems: at the end of February, a stone was discovered in his bladder requiring immediate surgery, followed by a lengthy recovery. He returned to work on 9 June and immediately resumed the interrupted thread of transatlantic cooperation, setting off twice across the Atlantic. Transatlantic travel in the bomb bay of an aircraft, with an oxygen mask at his mouth, must have been a considerable strain on a man who had not yet fully recovered. His lack of strength and his focus on cooperation with the US, which Denniston rightly regarded as a strategic problem, meant that he regarded turf wars as a nuisance unworthy of his attention. In presenting the background to later events, most sources emphasize that Denniston was also neglecting the elementary needs of the growing organization; new premises, staff and the necessary accommodation, equipment and transport. Serious organizational crisis was growing in the center, threatening its ability to capitalize on the successes achieved by the codebreakers. Their authors felt aggrieved by the disregard for the results of work done under difficult conditions and nevertheless yielding golden eggs. The initiator of a kind of codebreakers' revolt seems to have been Welchman. In the autumn of 1941, he perceived that the organization he had designed was missing opportunities as a result of deficiencies that could be remedied. Probably influenced by Churchill's recent visit, he decided that approaching the Prime Minister would be an effective solution. He discussed his idea among a few of his most trusted colleagues—Turing, Alexander and Milner-Barry—and together they sketched out a letter to Churchill. A message sent through official channels would have got stuck in the bureaucratic machinery, so the conspirators decided to deliver it in person and assigned Milner-Barry to this mission, deciding on a small psychological manipulation. They chose 21 October, the anniversary of the Battle of Trafalgar, for the delivery of the letter, the significance of which the Prime Minister could not remain indifferent to.

The fact that Milner-Barry managed to carry out his mission says much about the British political system. In the midst of the ongoing war, a young man appeared at the gates of the Prime Minister's residence at 10 Downing Street without even presenting his Foreign Office identity card, declaring only that he represented a secret structure, whose affairs he must present to the prime minister in a letter, the contents of which must not reach third parties. His statement must have caused confusion for Brigadier Harvie-Walker, the Prime Minister's personal secretary. Happily, the visitor invoked Churchill's recent visit to BP, which was known to his secretary and opened the door to a compromise: Milner-Barry agreed to leave the letter with the secretary, and the latter promised that it would get directly to the Prime Minister's desk. In the letter, the codebreakers drew attention to obstacles in the operation of BP, the solution to which is not understood by the center's management: "the work is failing and in some cases is completely stopped, mainly due to lack of staff. We are addressing you directly because we have done everything possible through the normal channels over the months and we doubt whether a rapid improvement is possible without your intervention". After all, they added, "thanks to the energy and foresight of Commander Travis, we are equipped with enough bombes to break the cipher". Pointing out Travis while skipping Denniston was a clear signal of who they trusted and, implicitly, who the codebreakers distrusted. The letter found its way to Churchill's desk, who the next day scribbled one of his famous 'Action this Day' on it, instructing the secretary of the Imperial Defense Committee to make sure that the codebreakers were "given everything they need with the highest priority". If the aim of the four was to resolve staffing problems, they achieved it: a few weeks later, Menzies was able to report to the Prime Minister that the situation in this area had improved. If they had hoped for Churchill's arbitration in resolving the management games, they were disappointed in the short term: Menzies criticized Welchman for bypassing the usual route and appointed an independent auditor to investigate the operation of BP.

In a thickening atmosphere, Major Curtis sent two memoranda to the Deputy Commander of Military Intelligence in which he ruthlessly attacked the current management of *Hut 3*. In both, he suggested that the management of hut should pass directly into the Army hands and that he himself should take charge of all military staff at the facility. A brigadier commissioned by Menzies to investigate the state of affairs also added fuel to the fire. Unfamiliar with the peculiarities of signals intelligence, he rather clumsily attempted to obtain the necessary information, distracting people from their work. In fact, at this stage the conflict took a fairly clear form: two of Denniston's deputies, Edward Travis and Nigel de Grey, made no secret of their intention to get rid of their boss. One of the factors that led to the exacerbation of the conflict was probably the attitude to the exchange of information with the Americans. Open reception of Sinkov's mission members and his departure for the US almost immediately after his convalescence indicate that Denniston was in favor of cooperation. The accounts of those who took part in the events confirm that de Grey and Travis were at best skeptical about it at the time. A memorandum from this period, which probably came out of de Grey's hand, leaves no doubt about the author's attitude: "It is perfectly understandable that the Americans would like to have a share in the success [… however] they have nothing to contribute to it". Denniston was unable to resolve the dispute between Army and Air Force representatives over control of *Hut 3*. But in the light of later events, the background to the codebreakers' revolt is somewhat different from the memories of its participants. During his convalescence and overseas travel that began shortly afterwards, Denniston was entitled to expect that Travis, who was replacing him, would take care of the needs of the center. The center's director and deputy director were equally responsible for solving just described problems. Welchman made no secret that "the thread of personal understanding that developed between Travis and myself became an important factor in the success of *Hut 6*". In doing so, he added that Denniston's deputy "possessed some of the qualities of Winston Churchill. He had the character of a bulldog and liked things to go his way". Most of the events that cast a shadow over the operation of BP during 1941 can be most easily explained by adopting the hypothesis of a palace coup aimed at Denniston and catalyzed, if not initiated, by the energetic and effective Travis.

By early 1942, it was clear that the situation in BP was ripe for resolution. Menzies, meanwhile, had received his envoy's report and decided to act. Being formally head of the GC&CS, he called a meeting of the Signals Intelligence Board for 5 February and delegated Travis rather than Denniston to attend. At the meeting he announced and pushed through a plan to split the GC&CS into two sections—civilian and military. The civilian section was to remain under Denniston's command and deal with diplomatic and commercial code-breaking (separating from BP and moving to premises in Berkeley Street, London). The Bletchley facility was thus becoming the military part of GC&CS and coming under the direction of Travis, who would gain deputy director status. The settlement was a blow to Denniston, who had led the GC&CS from the day it was founded and had guided his organization through the difficult days of budget woes in the early inter-war years. He felt betrayed by the people with whom he had worked for many years. On the other hand, something about the way he acted caused old friends to lose sympathy for him and withdraw support, while newcomers sided with Travis. Even if Denniston's ousting was the result of personal intrigue and a decision deeply unfair to him, it helped to improve BP's functioning. Travis was not a gifted cryptologist or even a good organizer. However, in addition to his stubbornness, he possessed two qualities that predestined him for his new position. The first was his ability to delegate tasks and powers and his talent for finding executors for his policies (Welchman was soon appointed deputy director for decryption mechanization). The second was the ability to make difficult decisions, which the protagonists of the war for control of Hut 3 were quickly to learn. Just four days after taking up his new duties, Travis attended a meeting, at which the RAF representative expressed the hope that under the new conditions his representative at BP, Humphreys, would be able to take up his position as head of *Hut 3* unhindered. Travis made it clear that he did not intend to tolerate the operation of private armies at BP and moved Humphreys to a position outside the hut. Despite this clear signal, Humphreys still attempted to fight back in June, alleging in a letter to Travis that *Hut 3* was notoriously failing to live up to its potential and suggesting that the remedy would be to appoint him its head. Spurred to action, Travis shifted *Hut 3*'s former head, Saunders, to oversee the bombe-building project, bringing in on 1 July 1942 an airman, Eric Jones, who, through skillful diplomacy managed fairly to shape a *modus vivendi* between Army and RAF representatives at *Hut 3*. After more than a year of wrangling and bargaining, the crisis within BP was resolved.

A series of changes to U-boot ciphers and procedures in late 1941 indicated that the Germans were unsure of the security of their communications. Premiere of the 4-rotor Enigma reinforced this thesis. A deadly danger hung over the secret of breaking Enigma ciphers: it was enough for the U-Boots to start sinking ships in convoys more effectively than in previous months for the Germans to deduce that the machine change had blinded the enemy. There was little the British could do to prevent this scenario from happening. The OIC's maps of the Atlantic had until recently been covered with a patchwork of symbols depicting submarine patrol lines, but after the introduction of the M4 they had reverted to their pre-August 1941 state. In the new situation, the OIC's knowledge of U-boat activities came almost exclusively from the network of direction finders. The Enigma secret was saved by decisions taken in Berlin. Four days after the Japanese attack on Pearl Harbor, Hitler felt obliged to declare war on the United States, and Dönitz sought to accentuate the Führer's participation in the new venture by sending several ships towards American shores. When the first of these left Lorient on 18 December, few observers believed that they were assisting in the birth of a murderously effective submarine offensive. Its beginnings were an improvisation on both sides. The Germans had no nautical charts of the American coasts; U-boot captains were equipped with travel guides. There was no data on Allied shipping on the western side of the Atlantic. The Americans transferred all ships worthy of the name to the Pacific, leaving almost nothing in the Atlantic. They attempted to remedy the situation by employing wealthy yacht owners with their vessels on patrol; their flotilla became known as the Hooligan Navy. The attitude of the US Air Force towards cooperation with the US Navy was similar to the relationship between the Luftwaffe and the Kriegsmarine. However, nothing influenced the shape of U-boot operations against Allied shipping off American coasts as much as the reluctance of US Navy commanders to take lessons from the Royal Navy. Admiral Ernest King recognized that in the waters surrounding US, organizing convoys would be an extravagance that lacked ships, men and time. Not even such elementary precautions as turning off the navigation lights or blacking out towns on the coast were taken. German captains did not need maps—they sank ships going with their position lights on, clearly visible against the backdrop of the well-lit coastline. In the history of the Kriegsmarine, U-boot operations off the American coast went down in history as Operation *Paukenschlag*. Rather, the ship's captains spoke of a "second happy period" or simply of "golden times". By August, when shipping protection was finally organized, the "golden times" had cost the Allies the loss of 609 ships with a total tonnage of more than three million GRT. The slaughter of Allied ships off the US coast and in the Caribbean Sea, however, drew the U-Boots away from the mid-Atlantic, preventing the Germans from noticing the difference in performance before and after the introduction of the M4 machine. The secret of breaking the Enigma ciphers was safe, although at a very heavy price.

The success of Operation *Paukenschlag* had a side effect: Hitler promoted Dönitz to the rank of full admiral. It was a stroke of luck that the order in which Dönitz announced the promotion poured some encouragement and confidence into the hearts of British codebreakers that the M4 ciphers could be broken. According to everything *Hut 8* knew about *Shark*, it did not raise new problems on the grounds of theory. The only problem was the performance of the bombe and the time required to find a solution. Breaking the *Shark* key seemed possible, assuming the availability of a certain crib and a lot of bombe time. The text of Dönitz's order was a crib of

100% certainty. The message was broadcast over virtually all Kriegsmarine networks and in all cipher systems. It was an ideal opportunity to test whether *Shark* hided any surprises. The experiment required almost all the bombes, so *Hut 8* first had to obtain the approval of the Y Board. Having received it, after several days' work, it was possible to break the keys in force on 23 and 24 February and 14 March. In the latter case, success cost seventeen days of work by six bombes.

American and British codebreakers were entering a second round of talks under changed conditions. The most important novelty was US participation in the war. During earlier talks, the British had sought to limit American participation in attacks on the Enigma ciphers, arguing that should the need for practical use arise, the US could rely on decrypts provided by BP. The tankers burning before the eyes of the people on the east coast of the US were a clear indication that the practical need had arrived, but at the most unfortunate moment the source of the promised stream of dispatches dried up. The British attempted to get the US Navy to adopt their tried and tested solutions. To this end, they sent the head of the submarine tracking section to the US in April 1942, but Roger Winn encountered a barrier of incomprehension, indifference and ill-will from the very first moment of talks. He had to remind his interlocutors that British ships were also being sunk off American shores and to firmly point out that "we are not going to sacrifice men and ships on the altar of your bloody incompetence and stubbornness" in order to get the Americans to agree to set up the equivalent of his section. In practice, it took several months; the American equivalent of Submarine Tracking Room did not start work until 27 December 1942. Perhaps the delayed launch was due to the realization that, for the Winn section, it relied mainly on decrypts from BP: without them the US equivalent of the Submarine Tracking Room could not make a significant contribution to US Navy operations. Another round of talks between BP and Op-20-G became inevitable.

Meanwhile, there was a minor personal earthquake on both sides of the Atlantic, resulting in a reversal of attitudes towards cooperation. Denniston, who favored transatlantic exchanges, could not reach agreement with the skeptical Safford and Driscoll. In 1942 the burden of negotiating with the Americans was taken over by Travis, who had previously made no secret of his reluctance to share BP's achievements with anyone. There was also a change of leadership at Op-20-G. The Americans had concluded earlier than the British that wartime required a different structure and leadership of a different type than peacetime. In February 1942, Op-20-G was split into three sections. Safford took over responsibility for naval cipher design under the banner of Op-20-Q, the Op-20-Y communications security section was separated, and Lieutenant Commander John Redman became the new head of cryptanalysis section. As it seems, Redman owed the appointment to his brother's position and had no cryptologic experience, so the actual leadership of Op-20-G was taken over by the Chief of Operations, Joseph Wenger. In 1926, Wenger was the first graduate of an accelerated cryptology course for naval officers organized by Safford and Driscoll. He was an advocate of the mechanization of cryptology and understood that with the US entering the war, cooperation with BP became a necessity. On the British side, awareness of this fact matured gradually. Once Travis had consolidated his position, resolved the immediate internal problems and decided to return to talks with the Americans, the very choice of delegate indicated the direction in which he intended to develop cooperation with the US. In April, Colonel John Tiltman, one of BP's most experienced cryptologists, flew overseas. His experience covered many areas, but at its center were Japanese ciphers and codes. If his superiors hoped that Tiltman would make an effort to steer cooperation towards Japan, they underestimated his pragmatism. During his conversation with Wenger, he was quick to recognize the rationale behind the Americans' claim to be involved in breaking U-boat ciphers and issued a memorandum indicating that "at present [the Americans] are involved in the war and have a vital interest in data on U-boot movements, they have the right to demand access to the decrypts or a detailed explanation of the reasons why the dispatches are not currently being read and the future prospects for breaking them. (…) If a quick and satisfactory solution to this problem is not found (…) their command will insist that their cryptologists duplicate our work on the Enigma ciphers". Neither Travis nor Menzies were prepared to accept the full and equal participation of Op-20-G in breaking U-boot ciphers, but they understood that they had to allow some concessions to the Americans. After exchanging several letters in late April and early May, Travis forwarded a note to Washington on 13 May stating that "the higher authorities have accepted our future policy regarding the Enigma (…) We will continue with the decryption, however, in August or September we will send you the device and provide you with a mechanic to instruct in its use. We will also provide you with a full manual and try to get someone to explain the methods we use". So, after almost three years of war and several months of transatlantic debate, the British decided to make a gesture and offer the Americans the gift they themselves had once received from the Poles without additional conditions.

The difference, however, was that Travis' promise was not kept. Fortunately, Tiltman was able to agree with his partners on another form of cooperation, which, over time, helped to create an atmosphere of trust and to undertake joint projects. On 1 July 1942, two US Navy lieutenants, Robert Ely and Joseph J. Eachus, arrived at BP and became the first Americans to be received at BP not as guests but as members of the *Hut 8* team. For Ely, coming to Bletchley was compensation for his disappointment at the time of the Sinkov mission. In the original plans, he was to head the group representing Op-20-G, but at the very last moment Safford and Driscoll decided that Weeks would replace him. The primary task given by Wenger to the representatives at BP was to get an in-depth understanding of how bombes worked. Despite a friendly reception, the barriers on the British side continued to operate. It was only after several days at BP that technical drawings of the devices were presented to the Americans, and conversations with Turing provided insight into the subtleties of their use. The Americans quickly developed at least two methods of overcoming obstacles and gaining information. The first was based on the fact that the American military bureaucracy classified them both as a detached unit, which allowed them to draw sugar in 100-pound bags, coffee in 20-pound cans, etc., from the American quartermaster's warehouse in London. In Britain, where food coupons had been in force for three years, a tin of coffee generally proved to be the argument advancing the discussion. The second mechanism, too, was one of the classics of the genre: Eachus won the sympathy of Denniston's former secretary, Barbara Abernethy, whom he married in 1947.

While cooperation was developing harmoniously at the purely human level, confusion and distrust continued among commanders. The British reneged on their promise to deliver the bombe, eliciting differing reactions from both American agendas. Friedman approached the problem calmly, pointing out that "the British have such a limited number of devices at the moment, and their needs are so considerable, that (…) even if they did deliver a machine, it would probably be of the earliest and least effective design – worthless as a model for replication". Wenger was convinced that the British were hiding the fact that they had broken the Enigma M4 cipher. Suspecting his partner of dishonesty, he drew more drastic conclusions than Friedman about the failure to keep the promise to deliver the bombe. At the end of August, a meeting was held in the Washington office of the British intelligence liaison officer, Captain Hastings, during which the parties attempted to clear up misunderstandings. Wenger learned that not only had *Hut 8* failed to break *Shark* so far, but the prospects for a breakthrough were slim: the British had admitted problems with the 4-rotor bombes. BP heads had lost hope of completing the *Cobra* project and decided to start parallel work on a prototype 4-rotor bombe by the BTM team led by Keen. The device has been given the working name "High Speed Mammoth". To this information, Hastings added a somewhat perverse interpretation of Travis and Tiltman's earlier promises, which implied that the British had in fact delivered on their commitments. If it was the conduct of this meeting that irritated the Op-20-G representatives, it was the irritation that saved the Allies. On 2 September, Wenger presented his commanders with a memorandum proposing that the US Navy undertake the construction of its own 4-rotor bombe. Faced with the bankruptcy of his strategy, Travis went to the personal Canossa. At the end of September, he travelled to Washington in the company of Frank Birch. On 2 October, after several days of talks, BP representatives concluded in Washington an agreement with Admiral King's communications chief, Captain Carl Holden, setting out the terms of cooperation. The agreement left the initiative in the area of breaking U-Boat ciphers in the Atlantic in British hands, situating the US Navy as a junior but full partner. After almost two years of unconstructive debate, including nearly a year of US participation in the war, the Allies finally created the conditions for the use of US resources in the cryptologic race. Given the obstacle-laden path that brought both sides to the table, the signing of the document in October proved to be a happy coincidence. The events that followed just a few weeks later brought a change of mood on the British side.

By this time, German submarines had abandoned the east coast of America, which had proved so kind to them, and moved partly into the mid-Atlantic and partly dispersed into peripheral waters where defenses were unprepared to face them. November 1942 became the worst month of the entire war for the Allies in terms of tonnage of lost merchant shipping—nearly 750,000 GRT. Dönitz's navy had achieved its strategic goal: a level of sinkings that, if they persist, guaranteed victory in the Battle of the Atlantic. The Allies were fully aware of the drama of the situation. In a surge of desperation, a letter was sent from the OIC to BP on 22

November in which Admiralty officers asked, with somewhat artificial politeness, if it would not be possible to pay a little more attention to the *Shark* problem. They reminded, as if anyone needed to be reminded of this, that the Battle of the Atlantic "is currently the only campaign in which BP is not making a significant contribution, and yet the only one in which the war could be lost if BP does not come to the rescue". Just as the OIC officers looked at the silent teletype from BP with a mixture of hope and helpless anger, the codebreakers at *Hut 8* had to clench their teeth waiting for news from TRE and BTM. But both British designs for the 4-rotor bombe were bogged down in technical difficulties.

Little did the British know that the mounting losses were only partly due to the *Shark's* silence, the increase in the number of U-Boots and the growing experience of their captains and crews. BP's codebreakers were not only losing the duel with Enigma during this period, but also with their counterparts from the *B-Dienst*, who had long been breaking convoys' signals. In June 1941, the Allies introduced a new code, *Naval Cypher No. 3*, which was used to coordinate the activities of the US, British and Canadian navies escorting the convoys in the Atlantic. Before it came into use, it was a reserve code, intended for temporary use in the event of any of the Royal Navy's main code systems being compromised. Worse still, the introduction of *Naval Cypher No. 3* took place under conditions of wartime improvisation. The most momentous error was adding to the new code of old superencipherment tables of only 5000 numbers. The Americans repeatedly pointed out the poor quality of the code: Safford, referring to its name, described it as a "third order communication system". Despite the warnings, the system, which became known as the convoy code, remained in use for more than two years. Its area of application coincided perfectly with the interest of the B-Dienst. The lively radio traffic associated with the organization of convoys of dozens of ships, escorted by rotating escort groups, gave the B-Dienst a unique opportunity. German Navy's chief cryptanalyst, Walter Tranow, focused all the efforts of his service on the new challenge. The results did not have to wait; by March 1942, the Germans had a reconstructed codebook. Should they crack *Naval Cypher No. 3* a little earlier, fragments of their decrypts would probably have appeared in the broken Enigma messages, signaling the threat. By March, however, communications on the *Shark* network were already impenetrable to British codebreakers. At the height of the Battle of the Atlantic, the B-Dienst was reading up to 80 per cent of the dispatches directed to convoys, including around 30 per cent in time for operational use. Rodger Winn complained that the image on the large map of the Atlantic in the Submarine Tracking Room seemed blurry; at the same time, a similar map in U-Boot Command drew the situation on the ocean routes sharply and precisely. Dönitz's son-in-law, Günther Hessler, who served on the U-boot staff as operations officer, remembered that in late 1942 and early 1943 "we reached a state where we needed only one to two days to read British radio dispatches. In some cases a few hours were enough. Sometimes we could guess when and how the British would try to exploit gaps in the U-boot lines. Our job was to patch these holes before the convoy arrived". As a rule, patching the holes succeeded, leading to an appalling increase in convoy losses in the summer of 1942. The second half of the year promised to be even worse. The British were intercepting signals coming from the Baltic from numerous new U-Boots, which were undergoing a training and crew integration in this area before entering line service. On the opposite side, the Admiralty was aware that, for the safety of the convoys carrying the invasion forces of Operation *Torch*, it would have to detach a significant number of escort ships from convoy cover in the North Atlantic.

When the OIC officers were sending their letter to BP on 22 November, they did not know that help was on its way. It had come from Egypt, where Montgomery had just pursued the enemy defeated at El Alamein. In a wave of panic that had erupted a few months earlier when Rommel's troops were expected to occupy the Nile delta any moment, Royal Navy ships left their base in Alexandria, moving to the ports of Beirut, Haifa and Jaffa, and U-boots followed. U-559, under the command of Hans Heidtmann, was among the first six ships diverted to the Mediterranean by Hitler back in September 1941. At the end of September 1942, the ship sailed from Messina on her tenth cruise with the intention of patrolling the coast between Alexandria and Beirut. On the evening of 29 October, Heidtmann was ordered to send a weather report on the morning of the following day and return to base. As the ship was broadcasting the report before dawn, the crew of the Sunderland flying boat from 47th Squadron RAF caught its echo on the radar screen and the report drew a squadron of five destroyers, *Packenham*, *Petard*, *Hero*, *Dulverton* and *Hurworth*, to the area. *Petard* and his eccentric commander Mark Thornton were to play a key role in the events of the coming hours. It seems that describing him as eccentric was a mild way of presenting his state of mind, bordering on neurosis. This manifested itself in a perfectionism taken to the extreme in the performance of his duties and a habit of subjecting his crew to tests whose severity bordered on cruelty. Thornton's main obsession was his ambition to capture a German submarine and secret documents on board; he could spin such plans for hours with his deputy, Anthony Fasson.

The British ships arrived in the area on 30 October at 12.20 PM. Assisted in their search by another aircraft, they located Heidtmann's ship and began the hunt. In shallow and clear waters, U-559 would only have had a chance of survival if the destroyers had other duties, but the hunters had plenty of time and a full supply of depth charges on board. Also, the attack tactics they employed did not give the hunted ship a chance: two destroyers maintained sonar contact with the submarine, homing in the attacking ship on the target. Heidtmann was an experienced captain and successfully avoided the worst for several hours. Meanwhile, early autumn darkness fell and his chances increased somewhat. Then, on board the *Petard*, one of the gunners suggested that the target had taken refuge on the seabed, deeper than the British charges were exploding. As the detonators could not be set to a greater depth, it was decided to plug their holes with soap to delay the explosion. The trick was successful—after a series of charges the destroyers' sonar again caught the moving echoe. At around 22.00 the situation became critical for U-559. By this time, the crew had counted 288 depth charges explosions. Finally, Günther Gräser, the submarine's chief engineer, signaled to the captain that the moment of decision had arrived—further delay would make surfacing the submarine impossible. At

22.40, the submarine appeared at the surface—Thornton could start implementation of his grand plan.

Most accounts of the event present the image of Anthony Fasson and Colin Grazier throwing off their clothes in a hurry and swimming towards the submarine. The reality was more prosaic: the three sailors jumped off the *Petard* and landed on the deck of the U-559 as the two ships stood side by side for a while. Three, for Lieutenant Fasson and Senior Seaman Grazier were joined by a 16-year-old cook's mate, Tommy Brown. When Fasson got inside the ship, he noticed that a shell had torn a hole at the base of the conning tower, and the water seeping in through was rapidly raising the level in the already partially flooded hull. With torch in hand, he made his way through room after room, partly using the keys hanging on the bulkheads, smashing with pistol shots those locks at which the keys were missing. He was followed by Grazier and Brown, transporting the lieutenant's chosen trophies to the ship's hatch. By this time, a whaler had also appeared at the scene: Ken Lacroix carried the prey out through a hatch in and handed it to the sailors waiting outside. Fasson, Grazier and Brown worked increasingly hurriedly inside the ship, rushed by the shouts of their colleagues, who warned that the U-559 was dipping dangerously. Tommy Brown had just made his third round with the prey when Gordon Connel, commanding the whaler crew, ordered the other two to be recalled immediately. Fasson and Grazier had just appeared at the ladder leading to the hatch when a wave of water washed Brown out to sea. Anthony Fasson and Colin Grazier were unable to extricate themselves from the sinking ship.

One of the rescued U-559 sailors, Hermann Dethlefs, watched with horror as the British threw secret documentation, including cipher keys and code books, into the whaler. He later attempted to convey the information about the loss of the documents in a somewhat naive manner in a letter to his family written from a PoW camp. He was not the only source of indiscretion. The crew of the *Petard* were obliged to keep the incident secret. But *Dulverton's* crew, who had been cheering on their colleagues' action from a short distance away, had not been warned; two of its members were heard describing the incident during bar chats in Haifa. Thornton's plan was only to capture the secret documents. In Haifa, he handed the papers to intelligence officers, who approached them with some indifference: the parcel containing the documents from U-559 did not reach BP until 24 November. The Admiralty had a bit of trouble with Fasson and Grazier. In the opinion of the commanders, their act merited the Victoria Cross. However, the justification for awarding it had to include a description of the combat action, leading to the secret of the capture of the codes being revealed. Citing the somewhat Machiavellian argument that the sailors had not acted in the face of the enemy, Admiralty led them to be awarded the civilian St George's Cross. The latest victim was Thornton, who was not relieved by the realization of his dream. *Petard's* doctor, William Prendergast, watched the commander's extravagances with concern until, three months after the events described, Thornton executed a group of surrendering sailors from the Italian submarine *Uarsciek*. At the doctor's request, Thornton was removed from command and transferred ashore.

When the trophy from U-559 finally arrived at BP it seemed that it would not excite the codebreakers. Among the documents, two seemed valuable; the short signal code book and the weather code. In the past, the dispatches in both codes had provided good quality cribs, but initial attempts to use them in an attack on the *Shark* had failed. The *Hut 8* faced a difficult task—messages in the short signals or weather code offered a good crib, but were generally too short to construct a menu for a bombe based on them. There was another problem associated with weather reports: determining the probable clear text of the message required breaking first the weather message sent by the DAN station in the *Germet* cipher. George McVittie's team, meanwhile, was able to establish that keys to this cipher were repeated at monthly intervals. When the documents from U-559 arrived at BP at the end of November, the first month of the cycle had just come to an end and the *Hut 10* team was having considerable difficulty breaking the key. On 8 December, however, the month of repeats began and, for good measure, it was only then that work on the cribs could start. Nearly fifty bombes were already available, but the testing of the cribs was moving at a sluggish pace, with each 4-rotor menu taking a considerable amount of time to check. It was only on Sunday morning of 13 December that one of the bombes found a possible solution. Shaun Wylie found out about it in the canteen, where he intended to have breakfast before returning home after a night on duty. But the news from the bomb section put both fatigue and the meal into oblivion. It turned out that the bomb had stopped in the position where the fourth rotor was set in the position A, where the M4 Enigma emulated M3 model. Confirmation of this result meant that the U-Boots' weather reports were being transmitted in 3-rotor mode. If it turned out that the only difference between normal dispatches and weather reports was the setting of the fourth rotor, *Shark* key could be recovered by attacking the menus created from the weather reports with the old 3-rotor bombe, and then testing at most 26 possible settings for the fourth rotor. Wylie ran to *Hut 6* to secure an allocation of six bombes, with which he began his search for the final element of the key. Later that afternoon, the first *Shark* decrypts since March came off the teletype in the Submarine Tracking Room. It did not have to wait long for the consequences. By December, losses were about half the tonnage of the ships lost during the fateful November, and by January had fallen to around 200,000 GRT, the lowest level since December 1940. Thanks to the mistake of the Germans using the M4 machine in 3-rotor mode, *Hut 8* was once again in the game.

It might seem that by breaking the dispatches of the Atlantic U-Boots *Hut 8* would detect the breaking of the convoy code by the B-Dienst. By a perverse twist of fate, however, the periods when the Germans read Allied dispatches and the British read German ones were intertwined. Until mid-December 1942, in dispatches encoded with *Naval Cypher No. 3*, the Allies were giving the superencipherment indicator in open text. Almost at the same time that *Hut 8* recorded a breakthrough in its attack on *Shark*, the Allies started enciphering the indicator, which blinded the B-Dienst for several weeks. The change helped the convoys, but it also delayed discovery of German breaking the convoy code. During the last weeks of 1942 and the first weeks of 1943, the rhythm of breaking the *Shark* repeated the achievements of the meteo section. During the first part of the new

Germet cipher table validity period, delays were considerable, up to several days. However, it was enough for McVittie's team to break the key or the reuse period to start for the delay to fall below 24 hours. Between 13 December 1942 and 10 March 1943, *Hut 8* managed to break 88 of the 99 keys in force at the time. After a few weeks of idyll, dark clouds began to gather on the horizon: references to changing the weather code appeared in the dispatches. The new *Wetterkurzschlüssel* came into force on 10 March, blinding *Hut 8* again. One Admiralty officer estimated that "the submarine tracking section would be blind to U-Boots' movements for a significant period, perhaps counted in months". An equally pessimistic report was passed to Washington: "The special intelligence source on U-Boots has suffered a severe blow. After 10 March it is unlikely that we will receive more than 2–3 pairs of days per month, and even these will not be delivered on a regular basis".

Had the Admiralty's assessments been confirmed, events in the Atlantic could have taken a disastrous turn for the Allies in the coming months. After relief in December and January, U-boots returned to convoy routes in February. The Allied invasion in North Africa and the beginning of the American troop transfer to Britain made the Atlantic the first line of resistance for the Third Reich in the West. This time, BdU had at his disposal a fleet of more than a dozen large ocean-going submarines, the so-called 'milk cows', designed as underwater tankers, capable of supplying combat ships with fuel, torpedoes and food. Each 'milk cow' carried 400 tons of fuel and 50 tons of food and ammunition, and had on board a repair shop, a doctor and a group of sailors ready to replace sick or injured crew members of the resupplied U-Boots. The length of a typical U-Boot patrol increased by two to three times and the saving in time spent cruising to base allowed more wolf packs to be kept in patrol areas. Between March and May 1943, the fiercest convoy battles in the history of the Second World War were unleashed. Pessimistic British commanders underestimated the ingenuity of their codebreakers. At the turn of the year, they opted for cribs based on the weather code, as these were the most reliable to use. After the change in the weather code, another asset remained in the hands of *Hut 8*—the short signal code. After the revelations of 13 December, codebreakers established that short signals were also transmitted in the Enigma's 3-rotor mode of operation. Their application was somewhat more complicated than that of the weather code. The content of an intercepted report could only be interpreted by knowing the circumstances of its transmission. In the new situation, contacts with the intelligence section at *Hut 4*, working under Edgar Jackson, became fundamental. His team, knowing the situation on the ocean, was able to link the intercepted ciphertext with, for example, the convoy escort's report that a ship had been torpedoed, allowing the codebreakers to prepare a menu for the bombe. The new approach had one advantage: there was no need to wait for the *Germet* code to be broken before launching an attack. It proved successful—between 10 March and the end of June 1943, it allowed 90 of the 112 keys of the Shark network to be recovered.

The resumption of *Shark* decryption by *Hut 8* was due to documents captured at the cost of the lives of two sailors. But an equally important factor was the recognition and breaking of various German cryptographic systems and the development of a system permitting to integrate data from various sources. Without breaking

Germet, a copy of the U-Boots' weather code would have rested on a shelf in *Hut 8*. Without the network of direction finders and the processing of the data derived from them, short signals would have been to the British just what they were supposed to be in the Kriegsmarine's view: unnoticeable and practically meaningless noise in the ether. BP was integrated into a global intelligence system that, thanks to US participation, covered all the Atlantic coasts, enabling every shred of radio transmission in its area to be intercepted and assessed. The success of the codebreakers came just in time. In the first phase of the campaign, the U-Boots gained the upper hand. In January and early February, Walter Tranow's cryptologists managed to break again into the convoy code. Dozens of dispatches starting with the words '*Geleitzug erwartet*', followed by the precise coordinates of the convoy, flowed to the U-Boots. The Allies originally paid no attention to the meaning of the term. It was obvious that the Germans had intelligence service that sought to establish convoy routes. The term "convoy expected" said nothing about the source of the information: an agent's report, a direction finding of signals transmitted by an escort, a U-boat report. However, on the night of 18 February, orders sent to the wolf packs *Ritter* and *Neptune* and the newly formed *Knappen* group were intercepted and deciphered. The dispatches were sent by BdU shortly after the Admiralty had ordered the convoy ON 166 to change course, and the deciphered orders directed the wolf packs precisely to the new convoy route. This was an alarming signal, indicating possible compromise of *Naval Cypher No. 3*. Contrary to Kriegsmarine in *Hut 8* and among the naval intelligence officers no one was looking for the treachery within their own ranks: a report was sent to the Admiralty that the code had probably been compromised. The Admiralty accepted the report but failed for months to replace the code with a new version. It was not until mid-May that the Allies obtained hard evidence confirming that the Germans were reading their dispatches. In the process, they also found out why the earlier signals bore only circumstantial evidence. It turned out that the Germans were sending the most operationally valuable information in *Limpet* key, the officer variant of *Shark*. In May, Op-20-G managed to break the key to *Limpet* and read a packet of dispatches addressed to the wolf packs *Rhein*, *Elbe* and *Drossel*. Lieutenant Knight McMahan, who was analyzing the dispatches, was alarmed by a sudden change in recently issued instructions, addressed to all ships: it indicated that the Germans had acquired important information. However, the codebreakers had no access to the instructions addressed to their own convoys. Fortunately, McMahan showed initiative and persuaded colleagues in another section to provide him with a copy of the original dispatches: the link between the orders to change the course of convoys HX237 and SC129 and the U-boot orders was obvious. This revelation was the ultimate proof of the compromise of *Naval Cypher No. 3*, but it probably came too late to play a decisive role. On 10 June, the Allies abandoned the old code, replacing it with Naval Cypher No. 5. Given the time required to print and distribute the code books, the change process must have already been underway when the Americans reported their discovery. This circumstance only slightly reduces the responsibility of the Admiralty for releasing insufficiently secure code for use, and keeping it in force for far longer than reason dictated. During the investigation, the Admiralty explained its delay by the complicated process of printing the code books and distributing them to bases and ships scattered around the world. However, the Admiralty was only partly responsible for the cryptologic blunder. Merit in breaking foreign codes and ciphers obscured the responsibility on the part of GC&CS. When the new institution was created in 1919, the construction of codes and ciphers of all three military types was excluded from its area of interest. However, the GC&CS charter required the School to "study all codes used by agencies of the British Government and the purposes for which they are used, primarily with a view to ensuring, and where necessary improving, their security". The GC&CS was also required to "advise on the principles of construction and limitation of the time of use" of codes and ciphers. There was a team of about 10 officers in the Admiralty to look after the security of its own communications. The GC&CS delegated two representatives to work on this team: one of the delegates was Travis himself, whose essential duties at BP hardly allowed for external commitments. The Bletchley codebreakers were aware of the weaknesses of the convoy code, for as late as 1941 they suggested to the Admiralty to supplement it with Tiltman's latest idea—a stencil for the subtractor table. The Admiralty ordered trials of the system, which dragged on until March 1942, after which they took an unhurried approach to printing and distributing the material. BP fell into the trap that nature sets for institutions of an integrated nature—success in one area of operation inexorably draws attention away from others. BP's tasks combined offensive signals intelligence, breaking enemy codes and ciphers, with defensive—monitoring the security of own communications. Successful attacks on the enemies' ciphers must have detracted interest from the second, equally important area of operation. As a result, during the crucial period of the Battle of the Atlantic, encompassing 1942 and the first half of 1943, the initiative and advantage in cryptological competition belonged clearly to the B-Dienst.

BP shared the responsibility for the delayed change of the compromised code with the Admiralty, but considering that the blunder took place in the main area of GC&CS responsibility, this is a stain, above all, on School's honor. The late change of the convoy code deprived the codebreakers of being a key factor in victory in the Battle of the Atlantic. At the same time that *Naval Cypher No. 3* was being decommissioned, the Allies had just settled the battle in their favor and the number of competitors for the role of decisive factor had become significant. The first candidate was aircraft. The appearance of long-range aircraft patrols over the *Black Pit* pushed the U-Boots underwater, making it difficult for them to follow convoys and use their surface speed advantage to take a convenient position to attack. The same role fell to aircraft taking off from the new escort carriers accompanying the convoys. The second candidate for laurels was radar. Reports from German captains noted with growing concern instances in which their ship was located and attacked in the open ocean at night. For a long time, the Germans could not believe that British technology allowed radar to be squeezed on board aircraft. Air patrols took away the U-Boots' advantage of invisibility during the day; radar also deprived them of cover at night. A third Allied advantage was ship-mounted HF/DF sets. Tracking from shore stations was subject to an error proportional to the square of the distance between the station and the ship being tracked. If it had been possible to mount the device on board the escort ship, the error in locating the U-Boat could have been reduced to a few tens of meters. Problems with the antenna stood in the way, but with help came a Pole, engineer Wacław Struszyński, who developed a solution suitable for mounting on a ship's mast. New anti-submarine weapons also appeared: the Hedgehog, throwing a volley of grenades with contact detonators in front of the ship's bow before the sonar lost contact with the target, and air-dropped torpedoes, guided automatically to the noise of the U-Boot's propellers. By mid-1943, the Allies had seen the cumulative effect of applying a number of innovations in the struggle against U-boots, which meant that it took only a few weeks to settle the battle. But for the same reason, it is impossible to point to any one of them as the decisive factor.

As the decisive convoy battles unfolded in the Atlantic, a race against time was underway in several engineering centers to develop the 4-rotor bombe. The one-year handicap given to the two British teams must have earned them priority. In March 1943, British codebreakers received a prototype of the *Cobra*. With no alternative available, BP management ordered a series of machines, but soon the first Keen-designed machines, which proved easier to use and more versatile than its rival, also arrived; production of the *Cobra* was discontinued after twelve devices had been delivered to BP. Mahon's memoirs reveal that the 4-rotor bombe prototypes suffered from childhood diseases, and codebreakers had no practical use for them for several months. The first success was the breaking of the *Limpet* key on 18 February, while still testing one of the High-Speed Mammoths at Letchworth. The next success had to wait until 28 May, but this time it took place in the USA.

The US Army entered the bombe-building race later than the Navy, but it set ambitious goals for itself and the project contractor, Bell Laboratories. As late as November 1942, Bell was to present a functional prototype of the bombe and, in the early months of the new year, deliver a prototype of the device, which became known in time as '003' or 'Madame X'. The device was a moloch several times larger than Keen's bombe, being the equivalent of 144 Enigmas. The typical test time for the three-rotor menu was 10 minutes, which was only twice the speed of the Keen bombe. The original version of 'Madame X' contained crossbar switches equivalent in number to the telephone exchange of a town of several thousand people. Attempting to build a 4-rotor bombe on the same principle would have required around half a million components—even the United States could not afford such an extravagance. In the end, at a cost of $1.5 million, 10 copies of 'Madame X' were built and used mainly for research purposes.

In parallel, the Navy was working on its own bombe model. Barely on 3 September Wenger submitted a proposal to build the device at a cost of around $2 million, Vice Admiral Horne had already approved the design and allocated a budget on 4 September, and on 15 September a detailed document outlining the design features of the device was submitted to Op-20-G. The short time between concept and decision was the result of actions taken by the Navy back in March and April 1942. Wenger's earlier experiments with tabulators had taught him to appreciate the importance of mechanization in cryptology. He decided to set up a section in Op-20-G, tasked with designing devices to aid cryptanalysis. He placed at its head the then captain of the naval reserve, Howard Engstrom, in civilian life a lecturer in mathematics at Yale. In the spring of 1942, the US Navy was looking among US corporations for a partner for several projects, of which automatic computing and electronics were a component. In theory, two US companies had the required expertise: IBM and NCR. As part of the wartime mobilization of industry, Vannevar Bush's council declared the automatic cash registers produced by National Cash Registers to be a superfluous product under wartime conditions, while ordering IBM to continue producing tabulators and sorters. So, in March 1942, the Navy somewhat out of necessity chose NCR, which had spare production capacity, commissioning the company's engineers to develop, among other things, a proximity fuse, a 'friend-or-foe' identification system and several communications systems. When Wenger was given the go-ahead to start the bombe project on 4 September, the choice of partner was a foregone conclusion. Especially since, in addition to its collaboration with the Navy on other projects, NCR's Dayton plant had other assets. The corporation's president, Edward A. Deeds, was an old and trusted friend of Vannevar Bush. The corporation he headed was a leader in electronics applications. Among other things, in 1940 it designed and produced an electronic tube calculator, later used for calculations in the Allied nuclear program. Finally, the NCR laboratories employed the man Wenger needed: Joseph "Joe" Desch. A typical American self-made-man, the son of the owner of a van manufacturing plant that went bankrupt during the Great Depression. He grew up running barefoot on the shores of Great Miami, then took a job as a ticket taker at a local theatre at the age of sixteen. The money he earned allowed him to graduate from the University of Dayton's electrical department. In his subsequent careers at Dayton Electric, Telecom Laboratories and Frigidaire, he became known as an open-minded experimenter with deft hands. In 1938 he took over as head of the newly established electronics laboratory at NCR, where he soon established himself as a leader in miniature electronic tube technology. Wenger was aware that the high-speed four-rotor bombe required electronic logic, which naturally brought him to NCR and Desch.

Desch had to be introduced to the secrets of cipher breaking even before the formal start of the project, otherwise it would have been impossible to prepare a study close to the final solutions within ten days. However, the concept he presented was a disappointment to the principals at Op-20-G, who had suggested a solution based entirely on electronics. Wenger and Engstrom's intention was to construct a universal cryptanalytic machine, capable of attacking not only the Enigma ciphers. Desch, however, knew the advantages and disadvantages of electronic technology too well to hijack the construction of a moloch containing tens of thousands of tubes. As a result, the concept described in his memorandum referred to the British original. His bombe was to have double-sided Turing mixers driven synchronously from a common drive axis. Desch's most important innovation was a thyratron-based memory, in which the device was to record the rotors' position corresponding the stop condition. The anticipated speed of the rotors prevented the bombe from stopping immediately after a stop was detected, so the author envisaged a mechanism that would independently retract the rotors to the position recorded in the thyratron memory. Desch and Wenger envisaged the construction of 336 devices, so that each device would test one of the possible combinations of rotors, without the need to reconfigure them between runs. The author closed the memorandum with a request for speedy "approval of the proposals presented, as plans are now being prepared based on our understanding of the functioning of the machine as described above". As a result of further agreements, a new naval unit, the Naval Computing Machine Laboratory (NCML), was established on 11 November 1942, based in Dayton, in the former NCR staff evening school building (Building 26). Joe Desch became its coordinator on the NCR side and Lieutenant Commander Ralph Meader became its head on the Navy side.

Building 26 soon began to populate with staff representing the Navy and NCR. On the Navy side, dozens of people arrived in Dayton, both representing Engstrom's team of officers—mathematicians and engineers—and the NCOs scheduled to handle the bombes. NCR delegated a group of dozens of engineers and technicians to work at NCML. However, the most numerous group was to become members of *Waves*, the women's volunteer naval auxiliary. Their numbers at Dayton were to exceed two thousand. The operation of this military-civilian conglomerate must have given rise to problems and friction. Officers found it difficult to accept orders given by engineers of Desch's team—in many cases an engineer's request had to be reinforced by an order from Meader. Naval officers attempted to stimulate the performance of civilian co-workers by methods from the drill yard. A wartime working week of 54 hours was introduced, but engineers would usually spend Saturdays and sometimes Sundays in the laboratory. In conflict situations, officers complicated relations by reminding those concerned that they had received a deferral of active military service solely because of their participation in the project. The navy stipulated that the project be billed at cost, with no margin. NCR, whose civilian production was halted, had to accept this condition, but looked for ways around its consequences. One was to outsource as much of the production as possible to subcontractors not subject to the zero-margin clause; Meader dramatized that the production of 93% of components and subassemblies was outsourced.

Organizational and engineering problems constantly intersected with issues of secrecy and security. In this sphere, the biggest concern for the project supervisors was 'Joe' Desch. He himself had been vetted even before working on the bombe, as part of other projects commissioned by the NDRC. In May 1942, he spent three days in Washington talking to counter-intelligence officers who tried to provoke him with accusations alluding to the most personal areas of his life. When, irritated by this treatment, Desch finally exclaimed that he did not care for a job involving humiliation, he learned that he had just been admitted to Navy secrets. But this was not the end of the problems for him or his supervisors. Not only was Desch's mother of German descent and had numerous social contacts among German émigrés in America, her close relatives in Germany were active Nazis. As a result, Desch had to accept severe restrictions on his private life for the duration of the project; he was expected to keep contact with his own mother and sisters to a minimum and to avoid his distant cousin altogether. His every move was tracked by agents who were not too secretive about their presence. The most astonishing precaution, meanwhile, was that Ralph Meader took up residence in the Desch's small house and occupied the bedroom opposite their own.

Meanwhile, the only incident that could have jeopardized the security and secrecy of the venture happened without any connection to Desch and his family and came to light thanks to chance. On 5 November 1943, an NCR employee got into a car belonging to a work colleague, James Montgomery, in the company car park. Montgomery was happy to drop off colleagues living outside Dayton from and to work. When a passenger wanted to light a cigarette while waiting for the driver, he looked in the glove compartment of the car for matches, where he noticed a bundle of papers containing Montgomery's correspondence with the German embassy and a long list of German and Japanese organizations in the United States. When news of the find reached Meader on the morning of the following day, it caused an uproar. Montgomery and his wife were arrested and subjected to several days of intensive interrogation. At the same time, agents searching their home found several schematics and electronic components that were part of the bombe. All the circumstances of the incident indicate that the compromising letters came from a period when the couple was in financial problems. Montgomery's imagination hinted at the possibility of offering service to a foreign power, although it is not really clear what kind of information he might have divulged. The electronic components found in his home were more likely to serve his radio passion than end up in enemy hands. But faith and trust are not virtues valued by counterintelligence officers: whatever his motives, Montgomery had to be isolated. He was convicted of stealing state property and was only released from prison after the war, and the Americans never disclosed the incident to the British.

Desch, meanwhile, was fighting a desperate battle against the vagaries of his own design. Trusting his engineering talents, he promised Wenger that the first bombes would be ready for use in early 1943. The Op-20-G began a search for buildings in Washington, D.C., in which the devices would be installed, and for staff to operate them. In January, Desch was only able to provide the blueprints. During the final period of work on the project, Alan Turing made a visit to Dayton. By late 1942 and early 1943, British codebreakers were becoming convinced that their own designs for a 4-rotor bombe had run aground. In these circumstances, it was worth looking into the actions of the Americans. Turing set off for America in November, with an extensive program of visits. The first stage was meetings with Army and Navy codebreakers in Washington. They must have taken place in a good atmosphere, for Engstrom decided to demonstrate to his guest the backstage of his projects to build cryptanalytic devices and took him on a trip to Dayton. The trip had to be impromptu, as neither Turing nor the Op-20-G officers accompanying him could find room in the overcrowded hotels. The Americans ended up spending the night on the floor of the hotel lobby, and Turing was given a seat in the living room of Desch's house, where it was certainly possible to discuss bombe concepts in an informal atmosphere. A surviving note of Turing's visit to Dayton confirms that he readily agreed with his interlocutors on the most important points. Turing questioned the idea of building 336 devices, pointing out that the rules governing the selection of rotors precluded the use of about two-thirds of the theoretically possible combinations.

Desch's original design envisaged rotors of different diameters; this concept could only work if the machine operated with a fixed set of rotors. The departure from the principle of one machine—one combination of rotors entailed the need for interchangeability of rotors. Under pressure from Engstrom, Desch accepted the recommendation, although only in subsequent copies of the bombe: the two prototype machines operated with rotors of different diameters. At the end of the meeting, Turing had to amuse his hosts with a remark about the secrecy attributes surrounding the project. Anything that might have suggested a connection to cipher-breaking was removed from the documentation, part names and organizational units in the project. The use of the number 26, associated with the 26-character Enigma alphabet, was banned: components were numbered from 0 to 25, bundles of cables connecting components consisted of wires marked with 28 rather than 26 colors. Turing, however, smilingly pointed out that work on the project was being carried out in building number 26.

At BP, life in *Hut 8* went on as usual—no one seemed to notice the absence of the section head. In later accounts, veterans hinted with a tinge of embarrassment that Turing had been the undisputed leader of the team during the formative period of the Enigma cipher attack theory. However, once the backbone of the problem was broken, his communication problems and somewhat antisocial manner made it difficult to continue his duties as head of the hut. Turing gradually moved away from the mainstream of events, which necessarily centered around the talented organizer, Alexander. Finally, when, during Turing's stay in the United States, someone approached Alexander to ask who was actually in charge of *Hut 8*, he replied with a tinge of hesitation, "Well, I suppose I am". His assumption of leadership was made in passing, as if in recognition of the *de facto* role he had been playing for some time. The circumstances of Turing's trip to the USA indicate that it might have been arranged, among other things, to bring about a change of leadership at *Hut 8* in a soft way. On 31 March 1943, Turing returned to the country. He also returned to BP, but not to *Hut 8*. He accepted the changes that had occurred in his absence. Those entering BP filled in a form in which one of the fields contained the details of the section head: Turing had put Alexander's name in it. It seems that without work on the Enigma, he could not very well have found a place at BP. His earlier successes and the recognition that came with them had boosted his self-esteem so much that, against preference, he became involved with Joan Clarke, to whom he was even engaged for a time. He broke off the engagement at one time, however, and when he attempted to renew their former relationship on his return from the USA, Joan let him know that she saw no future in the relationship. For the rest of 1943, Turing agonized in BP with no defined tasks or responsibilities. Apparently, he could most often be found in the canteen, where he would engage his friends in conversations on topics he was passionate about. For unknown reasons, however, he refused to participate in the team that had just undertaken the biggest challenge BP faced after breaking Enigma—breaking the German teletype ciphers. As a result, at the end of 1943, he bid farewell to BP and moved to a military facility in nearby Hanslope Park, where he worked on a speech secrecy system code-named *Delilah*. Alexander, who took over for Turing, summarized his predecessor's role: "Pioneering work tends to be forgotten when experience and routine make the task trivial; many of us at *Hut 8* felt that the magnitude of Turing's contribution was never fully appreciated by the outside world". While Turing was later successful, it seems that the four years he spent at Bletchley were the happiest period of his life. Perhaps it would have found a different and better ending had he not felt rejected by an environment that had previously shown him respect and which he had managed to claim as his own.

But let us return to Dayton and Desch. In March 1943, two bombe prototypes, christened *Adam* and *Eve*, stood in Building 26. Their operation was a nightmare for engineers and technicians. There were constant oil leaks. During operation, the rotors sparked, damaging their sensitive contacts and causing false stop signals. Rotors mounted in the fast position deformed under centrifugal force. When the bombe did manage to get it right, something invariably broke a few minutes after launch. In May, the patience of Meader and Engstrom's superiors ran out and they demanded results. Desch was informed that senior naval commanders would descend on Dayton at the end of May to observe a practical test of the machines. For the occasion, a secure teletype line was installed between Washington and Dayton to enable the results of the bombes to be transmitted to Op-20-G. On the afternoon of 28 May, *Adam's* operator, Phil Bochicchio, set up the machine according to instructions received from Washington and started the run. When the machine stopped, he was convinced that there had been another equipment failure. However, after a moment of stillness, the rotors began to turn in the opposite direction, where-upon a printer connected to the bombe printed a row of letters making up the solution. A skeptical Bochicchio asked a colleague operating Eve to repeat the test for the same data—the second bombe printed an identical result. Meader sent the result to Washington and soon received confirmation that the solution was correct. Phil Bochicchio recalled years later that, a few days later, Meader arrived enthused, announcing that the first success of the new bombes had already paid the cost of entire project. In fact, Meader was trying to boost the morale of the team: the first success of the American bombes involved an old U-Boot dispatch that had already been broken at BP a few weeks earlier. Success was at hand, but a few more crises had to be overcome on the way to it.

In May 1943, a new unit of the US Navy, the X Fleet, was created to become the main weapon in the fight against U-Boots. The most peculiar feature of the new unit was that it did not have a single ship. Its commander, Admiral Francis Low, was to focus on intelligence, training, new weapons and coordination. Effective signals intelligence was one of the key areas in his organization. Meanwhile, the equipment that was to be the X Fleet's main weapon continued to cause the constructors a lot of trouble. Overseas, the 4-rotor bombe program was slipping along with no hope of success. Nigel de Grey, who under pressure of necessity had converted to work-ing with the Americans, wrote to Wenger on 3 June: "The chances of breaking the Shark on the basis of short signals are at present dim. (…) in the light of this, your bombes have become more important than ever (…) What is the program for their production now"? Then came the most serious crisis during the entire project. On 18 June, Dayton was ordered to abandon the results of the work to date, to destroy the prototypes and to proceed with the design of a new, fully electronic version of the device. It appears, Engstrom could not accept Desch's rejection of his initial ambitious concept, and saw the problems with the bombe launch as confirmation of his earlier concept. The situation was saved by Meader. He had observed Desch's team's struggles with the prototypes and knew how much effort had gone into the design of the devices, and how little was needed for complete success. He supported the NCR engineers, and after a determined exchange of letters between Dayton and Washington, the project received additional credit. Desch climbed to the heights of engineering mastery. He saw that the cause of the problems lay in insufficiently smooth contact surfaces in the rotors. When the rotors were sanded, it became apparent that the bombes had started to generate false stop signals. This time, the cause was short circuits caused by a mixture of oil and dust left over from grinding. Desch came up with some ingenious dust removal tools and rigorous periodic contact cleaning procedures, after which the problems disappeared and the bombes began to prove their capabilities.

Assembly of the first bombe of the production series, identified as the Model 530, was completed on 4 July, Independence Day. By the end of the month, 15 had been assembled, but as the new facilities on Nebraska Avenue in Washington were not yet ready to accommodate them, the equipment worked in Dayton, with the results of their work being transmitted to Op-20-G by teletype. In August, Wenger ordered the equipment to be moved; on 11 September, the first six bombes made their way to Washington by train, and by the end of the month, seventeen machines were working in the new premises. By January 1944, eighty-four bombes had been delivered by NCR, reversing the balance of power in the cryptological duel. One run of Desch's bombe for the 4-rotor menu took about twenty minutes; the 3-rotor menu required a mere fifty seconds to test. Once the problems of its infancy were solved, the Model 530 proved to be a reliable machine: during its first year of operation, both machine maintenance and downtime due to breakdowns took less than 3% of its operating time. The machines and their female staff worked in three shifts, under harsh conditions. Each bombe contained about 2500 electronic tubes, generating more heat than a gas cooker. Powerful air conditioners were installed the rooms in which the bombes worked. Despite this, one of the operators, Beatrice Dunphy, recalled that she worked "in unearthly hot rooms. Salt pills were lying around at every water dispenser and the noise was just appalling". The heat emitted by the bombes also had a good side. The *Waves*, who were accommodated in makeshift rooms, did not find conditions in which to dry their laundry; as a result, they dried their underwear on the back panels of bombes. Desch bombes and the Waves operating them, as well as the Op-20-G codebreakers, quickly took on the burden of breaking U-Boots' ciphers. The British maintained that the Americans were trying too vigorously to emancipate themselves from their care. An efficient bombe was an important ingredient in the solution, but without a good quality crib it became an expensive pile of wires, gears and other parts. And finding good quality cribs required experience and constant contact with the network, both of which the Americans lacked. Mahon stressed that "the art of crib analysis was less about finding a good quality crib than it was about ignoring a bad one, and in this aspect Op-20-G was failing".

The Americans have had bad luck. Until recently, the weather code, followed by the short signal code, provided a supply of decent quality cribs. However, changes introduced by the enemy resulted in a decline in the quality of cribs in both codes, leading to a dispute inside *Hut 8* between proponents of proven methods and innovators. A source of hope for the innovators were the messages exchanged by U-Boots operating in Norwegian waters, encrypted in the old *Dolphin* system, broken constantly and without much difficulty. Knowing the way the German communications system worked, they argued that re-encodements between the *Dolphin* and *Shark* must happen. The argument was acrimonious, "the short-signal side had a large number of cribs of very poor quality, while the proponents of re-encodements had cribs that were few but absolutely certain". *Hut 8*'s problem was the lack of high-speed bombes on which these cribs could be tested. The dispute was resolved practically when, on 25 June, BP's first operational 4-rotor bombe, the *Trinidad*, succeeded on the basis of a crib derived from a re-encodement. This date marked the beginning of the end of the era of short signals and the dawn of the re-encodements period. However, their identification and use caused much trouble for inexperienced Americans. Thereafter, a reasonable compromise took shape whereby the Americans would send a list of candidate cribs to BP, and *Hut 8*'s experts would eliminate obviously bad proposals from the list. British-American cooperation also helped overcome another crisis when the Germans introduced the new *Caesar* reflector and a new narrow rotor, the *gamma*, to the Shark on 1 July. Had they done so with the debut of the M4 machine, they would have complicated the task of the Allied codebreakers. However, once the structure of the 4-rotor machine had been worked out, and the wiring of the *beta* and *Bruno* rotors was no secret, all the Allies needed to do to crack the new components was a long re-encodement from *Dolphin*. When they found them, they divided the work on the new challenge between the two sides of the Atlantic and after a few days were able to continue decrypting the *Shark*.

A measure of the relaxation that gripped Allied codebreaking agencies after the crisis surrounding U-Boot ciphers ended in the summer of 1943 was Wenger's prolonged leave. In November, with dozens of high-speed bombes in operation at the Naval Communications Annex, the Op-20-G head decided he needed to rest after the stress of the previous months. He went, accompanied by his wife, to Florida, where for six months he walked on the beach and engaged in his hobbies: drawing and painting. Fortunately, by this time, U-boots were no longer providing gruesome painting motifs to people resting on the beaches of the US east coast. As a result of the change in Allied naval strategy in the second half of 1943, the roles were completely reversed—the hunters became the game. The new rules of warfare came in part from the lessons of the convoy battles fought at the apogee of the Battle of the Atlantic, between March and May. The British attempted to use traditional evasive tactics, directing convoys around patrol lines set up by Dönitz. But during the crucial period, the number of convoys traversing the ocean and the number of submarines waiting for them only made it with the utmost difficulty to find a free piece of ocean; convoys routed around one wolf pack ended up in another. The US Navy commander, Admiral King, noted at the time that defensive tactics did not suit the new realities of battle—Allied superiority in ships, aircraft, men and intelligence.

The Allies were strong enough not to avoid the wolf packs but to draw German submarines into the battle around the convoys, where new weapons and better tactics gave the escorts an advantage. The number of ships at their disposal allowed the Allies to set up not only support groups independent of the convoy escorts, but also hunter-fighter groups, scouring the ocean freely for targets. So far Royal Navy commanders were inclined to agree with their American colleague, his further proposals went too far in their view. King believed that Allied superiority made it possible to direct the actions of hunter-fighter groups strictly and directly on the basis of deciphered German dispatches. The target he proposed were 'milk cows'. The calculation was simple—a submarine tanker would supply during a voyage up to a dozen line ships. Eliminating the 'milk cow' would force a fair number of U-Boots to return to base early. The British, for their part, pointed out that adopting offensive tactics would allow the Germans to realize the compromise of their ciphers and prompt a system change, the effects of which would also extend to the army and air force. 'Milk cows' generally operated maintaining radio silence, so it would have been difficult to justify the presence of Allied ships and aircraft in the area. However, King was determined to carry out his intention in the best "the devil with torpedos" style.[1]

After the nightmarish year 1942 and the fierce convoy battles of the first half of 1943, the Allies were finally victorious in the Battle of the Atlantic. Convoys carrying millions of tons of supplies and hundreds of thousands of troops to Europe and Africa were safe. Victory came too late for the thousands of sailors of the ships that fell prey to the U-Boots during the critical period of the Battle of the Atlantic. A significant number of them could have been saved had it not been for intra-Allied disputes over naval tactics and leadership in the cryptologic crusade. But as well as the unfortunate consequences of the codebreakers' difficult co-operation, they squandered at least two opportunities for the early cracking of the *Shark* key. The Americans were for many years inclined to consider their mistake a success: on 14 April 1942, the US destroyer USS *Roper* sank the submarine U-85, the first U-Boot sunk by an American ship in World War II. When U-85 reached the coast of Newfoundland in early April, it seemed that the greatest nuisance on its fourth patrol would be the unbearable heat. The crew spent their nights and days on the deck of the ship going on the surface, enjoying the sun and the sea breeze. On the night of 13/14 April, U-85 surfaced in the vicinity of Cape Hatteras. Despite the cooler weather, most of the crew were taking in fresh air on deck when an unidentified craft appeared nearby. It was the destroyer *Roper*, whose radar operator spotted echo more than a mile away. The speed of the target indicated that the *Roper* was dealing with a warship, and additional confirmation soon came in the form of a torpedo that narrowly missed the destroyer's hull. As the distance between the ships

---

[1] During the American Civil War, Admiral Farragut, commanding the fleet of the North, entered a minefield during operations in the New Orleans area. When this fact was brought to his attention, he said 'the devil with the torpedos there' (mines were described as torpedoes then), led his ships through the field and contributed to the capture of the port. The words he uttered on that occasion became a symbol of the US Navy's offensive tradition.

decreased to 250 meters, *Roper*'s searchlight illuminated the distinctive silhouette of the U-Boot, and the destroyer, in her combat debut, successfully fired on the enemy. *Roper*'s further actions also marked a debut. After several hits, U-85 submerged, leaving dozens of crew members floating on the surface. *Roper*'s crew were unsure whether the ship had sunk or submerged to continue the fight. Her captain had listened to stories about wolf packs and acted in the belief that other enemy ships might be lurking nearby. So, he did not proceed to rescue the crew members, but threw a dozen depth charges in the area where the ship was last seen. In the morning, a flying boat appeared over the scene and added her depth charges in the area where the oil stains signaled the presence of the U-Boot, and *Roper* soon added another few. When the retrieval of the German sailors from the sea began, there were only 29 corpses floating on the surface among the ship's wreckage and oil stains.[2] Even as they were being fished out, there was a momentary panic as the sonar caught a hard underwater echo, which ended with another series of depth charges. The U-85 was the most thoroughly sunk submarine in World War II.

U-85 came to rest on the bottom at a depth of only about thirty meters. As early as 15 April, the British trawler HMS *Bedfordshire* arrived in the area of the incident and easily located the wreck thanks to the air bubbles still rising from it, but due to the murky water, the divers' work on the wreck was unsuccessful. The *Bedfordshire* was soon replaced by the US Navy tug, USS *Kewaydin*, and by the rescue ship USS *Falcon*. Neither of these was successful. Hindered by unexploded depth charges that had to be detonated, and by the U-85's hulk being hollowed out and letting the air pumped in, divers only found that the U-85 kingstones were open: the crew had taken care to scuttle the ship. For the next 55 years nothing disturbed the ship's peace. It was not until August 1997 that Roy Parker located the wreck and extracted two wooden boxes containing four Enigma rotors each. During expeditions in 2001 and 2002, Jim Bunch and brothers Rich and Roger Hunting extracted two complete copies of the M4 Enigma and a large set of documents from the radio room.[3] Both the cipher machines and the documents were very well preserved. Giving up on rescuing the ship's crew and throwing depth charges among the survivors was not in keeping with the honor of the US Navy. It also harmed Allied interests. Had they not demolished the already sunken ship, they would have easily extracted important documents from it, enabling the *Shark* to be broken six months earlier than it actually was.

The Hut 8 codebreakers also scored a major stumble. Alexander spoke openly of an episode "over which even the least sensitive among us would gladly draw a veil of silence", Mahon described its circumstances as "… a regrettable story …, undoubtedly the worst episode in the history of the section". The redeployment of a group of U-boats to the Mediterranean in October 1941 resulted in a new German network appearing on the air, baptized in BP as *Porpoise*. Its messages were

---

[2] And two corpses massacred to the extent that American sailors were forced to leave them at sea. None of the U-85 crew survived the sinking of the ship.

[3] See http://home.comcast.net/%7Edhhamer/u-85.htm

recorded primarily in Malta, from where they were transmitted in mailbags to BP. When the first attempt to apply to *Porpoise* the methods effective against *Dolphin* and *Shark* failed, the *Hut 8* lost interest in those dispatches. Alexander assumed that the reason for the attack's failure must have been the bigram table, different from that of the other networks. As a result, bags of the *Porpoise* dispatches were handed over to *Hut 4*, where they attempted to extract any useful information from them, e.g. by traffic analysis methods. From these, specialists deduced that the new network actually represented three separate communication systems, separately for the Black Sea, the Aegean and the northern African coasts. By the end of 1941, the network was recording around 100 dispatches a day. When, by mid-1942, this figure had risen to 200–300 dispatches, the codebreakers decided to take a closer look. Without difficulty, they noticed that the message headers showed specific characteristics: if two messages shared the first letter of the header, their fifth letters were also the same, and similarly for the second and sixth, third and seventh and fourth and eighth letters. This was a behavior typical of the Kriegsmarine network until October 1937 and of the Wehrmacht and Luftwaffe until 1 May 1940, and the mechanism with which the Poles began their attack on the Enigma ciphers. The fact that the distinctive pattern was not recognized by anyone at *Hut 8* seems to confirm that the knowledge the Poles passed on was stuck at Turing's level. Mahon noted that "Turing, the only one of us who had come into direct contact with the double encryption of the message key, had by this time taken a modest part in the work of the section, and we lacked the theoretical background and experience to the extent that we were fixated on the idea that all naval networks functioned similarly to *Dolphin*". The problem was solved thanks to a happy coincidence. In August 1942, someone from *Hut 8* raised the problem of the *Porpoise* network in a conversation with a colleague from Hut 6. Hut 6 still remembered the double encryption of the message key and its breaking using Zygalski's sheets. It turned out to have been prematurely discarded. *Porpoise* was the first network after a long hiatus to signal a return to old habits and used this practice until June 1944. In subsequent years, other Kriegsmarine networks appeared using the system defeated many years ago by the Poles. German cryptologists used it wherever there was a need for a new network, and logistical problems made it difficult to distribute the documents necessary for *Dolphin* or *Shark*. The temporary nature of its use is evidenced by the fact that over time almost all networks abandoned this practice by adopting the general Kriegsmarine system. Alexander's note about 100 messages intercepted daily in late 1941 confirms that even then, breaking *Porpoise* was a relatively easy task. Had Rejewski and his colleagues worked alongside British codebreakers at BP, the specific and familiar features of the key would certainly have attracted their attention. But the Polish team at the time was devoting its experience, knowledge and time to secondary cryptological problems at P.C. Cadix in the south of France.

Chapter 20
Danger!

"The place was isolated, life there seemed easy, (…) we had means of transport available in case we needed to leave immediately". This is how Bertrand described 'Cadix', where the Polish and Spanish codebreakers started work in October 1940. Their stay at the center may have given the impression of an idyll, especially when compared to the hardships of war that were affecting much of the European population at the time. In the early days of the center, the supply of intercepts was symbolic. Initially everyone in the villa had too much free time, which they tried to manage in their own way. Bertrand insisted on not fraternizing with the inhabitants of the surrounding villages and towns. The isolation of the center was, in its own way, a protective mechanism, for rumors had spread among the local population that something strange was going on at Chateau Les Fouzes, with which it was better to avoid contact. But complete isolation was not possible; the codebreakers were keen to diversify their time with excursions in the area. The attractions of the south of France lay within easy reach of a bike tour; Nîmes with its Roman monuments, papal Avignon and the picturesque arches of the Roman aqueduct at Pont du Gard. Somewhat out of necessity, somewhat out of curiosity, the Poles adopted tastes and customs of their French hosts, for example by participating in frog hunts. Private and public celebrations were invariably popular, becoming an opportunity to spend a few hours in a more relaxed atmosphere, amidst chatter and chants.

There is an opinion among codebreakers that the best aptitude for this profession is a combination of mathematical and musical talents. In the Polish team, mathematical talents were strongly represented, but it turned out that musical tastes were not inferior to them. Both Różycki and Zygalski entertained their colleagues with their performances, and the showpiece number of Ciężki, who had a strong and clear voice, was Nadir's aria from "The Pearl Catchers". Occasionally, the Poles were joined by Spaniards whose songs added variety to the repertoire. Most members of the team tried to use the time to learn something useful. In some cases, it was a necessity: Ciężki and Palluth had to train several subordinates as radio operators. Others treated learning as a way to kill time or as an investment for the future.

© The Editor(s) (if applicable) and The Author(s), under exclusive license to
Springer Nature Switzerland AG 2025
M. Grajek, *Enigma Myth Deciphered*, History of Information Security,
https://doi.org/10.1007/978-3-031-65475-6_20

Rejewski, who had attempted to improve his French after arriving in Paris, changed his interests during his time in 'Cadix' and, with Bertrand's disapproval, began learning English. Many Poles took advantage of the period of stabilization to make contact with their families. Considerations of conspiracy did not allow letters to be sent under one's own name and from one's actual place of residence, but Langer and Bertrand arranged several channels allowing messages to be sent home without compromising the secrecy of the center's operations. As a result, families in the country occasionally received dispatches from the most unexpected places in the world (South and North America, Africa, Turkey or Switzerland), signed by people they did not know, but the familiar handwriting, the topics covered and the emotional charge made it possible to reliably recognize the closest people in the authors. Rejewski recalls that an operation to bring families to France was being considered, but had to be abandoned because of the obstacles that piled up.

After a period of idle run in late 1940 and early 1941, the center's activities began to take off, revealing the intricacies of its status in the process. On 7 March 1941, the Poles managed to establish direct radio communication with the Supreme Commander's headquarters in London. The exchange of information between the HQ and "Ekspozytura 300" was the result of the Polish group's position of dual dependency and took place entirely beyond the knowledge and control of their French hosts. As we shall soon see, information which did not arouse the enthusiasm of the hosts also reached London by this route. Bertrand vigorously sought contact with the British. His action represented a dissonance with the behavior of the vast majority of his compatriots, whose opinion was split at this stage of the war. On one side were the supporters of the Vichy government, which was almost mechanically associated with an anti-British attitude. The other side was made up of de Gaulle's supporters, who also had no sympathy for the British, but allowed them to provide assistance in the name of fighting their common enemies—the Vichy government and the Germans (presumably in that order). Bertrand maintained contacts on both sides of the barricade. While remaining a member of the Vichy apparatus, he tried to cultivate old contacts with the SIS. Denniston, when the proposal to continue cooperation reached him, presented a discouraging response to his superiors, asking the rhetorical question, who is paying for this operation and for what purpose? If this was to be a government that any minute might find itself in the ranks of Britain's wartime adversaries, it would be imprudent to cooperate. His response was understandable: in the first quarter of 1941, BP was quite adept at dealing with Wehrmacht and Luftwaffe ciphers, having just taken the first steps in attacking Kriegsmarine ciphers. The help of the Poles, who in previous years had been a factor in the attractiveness of cooperation with 'Bruno', no longer seemed necessary. BP had in fact rejected the offer of co-operation, but someone at SIS (with a high degree of probability it was Bertrand's old friend Biffy Dunderdale) reminded that, regardless of the practical utility of co-operation with 'Cadix', it must be remembered that the center's staff were privy to the secret of breaking Enigma. A brutal rejection of the offer of co-operation might cause a reaction leading to the secret being revealed to the Germans. It was decided to maintain somewhat pro forma contact with 'Cadix', but on the British side the role of correspondent was taken over by SIS. In March 1941, Bertrand travelled via Madrid to Lisbon, where he met Dunderdale in the local botanical garden and picked up the radio transmitter and ciphers that enabled the center to make direct contact with London, before forwarding them to Vichy by French diplomatic mail.

The 'Cadix' centre functioned as an organization of a rather schizophrenic nature and structure. It was part of the Vichy secret service, set up somewhat against the letter and completely against the spirit of the Armistice Agreement, and therefore illegal from the German point of view. These services, for the use of their French superiors, carried out activities directed exclusively against enemies of the Vichy state. The problem came down to the definition of enemy. In Marshal Petain's understanding, and even more so in Laval's, the enemy was Britain and equally General de Gaulle's partisans. When that country's secret services were reconstituted after the defeat of France, their members were firmly forbidden any contact with the British. Bertrand was one of perhaps four members of his service for whom an exception was made to this ban. However, this exception was accompanied by the comment that his aim was not to continue working for the British, but to spy on them.

In the center's mission so defined, the knowledge and experience of Polish codebreakers would have been of no use—the British did not use Enigma. Bertrand, however, hoped that by having the Enigma decrypts at his disposal he could induce the British to reveal the extent of their success in grappling with the cipher. The Poles were therefore to continue their work, even though the Enigma decryption was a clear breach of the Armistice Agreement and the mandate given to the service by the Vichy authorities. In their post-war memoirs, representatives of the Vichy secret service solemnly assured that the country's defeat had not broken their unequivocally anti-German stance and they were determined to continue the fight against their traditional adversary even in the new, difficult conditions. If their assurances corresponded to the truth, they must have been working in deep conspiracy in front of their political principals, who professed an entirely pro-German orientation at the time.

When the Polish team was placed under Bertrand's command in October 1939, its section formed part of the French Army's integrated intelligence service. With the war underway, the Enigma decrypts and the information they provided were at the very focus of the service in whose structures the Poles were functioning. After the defeat of France, however, the integrated intelligence organization was abolished, reverting to its pre-war, decentralized structure. The section headed by Bertrand became part of the Vichy counter-intelligence, a fact that its commander probably did not expose in his conversations with the heads of the Polish team. Among the intelligence services that could have entered the orbit of interest of the French counterintelligence, only the German Abwehr used Enigma, but this was a different model from the machine used by the German armed forces. Available documents show that Polish codebreakers were not even aware of the existence of the model referred to as *Zählwerksenigma*. As a result, the Poles' knowledge and experience of Enigma decryption was completely useless to French counter-intelligence.

Indeed, we will observe how, over the following months, the French transferred the activity of the team to simple hand ciphers used by German agents operating in areas subordinate to the Vichy authorities.

The orders received from London preserved the subordination of the Polish team to the French, against the interests of Poland, the Allies and the team members. Langer, Ciężki and their subordinates came to terms with the letter of the order, but never accepted its consequences, renewing their initiatives to move the team to the UK, where its work could offer more useful results. Perhaps to pacify their resistance, the Polish intelligence HQ in London proposed a bit cumbersome compromise: the Poles were to be subordinate to the French only administratively, but in operational terms they were directly subordinated to the intelligence headquarters in London as "Ekspozytura 300". Contacts with London and operations carried out at its orders were to be kept secret from the French, which added yet another layer of conspiracy to the operation of PC 'Cadix'. The subsequent fate of the team indicates quite clearly that the real reason for keeping the team away from London was animosity and friction within the Polish intelligence community. In time, its chiefs found a convenient and plausible justification for the cryptologists' stay in southern France. There was an effective network od Polish intelligence service operating in North Africa, from the codename of its creator and head known as 'Rygor'. However, the network had no direct radio communication with London. The intelligence HQ realized that 'Ekspozytura 300' could take over as a relay station for communications with Algiers. In time, Maksymilian Ciężki was transferred to Africa, where he informally acted as a signals officer for the 'Rygor' network.

If, to all the levels of internal and external conspiracy and distrust, one adds the isolation imposed by Bertrand between teams of Spanish and Polish cryptologists, and between groups of foreign mercenaries and French administrative personnel, 'PC Cadix' presented itelf somewhat like a caricature of the secret service organization.

Despite this, Polish codebreakers believed that little was missing to regain their former position in Allied cryptologial cooperation. A measure of Bertrand's optimism on the subject was a series of trips to Paris, which he undertook, among other things, to bring additional Enigma clones to 'Cadix'. Parts of the machines had already been ordered before the French defeat. In December, Bertrand transported Enigma components to 'Cadix' on successive trips between Paris and the unoccupied zone. This was a risky and not very prudent idea. The Germans carrying out checks on travelers on the demarcation line between Vichy France and the occupied zone were inquisitive; in fact, on the occasion of one of his journeys, Bertrand was said to have spotted on a list of wanted persons the name under which he had made a previous journey. An accidental check, culminating in the discovery of distinctive parts in his luggage, must have meant that the secret of the Enigma had been revealed. However, he managed to transport to 'Cadix' the components from which Fokczynski and Palluth assembled four Enigma clones. The final stage before proceeding with the decryption was the acquisition of raw cryptographic material, intercepted enemy dispatches.

During the period of work at 'Bruno', the Poles received messages registered by two French radio intercept services. Both were disbanded under the terms of the Armistice Treaty. In their place, a small civilian intercept service, the Groupement des Controles Radioélectriques de l'Interieur, GCR, was established. Its task was to monitor radio communications within Vichy France, detecting radio transmitters operating in contravention of the armistice treaty. A fortuitous circumstance for the work of 'Cadix' turned out to be the appointment as head of this service of Captain Gabriel Romon, an engineer, whose acquaintance with Bertrand dated back to pre-war times. Romon managed to fill the key positions with trusted people and to organize the work of the service in such a way that, in addition to its main duties, it intercepted various German messages. Its results were relayed to Bertrand, who visited the GCR headquarters in Hauterive once a week for this purpose. The information flowed both ways. Of the four radios installed at 'Cadix', two operated as liaison with London and the other two were being used as intercept stations. Their purpose was mainly to identify enemy networks of interest. Once the characteristics of such a network had been identified, technical details were passed on to the GCR, which took over operational listening. Interception of radio transmissions of German agents broadcasting from areas under Vichy control was entirely in Polish hands. Operators at 'Cadix' intercepted around 3000 such transmissions during the entire period of their work in the south of France. Later on, an additional source of ciphered messages was the French resistance in the occupied part of the country, which managed to tap the French telegraph network now used by the Germans.

The intercepted messages were encrypted in various systems and concerned a broad spectrum of issues. Despatches intercepted directly by the 'Cadix' belonged, as it later turned out, to German agents operating in the unoccupied zone of France. They were mainly deployed in ports on the French and North African coasts, with the task of relaying information on the movement of people and goods between the metropolis and its African dominions to Abwehr headquarters in Stuttgart. Working in field conditions precluded the use of cipher machines, so the agents were equipped with a manual cipher. Antoni Palluth sat down to solve it, recalling the forgotten days of his apprenticeship in the craft. He also broke it with great regularity, although the key for each period had to be recreated separately, with considerable effort. The reward for the effort was the breaking of a message in which headquarters ordered the agents to assemble at a hotel in Marseille, before being flown to Morocco. Informed of the fact, the French organized a raid on the hotel and imprisoned the whole company. It turned out, by the way, that all the agents were equipped with portable radios installed in identical suitcases. They attracted no attention scattered in various ports, but collected in the hotel they represented the agents' ID card. According to a later Polish report, 'almost the entire personnel [of the network] and six radios were then captured'.

The second type of dispatches came from tapping the telegraph lines. The German radio security service constantly monitored radio traffic coming from within and outside the occupied territories. The stations exchanged information on the time and frequencies of operation and the bearings of detected stations via a wired network, encrypted with the double Playfair cipher. Breaking the cipher was a labor-intensive rather than difficult activity, while the effect of the decrypts was extremely important for the underground organizations. Once the location of the station had been established with sufficient accuracy, a search team was sent into the field, relaying its instructions via the network overheard by 'Cadix'. In this way, many underground radio operators were warned in advance of the danger and avoided arrest. Another German hand cipher regularly broken at 'Cadix' was the one used by employees of the German armistice monitoring commission. The youngest member of the team, Kazimierz Gaca, for whom this was his first experience in cryptology, specialized in breaking it.

It is no coincidence that only in last place were German military and police dispatches encrypted using Enigma. Bertrand, in his detailed list, confirms that these accounted for less than 15% of the messages broken by 'Cadix'. Rejewski was right to admit that he could not remember "whether we had dealt with the Enigma cipher yet. I suppose, however, that we had too little cipher material and had to abandon dealing with this cipher". In fact, there were two reasons why the Polish team's involvement in breaking Enigma ciphers quickly faded. In the early stages of operation, the center did not have access to the cipher material. German troops in occupied France developed a wired communications network fairly quickly, cutting off the enemy's listening stations from the source of the messages. The situation was soon to change; first the German involvement in the Balkans, then the invasion of the Soviet Union provided many opportunities to break German dispatches, but then the codebreakers ran out of tools to work with. In the final phase of the French campaign and immediately afterwards, the Poles quickly adopted the methods developed in BP; cilia and herivelism. These were based on the mistakes and lack of radio discipline of the German cipher clerks and it was obvious that in time they would be eliminated by the German security service. Langer complained as early as February 1941 that 'the changes made by the cipher service had eliminated the errors on the basis of which at one time it was possible to solve the cipher'. The errors returned as soon as German units were back on the front line, but they happened relatively rarely and, according to Welchman's observations, had disappeared completely by early 1943. Enigma decryption had become entirely dependent on equipment: bombes, catalogues and tabulators, the use of which, under clandestine conditions at 'Cadix', could not even be considered. The fact that the Poles managed to break nearly 700 Enigma ciphers representing various networks during a two-year period of work in the south of France was a confirmation of their experience, knowledge of the enemy and brilliant intuition.

During this time, the British tried to maintain the the appearance of cooperation with "Cadix." Information preserved in the archives shows that during 1942 they handed over about twenty-five broken Enigma keys. Bertrand's recollections indicate that these came from networks that the French stations were not recording. Denniston consistently sabotaged the order to share the results of his work with the French. On June 15, 1941, in an internal memo he wrote: "It is true that we have not sent him [Bertrand] any keys since May 23; sending him current keys is clearly an action improper. I suggest that we respond by saying that we are encountering increasing problems with the breaking keys (…)". Documents kept by Polish cryptologists and found under astonishing circumstances many years after the war indicate that Denniston found an interesting solution to the dilemma. Among the documents was included a list of keys to the cipher, which ended up in the hands of the Poles. Juxtaposing it with information from British sources, it can be established that the British handed over to 'PC Cadix' only the keys captured in North Africa. Should the center be compromised and the keys fallen into German hands, they would not have raised the alarm; the Germans were aware of their loss, and its circumstances did not point to the systematic cracking of the cipher.

According to Bertrand's list Enigma decrypts involved an astonishing number of regions and countries: Austria, Belgium, Bulgaria, Greece (with Crete), France, Hungary, Poland, Libya, Syria, the Soviet Union and Yugoslavia. Several of these must have attracted keen interest in Britain; in particular, a package of messages confirming German preparations for the 1942 summer offensive and early evidence of atrocities behind the Eastern Front. These incidental successes were, however, unable to cover the truth: the work of the Polish group was more beneficial to the Vichy authorities than to the Allied cause, which the codebreakers wanted to serve. In a report written down after getting to London, Major Michalowski assessed that " the group's influence on intercept priorities was hampered, as the hosts, for their own utilitarian purposes, clearly favored certain areas at the expense of others. (…) The ciphers of the German intelligence network and the armistice commission were their main concern, focusing all the attention of their interception service".

The decreasing share of Enigma keys in the messages broken by the Polish team at 'PC Cadix' meant that the center's role in the plans and operations of Polish intelligence in London was changing. Headquarters treated the center increasingly as a transit hub for radio communications between London and Polish intelligence networks in southern France and the Mediterranean. This role was facilitated by the establishment of the African branch of 'Cadix' in early 1941, operating under the code name PO1. It was located in a villa situated in the Algier's suburb of Kouba, where, as Bertrand himself admitted, life flowed more normally than in the clandestine conditions of 'Cadix'. Interestingly, the new post was set up by the French with intention of listening to the dispatches of British troops rather than German ones. Algiers promised better conditions for receiving messages coming from the battlefields of Libya and Egypt. In addition to its function as a listening center, PO1 served as a resting place for codebreakers who were exhausted by the clandestine working conditions in 'PC Cadix'. It was agreed that members of both teams would take turns spending three months in Algiers. If the Poles were asking themselves why Africa did not become the main base for the 'Cadix' operations, they kept it to themselves.

French North Africa was an intelligence hotbed at the time, and among the many services, the Polish 'Rygor' network under Major Rygor-Słowikowski was one of the most effective. The Poles of 'Cadix' established contact with him in July 1941, and PO1 in Algiers became the network's main channel of communication with

London. Although the 'Rygor' network finally received its own radio set in February 1942, it continued to use PO1 as an intermediary until August. The operation of the Algiers branch was the cause of the first tragic loss suffered by the Cipher Bureau team during the war. On 6 January 1942, four codebreakers sailed from Algiers aboard the ship *Lamoricière*: Jan Graliński, Jerzy Różycki, Piotr Smoleński and Sylwester Palluth. They were supposed to return to France at the end of the previous year, but Captain François Lane, who was supervising the Poles, delayed their departure in order to spend the New Year in Algiers. Late in the evening of the following day, the ship's radio officer picked up an SOS signal coming from the freighter *Jumièges*, fighting a storm not far away. Captain Milliasseau was faced with a problem: due to wartime restrictions, his ship's machinery had been converted from oil to coal, reducing its power. In addition, the coal bunkered in Algiers was of inferior quality, making the storm dangerous for his own ship as well. Seafarers' solidarity prevailed; *Lamoricière* sailed to rescue the sinking freighter, but arrived too late at the scene of the disaster. Within hours, the situation had also become critical for the liner; the waves were flooding the decks, the soaked coal was unusable: at 5.10 PM on 8 January, *Lamoricière* herself sent out an SOS signal. In the morning, a glimmer of hope dawned when the ship *Gouverneur Général Gueydon* appeared nearby, but amidst the storm, attempts to retrieve passengers from the sinking *Lamoricière* failed. Half an hour later, the decision was taken to evacuate the people, which brought tragedy at the very start. When a group of children travelling on the ship were placed in the first lifeboat, a wave capsized it and all the children fell into the sea. The passengers watching the tragedy mostly refused to enter the lifeboats, so that when the *Lamoricière* sank at 12.35 on 9 January, only 93 people were rescued. Among those rescued, only Sylwester Palluth was found. The three Poles and the French liaison officer accompanying them, Capt François Lane, perished in the disaster.

Reports on the disaster shed some light on what the codebreakers were doing in Algiers—one or two copies of Enigma went down with the ship. The cedebreakers would not have risked travelling with dangerous cargo if rest was the only reason for their stay at PO1. It was late 1941/early 1942, and the British pursuit of Rommel's troops after Operation Crusader was just coming to an end in Libya. The fierce battles going on in the Egyptian and Libyan deserts left no doubt as to where the focus of the war had shifted to. The Poles attempted to do the only reasonable thing under the circumstances—to move their radio interception and decryption point closer to the main scene of the conflict; hence the presence of Różycki and the Enigma machines in Algiers. By this time, however, the methods available to the Poles for breaking Enigma had reached the end of their usefulness, which is probably why they were carrying both machines back to France.

The presence of Captain Jan Graliński and Piotr Smoleński in the team returning from Algiers reminds us of an aspect of the Cadix' operation that has been overlooked. The center not only worked on German ciphers, but also attacked, and successfully, Soviet ones. The latter were worked on by a small but experienced team consisting of Stanisław Szachno-Romanowicz and Jan Graliński. Szachno was a veteran and also a living symbol of the Cipher Bureau's Soviet section. In 1944,

Major Gaweł, the then head of Polish signals intelligence, wrote: "where Szachno is, there is also proper Russian codebreaking". Langer's 1941 report shows that Szachno and Gralinski had successfully continued their attacks on Soviet ciphers during their time in 'Cadix': "In the Russian section the work is going on, but there is a lack of intercepted material, which we are receiving only in modest quantities". Their work was probably more valuable to Polish intelligence during this period than the attacks on the Enigma ciphers. When the British felt confident in the field of German machine ciphers, they lost interest in Polish achievements in this sphere. They did, however, maintain a keen interest in everything the Poles had to offer about the Soviets. Even when, after the German attack on the Soviet Union, Churchill declared that the interception and decrypting of Soviet dispatches by the British services would cease, the GC&CS welcomed the continuation of the work by Polish outposts. A series of memoranda in late September/early October 1941 leave no doubt about this: "to save the British radio operators, equipment and cryptologic personnel, it was decided that the main part of the work on Russian communications systems would be carried out by the Poles. (…) The intelligence chiefs have decided to reduce activity in the Russian communications field to listening conducted by 2 receivers at Flowerdown, 2 at Cheadle and the Poles at Stanmore. It is assumed that the Poles will continue their cryptographic work. (…) The Poles are to relay to the GC&CS any intelligence they acquire from the listening carried out by Cheadle and their own". The results of this work must have been valuable if, around July 1943, the British radio intelligence chiefs asked General Klimecki to expand the team at Stanmore listening to and cryptanalysing Soviet dispatches: "we think that the present situation calls for the immediate strengthening of this valuable section of the Polish headquarters". The high rating of this area of Polish radio intelligence was also influenced by the work in 'Cadix'. Michałowski complained later that 'the amount of material decreased due to the abolition of a very well-working listening post in Algiers and closing the listening stations in Iran, Istanbul, Sofia and Budapest'. His report indirectly confirms that Gralinski and Smolenski's trip to Africa was linked to the search for better listening conditions of the Soviet communications.

Amidst the hard work, carried out under clandestine conditions, hidden emotions were bound to come to the fore and problems within the team to emerge. Langer was faced with a daunting task, obliged to enforce discipline in venture whose sense he himself questioned. Moreover, he was not dealing with the recruits, against whom corporal's methods from the drill square could be effective. Many years later, Rejewski was to admit in a conversation with Jerzy Palluth that tensions between the military and civilian parts of the team could be felt during his work in 'Cadix'. It seems that Langer and Ciężki tried to prevent their subordinates from discussing with the French the sense of their work. They were only able to do so by invoking secrecy and discipline, with no guarantee of effect. The codebreakers knew that they were in France against their own will, against the interests of their own country and with the thinly disguised disapproval of their own superiors. Bertrand, too, was aware of the mood in the Polish team, showing hypersensitivity to its outward manifestations. One of the team members recorded in his diary: "I have been voluntarily imprisoned here under a new name. I am staying in an area that is alien to me, without the right to move freely. Dry mathematical and linguistic work, the same company from morning to evening". The sense of imprisonment provoked a telling incident; the windows of the rooms on the ground floor of the villa were secured with thick bars. One day Piotr Smolenski allowed himself to be photographed from the outside as he looks out of the barred window with a sorrowful expression, apparently intending to send the photograph to his family. When the news reached Bertrand, he reacted with an indignation disproportionate to the seriousness of the episode. He forced Smolenski to apologize, in view of his ignorance of French made through Rejewski. The event itself says a lot about the atmosphere at the center, and the attention paid to it by Rejewski in his memories adds to its significance. The precision and honesty of his memoirs dictates that even seemingly secondary information is of considerable importance. In describing the story with Smolenski's unfortunate photograph, he clearly wanted to convey to the reader, in a discreet and synthetic way, his own assessment of the situation in which the codebreakers found themselves in 'Cadix'.

With all this in mind, Rejewski and his fellow mathematicians occupied a relatively privileged position. The vast majority of Polish-French disputes took place on the Bertrand-Langer line. The other members of the team (with the possible exception of Ciężki, who remained somewhat on the sidelines of life at 'Cadix' spending most of his time in Algiers) were unaware of the scale and importance of the disputes waged by the center's commanders. Bertrand did not accept that his Polish deputy was maintaining independent radio contact with Britain. He sensed that the results of the cryptologists' work, which he wished to pass on as his contribution, were reaching London by this route. In at least one instance, a dispute threatened not only the foundations of the center's operations, but even the secret of Enigma. When Rejewski broke the Swiss Enigma ciphers, Langer sent information about this to London, perhaps including a suggestion to warn the Swiss of the weakness of their ciphers. When Bertrand found out about this, he reacted with fury. Having financed the operation of the center with French intelligence funds, he felt he was the rightful owner of any results of the work. He maintained that the Swiss should have paid dearly for valuable information regarding the security of their communications. However, Langer was horrified when his interlocutor, in the heat of the discussion, began to hint that the only way to compensate him for his losses would be to sell the secret of breaking Enigma to the Germans. A threat thrown in the heat of discussion is not yet an act of treason. But Langer knew Bertrand's commercial attitude well enough to take his statement seriously. In a long and difficult discussion, he managed to reassure his partner and dissuade him from his fatal intention. After all, it seems that after this episode, their mutual relationship never returned to normal and trust was not restored. Later events indicate that Bertrand also divided the Polish team into a part worthy of his trust and affection and a group he merely tolerated. The first category included three mathematicians whom he respected for their achievements. The second group included the officers with whom he was forced to have difficult discussions: Langer, Ciężki and Palluth.

While 'Cadix' was experiencing internal problems, black clouds began to gather outside. In late August/early September 1942, the first indications of a planned Allied operation in the Mediterranean began to reach the center. Bertrand decided to go to Paris and take a closer look at the events as seen from the German side. For some time, he had been in contact with a certain 'Max', an official at the German embassy in Paris, who, for unknown reasons, assisted Bertrand in his missions (e.g. by offering a seat and diplomatic courier documents on the Paris-Vichy train) or occassionally shared information with him. News of these contacts must have reached the British, who, under their influence, began to regard Bertrand's organization as controlled or at least inspired by the Germans. This time, too, 'Max' proved talkative: according to his information, two German divisions were grouped in the Dijon area and ready to enter the hitherto unoccupied part of France at the first news of Allied operations. "Cadix" was in the area of their planned operations.

A clue that may point to the motivation of the mysterious 'Max' surfaced many years after the war, when the head of the Abwehr post in Paris, Oskar Reile, published his memoirs. He described an operation in which he introduced two agents into Parisian circles, whose task was to identify the structures of the French underground by pretending to be Germans disillusioned with Nazism. Reile introduced them only under pseudonyms: "Max" and "Moritz". The mere coincidence of pseudonyms in Bertrand's and Reile's memoirs is not sufficient to confirm that the 'Max' referred to in both of them was the same person. However, agent's described purpose and modus operandi, and the period of his contacts with Bertrand, indicate that the same person is involved. Bertrand's contacts with 'Max' must have been common knowledge in 'Cadix'. In Rejewski's private copy of the Bertrand's memoirs, the respective paragraph was annotated in the margin with "MAX!!!". The repeated exclamation mark indicates that Rejewski recognized the agent and considered his role to be important. If, as is quite likely, Bertrand was one of the French who swallowed Reile's bait and did not conceal this fact from those around him, one can understand that the British regarded his organization as being manipulated by the Germans.

The second worrying signal came in one of the messages intercepted from a cable tapped by the French and broken by the Poles. It appeared that the intense activity of the 'Cadix' radio station had attracted the attention of the German radio security service, which had begun a hunt for the transmitter. It was likely that German troops entering the unoccupied zone would be informed of an object to be investigated on the way. It turned out, however, that there was no need to wait for the entire German army to enter. On 25 September, during a visit to the prefecture in Nîmes, Bertrand learned that a German signals security team had appeared in the area. It had chosen the château of Bionne as its base and demanded the provision of some French number plates for its cars. Bertrand went to Bionne, where he took a close look at the vehicles of his alleged pursuers. He also soon found an opportunity to watch them himself as they relaxed in a hotel restaurant in Pont-Saint-Esprit after a day's work. Soon, cars belonging to the German team began to appear regularly in the vicinity of the Villa Les Fouzes—the Polish stations had switched to a regime of transmitting in blind, with no communication setup and no acknowledgement of reception, making the enemy's task more difficult for the time being. In the villa, preparations were underway to leave the place in alarm. Some documents were hidden in advance, while prepared hiding places awaited others and radio equipment. At the same time, the heads of 'Cadix' contacted London, asking for instructions. The reply came quickly and left no doubt—the center should be evacuated immediately to Algiers. Should Bertrand's capabilities prove insufficient, the British offered to organize the evacuation of up to twenty people by sea, from one of three designated points on the coast. The codeword was also agreed, the transmission of which by radio would mean that the Allied operation was imminent ('the harvest is plentiful').

The events described so far form a reasonably coherent picture, but in the next few weeks this will be replaced by mismatched pieces of a mosaic, fuzzy blobs and conjectures with which we will have to piece together the few available testimonies into a whole. The nature of the events will mean that the only sources will be the accounts of the participants. Their inconsistency, however, signals that not everyone was interested in an objective account of the facts. Let us therefore follow the trail of events aware that we are treading on uncertain ground. According to Langer's post-war account, as soon as clouds began to gather over southern France, he began efforts to redeploy his team to Britain, or at least to the safety of Africa, but was met by Bertrand's refusal to assist in the evacuation. The Frenchman argued that the coast of Africa could soon become a battlefield between French and landing Allied troops, making it impossible to seek safe haven there. Langer was dependent on Bertrand for almost all matters relating to his own and his subordinates' fates—he had to recognize an argument that he most likely believed to be false. When the recommendation for immediate evacuation to Algiers came from London, Bertrand, it is claimed, undertook a rather unusual consultation. Namely, he asked for instructions, but not from the commander, but from his adjutant. He justified this astonishing route on the grounds that he wished to avoid his contacts with London being revealed to the Vichy government. This was an astonishing argument given that prior approval for contacts with London had reached Bertrand through his immediate superior, Paul Paillole. Paillole's adjutant reassured him by saying that the command did not believe in any Allied operations in Africa in the near future, and therefore evacuation was not necessary. This friendly advice was presented by Bertrand to the Poles as an order and once again he refused to help with the transport to Africa. In this situation, clearly desperate Langer decided to resort to the assistance of the Polish organization, intelligence network F2, led by Lieutenant Colonel Romeyko. On 31 October, Langer and Palluth travelled to the Côte d'Azur in Bertrand's company to discuss the possibility of evacuation using the Romeyko's channels. The talks dragged on until 4 November, and on his return to 'Cadix' Bertrand found on his desk a message received in his absence which, when deciphered, turned out to be the agreed codeword 'the harvest is plentiful'. Bertrand's reaction was astonishing; he decided to make another trip to Paris to seek confirmation of the threat from 'Max'. In the meantime London confirmed on 5 November that 'the harvest is already very plentiful' and therefore there was no time for trips to Paris, and added that it would take the Germans about a week to introduce the planned retaliatory steps, so there was still enough time to carry out the evacuation.

At dawn the next day, everyone at 'Cadix' became convinced that the time to leave the villa has come. Langer spotted a German direction finding vehicle just a few hundred meters from the estate, preceded by a passenger car with several French policemen. The Germans had tracked the direction from which the last transmission had come, as they were interested in a small house positioned on the line between Les Fouzes and the vehicle with aerials. Within minutes, the villa had taken on the appearance of being completely abandoned—the gate, windows and doors were locked, and all movement inside froze. The masquerade came in handy after a few minutes, when the cars stopped in front of the gate of the property, the three French policemen got out of the car and circled around for a while, after which the caval-cade moved on. Later that day, the 'Cadix' resort was deserted. The Poles were transferred to the coast and deployed in secure accommodation. Bertrand and his wife and driver remained in Uzès for 24 hours, taking care to hide the center's equipment and documentation and watching the German team return to the site at 1 pm, 10 pm and 5 am the following day—usual transmission hours.

The Poles listened to the radio announcement of the Allied landings in Africa on 8 November already on the Côte d'Azur. It turned out that the British estimate of the German reaction to events was realistic; the villa at Uzès was not occupied until 12 November. By this time, the Poles found themselves under Italian occupation. The decisive period of a few days between the winding up of 'Cadix' and the entry of Axis troops into southern France was not used to evacuate the Polish group. The team, and with it the secret of the Enigma, were in deadly danger. Aware of this, the British made further attempts to pull the Polish team out of the trap. The radio sta-tion of 'Cadix' was not operational any more, but through the receiver of the F2 network in Nice, a proposal reached the Poles to pick up the group from one of the small islands off the French coast, presumably by submarine. Bertrand again made his assistance conditional on the approval of his command, but before he could obtain permission, the occupying forces imposed a ban on the movement of small vessels off the French coast, thus cutting off the planned escape route. This was the third consecutive attempt to evacuate the Polish team, which failed due to Bertrand's lack of cooperation. If there might have been periodic disputes between him and Langer up to that point, after this failure a sharp conflict had to break out. Langer threw Bertrand in his face the news from Africa confirming that the French had offered only token resistance to the Allies, and that many troops had gone over to their side. The catastrophic scenario that had been the pretext for rejecting the first offer to go to Africa had not materialized. Langer might have also asked why what Bertrand portrayed as impossible with regard to the Polish team was appropriate and possible with regard to the Spaniards? After all, the Spanish part of the team, at the very first sign of danger, was evacuated from the 'Cadix' and reached Algiers without problems in the first days of November. In his post-war book, Bertrand devoted to the departure of the Spaniards just a footnote in fine print. The evacuation of the Spanish group unambiguously confirms that he and his organization still in early November had the capacity to evacuate entire center from France to Africa. Langer would have been entitled to ask why that part of the 'Cadix' team, whose presence in territory occupied by Axis forces posed no threat to one of the Allies' greatest secrets, had been evacuated in the first place? Bertrand floated further plans for the evacuation, but Langer's anxiety reached its zenith when he learned that their author would not participate in the evacuation and thus would not have a personal interest in the success of the operation.

In the accounts of Langer and Bertrand, one can see traces of the emotions that drove them at the time. Langer accused his partner of negligence in the preparation of subsequent evacuation attempts and insufficient concern for the safety of the Polish team. Bertrand accused Langer of being uncooperative in the execution of the plans and of alcohol abuse. In retrospect, it seems that Bertrand tried to keep the Polish team with him at all costs, downplaying the danger to which he was exposing the team and the Enigma secret. Reporting on another episode of the Enigma story, Paillole justified Bertrand's exclusion from the operation as follows: "the peculiar thing, however, is what compelled us to take such a decision without hesitation. It had to do with Bertrand's character: he regarded H.E. as his personal property". He apparently extended this attitude to the Polish team as well. He failed to take advantage of the favorable time and missed opportunities to evacuate the Poles safely. When he finally realized that the stakes in the game had increased, the occupying forces had managed to seal the borders of France and the available earlier evacuation channels had been interrupted. Bertrand's superiors evacuated to North Africa on the evening of 9 November, leaving not only the Poles but also himself stranded. Bertrand claimed: "we had always assumed that I and my section would be part of the [evacuated] team, especially as Cadix was not too far from Istres", from where the planes with the French on board took off. He also remarked that when the evacuated officers had gathered on the airfield, it took them a full 24 hours to wait for the airplanes, whereas getting to 'Cadix' required only two hours by car. Not very discreetly, he shifted the responsibility to Rivet: 'Before his departure from Vichy, Delor spoke to Rivet about "Equipe Z", which had to be evacuated at all costs, but he replied: officers first'. As a result, 'there were many empty seats left in the three planes of the convoy'.

Paillole's memoirs show that the true circumstances of the evacuation of the Vichy secret services to Africa were somewhat different. Each of the several services, land army, air and naval intelligence and counter-intelligence, made independent decisions on how to proceed and how to evacuate to Africa. The evacuation by air from Istres airfield mainly concerned air intelligence staff under Colonel Ronin, who were joined by a group of officers from other services particularly threatened by the German entry into the unoccupied zone—it was probably to them that Rivet's term 'officers first' referred. Under Paillole's decision from the end of September 1942, the Travaux Ruraux, or counter-intelligence, was to remain in France and continue its work even under Italian-German occupation. This decision no doubt concerned Bertrand himself (hence his lack of interest in later evacuation plans), his French subordinates, and the Polish group, if no special exception was made for it. Such an exception would have complied with a request from British intelligence, addressed directly to Rivet still on 5 November. It seems, however, that the atmosphere of those days was not very conducive to gestures of Franco-British cooperation. In the course of instructing subordinate officers in the summer of 1942, Paillole expressed the opinion that even if Germany was the enemy number one, surely Britain was the enemy number two of his service. As early as November, he was to find that he was not alone in such an opinion; when he asked the Chief of Naval Intelligence, Captain Sanson, for help in evacuating the archives of his own service, weighing a total of some 40 tones, to Africa, he was told that 'there are no reasons why the archives should fall into the hands of the British rather than the Germans'. In the atmosphere described, Paillole might have 'forgotten' to issue separate instructions concerning the Polish group. Their absence, in turn, could have become a convenient pretext for Bertrand to try to retain the team under his own control.

Thus, having missed a convenient opportunity, further attempts to organize the escape of the Poles turned into dangerous improvisation. Langer, it seems, lost the remnants of faith not only in the competence but also in the goodwill of his French partners. He was in enemy-occupied territory. He was responsible not only for his team, but also for the security of the Enigma secret. He was entirely dependent on a man whom, under the pressure of facts, he no longer trusted: he had plenty of reasons to drown his feelings of helplessness in alcohol. But according to the theory promoted by Bertrand himself, not even a serious excess of wine could disturb the Polish officer's clarity of judgement. And the facts from Langer's point of view were clear: by blocking the possibility of the Polish team's departure, Bertrand took total and exclusive responsibility for its fate. In the coming weeks, evacuation plans shifted before the eyes of the Poles as if in a kaleidoscope. They were moved to successive clandestine locations in Cannes, Antibes and Nice. An attempt to pick up the entire group from the coast in the Le Trayas area on the night of 3/4 December failed—on the eve of the planned operation, the Italians began patrolling the cove from which the boat carrying the fugitives was to depart. Langer refused to repeat the operation in the same place on Christmas night, so the Poles were to sail away from the Cros-de-Cagnes area at the end of December—on the eve of the operation the intermediary arranging the operation was arrested.

An interesting question is why evacuation by sea, rejected by Bertrand before the occupation of Vichy, suddenly became an accepted solution. It seems that, thanks to Paillole's memoirs, we have a plausible hypothesis about the reasons for the change. Prior to the entry of German and Italian troops into the unoccupied zone, the evacuation was to take place aboard a British submarine. Thus, the success of the operation put the Polish codebreakers directly into the hands of the British. Paillole describes the story of the French submarine 'Casabianca', which managed to leave the port of Toulon before the French fleet scuttling itself and reached Algiers. There, her commander, Jean L'Herminier, placed the ship at the orders of the local French military authorities, who immediately used the vessel to make contact with the intelligence network left behind in France. The circumstances of the submarine's early, unsuccessful missions described by Paillole correspond quite closely to the evacuation plans presented by members of the Polish team and Bertrand himself. Evacuation aboard a submarine was an acceptable solution, as this time a French ship was involved, and thus a valuable team of codebreakers could be retained under French orders.

The failure of subsequent attempts at evacuation by sea led Bertrand to shift his hopes to Switzerland. Polish officers accepted this concept without enthusiasm. The Swiss took very seriously their duties of a neutral country directing all citizens of belligerent countries who crossed their border to internment camps. Langer and his subordinates wanted to get to Britain and continue their work, so the prospect of spending the rest of the war in an internment camp did not arouse their enthusiasm. The project collapsed when Swiss officials, who were hoped to make it easier for the Poles to bypass immigration regulations, could not be contacted in time. Spain remained the last option. When Bertrand notified Dunderdale of his intentions, the latter tried to dissuade him of the solution he was considering. But in a dispatch of 29 December, Bertrand insisted on pursuing it, writing that 'All that was left was the Spanish option, which we must try. (…) Here the danger grows every day and we must act quickly'. On 12 January the whole team was transferred to Toulouse, then in two groups to Perpignan. A mysterious in the light of later events reshuffling occurred at this stage, as Rejewski and Zygalski, who were to be evacuated in the second group, were separated from it and replaced by Paszkowski, accompanied by his pregnant wife. On 15 January, the whole group set off by bus to Arles-sur-Tech, from where they were to walk towards the Spanish border. But the bus was stopped on the way by French police, who picked out the Poles and arrested them on charges of using forged documents. During a friendly chat with the French gendarmes, it turned out that the organizers of the transfer had forgotten to warn the police about the presence on the bus of people whose papers should be viewed with a blind eye—the policemen made it clear that this was the standard practice. It was also the standard in the French reality of the time to bribe the police inspector in Perpignan, the guards in the local prison and to pay for a lawyer. Thanks to the bribes, after a month or so of detention the whole group left prison.

Meanwhile the French attempted to get Rejewski and Zygalski across the border. The first attempt also failed; the agreed guide failed to show up at the meeting point, forcing the two codebreakers to return to Toulouse. A few days later, they both again took the train to Narbonne, then Perpignan and Aix-les-Thermes, where they had to wait for contact from the guide. On 28 January, they were instructed to take the train to Latour-de-Carol, from where they would head towards the border on foot. The crossing was not without its emotions. First, the guide kept them waiting for a long time, then, already in the mountains, he threatened the Poles with a gun and robbed them of the rest of their money. But on the morning of 29 January 1943, the two fugitives crossed the border. On the Spanish side, they were quickly apprehended by a border guard patrol, escorted to a post in Puigcerda, and then sent first to the prison in Belver, then successively to Seo de Urgel and the district prison in Lerida. They were prisoners, but they were safe.

While Langer and his comrades were still stuck in prison in Perpignan, his group was followed from Cannes by another team including Michalowski, Szachno, Suszczewski and both Palluths. Having reached Toulouse, they learned from Captain Louis about the adventures of their predecessors and the need to return to Cannes.

But on 1 February they were back in Toulouse, this time without Antony Palluth, who was ordered by the French to act as rearguard and cross the border last. Between 2 and 4 February, Michalowski's group followed the route to Aix-les-Thermes and Latour-de-Carol. Its leader was disgusted with the French organization: "[the evacuation] was prepared very poorly and it was only by chance and my own forethought that the four of us did not fall into the hands of the German authorities". The guide customarily first exhausted the fugitives with a long march in snow and cold, then, threatening with a gun, robbed them of the rest of their properties. On the evening of 4 February, all four were already at the police station in Guardiola, from where they were transferred to Barcelona the following day.

On 12 February, Bertrand received a prison grep in which Anthony Palluth reported that he had fallen ill after eating too much of green beans. Green beans (haricots verts) was a term colloquially used to refer to Wehrmacht soldiers: it was obvious that Palluth had ended up in German captivity. The circumstances of his imprisonment remain a mystery. Detached from the group, he was to act as a rearguard and cross the border last. Indeed, he crossed the border unaccompanied by other Poles who could give an account of the circumstances of his arrest. However, the date on which the secret message reached Bertrand indicates that it must have fallen into German hands before or shortly after Michałowski's group crossed the border. Whatever the reason for his separation from the group, one of the key members of the BS4 team, with detailed knowledge of all the Enigma-related activities, was in German hands.

After leaving prison, the Poles in Langer's group were determined to avoid the smuggling network led by a certain Monsieur Perez, which had previously got them into trouble. All the more so when, while in prison, one of the local police inspectors whispered in their ear that Perez's agent, known to the Poles as Gomez, was suspected of collaborating with the Germans; he had been seen producing documents available only to those collaborating with the occupiers. Up to this point, Langer's and Bertrand's later accounts had remained fiarly consistent, but from then on they were to diverge dramatically. Bertrand maintained that the Polish officers were excessively flaunting themselves in the local bars and 'negotiating behind our backs with a Spanish guide who betrayed them to the Germans, according to the customs of the time in that area'. Indeed, on 12 March, Langer's group attempted to extricate themselves from the quagmire with the help of a guide recommended by a sympathetic police inspector, but on their way from Toulouse to Port-Vendres they had to flee the moving train when they saw Gestapo agents inspecting the passengers' papers in an adjacent carriage. They returned to Toulouse with the resolve to make another attempt the following day with the same guide. But on the morning of 13 March, before they could set off, mysterious Mr. Gomez found them in Toulouse and coerced them to rely on his services once more. Langer did not trust his network, but he found himself in a no-win situation: if Gomez collaborated with the Gestapo and managed to find them in Toulouse, they were at his mercy anyway. The agent tried to reassure Langer and his comrades—he signed a twenty-franc note and tore it into two parts handing one of them to Langer. The latter was to hand his part only after crossing the border to the guide, who was to be paid only upon presenting the bill. They set off around 7 PM, at Elnes they met Gomez, who handed them over to the guide, kissing everyone goodbye. They had managed to walk about three kilometers when the group was surrounded by German soldiers on motorbikes, shooting into the air. After arresting its members, they allowed the guide to move away unhindered; Gomez's farewell turned out to be a Judas kiss, and after Palluth also Langer, Ciężki, Gaca and Fokczyński found themselves in German captivity.

The story of the evacuation of the Polish team from France and the betrayal that placed part of it in German hands represent one of the most mysterious episodes in the history of Enigma. According to the previously quoted characteristics by his immediate superior, Bertrand apparently tried to treat the Polish team as if it were his personal property. He consistently torpedoed evacuation projects that would have resulted in the team being handed over to other hands. When it became clear that this could not be prevented, he largely lost interest in the fate of the team and the result of the operation. He had, moreover, new and unrelated tasks to carry out; after Paillole's evacuation to Britain and then Algiers, he gradually took over the organization and command of the "Kleber" intelligence network that remained in France. In effect, he delegated the supervision of the evacuation of the Poles to his earlier deputy, Captain Louis, whose talents he did not value very highly and treated him with disdain.

Nevertheless, the operation was not doomed to failure. The organization led by Paillole had at its service an effective smuggling network, ensuring the transfer of people and goods across the Franco-Spanish border. Its key organizers were the Ramonatxo brothers, based in Perpignan, with whom collaborated, among others, an epigone of the Catalan republican government, Manuel Valls de Gomis, for this was the name of the mysterious Monsieur Gomez described by Langer. Paillole himself used his services when crossing the border into Spain on the night of 28–29 November 1942. According to his own account, he and his companions had dined together at La-Tour-de-Carol. During the dinner they entertained the dog of German border guards sitting at a neighboring table, after which they were escorted to the house of a network member situated just hundred meters from the border. After nightfall, the smuggler led them without incident to the Spanish town of Puigcerda (if one does not treat as an incident the extortion under threat of arms of any cash in their possession—not even Paillole's status as head of the French network protected him from this). Paillole's account is extremely interesting, as it corresponds closely to Rejewski's report from his evacuation. Commenting on the betrayal and arrest of Polish group, Paillole assessed that 'this was a grave error on the part of our service, which had at its disposal absolutely certain routes to Spain'.

Absolutely certain routes were used in the case of Rejewski and Zygalski and the group led by Wiktor Michałowski, but the other two groups were apparently directed by other routes, with disastrous results. The question about the reasons for this decision is absolutely natural, and we have at least two possible answers. Bertrand, in his memoirs, maintains that the Poles rejected the services of the network controlled by his organization, and that the smuggler chosen on their own led them into an ambush. He apparently succeeded in convincing Paillole of this version as well, who noted that 'the Poles, left to their own devices, reached out, under circumstances unknown to me, for the services of a network controlled by the enemy'. Langer's report confirms to some extent the account of the two Frenchmen, contradicting it however on the key issue. The financial account of the evacuation funds presented by Langer in London, after his liberation from internment, includes the sum of 5000 francs as the remuneration of 'the smuggler who was to ferry us over the border, and who was put off by the French organization'. Getting a little ahead of events, let us note that Langer was given the opportunity to meet Bertrand face to face once more. Liberated from internment by American troops, he was passing through Paris in May 1945. He took the opportunity to find his former partner, presumably in an attempt to explain the circumstances of his imprisonment. In the course of the conversation, Bertrand reportedly had assured that he considered the evacuation of Langer and his group to have been successful after both halves of the note signed by the Pole had reached him. The experienced intelligence officer did not notice that he was thus compromising his own lie about the Poles using another smuggler. We do not know the other subjects and circumstances of the conversation between Langer and Bertrand. However, in a post-war letter addressed to his wife, Langer quite unequivocally accused the Frenchman of betrayal. Bertrand's lie received further confirmation when memories of his wartime past were published by Hector Ramonatxo, one of the main organizers of the smuggling network in Paillole's service. In them he raised the merits of Manuel de Gomis, or Monsieur Gomez, who offered Langer and his comrades a Judas kiss as a farewell. The story of the evacuation of the 'Cadix' hides many mysteries that cannot be explained on the basis of the available sources. Why did Bertrand block the evacuation of the Polish group to Africa when it did not yet involve any danger? Why did he organize the evacuation of the Spaniards efficiently and in time? Why were Rejewski and Zygalski, whom he had a fondness for, separated from the rest and successfully ferried across the border? Why did Palluth, whose position in the group placed him close to the disliked Langer and Ciężki, have to cross the border separately? Finally, could the head of a smuggling network operating on behalf of the French secret service have handed the group into German hands on his own initiative?

Permitting the capture of Poles who knew the Enigma secret was a defeat for French intelligence, which, nevertheless, paradoxically, had some justification for their action or neglect. In the period before the end of the center's work, information was reaching it from the UK, according to which the Enigma ciphers had become completely immune to attack. In such a case, the secret of breaking the Enigma ciphers was losing significance for the Allies. If the French followed this logic, it made it easier to accept another defeat. For, at the same time, another catastrophe was ripening in France, which might just as well have led the Germans to the truth about the security of their ciphers. Having installed themselves in Paris on 14 June 1940, the Germans energetically set about taking stock of their new holding. Local Abwehr representatives—Colonels Friedrich Rudolph and Oskar Reile—studied the archives of French military and police institutions left behind in Paris by their hosts fleeing south. On 22 June, their prey was augmented by the files of the French High Command, left carelessly in sealed wagons at La Charité-sur-Loire railway station. At the headquarters of the French Security Service in the Rue des Saussaies, too, nothing disturbed the order of folders deposited there. Studying the documents Abwehr officers gradually gained hope of unravelling the mystery of the leak from the highest circles of power in Germany, of which they had been aware since the story of the Hossbach Protocol in 1937. Soon, however, the Gestapo came to the fore in the race when Captain Kieffer found documents relating to 'Rex' Lemoine among French police files. The experienced officer was surprised by the scale of their subject's activities, commenting that 'it took a wheelbarrow to transport them'. Among the papers was a note from an Italian agent, dated 3 December 1938, stating that 'Rex' had tried to sell him a German code developed in ChiStelle that same year. The Germans knew the nature of 'Rex's' activities, for back in 1938 he had been detained by the Gestapo in Cologne, interrogated precisely by Kieffer and only released after declaring that he would henceforth also work for the Third Reich. At the same time, copies of reports compiled by the Forschungsamt were found among the seized documents of the French command. Consequntly, if there were two traitors, one of them worked at ChiStelle, the other at the Forschungsamt. However, if, more likely, it was one and the same person, the combination of facts allowed the investigation to be narrowed down to personnel working around 1938 with both institutions. As a result, around ten suspects were singled out in May 1941, among them Hans-Thilo Schmidt. The suspicions directed at the brother of one of the Wehrmacht's most revered commanders provoked Canaris' indignation, as a result of which the trail pointing to 'Asché' was temporarily abandoned.

However, it was not only the documents that remained on French soil, but also a man whose knowledge of Enigma was equally dangerous. In the days of the June defeat of France 'Rex' had as his first instinct tried to evacuate to Britain. He knew that once France was occupied, he would be wanted by the Gestapo for breaking a promise to cooperate made in 1938. On 20 June 1940, he appeared with his wife at the British consulate in Saint-Jean-de-Luz, where he was ordered to report the following morning to the port where a British ship was to be expected. Indeed, the next day, a Royal Navy minesweeper was moored at the quay, whose captain asked 'Rex' to return in the evening with provisions for the journey and a blanket. However, when the arch-spy returned to the harbor in the afternoon, the ship was already completely filled with evacuated Polish soldiers, and the British officer guarding order, when asked by 'Rex' to find a quiet corner for him, his wife and his secretary, responded with a lecture in naval language on the reasons for his dislike of the French. Lemoine, in whom Gallic inclinations had managed to prevail over an innate German matter-of-factness, countered that he would rather die on French soil than be insulted in Britain, and resigned himself to evacuation. Two days later, 'Rex' turned up at Bon Encontre, where the cream of French intelligence had gathered, including Paillole and Bertrand. Both realized that 'Rex' being within reach of German hands represented a ticking time bomb. Their dismay must have increased when, in the course of their conversation, 'Rex' opened the lid of his suitcase, in which they spotted a set of documents including secret code books and passport and identity card blanks from several European countries. After the Saint-Jean-de-Luz incident, suggestions that 'Rex' should take refuge in any safe country in the world, away from German eyes and hands, were doomed to failure. Lemoine was already, it seems, an old and very tired spy. All he could think about was living out the rest of his days peacefully in familiar surroundings and treating the contents of his suitcase as an insurance policy. Paillole merely persuaded him to abandon German-occupied Paris and move to Saint-Raphaël on the south coast, and to make sure that no material as compromising as that carried in the suitcase remained in his Paris flat. After all, if Poles knowing infinitely more about Enigma could work in the unoccupied zone, Lemoine could also find refuge there.

In April 1941, Paillole found 'Rex', who, as promised, had gone to live in Saint-Raphaël, but instead of living quietly without drawing attention to himself, he began to exploit the contents of the suitcase by selling the passports to people who, for various reasons, had to leave Vichy and had cash at their disposal. Paillole had to ask himself the obvious question: if the passports from 'Rex's' suitcase ended up on the black market, when would the other documents appear on it? He decided that the situation called for Lemoine to be permanently taken care of and persuaded him to move to the Hotel Splendide in Marseille, where his organization's associates could keep an eye on the old agent's activities.

Meanwhile, following the Gestapo's success in linking the traitor to the ChiStelle community, the Abwehr was back in the game. Admiral Canaris knew that the ongoing investigation was effectively diverting attention away from him. On his orders, the head of the Abwehr in France, Colonel Rudolph, commissioned Captain Wiegand to find 'Rex'. As late as September 1940, an observation of 'Rex's' Paris flat made it possible to intercept a postcard sent by him from Saint-Raphaël, but the first attempt to find him failed: Lemoine was absent. However, in February 1942, an Italian working for the Abwehr informed Wiegand that, on his way back from Tunis, he had accidentally come across 'Rex' in Marseille. Wiegand decided to act cautiously, assuming that better results could be achieved by getting the old spy to cooperate rather than attempting to act by force. When the Abwehr officer sent to Marseilles confirmed Lemoine's presence at the Hotel Splendide, Wiegand chose as emissary a French collaborator, Charles Marang, who knew 'Rex' from his pre-war business dealings. In June 1942, Lemoine informed the police inspector in charge of him, Osvald, that an agent working for the Germans had found him, in an attempt to persuade him to change his residence to Paris and his employer to German intelligence service.

In view of the impending German occupation of the south of France, this was alarming news in itself. Lemoine further raised the tension by refusing, in conversation with Paillole, to identify the German agent and to evacuate himself to Africa; the French officer understood that 'Rex' was trying to play both sides. Paillole and 'Rex' reached a compromise whereby Lemoine was to disappear again, holed up in a hotel in a small village in the Pyrenees, Saillagousse, from where he could easily cross into Spain in case of danger. But 'Rex' made a mistake unbefitting an experienced spy—he left his new address at the reception desk of the Marseille hotel. Thanks to this, in September, a message arrived in Saillagousse from the porter at Splendide informing him that his old acquaintance, Monsieur de Ry, wished to meet Lemoine. "Rex" must have been elated by the realization that he was still in the business and against all reason met his old acquaintance in Marseille. It turned out that de Ry had an Italian code he wanted to offer to the Germans; couldn't 'Rex' facilitate his contact with potential buyers? The rough-hewn German provocation, carried out right under Paillol's flank and in front of his associates, succeeded— Wiegand found Lemoine, whom the Germans had placed under constant surveillance after occupying the south of France. On 25 February 1943, Canaris gave the order to arrest "Rex": it was executed two days later. The affair was played out between professionals, so there was nothing of the violence in the relationship between Lemoine and the Abwehr officers interrogating him. "Rex" was placed in the Hôtel Intercontinental in Paris, where he was left plenty of time to take stock of the situation. After less than a month of reflection, he decided to use his insurance policy. Between 17 and 20 March, he briefed the interlocutors on his encounters with 'Asché'. On 20 March, Lemoine signed the minutes of his testimony, thereby passing a death sentence on Hans-Thilo Schmidt and revealing to the Germans the fact that he had betrayed the secrets of Enigma.

For the Allied intelligence chiefs, these must have been days of great anxiety. From the information Bertrand provided, it appeared that part of the Polish team was in safe Spain. The French officer concentrated his activity on getting Antoni Palluth, who seemed to be the only member of the team in German hands, out of captivity. To this end, he called upon on a person from Paris, who could indicate whom and with what amount to bribe to liberate Palluth. But before the efforts were successful, Palluth was transferred to Germany on 27 April, where Bertrand's hands did not reach. The news of the imprisonment of Langer's team reached the French almost simultaneously with the news of the arrest of 'Rex'. Menzies must have been shocked when the disastrous news reached London. All the more so because it did not end with them. In his memoirs, Paillole describes the sequence of events and negligence that led to the seizure of secret French intelligence documents by the Germans, the arrest of 'Rex' and the consequent revelation of Schmidt's treachery as a 'reprehensible and inexcusable negligence' on the part of French intelligence, dividing the responsibility equally between himself, Rivet, Perruche and Bertrand. But despite the harsh lesson, these same officers made the decision to leave Bertrand in occupied France. In this act of drama, there was little that British intelligence could do anymore: the secret of Enigma was tied to the fate and attitude of a few people who, completely defenseless, remained in German hands.

The hour of trial first came for Schmidt—on 23 March he was arrested by the Gestapo. At the time of his arrest, he was staying in a Berlin flat made available to him by his daughter's fiancé. It was not until a week later, when the daughter tried to visit her father, that she learned of his arrest. In desperation, she ran to the Gestapo headquarters, where she had to elicit sympathy from one of the officers, a certain Langemach, who made her aware of the seriousness of the situation and warned her against taking rash steps. The news of her father's arrest caused shock among all the family members. Gizela first decided to visit her brother, who had not yet fully recovered from his wounds sustained at the front in Russia, as a result of which his left arm had to be amputated. The son could not understand the betrayal of his father, the same man who had earlier encouraged him to join the Nazi party. During a conversation with his sister, he successfully acted out a suicide attempt, and after the crisis was resolved, they went together to their parents' home in Templin, near Berlin. There, in turn, they found their mother lying in bloodied sheets, drugged with opium and with her veins severed—but they arrived in time for Gizela to save her life. Faced with the evidence provided by Lemoine, Schmidt had no room for maneuver in the investigation and, judging by its swift conclusion, had to plead guilty to the charges, aware of the consequences. On the occasion of his daughter's visit to prison, he asked her to swap his warm winter coat for lighter clothing, while whispering in her ear a clue as to where he had hidden the letter. The contents of the secret letter robbed the family of hope—Schmidt asked to smuggle poison into the prison.

Gizela managed to concoct cyanide tablets and handed them over in the clothes she brought to the prison, but her Gestapo admirer, Langemach, warned the girl that the consignment had been detected. Only in time did it emerge that her brother had made an identical attempt. They never found out whether either of them had managed to deliver the poison to their father, or whether the Gestapo had decided to turn a blind eye to the consignment in an attempt to avoid the problem of executing the brother of a popular general. In September, Hans-Thilo Schmidt's family was asked to identify the corpse. There were no signs of torture on the body, and from the letters they found among the personal belongings of their father and husband, it appeared that Schmidt had written down his last words on the morning of 19 September 1943. Rudolph Schmidt buried his brother's body in a village cemetery near Berlin, not far from their mother's grave. When the story of the man known as 'Asché' came to light many years later, an anti-Nazi motivation was attributed to his actions, at least at the end of his life. His career as a spy began when no one yet regarded Nazism as a threat and ended before it had time to fully reveal its criminal face. Hans-Thilo Schmidt may have died as a man far from approving of Nazism, but anti-Nazi views were certainly not the motive for his betrayal.

The testimonies given by Lemoine and Schmidt meant mortal danger for the Poles in German captivity. After their arrest near Perpignan, Langer and Ciężki were first taken to the Paris Gestapo, but it seems that the interrogation by the officers there was only of a routine nature. They were then sent to Frontstalag 122 in Compiégne, which served as a transit camp, from where, in September, they were sent to the camp at Château Eisenberg. This was a somewhat surprising outcome. At the time of their arrest, Langer and Ciężki had to reveal their personalities and military ranks in accordance with international law. Despite this, they were not placed in an officers' prisoner of war camp, but in an internment camp for civilians. In March 1943, the ILAG IV at Schloss Eisenberg was reorganized as a special internment camp for prominent people (Sonderlager für prominente Persönlichkeiten), which brought together people of considerable social standing who had been arrested after the occupation of southern France. It housed, among others, high-ranking French officers who were not on active military duty at the time of their arrest (including General René Altmayer), representatives of the French aristocracy and other persons regarded as prominent, such as General de Gaulle's brother and Georges Clemenceau's son. By qualifying Langer and Ciężki for prominence, the enemies unwittingly showed them the respect they had not received from either their own command or the Allies.

Antoni Palluth did not achieve prominence status by his own choice. After the outbreak of war, he returned to active service, so he was able to invoke his status as an officer when arrested. He was probably afraid of being questioned about his military service record, which could have led the Germans to the Enigma traces, and decided to go into captivity as a civilian. Moreover, he decided to part with Jean Lenoir, whom he had pretended to be during the previous two years, and gave his real name at the time of his arrest. One can presume that his contact with his family in the past months had been somewhat one-sided: he probably did not know that he was being vigorously pursued by the Germans in Warsaw. Visits by German officers to the Palluths' home began from the very beginning of the occupation, and already one of the first ones could have spelled disaster for the family, as German officers found crates of radio parts in the flat. Later in the occupation, periodic visits by German officers became the norm in the Palluths' appartment. Jadwiga Palluth learnt to discipline her visitors by forcing them to leave their caps and gun belts on a hanger in the vestibule. In particularly persistent cases, Palluth's prize from his youthful years—a book with a personal dedication from Kaiser Wilhelm, whose name still made a magical impression on German guests—proved useful. When the Germans realized that they could not intimidate or surprise the brave woman, they resorted to other methods. In 1944, a certain Kazimierz Pilarski knocked on the door of the Palluths' appartment, invoking his pre-war acquaintance with Antoni and inquiring about his fate. Indeed, he had been employed as a radio operator at the Cipher Bureau before the war, but this position did not justify the degree of intimacy with Palluth that he had demonstrated during his conversation with his wife. Jadwiga had an additional reason to be cautious—she was involved in the forgery of documents allowing members of the underground to operate at the rear of the Eastern Front. So, she reported to her superiors about the suspicious behavior of the unexpected visitor: Pilarski was soon unmasked as a Gestapo agent and liquidated by sentence of an underground court. The cogs of the German bureaucratic machine must have been seriously worn out by this time—after the arrest of Antoni Palluth, the Germans did not recognize in him the person vigorously sought after in Warsaw.

Langer and Ciężki's first year of captivity passed surprisingly quietly. After interrogation at the Paris Gestapo, no one seemed interested in the wartime fate of the heads of the Polish Cipher Bureau. Their case had been closed before they were sent to the camp, so Langer was entitled to feel uneasy when he was unexpectedly summoned to the camp commander's office on 7 March 1944. The head of the Cipher Bureau sat in front of the three officers not knowing the cause and purpose of his interrogation. Doubtless he had been prepared for this conversation. Finding himself and Ciężki in German hands, they had to agree on a common line of testimony in case of questions about their service record and successes. The answer to the first question, which the German author had not even had time to fully formulate, came easily to Langer: no, he would not play the role of Colonel Redl, working for the Third Reich. The interrogators must have expected such an answer, because without delay they moved on to the next question—about the activities of the Polish

Cipher Bureau before 1940. In spite of carefully covering up all traces of their activities, Langer and Ciężki reckoned that the evidence of their service's pre-war achievements could fall into German hands. They decided to play a game with the adversary in which they intended to admit to isolated successes in Enigma decryption before the war, while emphasizing that changes made at the dawn of the conflict made it impossible to continue breaking ciphers. Langer suggested to the interrogators that technical knowledge of machine ciphers was Ciężki's domain, from whom they would obtain more accurate information. Ciężki, in his testimony, focused on the reasons that prevented the Poles from reading Enigma during the war, rather than on how ciphers were broken before the war broke out. The investigators accepted both officers' explanations and announced their return to continue their interrogations.

From the post-war testimonies of German signals intelligence officers, we know in outline the circumstances and the course of the interrogation at the Eisenberg castle from the opposite side as well. The investigation of the "Wicher" case has not progressed significantly in the period since 1939. From time to time, German cryptologists were summoned by the Abwehr or Gestapo to interrogate alleged members of the pre-war Polish Cipher Bureau, but as a rule these were false alarms. In July 1942, they were summoned to Warsaw, but it turned out that the local Gestapo had arrested random people bearing the same names as the wanted officers. That same month, in Stalag III, the Germans interrogated four officers, two of whom refused to give any information, while it was inferred from the statements of the other two that the Poles had managed to break the hand ciphers of the German army and police before the war. In 1943, Captain Adam Leja, the head of the listening station in Rivne before the war, was interrogated at Neuengamme camp, but it seems that the interrogation was mainly about the communications of the Polish underground, in connection with which he was arrested. According to the post-war testimony of the Wachmeister (and Doctor of Mathematics) Otto Buggisch, "much later, in 1943 or 1944, two Polish officers, a lieutenant colonel and a major, revealed certain facts during interrogation in a prisoner of war camp in Hamburg". Buggisch had second-hand knowledge of the interrogations from his immediate superior, Dr. Döring, head of the machine cipher section of the land forces' cryptologic agency. Under the circumstances, it is understandable that he mixed up the internment with the PoW camp and Castle Eisenberg with Hamburg. But the interrogation report he handed over is consistent with Langer's account.

Three officers participated in the interrogation: in addition to Döring, Dr. Hans Pietsch, head of the mathematics section of the OKH/GdNA, and presumably representing the Abwehr, the doctor Schneider. Buggisch reported: "(t)he conclusion was that the Poles had been reading Enigma for several years before the war (...). After a short campaign in Poland, they moved to France, where they probably tried to continue their work. However, the solutions stopped, and Buggisch only knew that it happened suddenly, as if due to a change in the system". On their return to Berlin, one interrogator expressed the opinion that the Abwehr had made a mockery of the matter by allowing the two officers to stay together for a long time, allowing them to agree on their testimony. As a result, contrary to their initial announcement, the German cryptologists did not return to Castle Eisenberg considering the case closed. The knowledge we have today provides an insight into the logic behind their conclusions. The key to their decision was probably the ovious consistency of the testimony given by Lemoine, Schmidt, and Langer and Ciężki. From the testimony of 'Rex', the Germans knew the period during which Schmidt had been in contact with French intelligence and the list of material he had provided. Schmidt had to confirm the picture the investigators had from Lemoine's testimony. They knew that he had regularly handed over the cipher keys to the French in the pre-war period. If both officers had denied any success of their service in the struggle with Enigma, the committee would probably have returned to castle Eisenberg with more inquisitive questions. Thanks to the clever tactics adopted by Langer and Ciężki, the facts known to the Germans formed a logically coherent picture which, in their view, did not require further explanation. Schmidt sold the keys to the French, who occasionally had to make them available to the Poles. These were able to read a few Enigma messages, but were unable to continue, both because the supply of keys was cut off and because of the change of the way the machine was used. Langer's and Ciężki's testimony was the decisive element; the security of the ciphers was confirmed, the investigation could be closed. If Schmidt had lived to see this moment and had indeed acted out of political motives, the outcome of the investigation would have been a consolation for him: the sacrifice of his life has been justifie, establishing the credibility of the Poles' testimony and contributing to Germany's defeat.

Three more insiders remained in German hands. In addition to Palluth, these were Gaca and Fokczyński. The Germans never associated Gaca and Fokczyński with the activities of Polish signals intelligence, so they were transferred to the Sachsenhausen-Oranienburg concentration camp immediately after their arrest. When Antoni Palluth also landed there and met his former associates, his first question was reportedly: "How can we continue to fight the Germans here"? Opportunity soon presented itself. The camp's inmates were used as labor force at the nearby Heinkel aircraft plant. Palluth became known as a talented engineer, and his knowledge of the German language helped to win the trust of the German supervisors, who often left technical inspection stamps in his hands. This opened the field for the organization of industrial sabotage on a considerable scale. Palluth and his associates turned many aircraft into flying coffins, masking the damage to mechanisms so cleverly that even the examination of the wreckage did not draw attention to the real cause of the disaster. Just when it seemed that a concentration camp offered Palluth effective protection from the Gestapo searching for him, tragedy struck. On 18 April 1944, American aircraft appeared over the Heinkel plant and one of the bombs took Palluth's life. A short time later, Edward Fokczynski died of exhaustion. The youngest and strongest of them all, Kazimierz Gaca, survived the war to bear witness of his fellow prisoners' attitudes and last moments.

The last participant in the Enigma secret to fall into German hands was Bertrand himself. The fact that he remained in occupied France confirms once again that neither he himself nor his superiors understood the significance of the secret. Bertrand became involved with the "Kléber" clandestine network, commanded from Algiers by Colonel Rivet, before replacing his former superior at its head in mid-1943. Bertrand's main asset was his contact with 'Max'. During his first visit to Paris in 1944 Bertrand was supposed to pick up a radio transmitter dropped in France on behalf of SIS. On 5 January, however, while waiting to meet a British agent, he was arrested. The Frenchman must have been suffering from a kind of polonophobia at the time, for his account contains several unclear and unconvincing suggestions that it had been the Polish members of the underground, who contributed to his capture. In Reile's memoirs we find a more plausible explanation of its cause. Continuing the story of 'Max' and 'Moritz' Reile reveals that "it was not a difficult thing to put Max and Moritz in touch with members of the [Allied] escape network. (…) They played their roles well enough in Lisbon that they were soon sent back to Paris, and after a few weeks a courier delivered the transmitter and ciphers to them". If, as we suspect, Bertrand's and Reile's 'Max' was the same person, the reason for the Frenchman's capture becomes apparent.

Bertrand was transferred to the center at 101 Henri-Martin Street and interrogated by a certain Christian Masuy, claiming to be an Abwehr officer, who offered to release him in exchange for his cooperation and to take up the radio game with London. Bertrand was, according to his own account, pretending to agree to the proposed terms, which, in a later meeting at the Intercontinental Hotel, were confirmed by the head of the Paris Abwehr, Colonel Rudolph. The test of the sincerity of Bertrand's intentions was a trip to Vichy made in the company of Masuy and Wiegand. During this trip, Bertrand encrypted and passed on to his radio operator a report to London, the contents of which amounted to treason, and to Wiegand a copy of report and the cipher key used. Having passed the test on 12 January, the French officer found himself at large, looking for a good hiding place and a contact with London that would allow him to disavow his previous message as transmitted under duress, and find help to get out of trouble.

SIS must have been in quite a dilemma when it was reported that Bertrand had been imprisoned and somewhat mysteriously released. In early 1944, with the approaching Allied landings in France, the secret of the Enigma was becoming even more important than ever. If it turned out that Bertrand had bought his freedom by revealing the secret, the Germans would gain the upper hand in the intelligence game. The British knew enough about the professionalism of the Paris Abwehr to approach Bertrand's release with extreme caution. The decision to evacuate him from France had to be taken after long and difficult discussions. When the signal of his release reached London, measures were taken to convince the Germans of the successful escape of the officer and his wife from France. To this end, the BBC broadcast several times on 27 and 28 January a message styled as information for the underground: 'The Bertrands have arrived in London'. After that, however, silence fell for weeks. Reason dictated the urgent evacuation of Bertrand, no matter how honorably or dishonorably he had behaved during his imprisonment. On the eve of the invasion, there was a flurry of activity between the British Isles and the secret airstrips in France: officers were arriving from Britain, agents returning to London with the latest information were waiting for departure. The Bertrands had to wait in hiding first until 27 April, when the first instructions were received, then until 3 June, when they were evacuated to Britain aboard the Westland Lysander

(Bertrand with disgust had to accept the presence on board of a Polish agent disguised as a Jesuit).

The welcome in London differed probably from the Frenchman's expectations. After initial hugs with old friends Dunderdale and Green, Bertrand was handed over to his former superior, Paillol. Menzies needed to get an assessment of Bertrand's actions following his January capture. He felt that knowledge of the environment predestined the former superior to question the ex-subordinate and provide a conclusion. Paillole had a long conversation with Bertrand, which yielded two conclusions. The first, positive for Bertrand, concerned the circumstances of his release. Paillole considered Bertrand's conduct towards the Germans to be honorable. He attributed the release in decisive part to a chance meeting at Abwehr headquarters with Hermann Brandl, to whom the Frenchman had saved his life in the first days of the war; the German officer had decided to repay a debt of honor. Paillole communicated his second conclusion exclusively to Bertrand, warning him that the mere mention in his report of the message he had sent to London under the Abwehr's control would compromise him in British eyes. Thanks to Paillol's conclusions, Bertrand was able to finally dine in the company of Menzies, reassured that the secret of Enigma remained safe in the crucial hours before the Allied landing in Normandy.

Bertrand's behavior, however, must have shattered this momentary relief—the French officer was surprisingly persistent in his enquiries about the date and place of the impending invasion, insisting on relaying them by radio to members of his network in France. Given the weight of suspicion that had just been lifted from him, Bertrand's questions must have awakened the vigilance of his hosts; Bertrand and his wife spent the days before and just after the invasion under a kind of house arrest. Despite issuing Bertrand with a certificate of morality, Paillole was surprised by his account, too. He recalls that "Rudolph and Wiegand, when talking about Lemoine and Schmidt, never mentioned the arrested Poles to him (…). There was never any mention of Enigma, of decryption methods, of the results achieved by the Allies… Amazing! (…) I don't know what to think about the fact that neither the Polish technicians, nor Perruche, nor Bertrand, were ever persuaded to reveal how the Enigmas (...) were exploited". To Paillole's questions, Langer or Ciężki could add another one: how was it possible that the Polish officers were not questioned immediately after their capture, that their case was closed, and that the inquiry committee appeared at Eisenberg Castle only a few weeks after Bertrand's arrest? Langer met Bertrand after the war. After the meeting Langer wrote down: "Bolek was arrested on the 9th of January 44, and the committee came to us on the 7th of March 44". Langer's note does not determine whther the connection between Bertrand's capture and the commission's appearance at Eisenberg Castle was only temporal or also causal. However, it is clear from post-war correspondence with his wife that he believed that Frenchman had bought his freedom pointing out to the Germans another target: Langer and Ciężki.

After crossing the Pyrenees, Rejewski and Zygalski for several months have been visiting a number of Spanish prisons. At the end of March 1943, they finally ended up in Lérida prison, which they left on 4 May thanks to the efforts of the Red

Cross. They spent the following weeks partly in Lérida, partly in Madrid, remaining at liberty, albeit under police supervision. It was only at the end of July that the Allies were able to organize the transfer of a group of internees to Portugal, where they were placed on a local fishing vessel, which met a British destroyer at sea and transferred the passengers on her board. Transported to Gibraltar, the codebreakers made their way to the UK on 30 July aboard the well-worn Dakota. After landing, they probably had to spend a few days in one of the Patriotic Schools, as the British counter-espionage centers were called, before finally arriving at the Polish military base in Kinghorn, Scotland. New assignments were already waiting for them; on 23 July, Major Gaweł reported to his superior that "6 German cryptologists from Ekspozytura 300, whom we are expecting shortly, we intend to install in Felden, attached to the Signals Intelligence Company".

Incidentally, the British and Polish military authorities in London were not the only services that wanted to use the experience of Polish cryptologists in attacking Soviet codes and ciphers. During the evacuation from France, the group led by Wiktor Michałowski included Stanisław Szachno-Romanowicz, one of the Cipher Bureau's top specialists in Soviet communications. While the members of the group were interned in the Spanish internment camp at Miranda del Ebro, he was summoned to the camp's headquarters, where two German officers were waiting for him. They declared that they were emissaries of the Japanese, who, highly valuing the Polish cryptologist's experience, were offering him a job in Japan. The Germans also hinted that they had found the officer's family in Warsaw, which the Japanese offer also included. Should it be accepted, the codebreaker and his family would be transported to Japan, where the officer would be able to continue his previous work. Szachno rejected the offer, which cost him months of mental torture when news of the outbreak of the Warsaw uprising and the fate of the city's civilian population reached him.

Before Rejewski and Zygalski were installed at Felden, they had to change their status. From the point of view of the military bureaucracy in the fourth year of the war, they were still civilians. So, they were conscripted into the army with the rank of private (it was only in October that they lived to see their promotion to the rank of second lieutenant) and assigned to serve in the signals intelligence company under the command of Captain Zieliński. They may have met their new commander before, as he had commanded Radio Intelligence Station No. 4 in Poznań before the war. Their work in Felden was not a continuation of their previous successes in attacking the Enigma ciphers. In the letter quoted above, Gaweł referred to Dunderdale's opinion, who 'stated that he was particularly concerned about Russian intelligence, as the radio intelligence company at Felden was their only source of information about Russia'. As a result, the new superiors of the two codebreakers considered the possibility of using the newly arrived specialists "to pull up Russian cryptology". The confusion caused by the discovery of the Katyń graves had been going on for several months. After breaking off diplomatic relations with Sikorski's government, Stalin's regime was relentless in attributing an anti-Soviet attitude to the Poles. Embarrassed by the split among the Allies, Churchill tried to persuade the Poles to be more cautious, not to say submissive, in their relations with the Soviets.

At the same time, British intelligence vigorously urged the same Poles to escalate their actions against the Soviet allies. As far as we know from the sources preserved, neither Rejewski nor Zygalski were assigned to the Soviet section, at least initially. Neither, however, were they able to return to their work on Enigma or resume their cooperation with British codebreakers, which had continued to a limited extent even during their stay at 'Cadix'.

Many historians have asked the question as to why the pioneers of the attacks at Enigma cipher were not allowed to join the teams at BP. An oft-repeated hypothesis is that, given the secrecy of the Bletchey Park facility, foreigners were not allowed to work there. By mid-1943, however, there were a large number of Americans working at BP and its outstations, both among the technical staff (radio interception, bombe servicing) and codebreakers and intelligence. But alongside Americans, BP was also keen to employ representatives of other Allied countries. In April 1940, a group of eight French naval officers (Mission DY) turned up at Bletchley and were assigned to Hut 4. After the defeat of France, both hosts and visitors faced a dilemma: the French were ordered to return to Vichy under threat of losing their citizenship, while the British were understandably wary of the prospect of officers privy to BP operations being transferred to enemy-controlled territory. In the end, the visitors were offered a freedom of choice, which they exercised in various ways. Four returned to France, but they only made the decision to return after the British action at Mers el-Kebir. One joined the Free France troops, another took refuge in the French embassy in London, but two linguists—Marc Vey and André Mirambel—remained and continued their work at BP. In the final phase of the French campaign, they were joined by two French Air Force cryptologists, Roger Baudouin and Felix Meslin; they continued working at BP until June 1942, when they were redeployed to SIS. It is reasonable to think that, under normal circumstances, the Poles could also count on a sympathetic reception at the British codebreaking center.

Another possible reason for the reserve against the Poles was the fear of introducing into BP people who had been in enemy-controlled territory for some time. Especially the period after the occupation of the south of France must have raised doubts for any counter-intelligence officer. The Polish team was divided into small groups, functioning without contact. In such a situation, no one could vouch for the fact that members of any of them had not fallen into the hands of the enemy and been compelled to cooperate. The only person who knew the fate of all the groups and could vouch for the security of their members was Bertrand. However, as a result of his not-so-clear contacts with 'Max', Bertrand's entire organization was considered by the British to be, at the very least, manipulated by the enemy. In the realities described, no counter-intelligence officer would have been willing to allow Poles into top-secret work.

The end of cooperation between British and Polish cryptologists was certainly not due to prejudice on the Polish side. In a memorandum dated 1 October 1944, Rejewski described the working conditions in Britain with considerable regret: "in view of the very small amount of intercept material (…), in view of the year-long break in work and in view of not receiving any help from the British, [the Polish cryptologists] limited their work to the only cipher which, under these conditions, offered some chance of breaking". He pointed out that "before the outbreak of the war, there had been Polish-British-French cooperation in the field of ciphers, during which, it should be stressed with all emphasis, the party giving, and giving generously, was exclusively the Polish side. It would be advisable now to remind the British of their debt (...) and to demand of them that they now cooperate with the Poles as loyally as the Poles worked and are working with them". Rejewski's note leaves no doubt that the Poles were interested in continuing their former cooperation, but did not meet with reciprocity from the British side.

Having eliminated many possible reasons for the cooling of relations between British and Polish cryptologists, two remain, the most likely. The first bears a purely practical character: by mid-1943 it seemed that most of the theoretical work related to Enigma had come to an end. The German machine was yet to provide the Allies with a few challenges, but after the victory in the Battle of the Atlantic, the prevailing feeling was that further struggles with Enigma would be the domain of engineers rather than codebreakers. Even at *Hut 8*, the longest defending bastion of cryptological theory, the number of researchers was just being reduced from a dozen to just four, and finally to two only. Moreover, starting in the winter of 1942/43, the Allies, hitherto pushed on the defensive on all fronts, went on the offensive. The tactical information provided by the Enigma decrypts had lost a big part of its importance; the Allies, imposing the initiative on the enemy, were rather looking for data on his strategic intentions. They found them in the decrypts of the teletype dispatches exchanged between the main headquarters of the German armed forces, bearing in BP the common codename 'Fish'. After the victory in the Battle of the Atlantic, and as the importance of the 'Fish' breaking increased, Enigma steadily declined in importance.

The fundamental reason why Poles were not included in the work on the new problems was probably the growing rift between the policies of His Majesty's Government and the Polish government-in-exile. The plane on which Rejewski and Zygalski arrived in the UK took off from Gibraltar airstrip less than four weeks after the plane with General Sikorski on board had taken off from the same runway for its short final flight. The General's death marked a breakpoint in Polish-British relations, and his successors were never to attain even a fraction of the standing and esteem that Sikorski enjoyed in the Allied camp. Fate spared him the necessity of facing a new situation in which the allies were vigorously shifting their sympathies to the side of the more numerous Soviet battalions. Cooperation in the sphere of cryptology became one of the first casualties of the British change of attitude towards the Polish allies; Rejewski was never to receive a reply to his memorandum.

Chapter 21
On the Other Side of the Mirror

The German investigation triggered by the revelation of the leaked meeting with Hitler had been ongoing since the end of 1938. From the autumn of 1939, it was accompanied by the 'Fall Wicher' investigation examining the possibility that the German ciphers had been broken by the Polish cryptology service. Both seem to have found a common end with the revelation of Hans-Thilo Schmidt's treachery and the interrogation of the heads of the former Polish Cipher Bureau at Castle Eisenberg. The issue of information leakage at the highest levels of the Third Reich was cleared up and, in the process, the security of the Enigma ciphers was confirmed. Despite this, historians and cryptologists alike are still debating how it was possible that the succeful attack at the Enigma ciphers, initiated in 1932 and carried out on a large scale since the outbreak of the war, did not come to the attention of German cryptologists, the Third Reich's intelligence service or the services using the machine? Any competent cryptologic service, and at least some of the German services in this area could not be denied competence, constantly analyses the security of its own communications, thoroughly investigating any events indicating even a slight probability of a breach. Why, in nearly 20 years of operational use of the machine, did none of its users notice the facts whose genesis could most easily be explained by the breaking of the cipher? Why did none of the cryptology agencies, which began employing mathematicians en masse at the outbreak of war, identify weaknesses in Enigma's design or use? Were the Allied measures taken to protect the secret so successful, or were the German cryptologists and intelligence services so ineffective?

Among the Western Allies, initiation into the Enigma secret was essentially equivalent to a life insurance policy. Strictly enforced rules prohibited anyone privy to the secret from getting so close to the enemy that there was even a theoretical risk of being taken prisoner. These rules basically served their purpose although there were incidental breaches by both the British and the Americans. German paratroopers searching General Freyberg's headquarters in Crete found a copy of a dispatch addressed from London personally to the general. The information it contained was

M. Grajek, *Enigma Myth Deciphered*, History of Information Security,
https://doi.org/10.1007/978-3-031-65475-6_21

said to be from the 'most reliable source', and the message itself was stamped with the urgency marker ZZZ; for those familiar with the organization of British military communications, an unmistakable sign that the source of the information was Enigma decrypt. In May 1941, the procedures for transferring and securing information from the BP to army field headquarters had clearly not yet been worked out in detail. Fortunately for the British, the nature of the information contained in the dispatch did not clearly indicate its source.

In the later stages of the war, US General George Patton became known for his rather cavalier attitude to the rules governing the military world. The rules on the distribution of Enigma decrypts forbade their transmission below army level; it was assumed that, in the frontline realities, even division or corps commanders could fall into the enemy hands. Patton, in many cases, justified the orders given to units of the 3rd Army with the knowledge derived from the decrypts, initiating thus sub-commanders into the secret, but it seems that these practices never had fatal consequences. A more complex situation applied to airmen, who, in the space of a few hours, could travel the distance between a secure headquarters deep in the rear and the skies above the enemy country. The rules forbidding participation in combat missions applied to them as well, but we know of several instances of their violation. In May 1944, one of the recipients of Enigma information in the RAF ranks, Air Commodore Ronald Ivelaw-Chapman, took part in a flight over occupied France. His plane was shot down and he bailed out. When it seemed that his adventure would end happily (he was initially taken into the custody of the French Resistance), he was taken prisoner by the Germans and was interrogated by the Gestapo, but the interrogators did not raise the issue of the security of German communications. Earlier, on 27 June 1943, US Air Force Brigadier General Arthur W. Vanaman had, against the rules, obtained permission from his superior, General Doolittle, to take part in a combat mission over France. His plane was also shot down, Vanaman was interned at Stalag Luft III in Żagań, where he lived in constant anxiety about the fate of the secret, but the Germans limited themselves to a superficial interrogation of the PoW. The contents of the captured German archives testify that the Allied errors that caused alarm in the German services were not so much resulting from the human intelligence, related to the breaking of ciphers themselves or the violation of security procedures, but rather to the use of the information thus obtained.

Of all the German services, the Kriegsmarine was the most sensitive to threats to communications security. Its problems with Enigma began almost simultaneously with the outbreak of war. In October 1939, three submarines were sunk in the English Channel area. The loss of the U-12 in particular seemed puzzling, its circumstances remain unexplained to this day (the only trace of the tragedy was the body of the ship's commander, dumped by the sea on the beach near Dunkirk on 8 October). Dönitz, aware that he had diverted his ships into the shallow waters off Dover, was already inclined to raise alarm on this occasion: British divers had, since the previous war, enjoyed an excellent reputation for searching wrecks. His communications officers managed to reassure their commander, and the only effect of the confusion was a ban on taking Enigma aboard ships carrying missions in shallow waters off the British coast (Prien's U-47 set off on her raid to Scapa Flow already without the Enigma).

Almost none of the incidents at sea described so far went without an echo on the other side. Each time, the Germans launched an investigation into the circumstances of the accident, looking into the possibility of compromising their own ciphers. After the loss of the U-33 in February 1940, Ludwig Stummel, in charge of the naval intelligence investigations, stated that the security measures in place with the Enigma made its cipher safe, even if the machine fell into enemy hands. Further confusion erupted when, in April and May 1940, the Kriegsmarine lost two ships in circumstances that might indicate a threat to communications security. The first cause for concern was the loss of the *Polares* on 26 April and the second was the sinking of the U-13 by *HMS Weston* on 31 May. Following the loss of U-13, Dönitz on 8 June forced the Kriegsmarine's communications security service to investigate whether the sinking of the ship might have compromised the security of the U-Boats' ciphers. Moreover, knowing the approximate position of the ship's sinking, the Germans decided to make it difficult for the British to explore the wreck and bombed the area around the wreck on 12 June. The crew of one of the aircraft involved in the raid noticed in passing that the area where U-13 wreck was believed to lie was marked with buoys. This disturbing fact was somewhat misinterpreted by German intelligence officers. The marking of the site with buoys was an obvious signal that the British were preparing for divers to examine the wreck. The Germans, meanwhile, were relieved to interpret it as confirmation that the ship had sunk rather than been captured by the enemy and escorted to one of the British ports. Also, the investigation into the loss of two trawlers, including the *Polares*, concluded on 21 May that, while it was likely that the vessels' crews had been taken prisoner, they had had enough time to destroy or sink the secret material. After these incidents, the Kriegsmarine remained unconcerned about the security of its own ciphers for nearly a year. It was not until April 1941 that Dönitz became concerned about the discrepancy between the precise information on convoy routes he was receiving from his own signals intelligence and the failure of his own ships to find these convoys at sea. The change in British naval codes on 20 August the previous year had blinded German B-Dienst for a time. But in the early months of 1941, the codebreakers managed to master the new code books sufficiently to be able to transmit data on the departure time, composition and planned courses of the convoys to the operations section. Basing on the information received, the U-boat headquarters at Kerneval deployed its ships in patrol lines in time to intercept incoming ships, but there were occasions in March and April when U-boat officers scanned the horizon in vain—the convoys missed the prepared traps.

Dönitz was convinced that the location of at least one wolf pack must have been known to the enemy, and he was probably right. After the 'Lofoten tip', *Hut 8* was taking first tentative steps in breaking the *Dolphin* network. But the BdU's reaction to the setbacks revealed a peculiar trait in his personality. Suspecting the activities of enemy agents, Dönitz caused Raeder to issue an order minimizing the number of officers and soldiers in contact with Enigma communications. There must have been certain amusement in *Hut 8* when the codebreakers deciphered the Raeder's order of 22 April: "The U-boat campaign makes it necessary to minimize access to ciphertexts. (…) In future, I will regard all transgressions of this order as criminal offences against national security". At the same time, Dönitz demonstrated conviction that treachery could have taken place in all Kriegsmarine staffs except the one headed by him: he demanded the separation of the U-boat communications network from the general Heimische Gewässer network, taking, incidentally, the first steps in this direction as early as April (the introduction of a starting position for the rotors being the inverse of the general key).

The first real alarm, relating to the Enigma security, was raised at the events surrounding the sinking of the Bismarck. The very loss of a ship being the pride of the German navy must have raised questions about the sources of the failure. We now know that BP's contribution to locating the battleship was modest and belated. Moreover, the Bismarck had used the radio often enough during that tracking her position was not a problem. What alarmed the Germans was not so much the course of the clashes between British ships and aircraft and the Bismarck, but its aftermath—the slaughter of tankers and supply ships sent to the Atlantic in connection with Operation Rheinübung. It was at this time that the British Admiralty received for the first time a steady and plentiful stream of Dolphin network decrypts, so one can understand its impatience and determination to exploit a new source of information. Sending its ships out hunting, the Admiralty was aware that intercepting all targets must arouse the suspicions, so it took care to provide a fig leaf by not pointing out one or two enemy ships to its hunters. As we recall, by coincidence the *Gonzenheim*, which was supposed to have escaped unscathed, was ambushed by two cruisers, and the *Gedania*, also encountered by chance, was captured by the auxiliary cruiser *Marsdale* and brought into Greenock harbor. *Marsdale* was quite fortunate in bringing his prize to port unnoticed, but the very fact and circumstances of the capture of the ship soon came to the attention of the Germans as a result of an incredible coincidence. Shortly after her action, *Marsdale* exchanged some of her crew with another auxiliary cruiser, *Malvernian*, which had been attacked by German aircraft off the coast of Spain on 1 July. The ship appeared to be so badly damaged that her captain took the decision to abandon the ship. 31 crew members, including the ship's captain, reached the port of La Coruña after three weeks of sailing in lifeboats; another group of survivors arrived a day later in Vigo. One lifeboat was picked up from the sea by the British sloop *Scarborough*, but the last of the rescued boats was found by German minesweepers on 17 July. Unfortunately, among the rescued sailors there were Lieutenant Keats and Seaman Blackburn, who had sailed on the *Marsdale* only a month earlier. In the luggage of Keats (who had commanded the prize crew on the *Gedania* en route to Greenock), German intelligence officers found photographs of a German ship and prisoners of war disembarking in a Scottish port. The results of the interrogation of the two sailors set off alarm bells in the German communications security services and led to the *Gedania's* captain being blamed for all the Kriegsmarine's failures over several months. In fact, the unlucky commander of the *Gedania*, Captain Paradeis, behaved perfectly sensibly: he first attempted to escape *Marsdale's* pursuit for two hours and, having recognized the hopelessness of the situation, led the crew to the lifeboats and detonated the charges that were to scuttle the ship. However, the British fished him out of the lifeboat, forced him back onto the ship, and then managed to patch a hole torn out by the explosion and get the *Gedania* to Scotland. The captain's mistake was to allow himself to be photographed in a situation which, when the photograph was found in Keats' luggage, was considered to be fraternizing with the enemy. German naval intelligence rightly concluded that the machine and documentation could have fallen into enemy hands on this occasion, wrongly attributed thereto most of the losses at sea during the tanker slaughter and in the weeks that followed, accused Captain Paradeis in absentia of treason, and considered the problem solved. The British capture of the machine and the secret documents sufficiently explained events by relieving investigators of the need to ask perplexing questions about the security of the Enigma ciphers. The capture of the *Gedania* was a failure of the disinformation plan surrounding the tanker slaughter. The capture of two *Marsdale* crew members was even more unfortunate event. After all, it turned out that the combination of the two cases had convinced the Germans of the security of Enigma! A further reduction of staff employed in the enciphering and deciphering of dispatches was ordered, the Abwehr organized a number of surprise inspections of cryptologic facilities, but nothing was done to change the ciphers themselves.

The late summer and autumn of 1941 had a few more surprises in store for communications security specialists. At the end of August, the British captured the U-570. Its commander, Hans Rahmlow, was later stigmatized by his captive comrades, but under enemy fire he kept cooll enough to destroy all the secret documents and equipment. The British boasted of their capture in the press, and the ship itself was put on public display in the Vickers shipyard at Burow-in-Furness. News of both the ship's fate and that of its first officer, who was shot as he attempted to reach the ship and sink it, reached Berlin. Dönitz on 5 November described the event in his combat diary as "depressing". But he had already commissioned Vice Admiral Erhard Maertens, the navy's communications chief, to investigate the possible consequences of the loss of U-570. The report he submitted on 24 October did not contain interesting findings. Maertens reasonably assumed that "we must reckon that U-570 could have fallen into the hands of the enemy without first destroying anything". Even if, however, the British had succeeded on this occasion in acquiring not only Enigma but also the cipher keys, they could only have broken the messages if any of the captured officers had revealed the *Stichwortbefehl* procedure password, transmitted only verbally. The final conclusions of Maertens' report were optimistic. When U-570 transmitted an open-text report indicating that it was under attack by aircraft and unable to submerge, and U-boat command attempted to extract more information, U-570 responded that it did not understand the incoming messages. According to Maertens, this may have meant (along with the transmission of the original open-text report) that the ship's crew had destroyed the cipher keys at the first distress signal. In the admiral's opinion, "everything indicates that the crew had the ability to destroy at least one of the secret cipher documents", thus guaranteeing the safety of the communications of other ships.

But Maertens' investigation also had to address other worrying developments. On 24 September, Dönitz noted in his combat log that "U-67 and U-68 have been assigned to the operational area around St Helena Island. At U-68's suggestion, acoustic torpedoes from the U-111 returning to base will be transferred to her deck. A rendezvous was set for 27.09 in quadrant EJ1696". The torpedo transfer operation required a location where no one would disturb the peace of the two vessels for several hours. U-67 wished to take the opportunity to consult a doctor sailing on board of the twin ship. One of its crew members was ill; if the matter were to prove serious, the U-111 returning to base could take him on board. Square EJ1696 corresponded to Tarrafal Bay on the island of Santo Antão in the Cape Verde archipelago. U-111 and U-68 were the first to arrive on the scene. At dawn on 27 September, they met at the entrance to the bay, then stopped in its calm waters, the captains had lunch together and the crew members engaged in torpedo transfer. At nightfall, both ships put to sea intending to return to the bay the following day to meet U-67. U-111's report confirming the date and location of the rendezvous was received and deciphered at BP. Tarrafal Bay was chosen by the Germans because of its location away from shipping lanes, sending British ships there must have aroused suspicion. The risk was probably accepted due to an opportunity to attack three enemy ships at the same time. In the area of the planned rendezvous, the Royal Navy had only one submarine, the *Clyde*, which had been on the lookout off Tenerife waiting for a German tanker. But the Admiralty's orders were unequivocal: abort the ambush, proceed at full speed to Tarrafal Bay to intercept the German submarines.

The Admiralty's plan had two weak pionts: a single submarine had little chance of sinking all the enemy ships, even in the event of a surprise. The second problem for *Clyde* commander David Ingram was time—he had no chance to arrive on the scene before the enemy and prepare an ambush. As a result, the operation ended in a blunder. The *Clyde* appeared at the entrance to the bay on 28 September, half an hour after midnight, and located U-68 fairly quickly. As the ship prepared for a torpedo volley, U-111 appeared in her vicinity and the British skipper decided to ram her. The German dived in time to avoid damage, the *Clyde* fired a volley of six torpedoes towards U-68, then dived rapidly to reload the torpedo tubes. The torpedoes missed, but two of them exploded on impact with coastal rocks, alerting a third German ship approaching the scene. As a result, when the *Clyde* surfaced two hours later, she herself was rammed by U-67, with both ships suffering minor damage. When news incoming in succession from the three ships somewhat clarified the initially confusing situation, Dönitz's first instinct was that 'it is likely that our ciphers have been broken or we are dealing with a breach of security. It does not seem impossible that a British submarine would find itself in such an isolated corner by accident'. Maertens drew a different conclusion from the whole incident: he felt that if the British had indeed known of the planned encounter and had attempted to organize an ambush, their operations would have been better prepared and less partisan. In the admiral's view, the way the British took up arms proved that it was they who had been surprised by the encounter with the three U-boats.

Even the creative unravelling of the mystery of the skirmish in Tarrafal Bay did not mean the end of Admiral Maertens' problems. His officers collected virtually all of the broken British Admiralty dispatches from the period under investigation and, having re-examined them, found at least one unexplained case. The Admiralty ciphertext contained information about a wolf pack operating in the South Atlantic. Usually, OIC officers tried to camouflage the true origin of the information in the messages they sent, with the standard alibi being a network of direction finders (D/F). This time the dispatch did not contain any identification of the source of the British knowledge. Worse still, German intelligence analysts easily confirmed that the ships concerned had not transmitted any signals since leaving port, making it impossible to determine their position via the D/F network. The simplest and most logical explanation for the situation was that the cipher in which the U-boats received orders directing them to the patrol area had been broken, but the authors of the report shied away from such a conclusion. Instead, they put forward a sophisticated construction: according to it, the group's ships were probably trying to transmit reports which, due to distance and interference, did not reach German stations, but were picked up and tracked by the British. In the report's conclusion, Maertens expressed the view that the British could only break U-Boats' ciphers in the unlikely event of simultaneously capturing the cipher machine and all elements of the cipher key, including only verbally transmitted *Stichwortbefehl* password. Even in such a case, they would only be able to read the dispatches for the period that the captured key elements remain valid. Maerterns concluded his report stating: "The anxiety about the compromise of our secrets is not justified. Our ciphers do not appear to be broken". Dönitz was willing to take the report's conclusions at face value. At the beginning of November, new cipher keys were coming into effect, which were certainly not on-board U-570. In December, he received an additional guarantee: a member of the U-570 crew managed to send in a dispatch from the PoW camp containing an agreed phrase informing that the secret documents had been destroyed.

However, before the reasuring letter could reach the staff, further puzzling information arrived from teh South Atlantic. This time it concerned the auxiliary cruiser *Atlantis*. In the autumn of 1941, Atlantis was returning home after the longest combat cruise ever undertaken by a Kriegsmarine vessel. She had just rounded the Cape of Good Hope and was heading north when U-boat command decided to use her to resupply several ships operating in the South Atlantic. U-126 received instructions to rendezvous with *Atlantis* on 22 November in quadrant FE7555. The instructions addressed to U-126 were deciphered and the cruiser *Devonshire* was directed to the area of the planned rendezvous. On the morning of 22 November, *Atlantis* was leading U-126 in tow, the captain and several members of the U-boat's crew having just had breakfast in the company of Captain Bernhard Rogge, when the cruiser's masts appeared on the horizon. Rogge knew from experience how ships under attack react to danger, so he ordered the radio operator to transmit the signal **RRR POLYPHEMUS** in open text, making it clear that she was the innocent Dutch ship *Polyphemus*, which had just been surprised by a raider (**RRR**). What Rogge did not know, was that as a result of his ship's activities, the distress signal from the raider had recently been changed to **RRRR**. Amidst the confusion of clarifying the identity of the ship, U-126 managed to dive, albeit without the captain, still aboard *Atlantis*. When a message from London ruled out the presence of *Polyphemus* in the South Atlantic, *Devonshire* opened fire sinking *Atlantis*, whose crew had managed to transfer to lifeboats, and promptly moved away from the scene so as not to risk contact with the submarine.

After some time, U-126 surfaced among the lifeboats, received its commander on board and took the *Atlantis* crew boats in tow, attempting to guide them towards South America. Dönitz, for his part, dispatched U-124 and U-129 to assist, while asking the Kriegsmarine headquarters for permission for the *Python*, a supply ship in the vicinity, to take over the survivors. *Python* arrived on the scene on 24 November and took over the 305 *Atlantis* crew members. However, this was not the end of the adventures of Captain Rogge and his crew. En route to French ports, *Python* was to supply several U-boats with fuel, including one of the participants of the skirmish in Tarrafal Bay, U-68. In light of recent events, the encounter between the supply ship and the U-boat was nerve-wracking. On 30 December, U-68 was just refueling when a second submarine, U-A, appeared off *Python's* side and its captain, Hans Eckermann, suggested moving the refueling to another area. The nervousness of the two U-Boat commanders went so far that they declined Captain Lüders' customary invitation to lunch on board his vessel. A hunch was not wrong for the experienced submariners; at around 17.00 the masts of the British cruiser *Dorsetshire* appeared on the horizon, and an hour and a half later only the lifeboats with more than 400 survivors from the two vessels remained on the ocean surface—Captain Lüders had scuttled his ship. The two German submarines had no choice but to take the lifeboats with the survivors in tow and move slowly towards the French bases, while Dönitz frantically assembled assistance for the unusual convoys, arranging a rendezvous with U-124 and U-129 and the Italian submarines *Corelli*, *Finzi*, *Tazzoli* and *Calvi*.

It was almost unbelievable that all the ships with survivors on board managed to break through the British blockade and reach French ports. This left more than 400 officers and sailors at home convinced that the enemy knew in advance the time and place of their planned encounters with the submarines. Particularly consistent in making his superiors aware of this view was *Atlantis'* captain, Bernhard Rogge. At a meeting held at his request at Kriegsmarine headquarters, Rogge presented crushing arguments supporting his thesis. He began by recalling that, while still in the Pacific, he had been ordered to resupply U-68 in the area of Saint Helena on its way back to France. The adventure in Tarrafal Bay must have been memorable for U-68's commander, Captain Karl Friedrich von Merten, because immediately after the ship surfaced near *Atlantis*, he suggested to Rogge that the refueling site be moved 300 miles to the south. Rogge, for some reason, also did not feel safe in the area, so he agreed to the proposal, both ships proceeded to the agreed location directly, without the use of radio, and completed the task undisturbed by the enemy. However, when first *Atlantis* and then *Python* attempted to rendezvous with the submarines at the locations whose coordinates had been radioed to them, a cruiser would appear on the scene sailing alone—as if making it clear that it was going on a special mission—sinking the supply ship, then moving away without delay, as if aware of the U-boats' presence. Not once was the appearance of the cruiser preceded by the passage of a reconnaissance aircraft on a routine patrol. Invited aboard *Atlantis*, the commander of U-126 attempted to reassure his host by informing him that prior to the scheduled rendezvous he had not transmitted a single dispatch in four consecutive days, not wishing to give away his position and course. Given the circumstances of *Atlantis'* and *Python's* surprise, each case individually must have been questionable—both together screamed loudly in alarm. It turned out, however, that the cry was heard only by people who had survived the sinking of their ships twice in the space of a few days, two thousand miles of travel in lifeboats and more than that distance aboard unmercifully crowded submarines. Rogge failed to convince his superiors of his theses during the discussion. What's more, he was persuaded to remove references to the alleged breaking of German ciphers from the ship's logbook. In such an atmosphere, it was difficult to suppose that the investigation led by Admiral Kurt Fricke would contribute to revealing the truth. Indeed, the report presented in mid-March 1942 was an apologia for the Kriegsmarine's cipher systems, 'superior to those used by any other country'.

The year 1942 was so successful for the U-boats that the events did not raise concerns about the security of their own ciphers. However, the situation began to change with the start of the following year. The most important instructions to the U-Boats were being transmitted in the officers' variant of the *Shark*, the *Limpet* network, the key of which was first broken by Allied codebreakers on 5 January 1943. By the end of January, the effects of this success had become apparent to an extent that prompted Dönitz to order another investigation. This time, the difficulties of the U-Boats in finding Allied convoys were taken seriously enough to involve the C-in-C's son-in-law, Günther Hessler, in the investigation. The chief of operations on Dönitz's staff, together with Lieutenant Adalbert Schnee, collated German submarine position data resulting from broken British Admiralty dispatches with the actual positions occupied by the U-boats over the previous weeks. A comparison of the data from the two sources yielded inconclusive results. Both officers concluded that 94% of the broken Admiralty dispatches added nothing of interest to the discussion of cipher security, with only 6% containing suspiciously precise data on U-Boat movements. A few worrying incidents were identified; a message of 29 January reported the presence of two U-boats in a particular area which were about to rendezvous with a supply ship. At the time the British warning was sent, both ships were only on their way to the rendezvous point and could not transmit any message permitting to establish their position. In other cases, Allied convoys sailing straight into a trap set by the U-Boats were ordered to change course to avoid the German patrol line. Some Admiralty dispatches gave the U-Boats' positions with an accuracy beyond the technical capabilities of the D/F system, while others specified number of ships that could not be determined by the D/F network.

On the other hand, officers noted that the positions of the U-Boats given in a significant, if not overwhelming number of enemy messages, represented at best an approximation of their actual position. British estimates of the ships' positions could, in most cases, be explained by the operation of the D/F network, air patrol reports and the use of radar. These reassuring conclusions were supported by Admiral Maertens, who used an argument as paradoxical as it was pertinent. He noted that many German dispatches contained clear references to information extracted by the B-Dienst from broken enemy ciphers. Maertens' question seemed logical: if the Allies were breaking German dispatches, they would have had to find in them ample evidence of weaknesses in their own codes. In that case, would they have used *Naval Code No. 3* for so long without changing it? Maertens added an additional argument to this—the most important orders to the wolf packs were sent in a double-enciphered, officer variant of the cipher. Allowing for the possibility of breaking the *Shark* key, it was difficult to imagine that the enemy could overcome the double veil of the *Limpet* cipher. The investigation traditionally focused at the treachery within the U-Boot headquarters or its communications service. Counter-intelligence officers questioned all members of both services who had access to the dispatches exchanged with the U-Boats, without finding a traitor. Only Dönitz him-self and his chief of staff, Admiral Eberhard Godt, were not questioned. Dönitz commented on this fact with a rare sense of humour, pointing out to Godt that in this situation only one of them could be the traitor.

As a result, on 5 March, the BdU wrote in his combat log that "a systematic assessment of the British U-Boat data for January and early February is reassuring from the point of view of the strong suspicion that the enemy has broken our ciphers (…). With the exception of 2–3 unexplained incidents, the British information on U-boat positions (…) can be explained on the basis of a logical combination of facts". The report attributed the role of the decisive factor to wolf pack tactics alone. The participating ships occupied a position in a patrol line in anticipation of a con-voy. During this time, they had to make occasional use of the radio, if only to relay weather reports, which allowed the British to build up an increasingly precise pic-ture of the pack. In the only case where the information contained in the Admiralty dispatch was exceptionally precise, the blame was put on the Italian ally as usual: the Italian ship participating in the pack was communicating with the base in an Italian code that was considered compromised. The case of the encounter between the two U-boats and the supply ship was explained basing on the linguistic criteria: both ships were ordered to circle (*schwabbern*) in the waiting area, while the British message described them as returning to base. Such a subtle difference in the description of the ships' missions was enough for Kriegsmarine analysts to rule out the possibility of breaking the cipher.

Following the catastrophic losses suffered by the U-Boats in May 1943 and their withdrawal from the convoy routes, Dönitz devoted the summer months to prepar-ing for the resumption of the naval offensive. This was to be aided by new types of equipment; a torpedo circulating in zigzags between convoy ships, radar warning devices, a new 'Greek' rotor and a new Enigma reflector. And it was during this period of relative calm at sea that the worst thing happened, feared by the British, who did not trust the secrecy on the other side of the Atlantic. On 10 August 1943, a report from an agent working for the US Navy Department, an American of Swiss origin, arrived at the Abwehr's post in Switzerland via Swiss intelligence service, stating that "during the past months the German naval ciphers used to transmit oper-ational orders to the U-Boats have been successfully broken. All orders are read in real time". The report caused understandable concern in the Kriegsmarine com-mand, which requested more information from the Abwehr, including an assess-ment of the credibility of the source. On 18 August, Dönitz received a supplementary report confirming that the agent was linked to a US military attaché and frequently visited London as part of a US Navy delegation. The report also added that "after the outbreak of war, a special office was set up to deal with codebreaking, which has had considerable success over the last few months". It was perhaps the most serious alert about Enigma security that reached Germany throughout the war. It was also treated with due attention—several investigations were implemented simultaneously, focusing this time not on possible security breach within their own ranks, but on the security of the machine and the ciphers it generated. The most serious attempt to assess the strength of the Enigma was implemented at the B-Dienst, where the so-called '100-day project' (*Hunderttagearbeit*) was launched. For one hundred days, the service's top cryptologists attempted to apply various attack techniques against actual Enigma messages from the period of the Norwegian campaign. It was clearly assumed that it would be easier to find cracks in the security shell of the weaker, three-rotor version of the machine. In parallel, an attack on the Enigma ciphers based on statistical methods was undertaken at OKW/Chi, and at B-Dienst another team attacked them independently of the '100 Days Project', with different methods, the details of which remain unknown. At least some of the projects brought less categorical assurances about the security of the ciphers than before. German experts announced that even if one considered the Enigma ciphers to be theoretically breakable, the effort, time and necessary mechanization level of the process required to do so made this practically impossible.

The mid-1943 investigations were, it seems, the last opportunity on which German cryptologists could and should have corrected their errors. The introduction of *Naval Cypher No. 5* meant that Kriegsmarine signals intelligence could no longer look over the shoulder of Allied navy staff officers. The many effective types of new equipment and weapons used in the Atlantic from the spring of 1943 onwards meant that the effect of breaking the German ciphers became overshadowed by other causes of losses suffered by the U-boats. The Allies' growing advantage allowed them to play the game calmly, without risky moves that threatened to compromise the Enigma secret. In the few cases where the circumstances of U-Boats and their supply units being surprised by the enemy raised doubts, the communications specialists consistently maintained their verdict: "our cryptological investigations have not provided evidence or even a trace indicating breakingthe ciphers in our major networks. (…) the strength of the [Kriegsmarine] ciphers was unequivocally confirmed by the available evidence". It would be interesting to know the frank and full opinion held by the chief cryptologist of the B-Dienst, Walter Tranow, on the investigations described and their results. We only have a snippet of it at our disposal, in which Tranow depreciates the efforts of his colleagues by acknowledging that 'the only effective action would be to do what we expect the enemy to do—launch a systematic cryptanalytic attack'. Tranow clearly did not regard *Hunderttagearbeit* as a systematic cryptanalytic attack, and he was probably right; another, very belated and far from determined German attempt at a cryptological attack on the Enigma cipher had more interesting results.

In June 1944, the Kriegsmarine department responsible for communications security borrowed Lieutenant Hans-Joachim Frowein from the Wehrmacht cryptanalysis department for six months. Around December of the same year, Frowein presented a concept for the identification of the M4 machine's connections, together with the plugboard settings. His proposed method of attack was based on somewhat optimistic assumptions: it required a probable clear text of 25 characters and only the machine's high-speed rotor moving during its ciphering. Even if it was not very practical for the M4, Frowein pointed out that its application to a machine-assisted attack on the M3 machine would have required the construction of a catalogue of enciphered phrases only seventy thousand punch cards long. In testimony after the war, Frowein stated that his findings had led to adoption of the proposed security measures. Only rotors with two notches should be used in the fast position. As a result, it would prove impossible to encipher twenty-five characters text with only fast rotor moving. It seems, however, that by 1944/45 the Third Reich's bureaucratic machinery had already become disorganized, which did not allow the codebreaker's proposal to be put into practice. Besides, the proposed solution would have been ineffective versus the key-breaking methods used by the Allies, anyway.

In the first half of 1943, the security of the Enigma ciphers also became a subject of study on the part of the Wehrmacht's cryptology service, OKW/Chi. In October 1942, 24-year-old Gisbert Hasenjaeger joined this service. A little earlier, in January of the same year, he had been severely wounded in Russia, and after his recovery, the German logician and young mathematician's mentor, Heinrich Scholz, protected him from returning to the front and caused him to be assigned to OKW/Chi. Hasenjeager became the youngest member of the service, being assigned to its youngest section, the just-established team dealing with the security of its own cipher machines under Dr. Karl Stein. The fact that such a service had only been established in late 1942 and early 1943 was astonishing. Equally surprising may have been the assignment of mathematicians who were undoubtedly talented, but only beginning their adventure with cryptology (Stein and Hasenjaeger had completed a short course taught by Erich Hüttenhain after joining the OKW/Chi team). What was most surprising was the objective set before Hasenjaeger—the security of the cipher of a commercial Enigma without a plugboard. The young mathematician tackled the task fairly quickly, identifying the internal connections and starting positions of the rotors from a 100-character message, but he was unable to develop his success into a military model, although he continued his research until the last days of the war. Even if German cryptologists failed to identify significant weaknesses in the Enigma or objectively assess the security of its ciphers, an undefine feeling of insecurity, hitherto felt mainly within the ranks of the Kriegsmarine, extended to the Wehrmacht and Luftwaffe. Arguably, the first major setbacks of the war contributed more to this than theoretical analyses; the search for the culprits of defeat is always conducted more vigorously than the search for the fathers of victory. Anxiety arising from the changing nature of the war resulted in a series of innovations in the German encrypted communications system that were to provide challenges for the Allied codebreakers until the final days of the conflict.

1942 went down as a time of BP's successes in dealing with Luftwaffe and Wehrmacht ciphers; it was even described in the post-war report as *annus mirabilis*. The campaign in North Africa provided an abundance of material for analysis, and the codebreakers coped with the enemy's ciphers far better than the British army did with the Afrika Korps. The number of new networks that appeared in the air after Rommel's troops landed in Tripoli meant that Welchman's color code was no longer sufficient; new networks were christened with the names of flowers, insects and birds. If the key of each network came to be attacked separately, the sheer number of them would become a problem for *Hut 6*. Fortunately, Reginald Parker spotted an enemy's cardinal error—the keys of certain networks predictably repeated key elements of other networks from earlier periods. The *Scorpion* network, for example, repeated the keys of the *Primrose* network from the previous month. *Scorpion* was a network for the air-ground liaison in Africa, so its messages retained operational utility over hours rather than days. Fortunately, an enemy error meant that the key to the cipher in most cases was already known from the previous occurrence, and the problem was limited to decrypting the intercepted messages. The German practice gained a name coined after its discoverer—*Parkerismus*. Parker's discovery allowed *Hut 6* to postpone or even avoid the crisis of having to break the growing number of keys. Postpone, for in the course of 1943, the repeats began to disappear; the German security service apparently recognized the reprehensible practice and succeeded in eliminating it. Before this could happen, however, the number of BP bombes had grown sufficiently to allow the keys of many networks to be broken independently. In the final phase of the war, when the Wehrmacht's administrative structure was clearly overloaded, key repeats reappeared, but by then they were no longer of any significance.

In the summer of 1942, the situation began to complicate itself. In July, *Hut 6* noticed that in all Wehrmacht networks the rotor order started to be cyclically changed three times a day. If, for example, at midnight the rotors were installed in order I-II-III, at 8 a.m. they were moved to position II-III-I, and at 4 p.m. to III-I-II. The practice did not come as a complete surprise, for as early as October 1941, the order of rotors began to be reversed at midday in Luftwaffe networks. During this period, the key of the *Greenshank* network was only occasionally broken in BP. One of its keys broken in 1942 made codebreakers aware that an even more sophisticated rotor reordering procedure was being used within this, the most protected of the Wehrmacht networks. All six possible permutations of rotors in force on any given day were used, with changes occurring at periods adapted to the varying volume of traffic on the network, so that the number of messages transmitted in each variant of rotor order was approximately equal.

At the end of 1942, some of Rommel's corps networks began to use a procedure identified by the Poles even before the war, adding words unrelated to the content of the message at the beginning and end of the message (so-called *Wahlwörter*, referred to by the British as Wahlworts). Some of the words and phrases used in this role (e.g. **HOTTENTOTENPOTENTATANTENATTENTAETER**), aroused the sincere admiration of British cryptologists. Others provided information about the mood dominating among the cipher clerks: **MUSIK** or **TANZ** on the positive, **HUNGER** or **WINTER** on the minor. Until now, *Hut 6* has been routinely breaking keys basing on the recipient or sender identification located at the beginning and end of the message. The use of Wahlworts did not put an end to the attack method, but it did complicate its application—several to a dozen different locations had to be verified for each message. In view of the limited number of bombes in BP, the new security measure could have proved fatal: "applied on a significant scale in 1940 (…) it could have knocked out the fledgling crib section before it was firmly on its feet". In particular if it had been applied in its more radical form, known among cryptologists as 'Russian splitting'. The practice, initiated by the cryptologists of Tsarist Russia, consisted of dividing a message into two parts of unequal length and swapping their places: in effect, the standard opening and closing elements of a message were located at a random place inside the message. The Germans used the described approach in only a few networks, including *Roulette*.

In the course of 1944, the previously popular *cillies* gradually disappeared from the Wehrmacht and SS networks. In October of the same year, the Allies obtained a document that explained the cause of this change—it was the use of so-called random message keys (random indicators). Back in 1938, the Kriegsmarine had prevented the possibility of the cipher clerk choosing a predictable message key, forcing him to take the next key from a list and cross it out after it had been used. This was a well-designed measure; Wehrmacht and Luftwaffe cipher clerks often abused their freedom, selecting predictable keys like **ROM-MEL**, **TOB-RUK** or **CIL-LIE** (hence called *cillies* in BP). It was not until 1943 that the German security service recognized the resulting danger and responded by implementing a system of random message keys. Its rules stipulated that the station commander would select a random unclassified text—a chapter of a book, a newspaper article, the lyrics of a song, etc., enciphering it at specific Enigma settings. He would then write down the resulting ciphertext, dividing it into groups of six characters each; successive groups from the list were used by the cipher clerk as individual message keys during the day. Given the randomness of the ciphertext characters, the procedure effectively eliminated predictable message keys. At the same time, its use did not produce any outward signs (other than the elimination of the cillies), so British codebreakers were unable to identify the start of its use.

At the same time, the Luftwaffe reached for another security measure, indicating that the concerns of its cryptologists were more about an attack with probable plaintext (in other words, the use of cribs). Cryptologists recommended that the plaintext of a message should be pre-coded using the so called Mosse code before encryption with Enigma. Rudolph Mosse developed his code in the early twentieth century, originally for internal use by his own companies—it was published in a book version after his death, in 1922. Mosse code allowed for significant compression of typical business phrases. For example, the sentence "We are not responsible for the delay in delivery. We recommend hiring another forwarder" in Mosse code took the form **SRUSUNYHBA**. Before the Luftwaffe started using the commercial code, it modified its dictionary accordingly. *Hut 6* began reconstructing it on the basis of broken messages, but before it could achieve significant results in September 1944 the Allies captured a complete copy of the codebook. The adversary must have been aware of the compromise of the code, because in early 1945 the dictionary was revised: by April of the same year, *Hut 6* had a full reconstruction. Using a commercial code to encode military messages was not a fortuitous idea. The use of a code gives the enciphered text an even more stereotyped structure than the clear text; where in natural language a paraphrase can be used, in a code there is usually only one code group equivalent to the term.

In the spring of 1944, in Luftwaffe messages references started to appear to a planned change in the plugboard settings. *Hut 6* took the warning seriously; changing all the pairs in the plugboard meant having to break the key of each network three times a day. However, when the threat materialized in early May, it was described as "the least sensible and most trivial of all the security measures" designed by the adversary. The novelty consisted of changing just one pair in the plugboard at 3 a.m., 3 p.m. and 11 p.m. Allied cryptologists struggled to make sense of the change, which did not significantly impede their work but apparently complicated the lives of the enemy cipher clerks; presumably the mistakes they were making caused the new procedure to be abandoned as early as mid-June. In the end, the opinion prevailed at *Hut 6* that the procedure known on the German side as *Zusatzstecker* had been introduced as a temporary solution, before the implementation of another novelty that surprised the Allies in July 1944.

In most cases, the earlier decrypts announced the implementation of changes to the Enigma cipher system, even if they did not provide details of the new solution. On 10 July, several of the broken messages of the *Jaguar* network caused a bit of a sensation in *Hut 6*, as the correctly deciphered header in the form of a number was followed by a gibberish in the rest of the message. It was obvious that the main text of the messages was hidden behind an additional layer of cipher. Another decrypt from the same network provided a clue that the puzzling messages had been encrypted using *Enigma Uhr*, whatever that was supposed to mean. *Hut 6* quickly dealt with the new challenge. Among the intercepted dispatches, a pair was found representing the same plaintext, encrypted with and without the use of the mysterious *Uhr*. It turned out that the key to the cipher in both cases differed only in plugboard settings. Having collected more messages of the new type, it was also noticed that the number starting the messages ranged from 0 to 39. It could be inferred that the Germans were using an Enigma attachment allowing them to select one of forty sets of distinct plugboard settings, prepending the identifier of the set used in the message header. Referring to the attachment as *Uhr* (clock) suggested the setting selection of by means of a rotary switch. The *Uhr* attachment was used initially in the *Jaguar* and *Cricket* networks, and over time its use extended to 15 Luftwaffe communications networks. In the process of implementing the attachment, the German cryptologists made at least two mistakes. The first was their use of machines equipped with and without the attachment within the same network. This meant that the same information had to be enciphered in two versions, providing *Hut 6* with an abundance of cribs. The second error was indicating the *Uhr* setting in clear text in the message header. The British were quick to identify the permutations implemented by *Uhr* in all forty variants, and decoding the number currently used, they continued to read the cipher messages as quickly as their proper receiver. The German cryptologists attempted to correct their mistake by introducing in November an uncomplicated enciphering of *Uhr's* position. A document later captured proved that their original intention was to implement position encryption simultaneously

Fig. 21.1 Enigma Uhr attachment. (cryptomuseum.com)

with the attachment use. Logistics failed, forcing the novelty to be implemented gradually, facilitating the task for the Allied codebreakers (Fig. 21.1).

The use of the *Uhr* posed a problem for Allied codebreakers in only one instance: when the key of a particular network had to be broken based on a message encrypted using the attachment. The original design of the Turing bombe was a failure—it was saved by Welchman suggesting the addition of a diagonal table using the reciprocity of the Enigma cipher. But some of the substitutions generated by *Uhr* were not reciprocal; if, for example, **A** was substituted for **T**, **T** could be substituted for **G**, and so on. As a consequence, the diagonal table had to be disconnected before the menu constructed from the *Uhr* encrypted message could be run in the bombe. Such a menu could only succeed by analyzing a crib of considerable length, and including several cycles. Such a problem was rare, however, and the use of the *Uhr* was appreciated by Allied codebreakers mostly in terms of an intellectual challenge, not a practical complication: "The Enigma Uhr was an ingenious device and provided a fair portion of entertainment for the experts at *Hut 6*".

In September 1944, several decrypts of the *Jaguar* network revealed an interesting phenomenon: a correctly deciphered beginning of a message was followed by a string of random characters. The last correctly read characters were always starting with the bigram **CY**, followed by two more letters of the alphabet, such as **DE** or **XY**. The codebreakers soon discovered that the rest of the message could be read if, after decoding the aforementioned four characters, the right-hand Enigma rotor was manually moved to the position specified by the character immediately following the bigram **CY**. For example, if the sequence **CYRS** appeared in the message, the machine's right rotor should be moved to the position **R** and the decryption could continue. The *Jaguar* was the only Luftwaffe network where the new procedure occurred. From the beginning of October, however, its application extended to all Wehrmacht and SS communications networks. In the middle of the same month, the

Allies obtained a document describing the procedure in detail. Its purpose was disrupting the regular movement of the rotors—a feature of Enigma that German cryptologists considered a weakness of the machine. According to the rules, enciphering of any message of more than 150 characters should be interrupted after 70–130 letters, where a CY marker should be inserted, followed by two characters defining the new position of the machine's right rotor. Thereafter message ciphering was continued. The new position could not differ from the current one by less than five letters. This procedure only complicated BP codebreakers' job if CY marker fell inside a crib or was used in a message being broken on the *kiss* from another network.

The complications in the Enigma cipher described so far caused only temporary excitement among Allied codebreakers. In late 1943 and early 1944, however, a novelty emerged that was perceived as a real, and this time serious threat to the ability to read German ciphers. Interestingly, it appeared in the Luftwaffe's networks. Up to that point, Luftwaffe ciphers had enjoyed an enduring popularity among Allied cryptologists: their users were known to be the most careless of all the German services. However, just as a sinner on his deathbed recognizes his misdeeds and declares atonement, late in the war the Luftwaffe cryptographers redeemed their earlier negligence. On 23 December 1943, *Hut 6* deciphered a *Red* network message announcing the introduction of a new reflector, designated *Dora*, at the beginning of 1944. The new reflector was not a problem in itself—*Hut 8* had managed to reconstruct the *Bruno* and *Caesar* wiring in the past. Their colleagues at *Hut 6* assumed that the new reflector would be used alongside the old one, allowing the cipher to be cracked through *kisses* between the new and old systems. They gave the new challenge the familiar name *Uncle Dick* (from *Umkerhwalze D[ora], UKD*).

The first day of the new year brought some emotional swings. The *Red* key was broken as early as 11.50 AM, and codebreakers began to wonder if concerns about UKD were premature. In the afternoon, however, it became apparent that the broken key could not decipher *Red* messages transmitted by Luftwaffe units stationed in Norway. They were attacked using the CSKO method, little known at *Hut 6*, having been developed and used so far only by *Hut 8*. Nevertheless, Oliver Lawn was able to cope with the challenge and by 1.30 AM the following day had broken the message, defining thus the UKD wiring. The reconstructed *Dora* caused no problems for the staff servicing the bombes. While designing the device, 'Doc' Keen provided the diagonal board with configurable connections that could be removed entirely from the device and replaced with another set (known in the jargon as QCU—Quick Change Uncle). When after Lawn's success *Hut 6* declared *business as usual*, on 11 January newly found reflector failed to decipher the *Red* messages. With the experience of the previous decade, *Hut 6* reconstructed the reflector connections again; this time speculations started to circulate that the design of the UKD might allow its wiring to be altered. Codebreakers had been anticipating the possibility of such a change for some time. As late as November 1941, a rotor with changeable connections had been used in the British TypeX cipher machine; one might have expected an opponent to come up with a similar idea. Between 1 January and 7 March, seven different UKD variants were found; the hypothesis of a variable-connection reflector was confirmed (Fig. 21.2).

Fig. 21.2 Enigma Umkehrwalze D, external view. (cryptomuseum.com)

Revealing the use of UKD brought *Hut 6* closer to explaining the earlier riddle of the *Greenshank* network. *Greenshank* was the new name for the *Green* network, first broken by the Poles, back in January 1940. Since then, it has acquired in *Hut 6* a status akin to the holy grail of cryptology. From just a few of its messages broken since then, the British knew that *Greenshank* was the communications network of the German military districts, used to transmit the most sensitive information. Consequently, it used the highest security standards in place and the cipher clerks adhered to them most rigorously. However, the importance of the network caused *Hut 6* to grasp every opportunity to renew attack on its key. In late 1943 and early 1944 several obvious *kisses* between the *Falcon* and *Greenshank* were used to attack the latter's key. Despite the considerable effort put into the attack (including the use of high-speed American bombes to try all possible rotor orders and reflector variants of both, Wehrmacht and Kriegsmarine, Enigma), it was not successful. In the new reality, codebreakers started to suspect that the *Greenshank* network was the first network to use UKD, back in 1943. The hypothesis was confirmed when, on 27 April 1944, Lionel Clark, after a week of feverish work, cracked the *Greenshank* key using a 200-character long *kiss* from another network (Fig. 21.3).

The rapid success in grappling with UKD did not reassure codebreakers. The Germans had only used the reflector in two networks so far, but it was conceivable that its use would become widespread over time. Reconstructing UKD connections after each change was not a serious problem from the point of view of theory, at least as long as the adversary used machines with and without the new reflector within the same network. BP developed several methods of attack (including CSKO, hand Duenna, and scritching), but these were labor- and time-consuming, so they

Fig. 21.3 Enigma UKD with visible rewireable connection. (cryptomuseum.com)

were only applied to networks of critical importance. Uncle Dick's connections were changed three times a month, their reconstruction by manual methods took about 5 days—half the period of their validity. As a result, by the spring of 1944, *Hut 6* was gradually giving up on breaking the less important Luftwaffe networks as they started to use UKD. This state of affairs was only acceptable during a lull in the war effort, but the impending Normandy invasion must have changed the priorities of the codebreakers' work. Firstly—a significant increase in the number of intercepted messages was expected. Secondly—networks that had been neglected could prove to be a source of key information. Thirdly—the increasingly widespread use of Uncle Dick was to be expected.

Consequently, a committee chaired by Alexander was set up at BP in April 1944 to plan a strategy for attacking the UKD reflector (Uncle D Committee). The first stage of its work took the form of an attack in the literal sense of the word: the intelligence services identified the factories producing the UKD and planned to bomb them (there is no confirmation as to whether the plan was carried out or had an effect). A more important action was the construction of equipment to automate the attack on the reflector wiring. The first plan was to construct an aggregate of four Turing bombes. The first copy of the device, dubbed *Giant*, was installed at BP on 9 June 1944. The effectiveness of the device left much to be desired: a run corresponding to one menu took three to four weeks. As a result, on 6 June it was decided to build another version, this time consisting of four four-rotor bombes. The new *Giant* was ready for factory trials at the end of August, but it had to wait until 2 December for its first success, when it broke the key of the *Puma* net used by *Flivos* on the Italian front. Although the four-rotor *Giant* completed a full run in about 16 hours, it was treated as an interim design, used to await the results of work carried out in the US.

At the same time, on the other side of the Atlantic, the two US services undertook UKD taming projects independently. The Army was working on a relay-based Autoscritcher device, planned as a proof-of-concept for an electronic Superscritcher. At the same time, the Navy commissioned a device called Duenna, which went down in the annals as one of the first cryptanalytic devices constructed entirely in electronic technology. The problem of the reconfigurable reflector must have absorbed Op-20-G's attention even before the Germans applied UKD: the first theoretical work on the problem is dated in 1943. New realities have led to an acceleration of work: the Duenna specification was ready on 25 February 1944 (Alexander, who was in Washington at the time, participated in its development). The new US Navy machine was in fact an attachment to a four-rotor bombe, so its implementation was entrusted to NCR engineers in Dayton. The prototype was delivered to cryptologists on 1 September and registered its first success on 26 December by breaking the *Puma* key. The Autoscritcher proved to be 2–3 times slower than the Duenna, so it played no practical role. However, when the fully electronic Superscritcher passed its first trials, it surprised even its designers with its reliability and speed: it was more than two thousand times faster than the Autoscritcher. Unfortunately, the Army codebreakers' innovative device did not play a practical role in the attacks on the UKD; its construction did not end until 1946. By the end of the war, various Allied cryptanalytic devices had been used to attack 62 Enigma keys. They were successful in 25 cases, of which 14 fell prey to the Duenna and four to the Autoscritcher.

Of all the security measures designed to improve Enigma security, UKD proved to be the most serious challenge to Allied cryptologists. The Germans had made the fatal mistake of introducing it in the networks where the old reflector was being used at the same time. However, the threat of widespread use of the UKD hung over *Hut 6* for a long time. This happened on 5 August 1944, when the UKD became the only reflector used on the *Wasp* network used by IX Luftwaffe Corps. From then on, "until the end of the war we saw the increasing difficulty of breaking most Luftwaffe keys". The Allied codebreakers, however, were somewhat fortunate in their misfortune. Uncle Dick was used exclusively in Luftwaffe networks and was never (excluding the case of the *Greenshank*) used by the Wehrmacht, SS or Kriegsmarine. It was only from post-war interrogations of German cryptologists that the Allies learned that the Kriegsmarine had developed its own version of the reconfigurable reflector, but had failed to deploy it operationally. The UKD epic found a somewhat ironic finale in late 1944 and early 1945. It proved troublesome to change Uncle Dick's settings, and the numerous mistakes made by cipher clerks prompted German cryptologists to abandon the use of the reflector. With a feeling of relief, the Germans abandoned a solution that could have defeated the Allies, if not on the ground of cryptologic theory, then at least in practice.

The Third Reich hampered Allied signals intelligence not only by modifying the cipher machine and the way it was used. Enigma-encrypted messages had to be transmitted to the recipient by radio, and it was by modifying wireless communication procedures that the Germans came closest to defeating the Allies. In September 1943, they introduced an innovation that seemingly did not relate to the Enigma cipher. In the autumn of 1939, Gordon Welchman worked out the structure of German communications, distinguishing several independent networks. At the time, every Wehrmacht and Luftwaffe message began with a five-character network indicator. The radio operator constructed the indicator by selecting one of the four triplets of characters specified in the applicable key (*Kenngruppe*), and adding two random letters before the selected group. If the key for the day provided, for example, *Kenngruppen* in the form of UCY, AMU, PWG and SKX, the network indicator could take the form of ASUCY, PKAMU, etc. The indicator was not encrypted, allowing each recipient to easily determine whether the received message was transmitted within his network. But it was just as easy for the opposite party's listeners to see if the message belonged to a network of interest. In the autumn of 1943, the Germans abandoned the network indicator altogether. German radio operators had no problem identifying the correspondent, using the call signs of the radio stations and their operating frequencies; the Allies lost the natural criterion of network identification. The crisis was eventually overcome, but "the existing team proved ridiculously small in comparison to the new problem, too small even as a nucleus that could provide training for new staff. Experienced staff stood on their heads working after hours, but still the challenge crushed them. (…) it was the most dangerous period in our work".

Overcoming the crisis caused by the abandonment of network indicators consumed the staff reserves being prepared for the future second front. Fortunately, this had already happened after the codebreakers' letter to Churchill: in early 1944, reinforcements in the form of 130 WRENs arrived at *Hut 6*. They could not have arrived at a better time. The messages broken in the early spring brought allusions to a planned change of the so-called '*F book*'—the list of German wireless stations' call signs. After the dropping of the network indicators, recognizing the structure of German communications was based entirely on the call signs, their change was bound to cause chaos. The expected blow fell on 1 April: the listening stations were unable to allocate the intercepted messages to specific networks. Fortunately, BP included a section whose tasks grew directly out of Welchman's pioneering work in BP's early days. Its name—Sixta—indicated that it originated from *Hut 6* and was concerned with traffic analysis (TA). Traffic analysis uses these features of an adversary's communications that do not require deciphering the messages to determine the structure of the communications network. By analyzing who corresponds with whom, at what times, when there is a peak of traffic on a particular network, etc., Sixta was able to reveal a surprisingly precise picture of the German war machine. This knowledge proved crucial in sorting out the temporary chaos after the '*F book*' change and reconstructing the list of new call signs. Once again, the foresight demonstrated by Welchman proved invaluable: "(w)hereupon what seemed a luxury from a codebreaker's point of view became a vital necessity, and had Welchman not fought for it early on, we would have had no hope of calming the storms that have almost overcome us in the last eighteen months".

German signals officers, however, had further surprises in store for the Allied intelligence. In November 1944, the Wehrmacht stations made in quick succession two changes that could have defeated Allied signals intelligence. The first was down to changing the frequency of all stations three times a day. The second was the encyphering of their call signs. The operating frequency of a station and its call sign were, at that time, the only criteria for classifying a station into a particular network. The moment of trial became at the same time a moment of triumph for the somewhat forgotten service of Allied signals intelligence: "it was at this time that [intercept station] Beaumanor achieved its crowning success (…). They fought the battle and won it entirely on their own, and sophisticated preparations on the part of [Hut 6] and Sixta proved superfluous". The success of radio interception proved doubly useful; on 1 February 1945, Luftwaffe stations followed the example of the Wehrmacht. *Hut 6* history noted that "the enciphering of Luftwaffe call signs resulted in a serious decline in decrypts, which we have not been able to fully make up for". This meant in practice a drop in the number of daily broken messages from 1.800 to about 1.000 and the keys from 150 to 104. But "the Germans made their usual mistake (…) by introducing (…) a change in operating frequency before they had implemented call sign enciphering", allowing each problem to be addressed separately.

The precautions implemented by the Third Reich's cryptologists were the result of their own assessment of the vulnerabilities of Enigma, the weaknesses in the procedures of its use and the mistakes made by the operators. In the light of our present knowledge, the Germans failed to identify any misuse of information resulting from decrypts that happened to the Allies despite all precautions. Virtually all the Allies made mistakes which may have led to the compromise of the secret of breaking the cipher. The game around protecting the secret of breaking Enigma was just as fascinating as the challenge of reading the cipher messages. It was risky and often fought on the brink of defeat. The British system of information distribution seemed to guarantee that content from the decrypts would not appear verbatim in the British own dispatches, and that commanders would not make decisions based exclusively on Ultra. But the Admiralty, in particular, was having trouble coming up with new fictitious sources of knowledge of the U-Boats' location, with the result that too many dispatches indicated the DF network as the source of information. Occasionally, decrypts yielded information so attractive that even top commanders could not resist the temptation to turn it into an operational use (cf. the 'slaughter of tankers'). Officers responsible for the security of decrypts were sometimes helpless when orders came from top-brass commanders or were taken independently by the American allies. The most common violation of rules was the quotation in operational orders verbatim of details that could only be derived from intercepted and deciphered enemy messages. The officer responsible for transposing information from the decrypts into an order for his own forces faced a difficult task. If he included too detailed information, he risked being accused of compromising the secrecy of the source. If he used too vague a description, he could make the mission difficult or impossible to carry out. Add to this the inevitable time pressures and tensions that accompany warfare, mistakes became inevitable. Among the classics is the juxtaposition of a broken dispatch informing of the departure of an Italian ship to sea with the order given by the submarine command in Malta:

Deciphered message	Operational order
S.S. *Bosforo* sets sail from Naples to Benghazi crossing 19 July at 0330 position 33°06′N, 20°16′E where 18 July afternoon will meet torpedo boat executing anti-submarine sweep. Expected time to drop anchor at Benghazi 19 July 0900.	Italian m/v *Bosforo* 3.567 BRT, length 347 feet, draught 21 feet, will soon set sail from Naples, crossing 19 July at 0630 position 33°06′N, 20°16′E. Torpedo boat would execute anti-submarine sweep on 18 July afternoon and meet Bosforo at the position given above. Drop anchor at Benghazi 19 1200.

Apart from the addition of the ship's technical data and the change of the time to the one used in Royal Navy, the two texts were no different. The commander of the Maltese submarine flotilla was cautioned for inappropriate action. But three days later, another Royal Navy officer gave the position of two U-Boats in the Atlantic repeating verbatim the data from the broken German message; worse still, the coordinates of the region he gave closely matched the corners of the squares on the secret German nautical chart. During the fighting in Tunisia that followed Rommel's retreat from El Alamein, Montgomery was warned by signals intelligence of an enemy offensive threat for his troops at Medenine. Despite the warnings addressed to him directly by Churchill, Montgomery unceremoniously pulled up artillery from the rear, deploying it precisely at the point of Rommel's planned attack. The Desert Fox's assault failed, he lost dozens of tanks, but he captured documents and took prisoners, whose testimony confirmed British knowledge not only of the fact of the planned attack, but also of its location and strength. Fortunately for the Allies, Rommel had the Italians on hand, to whom he routinely shifted responsibility for revealing any secrets to the enemy. The vast majority of threats to the security of the Enigma secret, however, were born at sea, where it was more difficult to mask the basis of one's decisions with information obtained from prisoners of war, air reconnaissance or one's own reconnaissance.

Knowledge of the Enigma secret was not limited to the circle of Polish, French, British and American codebreakers. By somewhat twisted routes, it reached also Moscow. Although BP's recruitment networks included both Cambridge and Oxford, Professor Frank E. Adcock, representing Cambridge, proved more successful than his Oxford counterpart, Hugh M. Last. As a result, Cambridge was represented in BP in far greater numbers than Oxford. What Adcock did not know was that his university's environment was also being effectively penetrated by Soviet intelligence. In these circumstances, it was almost inevitable that among the academics arriving at Bletchley there would be passive or active communist sympathizers and agents of Soviet intelligence. The best-known case is that of John Cairncross, who worked at BP for a period of about a year in 1942 and 1943. While at university, he was part of a group of left-wing students who later became known as the 'Cambridge Five'. In April 1937 he was recruited by Soviet intelligence and was given the pseudonym 'Moliere', later changed to 'List'. After graduating, Cairncross came first in the civil service examinations and joined the Foreign Office in 1936, from which he moved to the Treasury in 1938, finally becoming Lord Hankey's personal secretary in September 1940. Hankey oversaw the British intelligence services, as well as the work of several scientific committees set up by the Cabinet. Government documents relating to the empire's most secret affairs passed across his desk, and through Cairncross's hands. Among other things, it was through Cairncross that Soviet intelligence received the first official confirmation of Britain's nuclear bomb project. At this time, his controlling officer, Anatoly Gorski, "complained of an excess of material to be sent in cipher".

After Hankey was dismissed from his post, his secretary went to BP and was assigned to *Hut 3* in March 1942, where he was in charge of intelligence interpretation of decrypts. Also during this period, he was an efficient source of information for his Soviet supervisors. Gorski financed Cairncross's purchase of a car to facilitate his weekend trips to London, combined with the transmission of further batches of intelligence material. Thanks to Cairncross's activities, Soviet intelligence was aware of the operation of the Bletchley Park center (codenamed "Kurort" in Soviet files), as well as receiving from April 1942 onwards a considerable number of German dispatches relating to the Eastern Front. Among other things, the Russians received through him dispatches related to the planned Operation Citadel. Cairncross himself, however, was disappointed with his job at *Hut 3* and took the opportunity to move from BP to MI6 on 1 July 1943. Soviet sources stress that the organization in BP did not follow good counter-intelligence rules, facilitating the agent's task. Decrypts forming the basis of *Hut 3*'s work should have been destroyed with the assurance of document accountability. In reality, they landed in the trash bin, from where the Soviet agent would take them out by the handful and then deliver the source documents to his handler. The same Soviet source also states that, while working at *Hut 3*, Cairncross acquired information about the design of the device used to decrypt the 'Fish' teletype ciphers. On this basis, Soviet cryptologists were reportedly able to build their own version of the device and, for a time, read German messages independently. Sometime later, the Germans were to make changes to the design and Cairncross was no longer able to provide updates, which cut the Soviet cryptanalysts off from the source of the information. If the information provided corresponds to the facts, there must have been a serious violation of the *need-to-know* principle at BP; Cairncross could not have obtained this type of information as part of his standard duties at *Hut 3*.

Cairncross's transfer to SIS did not completely cut the Soviets off from their sources of information in the BP. Soviet military intelligence GRU had an agent, codenamed 'Dolly', located within the structures of the War Office. The agent, whose identity and even gender remain unknown to this day, passed on to his controllers a considerable number of decrypts relating mainly to Japanese and, later, German affairs. Amongst other things, it was 'Dolly' who in April 1943 communicated that 'the German Eighth Air Corps would be part of the operation being prepared [at Kursk], and that units forming the main striking force would be redeployed directly from Germany. The operation in question may form the main thrust of a future offensive towards Kursk'. After Cairncorss' transfer to MI6, 'Dolly' remained key Soviet source in British signals intelligence circles. Speculation has been rife that decrypts from the GC&CS diplomatic section were going to Moscow via another member of the Cambridge Five, Major Anthony Blunt, then working in MI5. Amongst other things, Japanese diplomatic dispatches originating from breaking the *Purple* machine ciphers would pass through his hands. After the war, the Russians announced that they had independently cracked the *Purple* ciphers in late 1941, without supporting this claim with any evidence.[1]

There may have been yet another, to this day unidentified Soviet spy, nicknamed 'Baron', working in or around BP. His activities came to light as a result of an unprecedented US signals intelligence effort, carried out from 1943 and not completed until 1980. The Americans attempted to crack Soviet spy dispatches, encrypted with a one-time pad, theoretically impossible to break. It turned out, however, that during the wartime crisis of late 1941 and early 1942, the service responsible for printing the cipher blocks had printed a batch of them twice, making it possible to read parts of the dispatches. It was not until 1962 that the May 1941 dispatch was broken, in which an anonymous agent quoted an Enigma decrypt containing a list of railway stations in Ukraine that were planned for use during Operation Barbarossa. The breaking of the dispatch caused a stir at CCHQ, the British agency continuing in the tradition of the GC&CS. The British had already learned of Cairncross's betrayal, but the broken message had been sent a year before the Soviet spy arrived at *Hut 3*. Nigel de Grey, who was still working at the former service, called a staff meeting and made a dramatic speech informing colleagues of the possibility of Russian penetration of GCHQ. Two years later, after Blunt's treachery was discovered, the British eased their consciences by writing off the alarming dispatch to his account. But was it right? Analysis of Venona's other dispatches indicates that the Soviet agent 'Baron', was rather linked to the émigré Czech intelligence community. Two suspects have emerged in this context. The first is František Moravec, whom we remember from Bertrand's account quoted earlier as 'Raoul'. His identity with 'Baron' could be confirmed by the revelations of General Pavel Sudoplatov. It is known that Moravec, in addition to heading the intelligence of the Czech government in exile, was also an employee of SIS. Sudoplatov maintains that Moravec had already been recruited in London to also work for the GRU. American and British intelligence analysts, however, are inclined to the opinion that the 'Baron' was rather identical with one of Moravec's closest associates, Karel Sedláček. Sedláček had extensive interests and contacts, working simultaneously with his own country, British, Swiss and Soviet intelligence. 'Baron's' access to the source of information at Bletchley is clear from the

---

[1] Soviet decrypts of Japanese diplomatic messages resulted most probably from the treachery of Kozo Izumi, clerk in Japanese embassies in Sofia and Istanbul, who provided his Soviet handlers with a key to the cipher.

Venona reports. On 3 April 1941, a cipher message was sent to Moscow containing the fragment: "This information comes exclusively from Baron, its origin should be well known to you, the intercepted message bears the Enigma mark". The Soviet agent was aware of the importance of the information provided, making further cooperation conditional on the establishment of a secure communication channel. A dispatch dated 17 May 1941 states: "On 17 May Dick met with Baron. Concerning the transmission to us of the intercepted message (…) Baron (…) announced that your answer to this question would determine future cooperation with him". Whoever passed the BP decrypts to Moscow, all the information leaks described confirm unequivocally that BP's counter-intelligence cover proved dramatically inadequate.

In the light of the available documents, the question of whether the Soviets were also able to break Enigma messages purely cryptographically, without the help of agents, remains unresolved. After the defeat at Stalingrad, the Germans themselves were concerned about the security of their communications. There were 26 copies of Enigma in the 6th Army cauldron, and the chaotic nature of the fighting in the final phase of the battle did not allow the Germans to make sure that they were destroyed or hidden. Russian historians claim that the first two copies of Enigma fell into Russian hands as early as December 1941, during the counter-offensive in the Klin area. The same source claims that three machines were captured in Stalingrad and that several cipher clerks were taken prisoner and persuaded to cooperate. Presumably the information they provided helped Soviet specialists to analyze Enigma's weaknesses. After almost a year of work, at the end of 1942, "scientists of the special group of the GRU cryptanalytic service, with the help of agents, proved the possibility of decrypting German Enigma-encrypted dispatches and proceeded to construct equipment to accelerate decryption". On 29 November 1942, the head of the GRU, General I. Ilyichev, signed a request for the award of decorations to a group of fourteen officers, of whom Colonel Malyshev, Lieutenant Colonel Tyumenev and Captain Yatsenko were to receive the highest Soviet decoration, the Red Banner Order. It seems, however, that apart from the decorations, the cryptologists' success had no significant consequences for the fate of the war: "(o)nly archival ciphertexts could be deciphered, because in January 1943 the Germans introduced new security mechanisms. Our cryptologists were unable to defeat these novelties because of the backwardness in electronic technology during the war". In the final analysis, the Russians proved helpless against the secret of Enigma, as did the British against the activities of Soviet intelligence.

Despite occasional errors, the stringent rules adopted by the Allies to protect the Enigma secret served their purpose. At various times during the war, cipher machines, secret enemy documentation, officers and soldiers privy to the workings of the German communications system fell into Allied hands. British, Americans, French and Poles aware of the Enigma secret fell into the hands of the enemy. Weighing up all the Allied mistakes in their approach to the secret, we are faced with a conundrum: what caused the Germans to rely unreservedly on the security of the machine's ciphers for nearly six years of war, despite numerous alarm signals? The reasons for this apparent blunder had several sources, and their significance changed over time. Immediately after adopting Enigma, the Germans were convinced of the cryptological advantage they had achieved over potential adversaries. The known weaknesses of the civilian model had been eliminated before Enigma was used in the services; the theory and practice of attacking machine ciphers was still in its infancy. Enigma had two qualities whose charms German cryptologists could not resist: it was a German invention and it was based on a scientific principle of operation. This scientific principle manifested itself in the magic of large numbers; it seemed to be possible to calculate precisely the time required for a successful attack on the machine's cipher.

All Enigma security assessments developed during the war were tainted by tunnel vision. Within the structures of the Third Reich, responsibility for communications security was dispersed, which was not conducive to the holistic view. As a result, each institution assessing the strengths and weaknesses of the machine did so through the filter of its own specific operation, knowledge and experience. Counterintelligence specialists looked for opportunities for betrayal within their own ranks, reducing the number of people in contact with secrets. Naval officers were concerned about printing codebooks with water-soluble ink. Cryptologists analyzed the cipher's resistance to compromise of individual key elements, reaching the risky conclusion that even the lack of a single element safeguards the security of the system as a whole. Cryptanalysts dabbled in the intricacies of combinatorics, presenting numbers of combinations that paralyzed the imagination. No one attempted to combine threats from different domains—the simultaneous acquisition by an adversary of secret documents or the machine itself and the application of an automated attack on the cipher. In the final stages of the war, German cryptologists were finally inclined to conclude that the Enigma could be broken, but they immediately reassured their principals that this would require time and effort beyond even American or British capabilities.

It seems that at this stage of the war they had no alternative to such a conclusion or to the Enigma itself. The machine was the primary cipher device used by all types of armed forces, the SS, the police and the railways. Questioning the security of its ciphers would have been perceived as undermining the foundations of the regime's security, with all the consequences for the perpetrator. Even if one allowed for the possibility of breaking the Enigma cipher, at this stage of the war the Germans did not really have an alternative to using the machine. Admittedly, back in 1939, a talented self-taught engineer, Fritz Menzer, designed for OKW/Chi the Schlüsselgerät 39 (SG-39) cipher machine. Its design combined elements of the Enigma and Hagelin machines, provided a significantly longer key period than the Enigma and offered operator convenience by printing plaintext and ciphertext and perforating both on teletype tape. Despite the device's advantages, only three copies existed in 1945, only one of which was complete. A technician employed in the production of the machine, interrogated by the Americans testified that the delay in the work was mostly due to army's indecision as to whether the device should combine both, the teletypewriter and the cipher machine, or act solely as a cipher device producing tape for the independent teletype.

This comment and the technical features of the machine indicated that the SG-39 was not supposed to replace the Enigma, but rather one of the teletype ciphering machines: T-52 or SZ/42. The successor to the Enigma was to be the SG-41 machine designed by Menzer in 1941. It was an ingenious extension of the design principles of the Hagelin machines, and its purely mechanical design favored its field use, analogous to the Enigma. The crank-powered machine earned the colloquial name of *Hitlermühle*, Hitler's grinder, but none of the services was determined to deploy the new machine. The situation changed when, in April 1942, as part of the OKW/ Chi reorganization, Menzer became advisor to the Abwehr. He harshly assessed the security of its ciphers, made *ad hoc* improvements and recommended the use of his design as a target solution. The Abwehr commissioned a thousand copies of the device from Wanderer Werke AG, an office machine manufacturer in Chemnitz, but it appears that it did not live to see the order delivered. Material shortages delayed production until the Abwehr was integrated into SD structures in February 1944. BP did not come into contact with the SG-41 cipher until October of that year.

In mid-1944, OKW decided to order 11.000 copies of the machine, and the German weather service included an order for two thousand copies of a variant encyphering numbers only. At this stage of the war, the Third Reich lacked both materials and a skilled workforce capable of meeting the order. Wanderer Werke managed to produce around 500 copies of the SG-41 machine. One can only specu-late on the impact of the potential replacement of the Enigma by the *Hitlermühle*. Certainly, the continuity of Allied decryption would have suffered. Mavis Batey, who took over from Dilly Knox at the ISK after his death, noted that 'we managed to break a few [SG-41] messages on depth, but during the war we failed to break [the] machine itself and it remained a mystery'. Even if it were possible to overcome the obstacles and implement the new machine Germans would be taking a risk by implementing a device sharing the working principle with Hagelin's equipment. As early as 1943, they were systematically breaking the cipher of M-209 machine, manufactured in the US under license from Hagelin. Despite the modifications introduced by Menzer, the functioning of the SG-41 and the M-209 was based on similar principles. Arguably, it was easier, quicker and cheaper to complicate the work of Allied cryptanalysts using another invention by Fritz Menzer—the so-called Luckenfüllerwalze. In the unanimous opinion of German cryptologists, the Enigma's greatest weakness was the regular movement of its rotors. Menzer designed a rotor whose perimeter had not a single stepping notch, but 26 notches with pins. An active pin caused advancing the next rotor, an inactive one caused no action. By configuring the pins, the movement of the Enigma rotors could be made more or less irregular. Heimsoeth & Rinke had developed the documentation for the Luckenfüllerwalze before February 1943. However, the decision to undertake the manufacturing the rotor was not made until a series of conferences organized by the last head of army signals intelligence (OKH/GdNA), General Gimmler, between November 1944 and January 1945. An order for 12.000 rotors was received by one of the Enigma manufacturers, the Ertel Werk, but did not manage to complete it before the collapse of the Third Reich.

The duel between code makers and codebreakers was one of the key elements of the intelligence battle between Allied and Axis services. Cryptologists on both sides demonstrated professionalism and achievements that, properly used, could influence the fate of the war. Neither side managed to avoid mistakes—triumph or defeat depended on the balance of their number and importance. In this sphere, the services of the Third Reich clearly failed, making mistakes on a larger scale, both in number and importance. German specialists could not be accused of being insensitive to communications security. Even before the war, employees of the Polish Cipher Bureau were concerned about how precisely changes made by the enemy would strike at their methods of attack. During the war, too, German cryptologists made improvements and modifications that were rational in terms of the methods of attack they themselves had successfully used against the enemy's ciphers. They were unaware that in the meantime, thanks to a handful of Polish mathematicians, a revolution in cryptology was taking place, subsequently developed by the British and Americans. A revolution whose effects were amplified by the use of sophisticated technical equipment and resulted in a paradigm shift in cryptology. In the new reality, any German assessments of the Enigma security must have been flawed, as they did not take into account the new methods of attack, unknown to Third Reich cryptologists. In this sense, it was crucial to preserve the secret of the pre-war success of the Polish Cipher Bureau, even under the most difficult circumstances.

But even limited measures, which were taken as a result of professional caution rather than under the influence of alarming signals from outside, could have produced far better results. The introduction of a fourth rotor in the *Shark* network cut *Hut 8* off from decrypts for almost a year, although it came as no surprise to the Allies. On the occasion of some of the innovations in the way Enigma was used, Allied cryptologists stressed that it was enough to apply two or more changes introduced separately at the same time for them to have to consider themselves defeated or, at the very least, experience prolonged interruptions in decryption: "the new German security measures could have, properly applied, virtually stopped the stream of intelligence information (…). If this result was never achieved, it was largely due to German mistakes".

Another source of Allied successes and German failures was a different organizational culture, resulting in a different approach to how the secret services interacted with the outside world. The British services had relied on the *'old boys'* system. Bringing in outsiders with the necessary qualifications and guaranteeing discretion due to a shared social and community background was standard in their organizations. The Poles had less experience in this area, but they were the ones who reached out for the help of mathematicians when the British still regarded them as guys out of touch with the real world. More importantly, apart from maintaining strict secrecy of operations, the Allied services did not insist that the recruits adopt the specific style of work of their new world. The BP huts more closely resembled the atmosphere of an academic seminary than the formal relationships of the militarized world of the secret services. For experienced Royal Navy officers, the manners of the codebreakers must have been shocking, but they accepted them while appreciating the results of their work. German military tradition prompted different priorities. One of Germany's most gifted cryptologists, Otto Buggisch, was sent to a reserve officers' course after mobilization, during which he probably learned to recognize the military ranks of his superiors, to fold his uniform properly in his locker and to salute. It was only after he had acquired basic military knowledge that he was given the rank of *Wachmeister* and was transferred to his proper duties as a codebreaker. Arguably, he was not the only mathematician whose talents were wasted on classes that were of no use in the codebreaker's job.

The lack of coordination was a natural and almost inevitable consequence of the organizational structure of the German services. Poland and Britain owed their successes to the concentration of the best people in one integrated agency. The United States struggled to overcome competition between Army and Navy codebreakers. In the Third Reich, signals intelligence activities were dispersed among at least six different services, pointedly ignoring each other's existence and operations. The British were cracking the keys of many networks on the basis of *kisses*, messages containing the same cleartext, encrypted in different systems. The organizational structure of the German services only rarely allowed agencies to exchange results and coordinate intentions. It was only in September 1943, under the influence of defeats on all fronts, that the OKW/Chi was entrusted with the scrutiny of new cipher systems. But even then, the other services did not easily relinquish their former independence, since in August 1944 a special coordination commission was created under the chairmanship of an OKW/Chi representative. The result of its activities was the much-delayed decision to deploy the SG-41 and Luckenfüllerwalze. With all the barriers of an organizational or psychological nature, the secret of the Allies' Enigma breaking would probably have been discovered by the enemy had it not been for the intervention of fate in several critical situations. The B-Dienst's reading of the Allied convoy codes was a drama for the sailors of the ships sunk by the U-Boats, but at the same time it became the key argument that prevented the Kriegsmarine from believing its own ciphers were broken. By coincidence, even the death of Hans-Thilo Schmidt made sense, making it easier for his compatriots to accept the not entirely true version of the story of the breaking of the Enigma by Polish radio intelligence.

Chapter 22
Decline

As late as the autumn of 1942, it seemed that no one and nothing, except nature, could stop the German war machine. In August and September 1940, despite poetic phrases about the few to whom so many were supposed to owe so much, Britain's most important defender had traditionally proved to be the sea. When German divisions stood at the gates of Moscow a year later, they were stopped more by space and weather than by the courage of Soviet soldiers and the prudence of their commanders. In other battles and campaigns of the first period of the Second World War, the initiative belonged to the Third Reich, and its armed forces exploited this advantage extremely effectively. Not even the entry of the United States into the war changed this picture; the mobilization of its industry and armies took time, and the U-Boats took care to make the route of the first units transferred to Europe as risky as possible. This picture of war changed dramatically in just a few months in late 1942 and early 1943. At the end of October 1942, Rommel's divisions stood at El Alamein, threatening to take Alexandria, Cairo and continue towards Palestine, Syria, Iraq and Iran. At the same time, the German flag was fluttering over the summit of Elbrus in the Caucasus, and Wehrmacht divisions appeared to have cleared Stalingrad of its last defenders, disrupting the artery of the Volga. Contact between the northern arm of the pincers, coming from the Volga steppe through the Caucasus, and the southern arm, led by Rommel, would have meant the loss of the Middle East with its fuel reserves, a threat to India, and the relegation of Russia to a passive role in the war due to fuel shortages. But by early November, the defeated remnants of Rommel's German-Italian forces were fleeing across the desert to the west. General Paulus commanding the German 6th Army had failed to stop the strategically pointless fighting to capture the ruins of Stalingrad. Within weeks or so, this was to be revenged on himself and his troops. After the ordeal of the winter siege, the 6th Army had to capitulate. If Dönitz was convinced that his U-boats ruled the Atlantic, the unhindered passage directly from US ports of an invasion fleet carrying divisions, which landed in North Africa on 8 November 1942, must have been a serious warning. He managed to mobilize his fleet for the winter campaign of 1942/43,

M. Grajek, *Enigma Myth Deciphered*, History of Information Security,
https://doi.org/10.1007/978-3-031-65475-6_22

inflicting heavy losses on the Allies, but from March 1943, luck began to turn against the 'grey wolves', and in May their commander had to withdraw ships to bases or calmer waters. In Africa, Rommel had to watch in amazement as Hitler, hitherto skimping on his every tank and soldier, generously threw divisions across the sea after the Allied landings in Morocco and Algeria, which only a few months ago would have guaranteed victory. Now, with the German-Italian forces in Africa caught in the pincers between the 8th Army pressing from the east and the British-American troops coming from Algiers, it was clear that the campaign was lost and every division thrown into its cauldron was sacrificed without justification. Whoever looked at the situation maps in the German staffs could harbor no illusions: in less than six months, on all fronts, the initiative had passed into the hands of the Allies, and the forces of the Third Reich had been pushed onto the strategic defensive.

The change in the nature of the war must have meant a change in the role played by Enigma codebreakers on the Allied side. Up to now, they had been a weapon of the weaker side, permitting to concentrate scarce forces at a key point on the battlefield, to avoid an attack by the superior enemy, to prepare own troops for a surprise attack at an unexpected time and place. Such a role is envisaged by the theory of war strategy for the intelligence of the weaker of the warring parties, but it seems that in the early stages of the war the British were completely surprised by the quantity, quality and timeliness of the data provided by their own signals intelligence, and as a result needed a considerable time to learn how to make proper use of the unexpected gift of cryptologists on the battlefields. During O'Connor's offensive, instead of concentrating their forces at a decisive point in the Libyan desert, they dispersed them by engaging in a hopeless Greek campaign. Adopting static, defensive tactics in the desert fighting, they allowed Rommel to beat their troops when and where he wanted. Operating usually in formations of brigade group strength, they almost always condemned their armored troops to fight an opponent who was numerically and technically superior. Even when Enigma brought precise information to exploit the enemy's weakness (e.g. during the fighting in the "Cauldron" near Gazala), the British developed their operations without any special haste, as if they hoped that the favorable situation would last forever. Learning to use the decryption data properly stretched over several stages. Arguably, this came most easily to the most conservative of the British services—the navy. There, after the initial mistakes and failures in the Norwegian campaign, an efficient mechanism for sharing information between BP and the OIC was organized, paving the way for the use of decrypts in war at sea. Probably due to this, a significant part of the analyses examining the impact of the breaking of the Enigma ciphers on the fate of the Second World War expose precisely the operations at sea. Indeed, the use of information from Bletchley in the land-based operations came with greater resistance. In the first phase, BP did not do badly in breaking the keys of the enemy's most important networks, but the system of distributing information to the field headquarters of the fighting troops was failing. In the second phase, of which the Battle of Crete was a classic example, information from BP reached the field commands reliably and on time, but the generals, who had gained combat experience in a different type of conflicts, were experiencing a problem using this information properly. The First Battle of El Alamein and, to an even greater extent, the Battle of Alam Halfa, were the first defensive clashes in which British commanders properly used Enigma decodes. However, when Montgomery felt strong enough to go on the attack, despite precise data on the enemy's strengths and weaknesses, he agonized terribly in the last of the battles at El Alamein before achieving victory—through brute superiority in men and tanks rather than subtle use of knowledge. Whereupon he promptly confirmed his own and his staff's inadequate preparation for the offensive use of information when, during the pursuit of Rommel's forces, he repeatedly allowed them to slip out of his trap, thus contributing to prolonging the African campaign by about six months.

It is time to look at how the changing nature of the war has altered the use of information from the broken German dispatches. Let us also take the opportunity to somewhat standardize the nomenclature used for the broken Enigma messages. So far, when talking about information from broken Enigma dispatches, we have used the terms 'decrypts', 'Enigma decodes' and so on. In the literature, it has become accepted to use a short and practical name Ultra to describe information derived from the decoded Axis messages. This convenient approach, however, is an anachronism; in the early days of the war, data emanating from BP was attributed to a fictitious agent codenamed Boniface, referred to as 'Special Intelligence' (SI), finally 'Most Secret Source' (MSS). The need to standardize the nomenclature arose as the United States entered the war and began providing regular intelligence to its representatives. Among the Americans a short and practical term *Magic* was used, referring to the deciphered Japanese diplomatic dispatches. It seemed sensible to coax an equally succinct designation for the decodes of German dispatches, which led to the birth of Ultra. Over time, the term Ultra came to encompass the entirety of Allied signal intelligence. The original meaning of the term was much narrower. Any documents containing decrypts were distinguished by a stamp signifying the highest level of secrecy in the then prevailing 7-level hierarchy in the UK—'Top Secret Ultra'. At the beginning of 1942, the Americans began to abbreviate the term, leaving only the last part, under which the whole of the activities surrounding the decryption of Enigma messages went down in history. From now on, when writing about breaking German ciphers, we will follow the accepted convention and refer to this activity as—Ultra.

However, the notion of Ultra also encompassed information derived from the messages encrypted in systems different from Enigma. Back in 1940, when police radio stations in Britain were scouring the ether for transmissions by German agents, they came across a particular type of messages whose sender could not be human—the characters in the messages were transmitted too fast and were not transmitted in Morse code. Having ascertained that it was not a transmitter of a German spy, the police officers abandoned further investigation. A year later, when the Germans were testing a new communications network between Vienna and Athens, the operator made a fatal mistake. On 30 August 1941, he transmitted a long message of about 4.000 characters and then, at the request of the recipient, retransmitted it again, enciphering it with the same key and changing a few details in the clear text. After two weeks' work John Tiltman reconstructed their open text. Having the enciphered and open text he could easily determine some 4.000 characters of the key, the analysis of which was entrusted to a young chemistry graduate, William Tutte. In December of the same year Tutte was able to repeat Marian Rejewski's success reconstructing a machine he had never seen and which the Allies were not to capture until after the end of the war; a so-called cipher attachment for a teletypewriter (Schlüsselzusatz) SZ40 manufactured by Lorenz. During the summer and autumn of 1942, at the height of the fighting in North Africa, a listening station in Malta periodically intercepted radio traffic of an unusual nature between Sicily and Rommel's bases in Libya. The operators of the new network were making the mistake of encrypting multiple messages with the same machine settings, which allowed Michael Crum to begin work on reconstructing it. This traffic involved a different model of ciphering machine, the T52C, made by Siemens. Incidentally, when British cryptologists embarked on an attack on its cipher, it had already been successfully broken in Stockholm.

Both the SZ 40 and its later versions, the SZ42 and SZ42A, as well as the T52 machine, later replaced by the T52A, B, C, D and E models, represented distant relatives of the Enigma, with which they shared the use of moving rotors. They were used to encipher teletype communications, carried out over both wire and radio links. The T52A and B were used from 1931 by the Kriegsmarine for communications between headquarters and large ships moored in the harbour. After the Luftwaffe was established, its staff reached for a later model of the device, the T52C. The same model was used by the Wehrmacht until 1941, when it was withdrawn from use and replaced by a device of Lorenz's design. While the Enigma was used at the tactical, up to army level, teletype cipher devices were installed at higher level headquarters, starting with the army group. In British nomenclature, teletype ciphers were collectively referred to as *Fish*. Within the *Fish* ciphers, communications encrypted by Lorenz equipment were referred to by the codename *Tunny,* and Siemens by *Sturgeon*. BP focused attention exclusively on *Tunny*, even though its ciphers presented a more serious challenge to codebreakers. Siemens equipment was mainly used by Luftwaffe headquarters, whose operations were well known through the regular breaking of Enigma messages. Meanwhile, BP's successes in breaking the Wehrmacht's Enigma were modest compared to those of the Luftwaffe and Kriegsmarine ciphers. Breaking the *Tunny* ciphers was a fair compensation for these deficiencies. Teletype cipher machines were installed in the Wehrmacht's most important headquarters. The messages carried fewer details useful at the tactical level, but more information of a strategic nature; assessments of the situation, directives for planned operations, specifications of the forces and equipment at hand. The decision to focus efforts on *Tunny* networks proved to be an expedient choice. When Tutte reconstructed the machine, the only intercepted network was used in training mode. But at the beginning of 1943 networks started to multiply and took on an operational character: by the middle of the year there were six, by the autumn ten, and in the first months of 1944 BP identified twenty-six Tunny networks linking the Berlin headquarters with the main Wehrmacht commands throughout occupied Europe. The story of the breaking of the *Fish* ciphers is just as fascinating as the Enigma story and certainly deserves a separate monograph. At the same time, it is a crypto adventure almost entirely separate from the Enigma story.

Let us leave the codebreakers for a while and return to the Allied troops we had left behind after victory at El Alamein and landing in French North Africa. After the victory Montgomery undertook an unhurried pursuit of Rommel's divisions. In the first few days after the battle, he was able to block their retreat and destroy them completely without much effort. However, he rejected the plans to bypass the German-Italian troops through the desert. He certainly remembered the experience of his predecessors, whose pursuit of Rommel turned into an equally swift retreat. The second motive for the slow pace of the pursuit was the belief that Rommel had nowhere to run. A British-American army was approaching from the east. It seemed that Rommel's forces were trapped with no way out. But the Allied plans clearly failed to take into account the irrationality of Hitler's decisions. When the Allies were primarily concerned with shaping some kind of political solution to unravel the withers of ambition and animosity in the French camp in Algieria, Hitler urged Kesselring to seize and defend Tunisia. On 9 November, the Allies learned from a broken dispatch from the Sicherheitsdienst's resident in Vichy that Marshal Petain had agreed to German intervention in Tunisia. On the evening of the same day, Ultra reported that the Germans were redeploying their air force to bases in the Tunis area and establishing a naval base in the city itself. Marshal Kesselring, who was in command of the Tunisian expedition, mainly used the Luftwaffe's communications networks, which allowed BP to keep an up-to-date and detailed eye on the increase in numbers and combat capability of the German troops being redeployed to Africa. Allied planning did not foresee Hitler committing significant forces to a campaign he could no longer win, and General Eisenhower did not have the forces with which to prevent a strong Tunisian Axis bridgehead. A failure in planning was to cost the Allies a six-month campaign amid Tunisian hills, winter rain and mud. In mid-December, the German forces in Tunisia were reorganized as the 5th Panzer Army under General Hans-Jürgen von Arnim. The consequences of Montgomery's error of judgement immediately after the battle of El Alamein became apparent in early February 1943, when Rommel's army, retreating from Libya, came into contact with German troops in Tunisia. Soon, too, the Allies were to find that the victorious conclusion of the African campaign would cost them a great deal of nerve and blood.

On its march west, Rommel's army reached the Mareth Line, a system of fortifications built at one time by the French as protection against an Italian invasion from Libya. Built in an inherently defensible location, the Mareth Line protected Rommel from the British 8th Army, allowing him to make an unexpected offensive turn towards the American troops occupying positions on the western side of the bridgehead. Rommel planned a joint strike by both armies against the Americans in the Gafsa area. Von Arnim, whose relations with Rommel remained cool, also wanted to go on the offensive, but in his own sector and under his own command. A conciliatory mission by Kesselring, who met with both commanders on 9 February, failed to reach a settlement—two separate and poorly coordinated strikes were agreed. The recent reorganization of command structures on both sides of the front meant that it was with the onset of February that the effectiveness of *Hut 6* against both the Fifth Army (*Bullfinch*) and Rommel (*Chaffinch*) networks declined. From several broken dispatches of the Luftwaffe network and the *Fish* network linking

Kesslering's headquarters in Rome with Tunis, the BP read the intention of both strikes, but without details or timing. As a result, the German attack on 14 February came as a complete surprise to inexperienced American troops and commands. Played out in difficult mountainous terrain, the somewhat chaotic series of clashes went down in history as the Battle of Kasserine Pass, which the US Army does not display in its glory gallery. But the initial chaos was brought under control and the German tanks retreated to their starting positions when one more coordinated attack could have broken through the Allied positions. The biggest loser in the battle was the chief of intelligence on Eisenhower's staff, Brigadier Mockler-Ferryman. Eisenhower was appalled by the surprise and attributed it to Mockler-Ferryman's excessive and one-sided confidence in Ultra. When the decrypts failed to provide timely warning, Eisenhower's intelligence did not reach out to other sources of information in time, putting American and British troops in a difficult position. Mockler-Ferryman's successor as Eisenhower's head of intelligence, Brigadier Kenneth Strong, was to persevere by his side until the end of the war, with Ultra rehabilitated already in the next clash of the African campaign.

As the Americans waged heavy fighting at the Kasserine Pass, Eisenhower encouraged Montgomery to advance towards the Mareth line, prompting Rommel to worry about his own rear. Montgomery responded deploying a battle group consisting of the 51st Division and part of the 7th Armored Division into the Medenine area. Somewhat contrary to the intentions of the Allied generals, Rommel treated the forces advancing towards his position not as a threat, but rather as bait. After he had to break off his attack in the west on 23 February, he decided almost immediately to throw his three panzer divisions to the east and destroy the advanced echelon of the 8th Army. Later that day, BP deciphered a dispatch from the Italian Commando Supremo informing that as soon as the retreat from Kasserine was completed, the armored forces of the newly formed Army Group Africa would strike the advancing British forces at Medenine. The further actions of the enemy were also closely followed by the codebreakers. On 28 February they broke Rommel's order to the Italians to carry out reconnaissance in the Medenine area. Montgomery's staff had originally planned to prepare troops to repel a possible attack by 7 March, but the intercepted order revealed that Rommel did not intend to wait that long, planning to attack on 4 March. Within days, the 8th Army managed to reinforce the XXX Corps standing at Medenine and, more importantly, to prepare an artillery ambush for the advancing tanks. Finally, the German attack began and ended early in the morning of 6 March. Having encountered concentrated British artillery fire and having lost around 50 tanks, Rommel realized that his attack was no surprise to the enemy and aborted the advance. The rest of the African campaign was not interesting either from a tactical point of view or because of the role played by Ultra. On 12 May it ended with two radio reports. Von Arnim reported to Berlin that his army had been forced to surrender after firing its last shells. Alexander meanwhile reported to Churchill: 'It is my duty to report to you that the Tunisian campaign is over. All enemy resistance has ceased. The coast of North Africa has been captured'. Ultra added an interesting footnote to both reports. More than a quarter of a million German and Italian soldiers were taken prisoner, but the Allies had no problem feeding this mass of PoWs—the captured food depots were stocked abundantly enough. Throughout the campaign, the precision of Utra reports allowed the Allies to sink ships delivering fuel and ammunition to Africa, letting units transporting soldiers and provisions through.

A few months before the fighting in Tunisia was resolved, in mid-January 1943, Churchill and Roosevelt met in Casablanca to determine priorities for future operations. Despite Stalin's hints about a second front, both politicians recognized that the forces at their disposal would not allow a landing in France during 1943. In order not to let the troops gathered in the Mediterranean languish for a year, it was agreed that Sicily would be their next target, followed by Italian peninsula. With an attempted landing in Sicily, more determined resistance was to be expected than from French troops in North Africa, so it was decided to confuse the Germans and Italians as to where the next move would be by planning Operation *Mincemeat*. A certain Major Martin, who was to become its protagonist, would have been outraged at the codename chosen by the staffers. He would have been, had he ever: many years later, the operation in which he played a central role was described in a book with the relevant title "The Man Who Never Was". The corpse of a tramp who had died of pneumonia was chosen to play the role of the alleged Major Martin in one of London's mortuaries. His body was disguised in an immaculate naval infantry uniform, a briefcase was attached to his wrist containing documents he was supposed to be carrying as a courier before his plane crashed, and then dumped in the sea off the coast of Spain on 30 April. Knowing the rules of cooperation between the Spanish authorities and the Abwehr residents scattered around the country, the British were confident that documents delivered in an unusual manner would not only end up in German hands, but would additionally be treated as credible. And the set of papers in 'Major Martin's' briefcase testified to Allied plans for a landing in the Peloponnese and a parallel operation to distract the Germans and directed against Sardinia.

Operation *Mincemeat* was the premiere of Ultra in a new role—that of a strategic deception support and monitoring tool. Every psychologist, or better still, every conman, knows that the best way to impose a belief in a certain fact on a victim is to ensure that this fact is as close as possible to the victim's own views. But where could one learn about the views of the enemy's commanders and staff? Ultra came to the rescue with reports containing periodic assessments of the situation prepared in Kesselring's headquarters or directives sent to him from Berlin. It was clear from these that Hitler had a similar interest for the Balkan area as Churchill. In this region lay the only deposits of certain raw materials of strategic importance to the industry of the Third Reich. After the failure of the advance towards the oil fields of the Caucasus, the Germans drew the bulk of their fuel from Romania; an Allied attack on the Balkans threatened the security of its supply. On the other hand, Hitler was aware that an invasion of Sicily and a further surge into the Italian peninsula would not lead the Allies to any goal worthy of the necessary sacrifices. The mountainous terrain, the numerous rivers crossing the peninsula latitudinally and the sparse road network allowed the defenders to delay the invading army's march north until it would find itself at a dead end at the foot of an impassable Alpine bastion. Hitler and his generals had ample reason to reckon with an attack on the Balkans rather than Italy, so the Allied planners expected them to swallow the bait willingly. But how to check whether the bait in the form of 'Major Martin's' briefcase had reached its destination? Ultra cleared up any doubts: on 12 May, an OKW dispatch addressed to Kesselring and other German commanders in the Mediterranean, passing on information 'from an absolutely certain source' about the Allies' planned parallel landings in the Peloponnese and the western Mediterranean, was broken at BP. Noel Currer-Briggs, then serving in a signals intelligence unit in Tunisia, later recounted how his unit stationed on the outskirts of Bizerta was visited by Alexander and Eisenhower asking for news. When the officer on duty informed them that the intercepted messages had confirmed the redeployment of one German division from Sicily to Sardinia and another to Greece, the two generals began to demonstrate their joy in a not very noble form, dancing on the beach: they have just received confirmation that 'Major Martin' had accomplished the mission.

When American and British soldiers jumped off the landing barges in the early morning of 10 July and dashed towards the Sicilian beaches, Ultra did not disappoint either. Even the previous day, Axis troops had sounded the alarm, but for Sardinia, not Sicily! On 11 July, a number of dispatches broken by BP in quick succession helped to warn the forces on the island of the German counterattacks being prepared. The course of the subsequent fighting on the island suggested that the game was not between Allied forces on one side and the Axis on the other, but rather between Montgomery's 8th Army and the Seventh Army under General Patton. As a result of the rivalry between the two Allied commanders, and the ignoring of Ultra's dispatches, German and Italian forces managed to withdraw from Sicily virtually unhindered, despite Allied superiority in the air and at sea. 75 ferries sailed unimpeded across the strait between Messina and Reggio for several days, evacuating some 60,000 German and 75,000 Italian troops from the island. Preparations for landing on the peninsula were taking place in the shadow of confidential talks with Marshal Badoglio's government about the terms of Italy's surrender, and the plans for the operation drawn up under these conditions represented necessarily a rather clumsy compromise. The Italians insisted on declaring their surrender only after the Allied landings at the tip of the Italian boot. Alexander therefore had to plan the operation as if he would meet the armies of both opponents on the beaches. In the first instance, the 8th Army was to cross the Strait of Messina and draw off the expected German counterattack. A few days later, when the German troops would be engaged in fighting with the British, General Mark Clark's American 5th Army was to land in their rear, in the Salerno area, and cut off their retreat northwards. In practice, nothing went according to plan. To cover the forcing of the straits, Monty staged one of the greatest artillery salutes of the ongoing war: for several hours his army's powerful artillery, accompanied by ship's guns and air force bombs, turned the area around Reggio into dust. When the first British soldiers landed on the beaches, they found that the enemy was simply not there. The Germans easily recognized the British intention and withdrew their units northwards. After the surrender of the Italians, they were willing to defend Italian soil only where and when it assured tactical advantage. As Daily Telegraph correspondent, Christopher Buckley, who observed the landing, noted, crossing the straits was 'an undertaking about as risky as sailing in peacetime to the Isle of Wight' off the British coast. The German evasion and giving up the attack at the British bridgehead cancelled the justification of the second part of the operation—the landing at Salerno. It was clear that the Germans would shield themselves with a token force from the 8th Army and throw all their energy against the landing Americans. On 8 September Ultra confirmed that the enemy had worked out the intentions of the Allied command. The Germans stated that the landing craft assembled in Sicily indicated the second Allied landing besides Reggio, and assumed that it would take place at a point delimited by the maximum range of air force operating from the local airfields. Broken messages revealed that the Kriegsmarine expected a landing in the Naples-Palinuro area and the Luftwaffe between Naples and Salerno. Military automatism worked, however, and the Americans landing on the night of 8–9 September encountered strong and well-organized enemy resistance. Only the heavy fire of the ships' guns allowed the Americans to survive the German armored counterattack, which reached almost to the beaches. But navy's role became complicated when the Germans effectively used their new weapon, radio-guided Hs 293 rocket missiles, dropped by carrier aircraft beyond the range of anti-aircraft cover. The Hs 293 bombs were no surprise to the Allies; the Luftwaffe's KG100 unit had already used them against shipping in the Bay of Biscay. Moreover, on 7 October, Ultra reported that the KG100 had been redeployed to a base at Istres in southern France, leading to the assumption that it would be used against landing forces. The fighting on the American beachhead was taking a dramatic turn. General Clark attempted to save the situation by ordering a night-time airdrop by the 82nd Airborne Division on 12 and 13 September. The courage of the paratroopers would have been of little use against the German Tigers had it not been for the problems on the German side. Wehrmacht divisions in central and northern Italy were busy disarming the Italians, unable to come to the aid of the forces fighting at Salerno. After the crisis of the battle between 12 and 14 September, on the sixteenth the leading units of the 8th Army linked with the Americans, and on the eighteenth Ultra delivered orders to the LXXVI Panzer Corps, signaling its intention to withdraw. The Germans not only escaped the trap, but took advantage of the Allied preoccupation with the fighting at Salerno to establish control over almost all of Italy.

During the fighting at Salerno, the Allies were able, thanks to Ultra, to watch discussions among German commanders about plans for further action. The first effect of the Allied landing in Italy was the withdrawal of German forces from Sardinia and Corsica. In the following days, Utra began to signal the German command's intention to surrender southern and central Italy without a fight and to offer resistance on the Genoa-Ravenna line, followed by a withdrawal to the Po river line. On this basis, General Alexander formulated on 21 September an assessment of the situation, predicting the capture of Rome by 7 November and the reaching of Livorno, Arezzo and Florence before the end of that month. Soon, however, further Ultra dispatches overturned his forecasts. In a dispatch addressed to Kesselring on 1 October and deciphered in BP the following day, Hitler stressed that it was extremely important to try to give as little ground as possible to the Allies, recommending active defense along the entire front line. A week later, a report from the LXXVI Panzer Corps revealed that the Germans were planning to take a fortified position south of Rome and hold out there through the winter. Even more surprising was a broken report from the Japanese ambassador, who quoted a recent conversation with Hitler in which the Führer announced that the Allies would be expelled from Italy. For a while, confusion reigned in Alexander's staff, for it seemed that instead of planning a rapid march north, care had to be taken to repel a German counter-attack. The staffers' main concern was that, in winter conditions, the Allied powerful air force would play little role in repelling a German attack. Fortunately, after two weeks of concern and hesitation, further reports from Utra cleared up the confusion: between 19 October and the end of the month, a number of enemy messages concerning the redeployment of more Wehrmacht troops to the eastern front were intercepted. By mid-October, Kesselring felt that the winter position of his troops had been prepared and abandoned his defences along the Volturno River, retreating to the Garigliano River line. The Allies trailing behind him found that the further route to Rome was baffled by a barrier of mountains, where all natural passages had been fortified by the enemy and manned by elite troops. The most important passage, leading through the valley of the Liri river, led through the charming town of Cassino, which lay in the shadow of the Benedictine monastery rising above it on the hill of Monte Cassino. Both names were to recur in staff bulletins and reports by war correspondents over the next six months.

Even before the Allied landing at Salerno, BP received a message that could have been an alarming signal regarding the security of the Enigma decodes. On 30 August, a dispatch announcing the arrival in Italy of the 18th Armored Division and the 14th Armored Grenadier Division, the spearhead of the XXXIX Panzer Corps, was broken. At the same time, a new network appeared on the air, used by the German Army Group B HQ stationed in northern Italy under Rommel's command. The network became known in BP as *Shrike* and resisted codebreakers' attacks until early October. It was only then that it came to light that its dispatches constituted deception: they were intended to confirm the presence in Italy of the XXXIX Corps, which was in fact deployed on the Eastern Front (as confirmed by the broken dispatches of other networks). Deception delivered via enciphered communications channel only makes sense on the assumption that the enemy can read their messages. Confusion arose among codebreakers—could it be that the Germans were aware of the breaking their ciphers and were taking advantage of this? Fortunately, messages coming from other networks were regularly confirmed by the facts: the game would have been pointless if it had only covered a part of the scene. Over time, it became apparent that the Germans were probably trying to influence the behavior of a wavering Italian ally through a trick with the *Shrike* network, suggesting that German forces in Italy were more numerous than they actually were. Later in the war, the Germans resorted to disinformation involving the creation of a fictitious communications network on more occasions. And the *Shrike* network itself became, in the winter months of 1943/44, one of the primary sources of information on German operations in Italy. *Hut 6* was breaking about 70% of the network's keys within a week, before traffic gradually died out in the spring of 1944.

   The static nature of operations on the Italian front in the winter and early spring of 1944 did not allow for much use of Ultra information. The exception to this rule was a few episodes in Operation *Shingle*, which was intended to breach the closed gates of Rome. General Alexander was aware of the strength of the German Gustav Line running through Cassino and the price his troops would have to pay for breaking it. In order to avoid a frontal attack on the fortified positions, he planned a landing on their rear to induce the enemy to abandon the Gustav Line for fear of being cut off. General Clark's 6th Corps of the 5th Army was to land on the beaches near Anzio and move inland cutting the roads linking Rome with the front at Garigliano. To disguise the operation and tie up divisions that might oppose the American landing, Allied troops launched an assault on German positions on 17/18 January. As the battle over Garigliano was dying down, American troops landed on the beaches near Anzio at 2 am on 22 January. This time the Allies dispensed with Monty-style firework show, left the ships in silence and made their way to the shore, where they encountered no resistance. Kesselring considered the possibility of an amphibious operation, but not where VI Corps had landed. The German command learned of the landing from a report made by a corporal from the track repair unit, who, seeing the Americans landing, got on a motorbike and set off to find an outpost where he could make a report. By chance, he came across Lieutenant Heuritsch of the 200th Grenadier Regiment, who managed to get a message to the mayor of Albano at 4 am, and the mayor relaying it to the German command of Rome. Kesselring passed the first information about the landing to Berlin in the afternoon. VI Corps thus achieved complete surprise, putting nearly 40.000 troops ashore on the first day of the operation, with no resistance from the enemy. But its commander, General Lucas, made an astonishing decision at dawn: instead of moving inland, he ordered his troops to consolidate the positions they had taken, running only a dozen kilometers from the beaches. A little later, Churchill summed up the events of the first hours of the landing by stating 'I thought we were going to throw a wild cat ashore, but in the end, we were left with a stranded whale'. Lucas certainly commanded his divisions disastrously, but the blame for the failure did not lie solely with him. During the first 48 hours of the operation, both General Mark Clark and Alexander himself appeared at the beachhead, but failed to influence Lucas and his tactics. Then again, Churchill's first reaction to reports of a successful landing also differed from his later assessment: "Thank you for all the reports. I am glad that you are consolidating on the territory you have gained, rather than establishing beachheads". The only commander who did not allow himself to make a mistake was Kesselring. In an impressively short time, he pulled reserves from garrisons in central and northern Italy and, surrounded the beachhead with a ring of field fortifications that resisted later attacks. Away from their garrisons the German troops had to switch to radio communications, and the codebreakers at *Hut 6* could watch the tightening of the German army's ring around the beachhead. Thanks to the sudden abundance of intercepts, codebreakers managed to save VI Corps from a probable defeat. When Kesselring rejected the much-delayed Allied advance on 30 January, he was already planning a counter-attack that could push Lucas' divisions into the sea. On 3 February, *Hut 6* broke his dispatch of 28 January in which Kesselring informed OKH of the details of his counterattack. On 5 February, further details of the planned German operation got to Alexander's staff, and in the following days the codebreakers kept on adding more elements almost daily, and finally on the 15th they reported that the enemy was ready to strike. As a result, a strike by 3 German divisions broke through the first Allied line on 16 February, but encountered further points of resistance deep inside the beachhead and stalled after five days without reaching the beaches. Shortly afterwards, BP detected and transmitted information that another attack was being prepared, the spearhead of which was to be *Hermann Göring* division. The German attack started, as expected, on 29 February and, although fighting continued into the first days of March, it failed practically at the start. On 9 March, a dispatch from the *Bream* network brought confirmation that Kesselring recognized his defeat and decided to withdraw most troops from the line, surrounding the beachhead with a line of fortifications sufficient to deter any Allied attacks.

At the same time, a bloody slaughter in several episodes was taking place on the frontline at Cassino. Its first act, the attack on the town of Cassino from 29 January to 4 February, fatally failed. Soon after the battle, a discussion began in the Allied headquarters about the role of the monastery in the ongoing struggle. In retrospect, one gets the impression that it bore a somewhat surrogate character. Faced with the stalemate at Anzio, the generals were not really able to propose a realistic scenario for breaking the German front. In Alexander's and Clark's staffs, the opinion was perpetuated that the key to the German position was the monastery itself, from whose walls German observers safely directed the artillery fire covering all approaches to the German lines. In fact, Kesselring had forbidden German troops to enter the monastery, and the commander of the XIV Panzer Corps, General Fridolin von Senger und Etterlin, as a devout Catholic, saw to it that this order was obeyed. Von Senger's subordinates convinced the Benedictine brothers to evacuate the library and treasury to Rome. Paradoxically, it was this decision, combined with Ultra, that might have contributed to the monastery's destruction. Along with the monastery's treasures, most of the 80 or so resident monks also went to Rome. However, the abbot himself, Dom Gregorio Diamare, decided to remain in the monastery, accompanied by ten brothers. The plan to bomb the abbey also raised doubts in the Allied command. Generals personally risked flying over the monastery aboard a reconnaissance aircraft to look for signs of the presence of German troops in its buildings. They came to contradicting conclusions; some spotted a radio aerial above one of the buildings and parts of uniforms drying on strings, others saw no one but monks. There exists a poorly documented version of events according to which Ultra may have provided a conclusive argument for the bombing. In a message broken by BP, the officer in command in the Monte Cassino area was said to have confirmed that the abbot remained in the monastery (*Abt bleibt im Kloster*). By an unfortunate coincidence, the dictionary of German military terms used by Allied intelligence expanded the word *Abt* as an abbreviation for *Abteilung*—detachment. If the supporters of the bombing had indeed received confirmation that a German detachment was stationed at the monastery this way, this was one of the most grotesque and unfortunate incidents in the history of Enigma. This course of events might be confirmed by the statement of General Maitland Wilson, who had just taken the seat vacated by Eisenhower as Commander-in-Chief of Allied forces in the Mediterranean. On 12 February, Wilson declared that he had 'irrefutable evidence' of the presence of German troops in the monastery. Two days later, Allied aircraft scattered leaflets over the abbey warning of the bombing, and the following day more than 200 aircraft appeared over Monte Cassino, their bombs turning St Benedict's monastery into rubble.

During the course of the fighting at Cassino, Ultra reliably reported important developments on the German side, including signaling the takeover of a key section of the front by the 1st Airborn Division. The nature of the fighting, however, limited the ability to exploit the decodes. Following the defeat of Operation *Shingle*, the Allies had to try a frontal breakthrough of a strongly fortified position. They were unable to achieve this goal during the offensive launched on 15 March, when they planned to carve a breach in the German lines using their powerful airforce. During the first three hours of the offensive, the RAF and US Air Force dropped nearly one and a half thousand tons of bombs over an area of a square mile, but the German paratroopers fortified themselves in the ruins of the town of Cassino and repulsed several assaults by New Zealand infantry. In the process, some Allied aircraft mistook Cassino and Venafro, a dozen miles away, smashing the mobile HQs of the 8th Army and the French corps stationed in the latter town. The final offensive began on 11 May. The Poles struck towards the ruins of the monastery, the British attempted to establish several bridgeheads on the Rapido river, the Americans fought heavy battles in the coastal sector, but the decisive events took place in the French sector. Moroccan troops in French service took advantage of the enemy's involvement in heavy fighting in other sections of the front and broke through the Aurunci Mountains, which the Germans considered impassable. The French advance into the Liri valley threatened to cut off the German divisions still fighting on the Rapido and Garigliano rivers, so on 16 May Kesselring decided that the winter position had served its purpose and began withdrawing troops. Success on the Gustav Line did not turn into the total victory it could and should have become. The plan was that, in parallel with the attack in the Liri Valley, the Americans trapped since the winter at Anzio, would break out of their beachhead and strike across the German lines of communication, cutting off the retreat of the divisions defending the Gustav Line. The start of the operation was promising. The Americans broke through several kilometers of terrain heavily fortified by Kesselring and took Cisterna. But then the commander of the American 5th Army succumbed to the temptation to play the role of conqueror of Rome and changed the direction of attack of his troops from east to north. He achieved his goal: on 4 June, the American spearhead of the 88th Division reached Piazza Venezia, and Clark himself was able to pose for a photograph under the plaque marking the entrance to Rome (the photographer had been instructed in advance to photograph only the left profile of the general, and the plaque itself went as a war trophy to the USA). Victory in the race to Rome, however, meant that the 5th Army left open the roads by which German troops were retreating from the south—the Allies were to fight many more hard battles against them in the remainder of the Italian campaign.

By this time, the Mediterranean operational theatre was just a side stage for world drama. All actors and spectators were aware of the imminent premiere of the second front, although discrepancies persisted among them as to its exact time and place. The anticipation was also felt in Italy. The landing at Anzio was carried out in unfavorable winter conditions, for the end of January was the last date on which it was possible to use the Allied landing craft in the Mediterranean: immediately after Operation *Shingle*, they were transferred to Britain in preparation for the Normandy landings. Along with the ships, a number of troops (including the Desert Rats and the American 82nd Airborne Division) and commanders (including Eisenhower and Montgomery) have left for the British Isles. The task of the troops remaining in Italy was to tie down as much of the German war machine as possible. But even engaging the Germans in Italy and supporting partisan operations in the Balkans did not change a simple fact: the Third Reich still had more divisions in northern Europe than the Allies could throw into the attack in France. If the landing was to be successful, it was necessary to keep the enemy, until the last moment, in the dark as to where, when and in what strength the Allied armies would land on the beaches of the continent. The Allies, in fact, set themselves an even more ambitious task. They wanted to keep the enemy uncertain about the location of the main strike even after the first wave had landed on the beaches of Normandy. The great disinformation campaign they prepared for this purpose was nicknamed Fortitude, and Ultra became one of its main tools.

The second tool of the Fortitude plan began to forge even before the war, when a certain Arthur Owen, a Welsh businessman, reported to MI5 with the information that he had been recruited by the Abwehr and equipped with radio equipment and ciphers to provide the reports from within the British Isles. After the outbreak of war, as a precaution, counter-intelligence placed him in an isolated location where Owen, as agent Snow, participated in a radio game with the Germans. Employees of the RSS, Britain's communications security service, monitoring the frequencies of a German station responding to Snow's calls, noticed that the same station was also exchanging on the same frequency other messages, cloaked in unintelligible cipher variants. Suspecting that they might relate to the activities of other Abwehr agents they attempted to interest GC&CS in the dispatches, but received a discouraging response. BP's attitude to the intercepted dispatches only changed when RSS's Hugh Trevor-Roper deciphered one of the secret dispatches proving that it did in fact relate to German intelligence activity. From then on, work on the family of German hand ciphers passed into the hands of the BP section headed by Oliver Strachey, known as ISOS (Intelligence Services Oliver Strachey). The messages of other agents broken by this section allowed them to be quickly and discreetly raked into the custody of MI5, where they faced a choice between a speedy trial and execution at dawn or continuing to work for the Germans, albeit under British control. In the vast majority of cases, those concerned chose the second option, which presented the British with the problem of defining the content they wished to impart to the Abwehr. This led to the formation of the XX Committee, also known as the Committee of Twenty or Double Cross Committee. When the Allies set about preparing plans for the invasion, Committee XX already had a substantial body of work in place. The first problem the double-agent controllers had to solve was to make sure they controlled the entire German agent network in Britain. Thanks to ISOS's methodical cracking of spy dispatches, the Committee was confident that it controlled the entirety of German intelligence operations within the UK. The second problem was to make sure that the British, for their part, were not involved in the German-led intelligence game. Almost all Abwehr operations in Britain were prepared and carried out in a grossly unprofessional manner. It occurred to members of the Double Cross Committee to think that the Germans had deliberately sacrificed ineptly prepared spies and carelessly planned operations in order to divert MI5's attention from a deeply camouflaged network of genuine agents. If this were the case, the value of controlled agents as purveyors of disinformation would be nil. The argument that reassured the British was provided by Dilly Knox's breaking the Enigma cipher used by the Abwehr. The contents of reports passed by MI5-controlled agents were regularly quoted in dispatches passed between Abwehr outposts (encrypted with Enigma and read at BP)—the Germans apparently took information from agents seriously. The final problem was to establish the reputation of the spies in the eyes of their German principals. This task was accomplished quite easily by occasionally transmitting through them information that was true but innocent enough not to harm British interests, or by transmitting it late enough to make its operational use impossible. The incoming dispatches from Hamburg announcing the award of the Iron Cross to individual agents provided confirmation of the effectiveness of this part of the game. The stage was set for disinformation representing a gigantic development of the aforementioned Operation *Mincemeat*.

As with the landing in Sicily, the first stage of Operation *Fortitude* was to establish what the Germans would most readily want to believe, given their own expectations. In light of the deep penetration of the Axis countries' ciphers, this was not a difficult task. In February 1944, the Luftwaffe reorganized the structure of its communications in the west, but the keys of the new networks did not resist attack for very long. Moreover, old *Red* Luftwaffe network retained its former importance, traditionally providing information of significance beyond the Luftwaffe's own field. In March of the same year, BP broke the key of the *Fish* network linking the headquarters of the C-in-C in the west, Marshal von Rundstedt, to Berlin. Thanks to the achievements of the Americans, the Allies also monitored communications between the Japanese embassy in Berlin and Tokyo, which was to prove extremely useful in planning the invasion. Information from BP flowed both to the London Controlling Section (LCS), the organization under whose innocuous name the entire deception effort was being conducted, and to SHAEF, Eisenhower's staff planning Operation *Overlord*. On 30 January 1944, an Enigma dispatch revealed Hitler's opinion that the landing at Anzio was a prelude to an invasion of Europe within the same year. During February, BP was breaking dispatches exchanged by German ships and aircraft conducting anti-invasion exercises at the mouth of the Gironde and in the Brest and Cherbourg area. March brought more interesting news: the Japanese ambassador to Vichy reported to Tokyo that, in view of the stalemate on the front at Cassino, the Germans considered an Allied invasion of southern France possible and were redeploying four infantry divisions and two armored divisions to the south. In the weeks that followed, the decrypts provided a good deal of chaotic opinions, the common tone of which was the belief that an invasion was not an imminent threat due to the Allied shipping shortages. But as early as 6 April, *Fish* delivered an opinion from von Rundstedt's staff to the effect that Allied preparations for an invasion 'were essentially complete' and that its target would be the northwest coast of France. Seemingly little was added to the matter by another dispatch from the Japanese naval attaché in Berlin, who, following a conversation with the Kriegsmarine's chief of operations, Admiral Meisel, reported that the target of the invasion would be northern France, although a simultaneous landing in the Bay of Biscay or southern France, accompanying commando operations in Norway, and possibly a landing in Greece to seize airbases, was to be expected. Admiral Meisel's opinion is a good illustration of the confusion that prevailed in the German staffs at the time. The consensus was that the main thrust would be directed towards the northern coasts of France, although there was no consensus as to the closer definition of the likely landing site: locations ranging from the Belgian coast to Brittany had supporters. The most partisan view was that the Allies would strike where the width of the Channel reduced both the risk of transporting troops across the sea and provided effective air cover—in the Pas de Calais. This group included Hitler himself, whose vote was decisive. But from the beginning of the war, the Führer had also been concerned about German positions in Norway, the control of which was to become particularly important to the Allies with the German attack on Soviet Russia: the ships and aircraft based in the Norwegian fjords were an obstacle to effective communication between the Allies. Knowing of Hitler's prejudice, LCS planners divided Operation Fortitude into two parts: Fortitude North and Fortitude South. The aim of the first one was to convince the Germans that the Allied forces assembled in Britain were preparing for the invasion of Norway, which would precede in time the landing in France. The second stage, Fortitude South, was to enter the decisive phase after the landing of Allied troops in France and to keep the Germans convinced that the Normandy landings were only a diversion before the actual strike, whose target would be the Pas de Calais coast.

We now know that the Fortitude Plan had the expected effect on the German command, drawing numerous elite divisions away from the field of the coming battle in Normandy. But over the course of April and May 1944, Ultra provided as much evidence of its effectiveness as it did alarm signals that the Germans had seen through the Allied game and were preparing a hot welcome for them exactly where Eisenhower and his staff had decided to land their divisions. In early May, the 3rd Luftflotte sent a report to the command, which included the following statement: 'In the light of the facts described, the repeatedly expressed view of the 3rd Luftflotte, according to which the landing is planned between Le Havre and Cherbourg, has been reaffirmed.' Five days later, the codebreakers broke von Rundstetdt's message, in which he had included an ambiguous opinion. He stated that 'the most endangered sector appears to be the section approximately from Boulogne to Normandy inclusive,' after which he reassured staffers concerned about this 'inclusiveness' by stating that 'given the importance to the Allies of the seizure of Cherbourg and Le Havre, Normandy and probably Brittany as well could become the object of strong air attacks.' His assessment seemed reassuring, but immediately afterwards BP received a series of dispatches reporting on the strengthening of defenses in Normandy. Between May 14 and 27, the experienced 21st Panzer Division was redeployed to the Caen area. To make matters worse, fresh divisions were garrisoned in the area where the plan for Operation Overlord called for the airdrop of American airborn divisions. The last-minute redeployment of the American drop zones represented the first direct Ultra contribution to the success of the invasion. Less than a week before the start of the operation, BP read a message from the Japanese ambassador, reporting to his government on a conversation with Hitler that took place on May 27. One can assume that it allowed the LCS team to breathe a sigh of relief: Hitler's statement sounded as if he was foretelling a well-learned lesson, composed of disinformation planted in his mind over the past months. Ambassador Oshima reported that Hitler expected diversionary attacks in Norway, Denmark, the Côte d'Azur and Brittany, which would be followed by a landing in Normandy or Brittany. The success of this operation was to determine the start of the real invasion, which the Führer was locating in the Pas de Calais.

Amphibious operations in the places and numbers envisioned by Hitler would have required the participation of troops far more numerous than the Allies had gathered in Britain, and above all—ships, vessels and landing craft that even the resourceful Americans lacked. However, Hitler and his staffers were convinced that the Allies had far more divisions than there actually existed. As part of the Fortitude Plan, the entire First American Army Group (FUSAG) under General Patton was created specifically for the purpose of deception. When the troops to take part in the invasion moved to isolated concentration areas, where they waited in radio silence for embarkation, transmitters belonging to FUSAG continued to work on the eastern coasts of England, convoys of American vehicles drove along the roads, inflatable mock-ups of tanks accumulated dust in the camps, and mock-ups of landing craft rocked on the water in nearby ports. At the same time, the Allied air force made sure that German reconnaissance planes appearing sporadically over Britain's shores saw only what had been constructed specifically for their eyes. Hitler was to believe that the Allies had enough forces to carry out two major landing operations.

Ultra was one of the key tools for planning the invasion, but when the landing craft touched the sand at dawn on June 6, 1944, and the soldiers of the first wave moved toward the beaches, even their commanders temporarily lost their influence over developments. Groups of soldiers were overcoming their fear, motivated to attack German positions by NCOs or lieutenants. Ultra could only watch their fate from a distance, but it did so effectively, thanks in part to the Germans' forced departure from the use of landline communications. In the weeks leading up to the invasion, the Allied air force made an effort to isolate the future battlefield, bombing railroad junctions in France and bridges over the Seine from Paris to its mouth. The vast majority of phone lines ran along the tracks and crossed the rivers by bridges that ceased to exist as a result of Allied bombing. Those wires that were not damaged by Allied bombs were severed by French resistance or Jedburgh sabotage teams sent to the continent before the invasion. On June 8, Hut 8 intercepted a report from the local Kriegsmarine command reporting that 'all lines to Berlin, Kiel,

Wilhemshaven, Paris, Brest, Aix and La Rochelle interrupted as a result of enemy action.' In the later phase of operations, when the war became maneuverable, most units naturally continued to use their radios. As a result, a plentiful stream of decrypts informing of enemy intentions flowed to Allied headquarters via BP. The new Enigma networks, broken on the fly, more than compensated for the loss on June 10 access to *Bream* network. On June 8 and 9, Ultra made it possible to locate the headquarters of Panzergruppe West, subsequently bombed so effectively that the group's command had to be taken over by the HQ of one of its corps. On June 12/13, Ultra's warning allowed the Americans to repel a counterattack by the 17th Panzer Grenadier Division. Immediately after the first Allied landings, Rommel was able to convince Hitler to place several armored divisions belonging to the neighboring 15th Army or the Supreme Commander's reserve at his disposal. As late as June 6, air reconnaissance confirmed that the 1st, 2nd and 12th SS Panzer Divisions, the 2nd and part of the 116th Panzer Division, the Panzer Lehr Division and the 17th SS Panzer Grenadier Division were being loaded onto trains with the obvious intention to reach the Normandy battlefield. Meanwhile, a counterattack by the only Wehrmacht panzer division present on the battlefield, the 21st Armored Division, broke through the vulnerable British defenses and would have reached the sea splitting the beachhead in two, had it not been stopped by fire from the ship's guns. Stopping on the spot, or at least delaying the German armored divisions heading toward the battlefield, was of paramount importance. At this point, the London Controlling Section intervened again. Alarmed by the air recce reports and their confirmation by Ultra, it tossed to the Germans, via double agent Garbo, the assessment that 'the Normandy landings, although undertaken on a significant scale, were only of a diversionary nature', preceding the actual attack in the Pas de Calais. This assessment reached Hitler's headquarters in the early morning hours of June 10, and Hitler immediately cancelled his approval to redeploy the tanks to Normandy, ordering them to remain in the Pas de Calais area. The scale of success of Operation Fortitude surprised the plan's authors themselves: for a long and crucial two weeks, the Germans did not reach for reserves whose presence in Normandy could have changed the fate of the campaign.

Ultra was to play an even more prominent role in this phase of the battle, which unfolded after the Allies had consolidated their beachhead and gathered enough forces to go into the offensive. Eisenhower's original plan was modified during the fighting as the situation developed. The British were expected to capture Caen, an important transportation hub, on the first day of the invasion. However, the Germans were aware of the importance of the city, concentrating most of their armored forces for its defense. Observing through Ultra the concentration of German armored divisions opposite the British, Eisenhower decided to change plans. Monty was henceforth to continue local offensives in the Caen area, more in order to force the enemy to commit more troops than to make a breakthrough. Meanwhile, the Americans on the western section of the front were to chop their way out of the hedgerows and into terrain favorable to the use of armored divisions. Having achieved this goal and taking advantage of the absence of enemy tanks, they were to break out of the beachhead and start two armored charges: toward the ports of Brittany and toward the

Seine. Monty's assaults in the Caen area served their purpose—they drew the bulk of the German armored forces to his section of the front. Meanwhile, General Bradley's American 1st Army broke through to the base of the Normandy peninsula near Avranches. Within a week Patton was able to abandon the masquerade as commander of the fictitious FUSAG and take command of the 3rd Army, at the head of which he moved through virtually undefended terrain toward the Seine. Even so, the offensive's objectives remained limited. The original plan for the invasion assumed that after breaking out of the beachhead, the Allies would, between D+120 and D+330 (almost a year after the Normandy landings) reach and stabilize a line running slightly south of the Loire, along the Meuse to the mouth of the Rhine. In a conversation with Churchill before the invasion began, Eisenhower showed a good deal of optimism, assuring the prime minister that 'the coming winter would see the Allies on Germany's borders.' Churchill, however, stuck to his plans, tempering the C-in-C's enthusiasm: 'I commend your enthusiasm, but liberate Paris before Christmas, we require nothing more from you.'

Patton set out on his raid towards the Seine on 1 August. On 3 August Hitler's order was intercepted at BP, changing all campaign plans and making a decisive contribution to the fulfillment of Eisenhower's optimistic forecast. Watching on the maps as successive divisions of the American 3rd Army poured through a breach several kilometers wide near Avranches, Hitler saw encouragement where his commanders saw the specter of doom. During the 1940 French campaign, Guderian's and Rommel's divisions were ahead of any plans the Allies made, not to mention actual troop movements. In 1944, however, German divisions were far less mobile than their Allied counterparts: the situation dictated that all units be withdrawn as urgently as possible from the Normandy front and attempt to reconstitute a front line on the Seine. Hitler, however, thought that he would be able to change the fate of the campaign with one brilliant maneuver. On 3 August he ordered the withdrawal of armored divisions opposing Monty, forming them into an armored fist that would set off from Mortain and reach the sea in Avranches, cutting off and then destroying Patton's divisions. The Führer's order represented such incredible bravado or stupidity that Eisenhower's deputy personally called Winterbotham to make sure there was no mistake. Marshal von Kluge, who had in the meantime replaced von Rundstedt as Wehrmacht commander in western Europe, tried to convince the Führer to give up an operation he considered as suicidal. The memoirs of Allied commanders were mostly written down before it was possible to talk openly about Ultra and its impact on the fate of the war. However, those who lived to see the moment recalled the excitement in Allied headquarters, where the discussions between Hitler and his commanders were being watched. In the end, on the night of August 6–7, an onslaught of several German panzer divisions moved in. It encountered the tough defense of the American 30th Infantry Division, whose regiments fought in encirclement for some time, supplied from the air. At dawn, aircraft of the 9th Army Tactical Air Force appeared over the battlefield, whose attacks in the early afternoon stopped the German advance. Meanwhile, a race was underway on both wings of the advancing German divisions toward their rear. Patton directed his divisions to the northeast, toward Argentan. Montgomery, aware that few armored units remained in his section, moved toward Falaise in an attempt to close the pocket in which the German 7th Army was stuck. The race would probably have yielded better results had the rivalry between Monty and Patton not surfaced again. Monty led the chase in his methodical style. Irritated Patton at one point asked Eisenhower for orders to advance north, promising to throw the Germans and British into the sea together, repeating the Dunkirk for the latter. Finally, on August 14/15, the exit from the Falaise packet was sealed by the Polish 1st Armored Division. Hitler proclaimed that August 15, 1944, was the worst day of his life. The Germans had lost much of their 7th Army, had to withdraw from almost all of France and Belgium, and to abandon much of the Netherlands.

The march of Patton's 3rd Army through France was one of the best examples of the role that timely delivered and intelligently used Ultra decrypts could play in Allied operations. Patton's army launched its raid on 1 August. It did not meet the American forces advancing from the south of France, as part of Operation *Dragoon*, until 10 September. For more than a month, Patton's troops charged with their right wing completely exposed, with an absolute disregard for the fundamental principles of military strategy. Patton's command style, based strictly on Ultra's decrypts, raised doubts or even protests from his subordinate commanders. General Irwin, who commanded the 5th Division, was irritated that the orders arriving from Army HQ made no sense, forcing him to constantly change the direction of the attack, without any justification and, above all, without contact with the enemy. Irwin did not know that Patton was deliberately avoiding the centers of German resistance identified by the decodes, in an attempt to move his divisions as quickly as possible to the deep rear of the German front. When General Manton Eddy, commanding his right-wing XII Corps, drew his superior's attention to the information provided by his own staff, according to which 'I have 90.000 Germans on my right wing, 80.000 on my left, and I have only four divisions at my disposal,' Patton bristled at the question in his style: 'Ignore the bastards and roll forward'. Major Melvin C. Helfers, who served on Patton's staff in a role analogous to British SLU officers, shared his opinion years later: '[Patton] knew from Ultra that there was no threat to his southern flank north of the Loire'. The march of his army through hostile territory without flank protection was unprecedented in the history of wars'. In fact, south of the river the Germans had troops with a total strength of about 36.000 men that could have been thrown into an attack on the flank of Patton's troops and disrupt the Third Army's march east. From Ultra dispatches, however, Patton knew that General Erich Elster, commanding these forces, had been ordered to bring them to the Belfort area. This meant that his troops would be racing eastward, marching parallel to Patton's axis of advance and trying to avoid being locked in a cauldron. The Germans systematically informed von Kluge's staff of the points reached and the directions of the march for the next few days. Their reports went not only to Patton's staff, but also to General Otto P. Weyland, commanding the 19th Tactical Air Army assigned to Patton's forces. The August weather favored the operations of American aviation, which systematically destroyed bridges, railroads and roads on the route of German troops. Finally, on September 9 General Elster sent liaison officers to the nearest American outpost with an offer to surrender his troops on condition that

American aircraft allow them to reach the site of the planned surrender unharmed. The Germans not only demanded the presence of the air commander at the surrender ceremony. The highest-ranking American commander on the ground was General Robert C. Macon, commanding the 83rd Infantry Division. General Elster, however, violated protocol by surrendering his weapons to General Weyland. In doing so, he indirectly paid honor to the Allied codebreakers whose work was hiding behind the precision of the air force's strikes.

As announced by Eisenhower, the Allied armies stood on Germany's borders well before the winter. But before they moved the war into the enemy territory, several bloody battles came to be fought. The earliest and most controversial has been remembered over time as the operation that went one bridge too far. When the Allied armies reached the borders of Germany in early September, and entered Belgium and the Netherlands in the north, it became obvious that they had surpassed their own success. Supplies flowing from the distant beaches of Normandy in an increasingly drying stream were not enough to sustain the pace of the offensive, even threatening to stop it. Potential pause in hostilities meant time for the German armies to reorganize and replenish equipment from centers in the Ruhr. There was only one way out of the deadlock: one arm of the advance must be halted and all available fuel and ammunition diverted to the other. Both Patton and Montgomery saw themselves in the role of that arm of the offensive to continue the march. Eisenhower was to play the role of arbiter in the dispute between the ambitions of his subordinates, and Monty assumed that the C-in-C would bet on his compatriot. As a result, he devised and presented Eisenhower with a plan that hit several of his weak points.

It can be assumed that if Montgomery, known for his devotional religiousness, had entered Eisenhower's staff wearing a ballerina skirt, with a bottle of Bourbon in hand and accompanied by a group of scantily clad young women, he would have made less of an impression than he did by presenting his blueprint for the operation. It contradicted absolutely everything that observers identified as Monty's principles of warfare. Considered a war materials general, moving forward only when he had decisive advantage in men and equipment, and even then taking care to first batter the enemy with hours of aerial and artillery bombardment, this time Montgomery presented a plan that was extremely risky, not to say adventurous. It envisaged an attack more than 100 kilometers deep into enemy territory, conducted along a corridor narrowing in places to a single road or bridge, assuming the crossing of several water obstacles along a carpet to be formed by airborne divisions. Hearing of his rival's plan, Patton must have murmured, and not without reason, that the Allied armies' supply problems would have been solved if Monty had bothered to take a look at a map before ordering his divisions to seize Antwerp. No one on his staff thought to capture the mouth of the Scheldt as well. The German 15th Army, holding the river mouth and Beveland Island, still controlled traffic on the waterway leading to Antwerp for weeks, preventing the Allies from using the port. Patton's reservations were valid, but Eisenhower was impressed by Monty's plan. After the Normandy victory, he felt that the end of the war was within reach, achievable as late as 1944. In the pursuit of the Germans, he had so far failed to take advantage of all his assets, among them the powerful Allied airborne army. SHAEF staffers had successively designed 17 operations in which it could have played a decisive role, but so far events had always overtaken the plans—before the operation could be completed, ground divisions were reaching the projected target. Monty brought Eisenhower a plan that would allow him to spectacularly demonstrate the power of Allied airborne troops and, if successful, break through the Rhine barrier, the last major natural obstacle before the Ruhr and the heart of Germany. The vision of ending the war before winter proved stronger than American solidarity and Patton's objections.

To make his vision come true, Montgomery and his staff went so far as to manipulate the contents of intelligence reports. Widely known is the story of aerial recce photos commissioned by Major Brian Urquhart, proving the presence of German armored divisions in the Arnhem area, the target point of the planned offensive. When he tried to reach the top commanders with his findings, he was ordered to take a forced leave. The plan for Operation *Market-Garden* was an expansion of the earlier, unrealized Operation *Comet*. Compared to the original plan, the number of participating airborne divisions was increased from two to three, motivated by 'the increase in German forces in the Arnhem-Nijmegen area.' This assessment was presumably based on a message broken on 5 September, reporting the departure of uncommitted units of the 2nd and 116th Panzer Divisions and the 9th and 10th SS Panzer Divisions to the Venlo-Arnhem-s'Hertogenbosch area for rest and refit. Subsequently, the two SS divisions disappeared from Ultra for several days, only to reappear in SHAEF's intelligence summary of September 16, which mentioned the withdrawal of the '9th, and with it presumably also the 10th SS Panzer Division, to the Arnhem area of the Netherlands, where they are being equipped with new tanks from the Cleves depot'. Confirmation of the presence of German tanks in the area of the planned landing came from Dutch resistance sources. The convergence of information from Ultra, air recce and the Dutch underground should have alerted the authors of the plan. These were not the only warning signals. As of September 15, the Allies knew from Ultra that Field Marshal Model's staff was installed in the immediate vicinity of the British 1st Airborne Division's drop zone, at Oosterbeek, which signalled the vigorous command of the coming battle. A 10 September decode reported that Model's Army Group B HQ ordered the 3rd Luftflotte to conduct an air recce to determine if the Allies were preparing for an airborne operation in the Aachen or Arnhem area, and again three days later to check if gliders had been spotted at airfields in Britain. In an airborne operation, the surprise is an asset that compensates for the light armament of the troops. Ultra dispatches signaled unequivocally that the enemy's surprise cannot be reached. Information from all sources converged on the desk of SHAEF's intelligence chief, General Kenneth Strong, who did not remain blind and deaf to the contents of bulletins promulgated by his own service. He managed to convince SHAEF's chief of staff, General Bedell Smith, to pay a joint visit to Montgomery's HQ and urge him to reconsider the rationale for the operation. The four eyes conversation between Montgomery and Eisenhower's chief of staff had no effect. Bedell Smith recorded in his memoirs: 'We were stuck. Montgomery simply swept my doubts aside with a smile.'

Despite its boldness or even adventurism, Operation *Market-Garden* was not a bad plan and was vigorously implemented. Thereafter Montgomery, making a good face for a bad game, stated that 90% of the operation's objectives had been achieved. The British and Polish paratroopers at Arnhem held out longer than anyone could have expected. The attack by American paratroopers across the Waal River at Nijmegen was an example of the superb combat training of the army's new elite. Tankers from the Irish Guards and infantry from the 43rd Wessex Division fought bravely breaking through and repelling German counterattacks on the front one tank wide. The plan for any battle must include a margin of uncertainty and risk. Disregarding numerous warnings from Ultra and other sources before the start of Operation *Market-Garden* widened that margin so significantly that an undertaking risky in nature turned into a precisely planned disaster. After Arnhem, both sides recognized that they were exhausted by the campaign so far that they could only conduct local operations. The Allies busied themselves organizing supply lines. The Wehrmacht frantically reorganized the armies that had emerged from France as a mass of men and equipment void of structure. Allied successes in the summer of 1944 meant that the front line was not far from German factories and arms depots. The German armies also gained another advantage: operating on their own soil, they could take advantage of a landline communications network, despite Allied bombardment. The Germans must have had unlimited confidence in the security of their ciphers, for in the last phase of the war the number of messages broken daily in BP steadily increased, reaching a maximum in March 1945. However, the ability to use the landlines was to play a key role in the late autumn and winter of 1944.

Hitler quickly regained his stamina after the defeat in France. Just four days after the critical 15 August, he announced at a staff conference his intention to regain the initiative in November, when the autumn weather would deprive the Allies of the advantage of a powerful air force. Intercepted by the Allies just before Operation *Market-Garden* orders to withdraw armored divisions from the line and move them to the rear to replenish men and equipment were the first practical manifestation of this plan. On 16 September, the Führer revealed to members of the staff that he intended to strike through the Ardennes towards Antwerp. Hitler was apparently convinced that he had found a magic way to win the war in the west. The operation he was planning was a copy of the strike that had determined the outcome of the French campaign in 1940. The stage of the German offensive was again to be the Ardennes, but this time the turn north to cut off and encircle the northern group of Allied armies was to take place earlier, after reaching the Meuse. Hitler wanted to achieve two effects at the same time: to separate the British in the north from the Americans in the south and to capture their logistics base at Antwerp. Work on the plan for the operation, which was to become code-named *Wacht am Rhein*, was carried out in a hermetic circle of staffers and field commanders, so only scraps of information were reaching the Allies. At the end of September, Ultra brought news of the withdrawal of four panzer divisions behind the Rhine with accompanying heavy tank detachments and their amalgamation into the 6th Panzer Army, with Sepp Dietrich in command. In the days that followed, information arrived about the systematic reinforcement of the new army by further divisions and its transfer to the

OKW reserve. On 18 October, Ultra explained that orders to replenish equipment of the 6th Panzer Army had come from Hitler himself. On the same day, information arrived from BP that the 5th Panzer Army had been withdrawn from the southern part of the front and subordinated to Army Group B, fighting in the north. In time, decrypts brought news of the withdrawal from combat of three elite airborne divisions. In early November, the Allies intercepted and deciphered dispatches ordering these divisions to maintain strict radio silence. In the run-up to the German offensive, the Allies managed to crack again the key to the Enigma used by German railways. This enabled them to identify numerous trains being unloaded in the Ardennes, just a dozen kilometers from their own positions held by the weak and overstretched divisions of the American 1st Army. Ultra reported on the Luftwaffe's preparations for offensive operations. Fighter aircraft stationed in the west were to be prepared urgently for fighter-bomber operations, and the use of the term 'assault operations' (*schlagartiger Einsatz*) in the order left no doubt that the aircraft were preparing for offensive tasks. From the last week of November onwards, Army Group B command regularly ordered the Luftwaffe to carry out aerial reconnaissance over Liège and the crossings of the Meuse between Givet and Liège. Moving the most experienced armored divisions out of the front line, merging them into new armies, combined with the preparation of elite infantry and airborne units, signaled the German command's intention to undertake an offensive of considerable scale. Two views emerged among the Allies regarding the likely time, place and purpose of such an operation. The first assumed that the Germans would attempt an offensive as early as November, with the aim of disorganizing and delaying the anticipated Allied advance towards the Rhine. Representatives of the second option believed that the units pulled out of the front line formed a reserve army group that would be used to launch a counter-attack should the Allied spring offensive manage to penetrate the German defenses. Among intelligence officers there was a belief that the Germans were preparing for a local offensive, but any conjecture as to its time and place was distorted by knowledge of planned Allied operations and the belief that the enemy was already incapable of attempting to regain the strategic initiative.

In mid-December, Monty asked Eisenhower's permission for a Christmas holiday at home. The American generals, whose homes were too far away, planned to spend Christmas amid the luxuries of Paris. Grounded by the December weather, Allied reconnaissance aircraft remained at airfields. In the Ardennes, a frail screen of US divisions recovering after heavy fighting, or the green units fresh from US training camps held the front. Strict radio silence, the transport of troops and supplies at night, the preparation of operations on home territory, the proximity of supply bases, the winter weather limiting the use of aerial reconnaissance—all combined to take the Allies completely by surprise when, on the morning of 16 December, hundreds of German tanks did not so much break through the American lines as drive past completely surprised American soldiers, moving towards the Meuse and Antwerp. The course of subsequent operations confirmed the opinions of those Allied intelligence officers who maintained that Hitler could not afford to undertake a strategic offensive. German tanks went into the attack with a meagre supply of fuel and the hope that it could be replenished in captured Allied depots. The Americans, however, took care to remove fuel stocks from the range of the German offensive or destroy those that could not be evacuated. The German armored grenadiers and paratroopers thought that the American soldiers would surrender at the mere sight of them. However, the Americans, after the initial surprise, spontaneously joined into groups, occupied key points and defending their positions as fiercely as effectively. When, after several days of fighting in the midst of a snowstorm, the sun came out, hundreds of Allied aircraft appeared in the sky, preparing the German tanks for the same fate as at Mortain and Avranches. Although the battle was decided, it was not until the end of January 1945 that the Allies managed to push the German divisions back to their starting point.

Once the situation on the front was stabilized, the hunt for those responsible for the initial failure has begun in Allied headquarters. Many Ultra-initiated staff members attributed the surprise to its silence on the impending German operation. The echoes of these opinions must have reached the codebreakers at BP, for they undertook an investigation on their own to find out whether a more categorical warning could have been formulated on the basis of the broken dispatches? The results of this investigation were ambiguous. No explicit reference to the time, place and purpose of the operation was found in any of the German dispatches. At the same time, the codebreakers confirmed that, knowing later developments, they were able to identify numerous dispatches which, considered in a wider context, should have alerted Allied intelligence officers and provided sufficient indications to work out the enemy's intentions. This is where the BP's conclusions end, but they need to be supplemented by observations made by later commentators. Analyzing the work of Allied intelligence prior to the Battle of the Ardennes, military historians have pointed out that SHAEF intelligence bulletins during this period focused more on the Allies' own plans than on enemy's intentions. Despite the failure of Operation *Market-Garden* and the need to fight a winter campaign on Germany's borders, the Allied HQ viewed the opponent as defeated. The generally accepted assumption that the reserve army group created by the OKW would be used for a counter-attack during the Allied spring offensive completely obscured other options and meant that any information pointing to the contrary was dismissed just as it had been during the Arnhem operation.

While operations at Arnhem and the German offensive in the Ardennes marked the twilight of Ultra's role in ground operations in Europe, it retained some importance in the war at sea. After the first defeat of the wolf packs in May 1943, Dönitz planned a rapid return of his ships to the Atlantic, equipped with new weapons and equipment. However, between May and the autumn, two significant developments influenced the struggle the codebreakers. In June, the Allies abandoned the use of the utterly discredited Naval Cypher No. 3, blinding the B-Dienst. In the autumn of 1943, Tranow's team briefly gained a glimmer of hope cracking a handful of dispatches in the new code, but the break-in proved to be an episode without sequel: for the rest of the war, the Allied convoy code resisted German attacks. At the same time, in the summer of 1943, Joe Desch came to grips with the vagaries of the machine he had created, and NCR began serial production of Type 530 bombes. The nearly one hundred devices built by the end of the year marked a breakthrough in the attack on the *Shark* network keys. When the technical capabilities of the American bombes were combined with British experience in selecting effective cribs, breaking the M4 Enigma became nearly a routine activity. *Banburismus* became the first casualty. New bombes permitted to break the key analyzing all the rotor orders, before the banburists had managed to assemble a packet of dispatches allowing them to proceed. As a result, *banburismus* was abandoned in September. 13 October 1943 was the last day during the war when the daily key of the *Shark* network failed to be broken. From that point until the victory over the Third Reich, the chronicle of the secret struggle with the ciphers of the naval Enigma resembles more closely a management report of an industrial corporation than a battlefield dispatch. In the early months of 1944, the effectiveness of US bombes caused the entire struggle with the four-rotor Enigma ciphers to be shifted to Washington. Wenger, Engstrom and Meader were able to write with satisfaction in the report that 'in the original discussions the US Navy was only to assist the GC&CS in grappling with German naval ciphers. Now practically the whole problem is being solved with US Navy bombes'. American codebreakers were gaining increasing knowledge of the subtleties of German ciphers and, with fresh eyes, were seeing features having escaped the attention of their British colleagues for years. The Americans compiled statistical summaries of the keys they had broken in the past and, around February 1944, noticed regularities in the settings of the Enigma rotors in force. The most important among the rules discovered was that never did the same rotor remain in the same position on a subsequent day, and that the that rotor selection was made from 8 groups determined by the right rotor, selecting a rotor from as far apart a group as possible on consecutive days. If, for example, on a given day the left position was occupied by rotor II, on the following day there was a negligible probability of rotor I or III being used in the same position, while the probability of selecting rotors VI, VII or VIII increased. The observations of the Americans, collectively referred to as the rotor order rules, represented a new and more effective equivalent of *banburismus*.

Appreciating the experience of the British with regard to routinely broken keys, the Americans concentrated their interest on systems that BP had hitherto treated as less important, abandoning attempts to break them. Several of these proved to be relatively easy to break and provided interesting information. The first cipher to be broken was *Sunfish*, which the Americans cracked in August 1943. In September, *Seahorse* was broken, which was of interest to the Americans because it was used for communication between the German naval attaches, particularly between Berlin and Tokyo. It turned out that both networks used a repeated message key system, vulnerable to attack by methods developed still by Polish codebreakers, the only difference being the use of a four-rotor Enigma in the *Seahorse* network. It is interesting to note that *Sunfish* retained the repeated message key until the end of the war, providing a somewhat ironic commentary on the British abandonment of the methods developed by Rejewski. Op-20-G decided to make appropriate modifications to several copies of the bombe, essentially reverting to Rejewski's original version of the device. The decision proved to be forward-looking, as the following months saw an unexpected renaissance of ciphers based on the this system, including several 3-rotor ciphers. In February 1944 *Porpoise*, a veteran among BP's broken ciphers with a repeated message key, switched to 4-rotor but retained its previous structure. It soon became irrelevant, however, when the Allies seized all the stations providing material from this network. In April, the next representatives of the group appeared: *Trumpeter* and *Bonito*. It is difficult to identify the reason for a phenomenon equivalent to weakening of the security of German communications. Three-rotor ciphers and ciphers based on repeated message key were used mostly in training networks and in theatres of war of minor importance (e.g. *Grampus* in the Black Sea). It can be speculated that the German communications security services were primarily concerned with securing networks transmitting operationally significant information, using a standard version of the cipher machine for training purposes and accepting the use of procedures adequate to the capabilities of inexperienced cipher clerks. There was a danger in the policy of graded network security—many of the most important German communications networks were being breached by the Allies using *kisses*—messages with identical open text, encrypted in two different keys. Alexander assessed that 'Bonito was another striking example of a network that we would never have solved if it had not been for its use of the repeated message key system'.

In the summer of 1943, *Shark* keys were mostly broken on *kisses* from *Dolphin*. The downside of this approach was the inevitable delay in reading the messages of both networks. As a rule, it took *Hut 8* codebreakers about 48 hours to break the key of the three-rotor *Dolphin*. Only after this time could an attempt be made to crack the *Shark* key as well, and even this assuming the presence of *kisses* from *Dolphin*. *Hut 8* knew other potential cribs, but these however appeared in too many variants. One of these was a daily report from the weather station at Boulogne, which relayed the weather forecast for the Bay of Biscay. By October 1943, the *Shark* key had been broken only once on its basis. But at the end of the month, the weather forecast unexpectedly took a constant form: the report regularly began with the phrase **WETTERVORHERSAGEBISKAYA**. Allied codebreakers could hardly believe their eyes as they watched day after day and month after month the bombes breaking the *Shark* key based on the same menu. In January 1944, they experienced moments of unbearable tension when they decoded a reprimand addressed to the sender of the forecast. The German radio security service pointed out that 'the message for the Biscaya is broadcast at almost the same time every day, the greater part of the text remains constant, the length and beginning of the message is also constant, which is then repeated on the Irish frequencies'. The German weather station staff took only the last element of criticism to heart and began broadcasting the message intended for U-Boots in the Irish area in a different, though still constant, form. However, they did not change the content of the message for the Bay of Biscay, making breaking the *Shark* a rather dull job. Op-20-G did not find enough challenges of Kriegsmarine ciphers to keep their bombes busy in 1944. Travis and Wenger entered into a tacit agreement whereby the British would send ground force and air force Enigma cipher menus to Washington (such menus shared the code name *Bovril*). Wenger's agreement solved a significant BP problem. In the run-up to the Normandy landings, the Germans had multiplied 3-rotor keys, stretching British bombe schedules to the limit. An American four-rotor bombe took only about 50 seconds to run a three-rotor menu. The goodwill towards the British caused a bit of a row in the American camp. When the SIS codebreakers learned of their naval colleagues' deal with BP, they began to murmur unkindly about the Navy's intrusion into the Army's exclusive area of interest. Wegner had to explain to his colleagues that American boys would also be landing on the beaches of Normandy, which deserved a good information on the enemy regardless of who provides it.

Largely thanks to Ultra, the Allies did not allow themselves to lose the advantage they had won in mid-1943. When the wolf packs returned to the convoy routes in the autumn of that year, their captains saw how much the war at sea had changed over the previous months. Most convoys continued to successfully evade their patrol lines, but in October the OIC decided to make an experimental change in tactics, which was tried on convoy SC143. When the British became aware of a wolf pack threatening the convoy, they reinforced its escorts by transferring additional ships from another convoy, then deliberately directed SC143 towards the enemy. The clash that ensued saw the result close to a draw: three sunk U-Boats at the cost of one merchantman and one escort. Dönitz did not know that the purpose of the maneuver was diverting his attention from a large and poorly defended convoy sailing nearby. The following January, he sent some 20 U-boats into the waters south of Ireland, providing them with strong and constant aerial reconnaissance. Despite this Ultra allowed the OIC to run 12 convoys in succession past the U-boat patrol line. The balance of the operation was depressing: at the cost of 11 submarines lost, the Germans sank one escort, one straggler merchantman and shot down two aircraft. The Kriegsmarine commander quietly admitted his tactics had suffered the ultimate defeat. The operations of January and February 1944 were the latest instance of the use of wolf pack tactics.

Hitler ordered the large surface ships having survived the losses of the early years of the war to the Norwegian fjords, where they could play several roles at the same time. Given his constant fears of an Allied invasion of Norway, the *Tirpitz*, *Scharnhost* and *Lützow* posed a threat to any invasion fleet that might appear off the Norwegian coast. Earlier forays by German battleships and cruisers into the Atlantic had forced the Royal Navy to maintain a considerable force in Scapa Flow. Besides, German ships did not have to go out into the Atlantic at all to threaten Allied shipping. A route of convoys delivering aid to Soviet Russia led along the Norwegian coast. The horrors faced by the participants of the Atlantic convoys paled in comparison to those common for the Arctic ones. Hitler insisted on maintaining a sizable flotilla of U-Boots in Norwegian waters as a precaution against an attempted Allied invasion. In the Atlantic, Dönitz's ships suffered from a lack of aerial reconnaissance, while the waters in which the arctic convoys sailed lay within range of German aircraft taking off from bases in northern Norway. Not confined to reconnaissance, they frequently and effectively bombed or torpedoed convoy ships. Behind the horizon lay the constant threat of the enemy's capital ships. Should *Tirpitz* or *Scharnhorst* managed to slip out of Altafiord unnoticed, the close escort of the convoy would be helpless against the power of their guns. The only source that could bring timely warning of German capital ships leaving their bases was Ultra.

The Allies never cracked the key of the network used by the Kriegsmarine to communicate with its capital ships. However, every operation by a *Tirpitz*, *Scharnhorst* or other large ship triggered an avalanche of dispatches on other communications networks. Luftwaffe networks brought information about the aerial reconnaissance being carried out of the areas of interest. Battleships and cruisers were escorted by destroyers, which received their orders in the *Dolphin* network. German submarines operating on their planned route were warned of the presence of German surface ships in their patrol zones. Even if Ultra dispatches did not clarify the enemy's intentions, they served as a warning. Ultra's crucial limitation in its role as guardian of the safety of Arctic convoys was the time required to break the key. In at least one case it proved to be too long to allow the Admiralty to make a cool assessment of the situation. This ended in the premature dispersal of convoy PQ-17, which was in effect massacred by combined U-boot and air attacks.

But, as if in revenge, Ultra also determined the fate of Germany's largest ships over time. On 22 October 1943, British midget submarines managed to damage *Tirpitz*, and Ultra confirmed that it would take nearly six months to repair the damage. In January 1944, another Ultra signal brought a warning that *Tirpitz* would return to service on 15 March. *Lützow* was transferred to bases in the Baltics at the same time, so temporarily the *Scharnhorst* remained the only threat in northern waters. In December 1943 Ultra reported her put to sea, where she was intercepted by a Royal Navy team under Admiral Fraser and sunk by gunfire from the battleship *Duke of York*. Until the time *Tirpitz* returned to service, the threat to the arctic convoys from the capital German ships had temporarily ceased. In late March/early April 1944, the Royal Navy planned Operation *Tungsten*, in which aircraft taking off from three aircraft carriers were to bombard *Tirpitz* undergoing technical trials at Kaafiord. As the carriers were on their way to their positions a dispatch was broken at BP announcing the battleship's planned departure to sea for speed trials. The information was relayed in time for the carriers to increase speed and get within effective striking distance at the right time. On the morning of 3 April, Allied aircraft came over the fjord just as *Tirpitz* was weighing anchor and damaged the ship again. Soon Enigma decrypts brought confirmation of both the nature and extent of the damage and the expected repair time. The hunt for the *Tirpitz* dragged on into the autumn. Attacks by aircraft taking off from carriers in July and August 1944 were unsuccessful, but in September the British were able to persuade the Russians to grant RAF aircraft permission to land at bases in the Russian north. Taking off from Russia, British aircraft again severely damaged the ship. Originally, Utlra decrypts indicated that the battered *Tirpitz* would remain in Altafiord, but in October another dispatch brought word that she would be moved to Tromsö, where she would undergo necessary repairs. She came thus within range of bombers taking off from the Shetland Islands, which carried out an attack on 12 November. Codebreakers at *Hut 8* waited impatiently for German dispatches to be transmitted after the bombing to assess the impact of the attack. The first message intercepted was transmitted in an officer's variant of the key, which had not yet been broken. Someone in *Hut 8*, with a fair amount of optimism, suggested that the phrase `TIRPITZXGEKENTERT` (Tirpitz capsized) be used as a crib. To everyone's surprise, the menu based on the proposed crib yielded the correct solution: the threat from capital ships to the arctic convoys had finally been eliminated.

At least, this is what the hunt for *Tirpitz* looked like as observed through the eyes of BP codebreakers and the recipients of their reports in the Royal Navy. Its story illustrates one of the key limitations of the Ultra system, and of radio intelligence in general: even intercepting and deciphering all the enemy's dispatches never gives a complete picture of the situation. Some decisions are made and communicated via communications networks outside the range of the interception service. This is exactly what happened with *Tirpitz*. The ship's transfer to Tromsö was not the result of an intention to make repairs there that would allow her to go to sea again and threaten Allied convoys. When Admiral Dönitz took over from Erich Raeder's at the head of the Kriegsmarine, he accepted without resistance Hitler's thesis that the heyday of the mighty battleships was irretrievably over. As a result, he decided to relegate the hitherto pride of the German navy to the role of a floating artillery battery, protecting the Norwegian coast from Hitler's eternal bane—the Allied landing. Part of her crew were removed from the ship, leaving the gunners, and she was anchored in waters shallow enough that even damage below the waterline would allow her to settle to the bottom and continue to use her powerful guns. However, the sea where the ship was moored proved to be deep enough that it managed to capsize after being hit by bombs, becoming a trap for the crew.

The *Tirpitz* was not the last weapon of this name the Allies were to counter. Following Japan's entry into the war, Germany signed a new military alliance with this country on 18 January 1942. It was followed by a naval communications agreement, signed on 11 September. The broken dispatches of the *Seahorse* network confirmed both sides planned to secure the exchange of information using a new variant of the German cipher machine, the so-called T-Enigma, bearing the designation Tirpitz (transcribed as Tirupitsu by the Japanese). The new machine was also to take over as the main cipher machine in the IJN, which placed an order for 800 copies of the device. A separate key list, designated *Gartenzaun*, was developed for communication between the navies of the two countries. On 22 August 1942, the Japanese submarine I-30 left the Kriegsmarine base at Lorient with 50 Tirpitz copies on board. It had only managed to hand over 16 to the IJN units it encountered along the way, when it ran into a British mine in the Singapore area and sank with all its cargo. The copies that reached Japan must have served as models for the start of local production—two of the Japanese copies of the Tirpitz survived the war. Shortly afterwards, the Japanese informed Berlin that they were cancelling the order and the right to manufacture machines. It is possible that the initiative to abandon T-Enigma production came from the Germans themselves. Growing material problems prevented them from completing the order quickly; moreover, German cryptologists had concluded that the Tirpitz was not a secure system with the high volume of information exchanged. On 15 November 1943, the Germans informed the Japanese of their intention to design another variant of the machine, combining higher security with the use of more readily available materials—thus the A-Enigma concept was born. The T-Enigma was a 4-rotor machine without a plugboard. Its security was based on the rotor design, equipped with five notches causing an irregular movement of the rotors. The Tirpitz entry drum connections also differed from the standard Enigma—they had a random structure. Judging from the references in the deciphered exchange, the design of the A-Enigma was intended to copy the standard machines used by the Kriegsmarine, with different rotor connections and rotor stepping mechanism. By the time the new German proposal was presented, the Japanese must have already produced a number of Tirpitz copies, as they were concerned about the cooperation between the T-type and A-type machines. The reassuring German response suggested that the A-Enigma, with no plug connections, was functionally identical to the Tirpitz. The first batch of 10 new machines arrived in Japan on 16 April 1944 aboard the submarine I-29, along with prototypes of radar detection and jamming equipment and sonar decoys. Subsequent deliveries were less fortunate, and after the Normandy invasion the Allies found dozens of T-series machines awaiting shipment in Kriegsmarine warehouses near Lorient. The effectiveness of the Allied blockade of the sea routes between Germany and Japan was a fortuitous circumstance for the course of the war in the Pacific. Japan's use of new cipher machine would have outdated the methods developed by Op-20-G for breaking Japanese naval codes. At the same time, the cipher breaking methods known from the European theatre of war would might proved ineffective against a new version of the machine.

In the eyes of *Hut 8* codebreakers, the last year of the war was dominated by the feverish activity of German cipher services. Frequent changes in operating procedures, splitting of communications networks and the introduction of individual keys gave the impression that German signals service wanted to contribute at all costs to the strengthening of the creaking fronts. In June 1944, *Porpoise* and *Grampus* gave up the repeated message key, with *Bonito* following in September. *Sunfish* remained the only network using this system until the end of the war, but as it passed into American hands, from September '44 BP was breaking all the keys exclusively on cribs. By the time the repeated key system vanished, it had allowed the *Porpoise*, *Grampus*, *Trumpeter*, *Sunfish*, *Seahorse* and *Bonito* networks to be broken in succession, and facilitated the breaking of several other keys on the basis of *kisses*. In September 1944, a Luftwaffe aircraft dropped on the positions of Americans besieging Brest a parcel intended for its defenders and containing the keys to the *Dolphin* network for October. This error forced the German services to improvise, whereby *Dolphin* was split into three sub-networks using different keys. The resulting inconvenience to the Allies was largely offset by sending to the German garrison in the Channel Islands *Dolphin*'s settings for October encrypted with the September key. It can be assumed that problems with key distribution in areas where German influence was rapidly fading were the reason for the split of the *Porpoise* network, from which *Catfish* (for the Aegean area) and *Bloater* (for the Asian waters) were branched off.

In November, the German service implemented several innovations simultaneously. Firstly, *Dolphin* and *Plaice* started to use a 4-rotor machine, requiring to recover new bigram tables. The traditional method based on the **EINS** catalogue was impractical for the 4-rotor machine. Machines previously developed at BP (Jones' Dudbuster) and Op-20-G (Grenade), testing the message text for the presence of encrypted versions of the **EINS** and message urgency markers (**KRKRs**), also proved too slow. Faced with the new challenge, BP developed a modified version of the bombe, called *Fillibuster*, which simultaneously checked for the presence of the **EINS** phrase and several versions of the urgency marker—**KRKR, BINE, NUKE** and **WESPE**—allowing the rapid reconstruction of new bigram tables.

The second of the innovations implemented in November was the proliferation of a system rarely used in the past, which involved providing the ships and isolated garrisons with individual keys (Sonderschlüssel). Ciphers with individual keys were broken occasionally. In January, *Shark* dispatches using an individual key were broken twice (thanks to *kiss* from *Limpet*); in April, Op-20-G broke the *Sonder-161* based on *kisses* from *Sucker*. Individual keys posed a serious challenge to the Allies: breaking *Sonder-161* cost 5300 bombe hours and took six weeks. Since the Kriegsmarine began assigning individual keys to virtually all operational U-boots in February 1945, the Allies had only superficial information about the enemy's underwater activities until the end of the war. The counterpart to the Sonderschlüssel used in parallel by the Wehrmacht was the so-called emergency key (Notschlüssel). German ground troops were increasingly faced with the problem of communicating with garrisons or field units that had been cut off from their own lines and there was no way of safely delivering new keys after the old ones expired. The Notschlüssel procedure allowed complete machine settings to be constructed from a long word or phrase transmitted in open text or in another cipher system.

The third news of the month was the replacement of *Bonito* by the 3-rotor *Bounce* key, designed for German forces in northern Italy. Somewhat surprisingly, the evolution of the cipher gradually extended its coverage to all river areas in Europe, and in this role it received its moment of glory during the battle around the Rhine bridge at Remagen. Among the minor changes to cipher systems, the Germans managed to prepare a significant surprise before the end of the war. In April 1945, the *Plaice* network discontinued the use of a single Enigma base setting for an entire day, allowing the operator to select one of 288 settings and transmit it to the receiving side in the form of a fixed trigram. The recently developed Fillibuster proved useful in this case too, allowing the trigram table to be reconstructed, but this was changed again in May to include *Dolphin*. Such frequent changes could have presented Allied cryptologists with a serious obstacle to breaking the dispatches in a reasonable time, but amidst the chaos of the final phase of the war, the new system did not manage to cover a significant part of the Kriegsmarine's communications. Despite all the setbacks, *Hut 8*, which processed an average of 460 messages per day in 1941 and 1942, regularly increased its output to 980 messages in 1943, 1560 in 1944 and 1790 in 1945 (these figures do not include *Shark* messages and other keys broken by Op-20-G).

From the autumn of 1943, Enigma, like the German armies on all fronts, was on the defensive. The Allied codebreakers could look back with satisfaction at their achievements to date and with hope for the future. Fundamental theoretical problems had been solved. Recent graduates and students of mathematics, who had

matured into full-blooded codebreakers within a year, were gradually being redeployed to tasks unrelated to Enigma. Since the abandonment of *banburismus*, the
system they designed for breaking Enigma ciphers required industrial planning of
resources dedicated to different keys rather than solving new theoretical problems.
Hut 8, which encountered the most numerous challenges during the first period of
the war, employed a dozen codebreakers at its peak; by August 1944, only four were
working in the *sahib* room, as the room in which the section's collective brain functioned, and on the day the war ended, the then head of the section, Patrick Mahon,
and Joan Clarke were toiling there alone. However, that somewhat sleepy landscape
of BP sections toiling in the final days of the war to attack Enigma ciphers was
misleading. In the light of everything we know about the intentions and designs of
the Third Reich's cryptologists, they had the resources to be able to suddenly revi-
talize the atmosphere of the center, and the Allied cryptologists were fully aware
that any moment now they might be presented with a challenge that would force
them to the utmost or even outstrip them.

That none of the catastrophic scenarios came to pass is a mystery of equal measure to that of Enigma itself. Since the secret of the struggle with the Enigma ciphers
came to light many years after the war, historians, cryptologists and amateurs have
been puzzling over the reasons for the German failure and the Allied success in the
cryptologic race. If, despite decades of study, no single, universally accepted cause
has yet been found, there were probably more. Here are a few candidates for
this role.

Among contemporary historians and cryptologists alike, the common opinion is
that the most important factor of the strategic failure of the Third Reich in the cryptological race was the dispersion of competences among several independent cryptological services. Not only did these services operate independently of each other,
without a centrally coordinated plan of research and action. The nature and power
structure of the Third Reich presupposed their mutual competition. In the first half
of the twentieth century, cryptology did not function as an open field of knowledge
that could be studied independently and beyond the isolated world of military,
diplomacy, and secret services. All of these services maintained the utmost secrecy
concerning both the ways in which they secured their own communications and the
attacks on their adversaries. The natural consequence of this practice was to mini-
mise the circle of people familiar with it. The scarce availability of know-how and
qualified and experienced staff dictated the need to concentrate it within the integrated crypto service. This solution was adopted by the services of two countries
that were later to make key contributions to the struggle with the Enigma ciphers:
Poland and Great Britain. During the Weimar period, it seemed that German cryptology was also moving in a similar direction—experienced staff was concentrated
in the crypto services of foreign ministry and the army. However, the functioning of
the Third Reich was based on different principles, involving the parallel operation
of several services with overlapping competences. This inevitably led to the dispersal of scarce competences and mutual competition between the services. Efforts to
coordinate their activities were only made late in the war. Too late and with to too
limited scope to affect the outcome of the cryptological duel.

On the other hand, virtually all the breakthroughs in the history of Enigma can be attributed to individuals, not the collective efforts of complex organizations. Mathematics was not a part of the organizational culture of Polish military intelligence when Lieutenant Maksymilian Ciężki decided to enlist the help of its adepts in grappling with the Enigma challenge. The heads of his service deserve respect; they accepted Ciężki's pioneering idea, provided the money and time necessary to implement it, without any guarantees of success. The very idea of using mathematics to attack ciphers predated the development of cryptology in other countries by at least ten years and was Ciężki's own contribution. Years later, Marian Rejewski emphasized the team nature of the early work on Enigma as one of the reasons for success. Ciężki invested several years of hard work to turn young mathematicians into professional cryptologists. In this, and this alone, his plan failed; when Rejewski undertook his earliest attack on Enigma, he was fortunately still more of a mathematician than a cryptologist. Had he attempted to attack the cipher using any of the methods learned during his cryptology course and his apprenticeship at the Poznan branch of the Cipher Bureau, he would surely have failed. When, at the end of 1932, he transformed his knowledge of the machine into mathematical equations, he was unaware that their solution meant not only a triumph over the Enigma, but also the cryptology's rite of passage into a new age of its development. He had made a theoretical breakthrough relying solely on his own mathematical knowledge, and his solitude at this stage of his work was, it seems, an essential ingredient in his success.

The baton passed on by Polish codebreakers was taken up by Alan Turing. In theory, he was doing so on behalf of British cryptology. It seems however, that Turing lived in his own world, paying only token attention to his surroundings and not much regard for their opinion. His attitude systematically deepened the sense of loneliness that was to become his nemesis over time.

The Enigma decrypts would probably not have played a significant role without Gordon Welchman's designing an organizational structure permitting to reach their potential. However, he did not do so as part of nominal occupation at BP, but rather as a result of his conflict with Knox and the resulting sense of alienation.

The most important theoretical and practical breakthroughs in the struggle with Enigma represented the achievements of individuals working in isolation, sometimes on or beyond the margins of the organizational structures, even contradicting their official policies. So maybe centralization on the one hand, and dispersal on the other was not a key success factor? Talent or genius can manifest itself just as effectively in a small group as in a service of complex structure. Moreover, a novel idea is easier to spot and develop in a small team, hence the role of start-ups in modern entrepreneurship. Crossing the barrier of anonymity is more difficult in a larger group, also, or even especially, for individuals exceptionally gifted. The laws of social dynamics are at work here, giving rise to competition, animosity and conflict. A diagnosis that reduces the success of one side and the failure of the other to the dispersal or centralization of cryptologic services seems an oversimplification.

Perhaps, then, the Allies had a special competence that the Third Reich's cryptologists lacked? Historians and cryptologists agree that the Allied success in attacks on Enigma found its beginning in two revolutions in cryptology enacted by mathematicians and engineers of Polish Cipher Bureau. The author of the first one was Marian Rejewski, who as the first in the world successfully applied mathematical approach to cryptanalysis. This approach allowed him to recreate (today we would say re-engineer), solely by mathematical inference, the design of a machine, the original of which he had never seen in his life, and to identify a feature of its cipher which allowed him to recover the key. When changes in the way the machine was used made the key recovery too time-consuming, engineers stepped in. The Poles were the first to realize that the natural adversary for the cipher machine was another machine, assisting the key breaking process. Rejewski developed the idea of his bombe, and Antoni Palluth and his colleagues at AVA factory transformed his theoretical concept into a functional device. And although the practical career of the Rejewski bombe proved to be short, it became the prototype for most of the devices constructed by the Allies during the war.

Neither the availability of talented mathematicians nor the ability to construct and manufacture sophisticated devices, however, constituted an Allied monopoly. In the interwar period, German was the official language of mathematical literature, and Marian Rejewski, after graduating from Poznan University, continued his education in Göttingen, then regarded as the global Mecca for mathematicians. Germany was one of the first countries to start employing mathematicians in cryptology agencies. Werner Kunze started working in the cryptology section of the German Foreign Ministry as early as December 1918, in time assuming the duties of head of its mathematical section. The mass recruitment of mathematicians did not start in earnest until the eve of the outbreak of war, and the process accelerated after the outbreak of war with Soviet Russia. In the light of Erich Hüttenhain's post-war memoirs, however, the sudden popularity of mathematicians was not entirely due to the recognition of their usefulness in cryptology. Rather, it was an expression of professional solidarity; mathematicians involved in the cryptological services invited colleagues to join them, viewing their employment as protection from front-line duty. A source book on the mathematicians in the service German cryptological agencies lists the names of more than 30 academic mathematicians and actuaries employed in several organizations. The lack of mathematicians was certainly not a reason for the Third Reich's failure in the cryptological race.

If the mathematicians had not failed, perhaps the problem lay with the engineers, incapable of turning theoretical ideas into practical solutions? The Allies did indeed acquire some extremely talented engineers who contributed as much as mathematicians to the victory over enemy's ciphers. In Poland, Antoni Palluth, designer of the Rejewski bombe, played such a role. The successes of British codebreakers would have been probably less spectacular had it not been for the engineering talents of Harold 'Doc' Keen, designer of the Turing/Welchman bombes, and Thomas Flowers, creator of Colossus. In the US, Joseph Desch, designer of the four-rotor bombe, played an equally important role. German codebreakers did not have such advanced and effective equipment, but this was certainly not due to a lack of people who could construct such devices. As early as 1937, Konrad Zuse had obtained a patent for a computing device with a structure that today, not quite rightly, we call von Neumann architecture. The importance that the Third Reich authorities attached to research into the automation of computing is illustrated by the fact that prototypes of his early designs were created in the private apartment of Zuse's parents. When he approached the authorities for financial support to convert the relay-based design to electronic one, his request was rejected as 'strategically irrelevant'. During the war, engineers working for the German crypto services constructed several ingenious devices to support the cryptanalysis of Allied codes and ciphers. If devices as groundbreaking as Rejewski's and Turing's bombes were missing, it was not the engineers' fault. Rejewski, and in his wake other mathematicians on the Allied side, manage to reduce the complexity of the problem of cipher/key breaking, but its size was still beyond the capacity of the unarmed human mind. In the next step, however, they designed an algorithm for the operation of a machine capable of overcoming the remainder of the problem in a reasonable time and handed it over to the engineers. To the best of our current knowledge, no cryptologists working on the opposite side have reached for an equally pioneering combination of mathematical theory and engineering.

In the author's opinion, probable solution to the mystery of the Allied success and Third Reich's failure is to be sought in the differences in the organizational and human culture of the services. Despite the war going on around, Rejewski and his colleagues remained civilians until 1943, and their superiors had enough common sense not to try to impose military discipline on them. When a group of mathematicians received their mobilization cards on the first day of the war, Colonel Langer promptly caused his subordinates to be relieved of active duty, as if intuitively sensing that military discipline was not conducive to the work of a codebreaker. The concerns surrounding the recruitment of 'professor types' to serve in the GC&CS indicate a certain level of conservatism within this organization. Fortunately for the mathematicians, it was formally subordinate to the Foreign Office and a style reminiscent of the world of diplomacy rather than the military prevailed in it. Army and navy officers working at BP were probably appalled by the manners of the university eccentrics, but they did not give expression to their feelings. As a result, wartime codebreakers from Cambridge, Oxford and other universities formed a miniature copy of the republic of scholars, with its free discussion, intellectual openness, banburismus, dummyismus, herivelismus, yoxhalismus and other carbon copies of the language of science. And when Churchill's 'geese' began to lay golden eggs, reason dictated acceptance of their dissent. Especially after scientists gained the position of being able to address the Prime Minister directly.

At the same time, mathematicians joining Kriegsmarine were sent to a reserve officer's training, where they learned to recognize military ranks, report appropriately and generally behave in a manner consistent with the traditions of the service. There were cases where experienced codebreakers were transferred from the Berlin offices directly to the front, where they were sure to do well as platoon or company commanders. The military tradition prevailing in German society of the time required that scientists who were conscripted into the army should feel in the first instance that they are soldiers or officers. Many talented scientists served in the Wehrmacht, Kriegsmarine and Luftwaffe, including some excellent candidates for the codebreakers. They did their job to the best of their ability: they constructed procedures and equipment that, properly applied, could have caused the Allies a great deal of trouble. However, in a hierarchical, militarized environment imbued with Nazi propaganda, their voice had no chance to be heard. The prevailing cult of infallibility among officers made it difficult to acknowledge past mistakes, especially if they were to be exposed by civilians. In what they considered a critical situation, British codebreakers did not hesitate to appeal over the heads of their superiors directly to Churchill. It is difficult to imagine officers of the B-Dienst, OKW/Chi or Inspk. 7 OKH appealing to Hitler, bypassing their own commanders.

World War 2 was a conflict of industrial blocks in which the availability of raw materials, industrial and logistical efficiency are perhaps more important than the number of divisions, ships and aircraft. Against this backdrop the duel between the two sides' cryptologists takes an almost symbolic significance, looking like an island of modern age, harbinger and antechamber of the current knowledge economy. But in the final months of the war, the Allied codebreakers were certainly unaware of their role as harbingers of tomorrow. Day by day, they anxiously awaited German ciphers, whose new and unusual form would have induced a rapid heartbeat and flurry of activity among BP, Op-20-G and SIS staff. Instead, the Allied listening stations one day received an open-text radiogram broadcast containing the phrase **DERXFUEHRERXISTXTOT** (the Führer is dead). The war in Europe, and with it the Enigma adventure, was coming to an end.

Chapter 23
Postscript—People, Machines, Ideas

uring the five years of the war, a small town of barracks housing the various sections of the GC&CS grew up next to the Victorian mansion that had become the centre of BP. By the end of the war, the buildings and park resounded with the voices of nearly 10.000 people. It took only a few weeks of peace for the hitherto bustling place to become almost deserted. Certainly, most of the staff left with a feeling of deep relief. The bulk of the BP staff were young women who had enlisted in the auxiliary formations of the British armed forces, some wanting to experience something heroic, others hoping to find even a little romance in war. After the conflict, they must have found that there was about as much heroism or charm in their wartime occupations as in the work of a lathe operator. Subjected to military discipline, they performed day and night a monotonous job, requiring concentration and patience, the meaning of which they did not know.

Sometime later, Poles arrived at the BP, that the original staff had abandoned. They were not, however, codebreakers, but DPs, displaced persons, people who had been thrown far from their homes and families by the war and the German occupation of their country. They were the first Poles to reach BP, to whose organisation their countrymen had contributed so much. The fact that they came to BP as exiles was a poignant commentary on the relationship between the Allies late in the war. This was not the only disappointment to be experienced by the Polish participants of Enigma story. In the first days of May 1945, Langer and Ciężki were liberated by the Allies. In their own way, both officers were lucky, as Castle Eisenberg was in a zone to be occupied by Soviet troops. Fortunately, the train on which the internees were evacuated reached the furthest town liberated by American troops. They thus avoided being 'liberated' by Soviets: given the fate of the officers of the pre-war Polish intelligence service who found themselves in Soviet hands, this circumstance probably meant a chance of survival. However, when they reached London, they had to experience a surprise. Arriving among the many soldiers and officers liberated from PoW camps, they probably did not expect a triumphant welcome. Perhaps they were hoping for a symbolic interest on the part of the British. Before the

M. Grajek, *Enigma Myth Deciphered*, History of Information Security,
https://doi.org/10.1007/978-3-031-65475-6_23

outbreak of war the decision to share the secret of Enigma with the Allies was made at the highest levels of Polish intelligence service, yet the initiative must have come from Langer and Ciężki. If the lack of interest on the part of former Allied partners was a surprise, the indifference or even hostility on the part of Poland's own intelligence community must have come as a shock. For a long time after their return from captivity, Langer and Ciężki could not get an opportunity to report to Colonel Gano, their pre-war friend and now head of the service. They gradually realised that the course of events leading from rejecting their service in the Polish Army in October 1939, and placing the entire team under French orders, through General Kleeberg's subsequent order to remain in unoccupied France, was neither an accident nor the result of wartime necessity. The leadership of the Polish intelligence service in London was decidedly unfriendly towards their former colleagues, and the officers' several years of separation from that milieu allowed an intrigue dictated by political or personal motives to grow to caricature proportions. The resentment towards the former heads of the Cipher Bureau must have extended to its staff as well: we know a copy of the report submitted by Major Michałowski after getting to London, peppered with ironic remarks by Major Langenfeld. This was not the only vendetta within the Polish intelligence community in London. According to Langer and Ciężki, at least part of the responsibility for their reception in London fell on Bertrand. The Frenchman had arrived in Britain a year earlier and had ample time to disseminate and consolidate his version of the events that unfolded in the south of France in late 1942 and early 1943. Given the hostile attitude of Polish intelligence chiefs towards the two officers, his account attributing to them responsibility for indecision and alcoholism leading to the fiasco of evacuation, fell on fertile ground. Langer and Ciężki eventually returned to the ranks of the Polish army, but remained sidetracked. Both were assigned to the Signals Training Centre in Kinross, Scotland, where Langer died of heart failure on 30 March 1948. Ciężki lived to see the day of demobilisation as head of the Signals Intelligence Training Centre. Knowing that his return to the country meant deadly danger, he improvised plans to get his family out of Poland. His letters found his wife, Bolesława, in Sochaczew. She had bad news concerning their three sons. She knew the fate of only one of them, Zbigniew, who was captured by the Gestapo in June 1942 and taken to Auschwitz, where he soon died. The other brothers were, like Zbigniew, members of the underground and disappeared while on duty. Zdzisław was employed in a cartographic office where he participated in sabotage. In September 1943, he was warned of his impending arrest and fled Warsaw, hiding for several months in the village of Kowaniec in Podhale region. He returned to Warsaw in April 1944 and in the same month accidentally fell into an ambush set by the Gestapo on someone else. The youngest, Henryk, took part in the Warsaw Uprising, after its fall was taken PoW and went through several camps before being liberated from captivity in April 1945. He ended up with the Polish Army in Italy, from where he managed to make contact with his father in Scotland. They were soon reunited in the UK, but not for long, as Henryk was determined to return to Poland. After demobilisation, Ciężki's road to family reunification was closed. A polite clerk informed him that he is entitled only to a reduced social assistance, which is not enough to support a family. However, HM government is offering to the former military a free training in civilian occupations; would Mr Major not like to take a weaving course? Living on half of normal allowance, Ciężki decided to move to the provinces, ending up in Par in Cornwall. There, lonely and broken by life, he died on 9 November 1951. Before his death, he did not learn that the second of his sons, Zdzisław, was executed by the Germans in the ruins of the Warsaw Ghetto on 28 April 1944.

Rejewski and Zygalski were also faced with a choice—return to the country or emigration. Their decisions were dictated by the most personal motives, namely responsibility and concern for the nearest people. Leaving Warsaw, Rejewski left behind his wife and two children, while Zygalski remained without family. Fortunately, they did not have to decide on their future immediately after the end of hostilities. In the last days before the German surrender, both went to the south of France, where they had left personal belongings that could not be carried with them when escaping across the green border. Giving them guidance for their journey, Bertrand was embarrassed to admit that his guardian had abused his trust and ransacked the luggage left in his care. Once there, Rejewski and Zygalski repacked the books, which were the only valuable part of their possessions, and set off back to Felden, celebrating the end of the war in Europe in Paris on the way. In the autumn of 1945, the codebreakers were sent to Scotland to participate in the Officers' Military Administration Improvement Course. Under this innocuous name was hiding a training in intelligence techniques. They probably ended up there by virtue of the inertia of the Polish military structures, but the fact that they had attended the course could have proved a serious problem should they return home.

The day of decision came in the autumn of 1946, when demobilisation began. Rejewski, without much hesitation, refused to join the Polish Resettlement Corps, which was to be a transitional stage for soldiers and officers choosing to emigrate, and returned to the country on 21 November 1946. His flat in ruined Warsaw was occupied by squatters, so he settled in his hometown of Bydgoszcz in the house of his in-laws, in time taking up a job in the invoicing department of the Kabel Polski factory. Quite quickly, he found out that his past at the Cipher Bureau was following him. First, he sensed signs of the security service's interest in his person, then in March 1950 he lost his job. He fell into a bizarre spiral of fate: in successive jobs his intelligence, diligence and competence predestined him to occupy higher positions. At the same time, whenever the prospect of promotion arose, it was followed by pressure from the communist political police, which insisted that he be dismissed. Rejewski grasped this mechanism quite quickly, and realising the fragility of his own existence in the police state, he gave up his ambitions in a desire to protect his family. In a letter to an old friend, he wrote: 'in a very short time I realised (…), where the wind was blowing from, in view of which I abandoned the resumption of any old contacts and reduced the existing ones to the necessary minimum'. Until the end of his professional career in 1966, he held subordinate positions in the accounting departments of various institutions in Bydgoszcz. Years later, his family was to learn from declassified security files that Rejewski's caution had probably saved him from serious danger. During the 1950s, the political police watched his every step, attempting to implicate him in investigations conducted against groups undertaking alleged anti-communist activities. It was not until he retired at the beginning of 1967 that Rejewski sat down to write up his memoirs, which he submitted to the Military Historical Institute around April of that year. There, his typescript attracted the attention of Władysław Kozaczuk, who wove the theme of breaking the Enigma ciphers into a book published the same year on the activities of pre-war Polish intelligence.

The book failed to attract interest in the world of historians. Rejewski had to wait for until 1973, when a wartime memoir entitled 'Enigma ou la plus grande ènigme de la guerre 1939–1945', was published by Gustave Bertrand. His book presented the story of the breaking of Enigma from a rather subjective point of view, and the British, understandably irritated by his version, responded with books by their own authors in which, taking up the polemic with Bertrand, they presented the role of the Poles in a completely distorted way. Rejewski joined wholeheartedly in straightening out historical inaccuracies, giving endless interviews to historians and journalists from all over the world and complementing his own memoirs. In the small Warsaw flat to which the codebreaker had moved in the meantime, the door would not close for some time and the telephone would not go silent. Rejewski, however, did not manage to unravel the confusion surrounding the breaking of the Enigma—he passed away suddenly on 13 February 1980.

Zygalski, who had left the country as a bachelor, had a wider margin of discretion in choosing the way forward and decided to stay in Britain. Around 1950, the most important person in his life became Berta Blofield, with whom he shared a passion for music. Fortunately for their future life together, Zygalski's mathematical knowledge and talents proved to be an asset that remained valid even after the war. It was not, however, without obstacles. The British authorities refused to recognize Zygalski's degree awarded by the University of Poznań, forcing him to repeat his studies and obtain a British degree. Thereafter Zygalski was able to earn a living as a lecturer in mathematics at several British schools and universities. After his retirement, he and Berta moved to Liss near Portsmouth. Disclosure of his involvement in breaking the Enigma ciphers earned him an honorary doctorate from a Polish university abroad. He received it, however, already broken by illness, and died on 30 August 1978. Notwithstanding his mathematical and musical talents, he must have been a man worthy of great and beautiful affection, for soon after his death, Berta decided that a life without Henryk by her side represented no value.

The two surviving mathematicians of the team which played a pioneering role in the Enigma story had no contact with cryptology after the war. Arguably, this was not a conscious choice by people tired of several years of working under conditions of extreme mental concentration. As a result of controversial, to put it mildly, decisions by the Polish military authorities and then the French and British allies, they were sidetracked during the war. After the end of the hostilities, the story of the breaking of Enigma seamlessly changed its status from the greatest secret of the Second World War to one of the key secrets of the Cold War. For a long time, none of its participants could claim their share in a success that did not officially exist.

A pity, as the post-war world offered ample opportunities for the use of their knowledge, so both would have been able to find occupations more suited to their knowledge and experience. Gustave Bertrand, for example, easily found his way in the post-war world. In August 1944, he finally gained an assignment in active service, taking up several positions in French intelligence in succession: he was able to embark on the rebuilding of his service, which under the SDECE became known as the Service Technique de Recherches (STR) or, for a shorter time, Service 28. The easiest way to rebuild its former position was to take in castaway, of which there was no shortage in post-war Europe. This is how six members of the former 'Camazon seven' (one of whose members emigrated to Mexico) returned under Bertrand's wings.

In time, two members of the former Polish team also joined Bertrand's organization. Liberated from the concentration camp, Kazimierz Gaca arrived in Paris and knocked on the only door he knew in the French capital, at 2 bis Rue Tourville. His former boss received him with joy, which, however, was no match for the enthusiasm of his driver's daughter. It turned out that during their time together at Uzes, the then-teenager had a secret crush on the youngest member of the Polish team. When Gaca appeared in Paris, emaciated by his camp experience and in need of loving care, she was definitely no longer a teenager and the girlish fascination turned into an affection leading to marriage. Gaca worked at SDECE until the end of his career, died and was buried in Paris. For a time, his working companion was also Sylvester Palluth, who, reportedly, was to suffer a crisis of conscience observing the activities of the French in Algeria, parting with his former employers.

Bertrand continued to develop his cryptologic service for some time, recruiting Lithuanians, Latvians and Estonians who had previously served in German or Finnish radio listening posts, some Italian cryptologists whom the French had taken prisoner in North Africa, and even Japanese. The British demonstrated an ambiguous attitude to his activities. Some of his operations were carried out in consultation with and even co-financed by MI6: traditionally Bertrand's partner on the British side was *Biffy* Dunderdale. But the codebreakers consistently distanced themselves from their former ally. When John Tiltman was due to represent his service at the Allied Radio Intelligence Conference at Versailles in mid-1945, Menzies banned him from discussing any professional issues with Bertrand, should they meet. The Frenchman was aware of the reserve on the part of the British. Meeting Tiltman at the hotel lobby he quipped 'I'm sure you wouldn't want your boss to find out about our meeting'. It seems that Bertrand's post-war career also encountered obstacles in France. Around 1950 he parted somewhat mysteriously from the SDECE, retiring and settling in Theoule near Cannes, where in time he became a councillor and then mayor. Following the publication of his book in 1973, he took very personally the polemics by British authors. Paillole records that: '(…) he rejects the discoveries about the Enigma machine announced off his back in England and questions their veracity. He still considers himself the sole depository of the secrets'. Bertrand died in Toulon in 1976 and the French gendarmerie sequestrated his sizeable archive. Parts of it were declassified only in 2015, providing new information on the history of breakieng the Enigma ciphers.

When the demobilisation of BP began after Germany's surrender, the centre's section chiefs set about the privilege and duty of the victors—to write up the official history of events. For many of them, this last task was a source of frustration. They were aware that they were probably writing for the drawer only: given the secrecy surrounding BP's activities, it was unlikely that their accounts would ever see the light of day. Fortunately, they treated the final chord of their BP work with as much seriousness as the purely cryptological challenges of wartime. As a result, we have a vivid and juicy picture of how the centre functioned. Some sections continued their earlier activities, breaking previously intercepted messages that did not fit into wartime priorities. Bombes are known to have attacked the keys of previously unbroken networks as late as March 1946. Several Colossus machines also escaped dismantling and scrapping. They were moved to Eastcote after the war and used in attacks on Soviet teletype ciphers, knowledge of which the Allies owed to German prisoners of war. Having fulfilled their chronicling duty, the codebreakers took it in turns to sign a reminder of their obligations under the Official Secrets Act and return to pre-war life.

In nothing has the unique character of the team assembled at BP during the war manifested itself so intensely as in their post-war lives. The most interesting personality of *Hut 8* and arguably of the entire BP of the war years, Alan Turing, had already been working at the Hanslope facility since 1943 on the design of the prototype of *Delilah,* device encrypting conversations. There, in mid-June 1945, he received a phone call from J.R. Womersley, head of the newly created mathematics department at the National Physics Laboratory (NPL). The government-funded scientific institution tempted Alan with the prospect of a continuation of his deliberations on a universal logic machine, the practical realisation of which had been made possible by the wartime development of electronics. Turing took up the offer and already on 16 February 1946 had presented a design for earliest electronic computer in the present sense of the word. After a brilliant start, however, came disappointment. The NPL was a government institution, subject to the dictatorship of budget cycles and limits, so work on practical implementation was delayed. Annoyed by the dictatorship of the bureaucrats, Turing withdrew from the project just as it began to take shape. 18 August 1947 was declared the official start date for the project, but Alan renewed on 30 September his membership of King's College. He probably did not yet know that computers were his destiny. As he struggled with government bureaucracy, his former Cambridge mentor and BP colleague, Max Newman, was undertaking a competitive computer building programme at the Manchester University. In 1948, he induced Turing to move to Manchester, where he reserved for him the position of lecturer and then deputy head of the computing laboratory. In his new role, Turing's main task was to develop software for the yet-to-be-developed Manchester Mark I computer. Alan was as ambitious as ever: he set about creating a program imitating the logic of a chess player. Ultimately, both projects he was involved in were successful. On the night of 16–17 June 1949, the prototype dubbed Manchester Baby determined one of Mersenne's numbers in nine hours of continuous operation. If anyone was trying to define the beginning of the information age, that day is a good candidate.

ACE, Turing's project at NPL, performed its first computation on 10 May 1950. Meanwhile, Alan again shifted his scientific interests into new areas, first artificial intelligence, then mathematical biology. He passed away in tragic circumstances before he had time to fulfil his intellectual potential. In 1952, while reporting a burglary at his home, he admitted in passing that his homosexual partner may have been an accomplice in the crime. After this confession, officers lost interest in the burglary, focusing instead on Alan's homosexuality, an act prosecuted by law in Britain at the time. Sentenced by the court, facing the prospect of prison, he chose forced hormone treatment. This had deplorable side effects, both physically and mentally. As a result of his sentence, he also lost his security clearance to participate in confidential projects. He also had to lose the will to live: on 7 June 1954, he dipped an apple in a cyanide solution and took a sizable bite. His former colleague at *Hut 8*, Shawn Wylie, said of Turing years later: 'I don't think people were an important part of his life. (...) Obviously there must have been people who meant something to him. But I think devices and ideas were his only real love'.

Gordon Welchman emigrated to the USA in 1948 and joined the MITRE corporation, working in communications security. He became a US citizen in 1962 and retired in 1971, maintaining his association with MITRE as a consultant. In 1982 he published memoirs of his time at Bletchley. Although BP's wartime activities had been publicly discussed for nearly 10 years at the time, and the breaking of the Enigma ciphers widely known, with the publication of the book he drew the ire of the British continuation of the GC&CS tradition, GCHQ, on his head. The British took advantage of their special relationship with their American counterpart, the NSA, to revoke Welchman's security clearance (which put an end to his work with MITRE) and to prohibit him, by court order, from publicly discussing the details of his former work. He died on 8 October 1985.

C.H.O'D. Alexander returned to his position as head of research at the John Lewis Partnership for a time, but the challenges of codes and ciphers continued to absorb his attention. In the summer of 1946, he returned to his former workplace, which had since been renamed Government Communications Headquarters (GCHQ) and moved to buildings erected during the war to house one of the bombe stations, at Eastcote. Alexander became head of Department H, responsible for cryptanalysis. It seems that inventing new ways to break other people's ciphers was still his favourite pastime, along with chess, as he refused several times offers of promotion to positions of a more prestigious but only administrative nature. He continued to play chess at the highest level, winning the title of international champion in 1950 and recording victories over Botvinnik, Euwe and Bronstein, among others. He retired in 1971 and died on 15 February 1974.

The last of the four codebreakers to co-author the letter to Churchill was Stuart Milner-Barry. After the war, he moved into the civil service taking up successively several posts in the Treasury. After a two-year tour at the Department of Health, he returned to the Treasury, ending his public career in 1966 as Under-Secretary. Alongside Alexander, he represented Great Britain in several chess events, passing away on 25 March 1995.

Alongside them, many people worked at BP, who in later years made their mark on public life in Britain, Europe and the world. Former Testery employee Peter Benenson founded Amnesty International. His section colleague, Donald Michie, was one of the pioneers in the field of artificial intelligence. Yet another Testery member, Roy Jenkins, reached first the position of Chancellor of the Exchequer, then Home Secretary (in this role he implemented the metric system in the UK), to chair the European Commission from 1977 to 1981, preparing the then Community for economic and monetary union.

The post-war careers of former codebreakers on the other side of the Atlantic provide an interesting commentary on the links between the wartime attacks on the Enigma and Fish ciphers and the origins of computer technology. The head of Op-20-Gm, Howard Engstrom, had already become known as a proponent of electronic technology during the war. Then he nearly derailed Desch's bombe design by trying to replace it with an all-electronic design. But by early 1945, the campaign against the U-Boots was settled and the end of the war in Europe was near. In Op-20-G, everyone was thinking about the post-war future. Joseph Wenger was concerned about the prospects for his organization. The majority of staff members called up for the war did not intend to remain in the ranks of the navy after its end, his research team was about to disband soon after victory. Also, his current business partner, the NCR, was hinting that cost-based calculations of wartime contracts were not an attractive peacetime proposal, so it intended to return to profitable production of cash registers. The gusts of the coming Cold War could be sensed in the corridors clearly enough to reckon with the need to maintain the codebreaking agenda, meanwhile Op-20-G was in danger of losing its most important assets.

At the same time, his subordinates considered how their experiences of war time could be used in civilian life. Even Engstrom, who had moved to the Navy from prestigious position as a mathematics professor at Yale, was less than thrilled at the prospect of returning to academia. His section colleague, William Norris, was also reluctant to think about returning to his role as a travelling salesman selling yet another X-ray machine around small towns in Illinois, Iowa and his native Nebraska. Out of their conversations came the idea of setting up a company to design and build specialised cryptologic equipment for the Navy. Their idea prevented the dispersal of a group of uniquely qualified professionals, so it gained the support of Wenger and, thanks to his energy, also the Navy top brass. Engstrom and Norris, still in their smart Navy uniforms, began touring US corporations looking for potential customers and those willing to invest in the new business. However, their position in the talks was rather difficult: information about the devices they intended to manufacture was still classified. After dozens of fruitless meetings, however, they found, thanks to the personal involvement of Admiral Nimitz, a partner willing to invest his money in the venture. For John Parker, his interest in a business was a form of big rush ahead. He owned the Northwestern Aeronautical Corporation, a company located in St Paul, Minnesota, which manufacured gliders for the army during the war. With the end of the war, orders stopped abruptly and the plant was threatened with bankruptcy. Trying to switch the company from manufacturing wood-and-canvas gliders to electronics was risky, but it seemed the only solution to Parker's problem. In January 1946, the partners registered a new business—Engineering Research Associates (ERA). Engstrom, Norris and Parker were also joined by Ralph Meader, whom we last met in one of the bedrooms of Joe Desch's house in the role of guardian of his conscience. Among the founders, Norris was the only person with engineering experience, so he took the role of CEO.

Aside from minor projects, ERA's first significant achievement was the construction in 1947 of a cryptanalytic machine called Goldberg, whose major innovation was its magnetic memory, the precursor of the modern hard disk. The company's next creation, the Demon, lost significance before it was completed. It was to be used against one of the Soviet ciphers, but before it could be completed, the day referred to in the history of American cryptology as 'Black Friday' arrived on 28 October 1948. The Soviets, alerted to the reading of their ciphers by ASA employee William Weisband, changed most of the systems within a day. More fortunate was the next ERA project, started in 1947 and code-named Task 13. In 1950, ERA delivered to the Navy the first universal computer, which in its final version was referred to as Atlas. The design was so successful that ERA decided to sell the device also to the businesses, under new name ERA 1101 (1101 was the binary form of the project's original designation, Task 13).

One sometimes encounters the opinion that some cryptanalytic devices, constructed during the war for attacks on the Enigma and Fish ciphers, were in fact an early form of computers. This argument is about as accurate as the claim that an abacus or a mechanical adder represented an early form of a computer. There are two determinants of the modern notion of the computer: electronic technology and the universal use resulting from storing the programme in the same memory as the data. Devices constructed before and during the war by the Polish Cipher Bureau, BP, Op-20-G and SIS grew out of a common root with the machines whose ciphers they were supposed to break: they were electromechanical. The need to speed up information processing and the parallel development of electronic technology led to the birth of hybrid devices, some of which were constructed in mostly electronic technology (Colossus, Duenna, Superscritcher). However, their structure was based on a specific method of attack at particular cipher. In some cases, they could be used to analyse another, similar cipher (Colossus was used to break Soviet ciphers after the war), but they could not be reprogrammed to perform universal mathematical operations, for example. Their 'programme' of operation was contained in their structure, and thus they did not meet the condition of universality that we require of a computer. The link between the wartime cryptanalysis and the development of computer technology obviously existed, but it was abot people and not machines. In Europe, the post-war pioneers of universal computer design were BP veterans Max Newman and Alan Turing. On the other side of the Atlantic, the initiative belonged to Op-20-G engineers and scientists William Norris, Howard Engstrom and Ralph Meader.

The dozen years between Marian Rejewski's first success in the attack on the Enigma cipher in December 1932 and the end of the Second World War transformed cryptology. When the Polish Cipher Bureau embarked on its attack on German machine ciphers, the functioning of most cryptologic agencies in the world was not significantly different from the picture that prevailed during the years of the First World War and many years before it. From a contemporary perspective, it can be said, without a tinge of disparagement, that cryptology before Enigma was a craft rather than a science. As in a craft, key skills could only be mastered by apprenticing alongside a master. The fact that masters were only ever employed in the Black Cabinets, working for rulers and state governments, further strengthened the monopoly of practising cryptologists on their secret knowledge. Not even the publication of the first textbooks on cryptology changed this picture of the discipline. Like manuals for goldsmiths, watchmakers and shoemakers, these contained not so much a systematic lecture on the theory of the discipline as a description of the classic known types of codes and ciphers and the methods of attack that were effective against them. When the breaking of the Enigma cipher proved the effectiveness of mathematical methods, ten years after the Polish Cipher Bureau, other countries also began to employ mathematicians in their cryptologic agencies. This was not an immediate or easy process. When the British decided to recruit future cryptologists from among university students and graduates, they originally preferred physicists, who were supposed to present a more practical attitude to reality. After the war, no one raised this type of objection. Cryptologic agencies, within a few years, ceased to be the domain of classical philologists and linguists, turning into the almost exclusive domain of mathematicians and engineers. The work of the three Poznań mathematicians and their successors made cryptology into fully-fledged branch of applied mathematics, a science with a solid theoretical basis involving the keenest mathematical minds.

The breakthrough in understanding the nature of cryptology came just before the Second World War, which encouraged the development of new ideas but not their propagation outside the world of military applications. But the discipline's new scientific status was to produce surprising results as soon as the guns fell silent. Demobilised mathematicians returned to universities, departments and institutes inspired by the new field of application of their discipline. Cryptology, a field hitherto reserved exclusively for the secret services, diplomacy and the military, embarked on a two-pronged development. One path remained state agencies hidden behind the doors of secrecy. The other became universities, soon followed by business. The openness of university research meant that cryptology opened up to the world, gradually emerging from its military-diplomatic underground and embracing new areas of application. The two streams of development of its methods meant that there was inevitably duplication of work and repetition in the civilian sector of discoveries whose secrecy prevented scientists working for the military from revealing. However, the proportions of the two sectors' contributions to the state of the art gradually evened out as applications of cryptology moved into the civilian sphere— telecommunications, the financial world, privacy and the Internet. The vast majority of us have not noticed that cryptology has made its way into our daily lives: when paying using a credit card, talking on a mobile phone or ordering goods from an online shop we are using its services. Perhaps reaching for the mobile phone we should remember the people whose work and contributions influenced not only the fate of the Second World War, but also our everyday lives.

Chapter 24
Post-postscriptum

A summary of the Enigma history would be incomplete without an attempt to assess its historical significance, in particular its impact on the fate and outcome of the Second World War. Evaluations of this kind always carry a flavor of subjectivity, so the comments presented below differ fundamentally from the fact- and source-based narrative of the previous chapters. Also, the resulting picture of the role that Enigma decrypts played in the various stages of the war differs from the conclusions presented by earlier commentators. Their views are mostly situated between the extremes set by Harry Hinsley and John Keegan. Harry Hinsley, an active participant in the events with Enigma in the background himself, assessed that without the decrypts provided by Allied codebreakers 'the war would probably have lasted two or three years, perhaps even four years longer'. Keegan estimated that the decodes shortened the war by only three months. In their absence, the first American nuclear bombs would have fallen not on Japanese but on German cities, bringing the war to an end. Any attempt to assess which opinion is closer to the truth, and thus how many months or years of war and how many lives were spared by the work of cryptologists, is situated dangerously close to the judgements quoted above. However, on the basis of the available information, it is possible to assess the extent to which Allied cryptologists, intelligence analysts, military commanders and politicians managed to exploit the opportunities they gained from being able to systematically look into the enemy's cards during the conflict.

Before proceeding with such an analysis, it is necessary to comment on, or even correct, a number of questionable opinions often repeated by historians of Enigma. Let us begin the review with one of the often-repeated assumptions about this history. Many authors tacitly assume that the machine's ciphers would also have been broken without the pre-war success of the Poles. In doing so, they continue the narrative that constituted Dilly Knox's initial, very emotional reaction to the news that they had broken the cipher; the Poles had managed to break the cipher thanks to a stroke of luck; they had managed to guess the structure of the machine's element on which his own attack had stopped. Indeed, Rejewski admitted that during his first

M. Grajek, *Enigma Myth Deciphered*, History of Information Security, https://doi.org/10.1007/978-3-031-65475-6_24

attack he assumed the alphabetical order of the entry drum, and this assumption yielded the correct solution. But he also described a purely mathematical method of reconstructing the connections of this element. When an accurate conjecture allowed the problem to be solved, the Poles were happy to use it. However, they were ready to replace intuition with mathematics at any time. Dilly corrected his judgement over time. Some historians, however, have chosen to refer to his earliest misguided reaction.

Another commonly reproduced narrative concerns the role of documents provided by French intelligence. In his post-war memoirs, Rejewski stressed that he made use of only one of them; the list of cipher keys for September and October 1932. Knowledge of the plugboard settings for these two months allowed him to eliminate one unknown from his equations, which was enough to solve them. In the same memoirs, however, he described a method, which would allow him to solve his equations also without the keys. Contemporary cryptologists confirmed the correctness and effectiveness of the approach described by Rejewski. As its application required cipher material from about a year, involved a considerable amount of work, and its expected result had previously been achieved by other means, it remained a historical and mathematical curiosity only. If, however, there had been a need to break the cipher without the material provided by Hans-Thilo Schmidt, the Cipher Bureau team would simply have achieved the solution a year or two years later, still long before the outbreak of war.

Many historians make it clear that the Polish contribution was limited to handing over a copy of the machine to the Allies. Let us not comment the particularly pervert versions, attributing to the Poles the theft of Enigma documentation from the factory manufacturing the machine or the hijacking of trucks delivering the machines somewhere in the back of the front line. Copies of the machine alone, without knowledge of the daily changed key to the cipher, would be an intelligence curiosity devoid of practical significance. Therefore, the package of information provided by the Cipher Bureau to the Allies included items of far greater importance. The first and perhaps most important of these was the sheer certainty that the Enigma cipher, hitherto regarded equally by German, French and British cryptologists as unconditionally secure, could be broken. According to the memoirs of GC&CS codebreakers from the pre-war period, extreme pessimism about the prospects of breaking the machine cipher was dominating in their circles. Overcoming this attitude and replacing it by an activity typhoon would not have been possible without a transfusion of hope from Warsaw.

The second key element was a description of the methods permitting to reconstruct the cipher key. Some authors suggest that the role of the Poles came to an end in September 1938, with a change in the Enigma operational procedure. From a Polish report declassified in 2015 by the French secret services, we know that the last day before the outbreak of war for which the Cipher Bureau managed to break the key was 26 August 1939. In fact, the period between September 1938 and the end of the year was the period of greatest glory for the Polish team. After changing the cipher procedure, the Poles developed and implemented two methods of attack that enabled them to resume decryption in just two months; Rejewski bombes and

Zygalski sheets. The introduction of two additional Enigma rotors reduced the practical effectiveness of both methods, but did not influence the theoretical foundation of their operation. Rejewski bombe became the prototype for big part of the cryptanalytic devices constructed by the Allies throughout the war. Zygalski sheets, after the British developed a set including the new Enigma rotors, made it possible to break the cipher until May 1940.

However, the most important element contained in the report handed over to the Allies at Pyry, in July 1939, was the new definition of cryptology developed by the Poles, who were about ten years ahead of the rest of the world in this respect. The employment of mathematicians in the section breaking the Soviet ciphers in 1920 may have been the result of lucky coincidence, but the Poles learned the lesson well. When, in 1925, the Japanese army was to send a number of officers for training at the Polish Cipher Bureau as part of the two countries' signals intelligence cooperation, it asked about the preferred characteristics of the candidates. The answer from Poland was that they should have a good knowledge of Russian language and mathematics. The appreciation of the role of mathematics by Polish codebreakers makes fully understandable the decision of Lieutenant Ciężki, who in 1928 decided to look for candidates for future Enigma breakers among students of mathematics. In this way he assisted in the birth of a new face of cryptology, in which mathematics replaced the linguistical skills. Even ten years later, when the heads of GC&CS had decided to recruit candidates of a 'professor type', they quite emphatically denied the need to recruit mathematicians. Luckily, Alan Turing was among the first recruited candidates for codebreakers.

As a result, British delegation to the Pyry conference consisted of old-school cryptologists with purely linguistic experience. They were not prepared to take on all the know-how that the Polish codebreakers wanted to pass on to them, which contributed greatly to the misunderstandings at the conference. On his return to the UK, however, Knox invited Alan Turing for a weekend to his country house, where together they reviewed the material brought from Warsaw, and the mathematician confirmed the correctness of the Polish colleagues' reasoning and conclusions. The result was a revolutionary change in the attitude of the GC&CS management. Earlier doubts were dismissed and the recruitment of mathematicians went into full swing. In this context it is appropriate to comment on an element of Gordon Welchman's post-war account. In his memoirs he noted that in his early period at GC&CS, working without knowledge of Polish success, he repeated to some extent the achievements of the Poles by constructing an equivalent of Zygalski sheets method. This statement gave probably the basis for claims that British codebreakers would also have been able to cope with the Enigma challenge without the knowledge transfer from Warsaw. However, let us note that without the earlier Polish success Gordon Welchman would probably not have been a GC&CS employee in late 1939 and would not have had the opportunity to face the Enigma challenge. Moreover, he designed his attack with full knowledge of the machine's design. British troops did acquire copies of the machine over time, but this happened at the earliest during the Norwegian or French campaigns. Only then would Welchman or anyone else have been able to begin work on a key recovery method. Perhaps this is why Welchman assessed himself that without the Pyry conference the breaking of the Enigma would have been delayed by at least a year.

The mathematical revolution in cryptology was a crucial, although not the last coup that the Polish codebreakers delivered. Mathematics made it possible to reconstruct a machine, and to reduce the scale of the key recovery problem to a more manageable size. However, the military codebreakers were not interested in theoretical breakthrough only, it was necessary to read the cipher in time to use the information obtained. Their success so far was only half-hearted and required a practical complement. Trying to meet the expectations of their superiors, they were the first to realize that the natural adversary for the cipher machine was another machine, assisting the attack at the cipher key. Rejewski's bombe became the prototype for most of the cryptanalytic devices constructed by the Allies during the Second World War. British delegation returning from a meeting in Pyry brought with them the foundations of the later Turing-Welchman bombe. It consisted of the practically proven effectiveness of Rejewski bombe and one of Rejewski's earliest insights into the structure of the Enigma cipher—the independence of the cyclic structure of the cipher from the plugboard settings. Both ideas landed in good hands; Turing managed to combine them in a construction more flexible than Rejewski bombe, Welchman added the crucial diagonal board, and 'Doc' Keen—a solid dose of engineering. Their construction was a gigantic success, becoming the basis for the wartime activities of the BP. In light of this success, historians tacitly assume that the emergence of the Turing-Welchman bombe in the first year of the war was a foregone conclusion. But was this in fact the case? Without the ideas brought from Warsaw in Denniston's luggage, would the device really have been constructed so quickly? Would it have been created at all?

The revelations brought back from Warsaw, on the one hand, gave direction and impetus to the work of the codebreakers, but on the other hand, created a great deal of confusion in the sphere of military intelligence and operational command. Harry Hinsley, a BP veteran and post-war chronicler of British intelligence, formulated the elementary conditions for successful signals intelligence operations. He stated that the enemy's message must be intercepted, deciphered, analyzed and interpreted in the context of other sources of information, and the resulting conclusions must be communicated to the recipients. This scheme illustrates the operation of signals intelligence as seen from its internal perspective. The information it provides will be of no use unless appropriate action is taken on its basis. Hinsley's list needs to be supplemented by several conditions. Firstly, the information provided must be credible to the receivers. Secondly, this information should, on the one hand, protect the secret of its origin and, on the other hand, allow recipients a degree of freedom to make decisions based on it. Thirdly, recipients must have the means to implement such decisions. During the first two or even three years of the war, Britain encountered problems in each of the areas indicated. It entered the war with a unique asset—an integrated cryptologic organization. On the other hand, for the past almost 20 years, the GC&CS has been under the authority of the civilian Ministry of Foreign Affairs. Its communication and cooperation mechanisms with the Navy, Army and Air Force existed in embryonic form. News from Warsaw accelerated the wartime expansion of its codebreaking arm that the recipients of its reports were unable to keep up for some time.

Capitalizing on the codebreakers' success was a difficult task. Few people realized this as much as Churchill did. In the early days of the conflict, he took up the post of First Lord of the Admiralty, the same post he had held in the early days of the First World War. Britain then received a gift from its Russian ally, the German Imperial Fleet codebook. Churchill personally designed the rules for the use of the decrypts in an attempt to protect the security of the source in the first place. The extremely restrictive nature of these rules severely hampered the operational use of information gleaned from the decrypted messages. It was the consequence of their application that allowed the German Hochseeflotte to avoid the Royal Navy's trap during the Battle of Jutland and even inflict losses on Admiral Jellicoe's fleet that exceeded its own. Mindful of this setback, Churchill tried to make sure that the rules for the use of Enigma decrypts guaranteed both the security of the source and its practical use on the battlefield.

Most of the available descriptions of the Ultra system refer to its mature form, which became operational in the second half of 1941. Earlier episodes of its history presented improvisation of somewhat limited effectiveness. The problems were largely due to the very nature of the Enigma messages. Enigma was used to secure communications at the tactical level, in the division-corps-army (and their equivalents) relations. Information of a tactical nature retained operational utility over a period of a few hours to a few days. In the early days of Ultra, breaking the network key, decrypting the dispatches, processing them, encrypting them back with own cipher, transmitting to the field command, and decrypting them back took a total time exceeding the useful life of the information. Moreover, in the early days of Ultra, the security of the source was ensured by tagging the decrypts with the signature Boniface, suggesting their origin from an agent. For quite a time recipients treated them with certain distrust, consequently failing to act on them. One can only speculate on the reasons for this behavior. British intelligence enjoyed varying degrees of prestige in military and civilian circles. SIS was traditionally held in high esteem both inside the country and abroad. But in the military environment, only candidates who could not be meaningfully deployed in other areas were detached to intelligence. One BP codebreaker recalled that, during an interview investigating his suitability for the army, he managed to get a sneak peek at a conclusion. It read: "No good. Not even for intelligence." Many commanders of the Second World War had experience from the trenches of the Great War. They brought from them a belief in the low value of information provided by agents. Perhaps influenced by such observations the unfortunate Boniface was replaced overtime by the more abstract term CX/MSS (Most Secret Source). In the early days of Ultra, decrypts did not significantly affect the military operations, but gradually built up the credibility of the source.

By becoming involved in the design of the signals intelligence structure, the codebreakers went far beyond their nominal role. In theory, their involvement should have and ended with the passing on the raw decrypts for further intelligence interpretation. Probably due to his previous business experience, Gordon Welchman was able to see the broad picture of the structure of which the codebreakers' job was a part. He was able to impose priorities on the work of interception, ensuring the double-banking of the networks that offered best prospects of breaking or were operationally important. To this end, he had to overcome the partisanship of the services, which found it difficult to accept that, for example, the Royal Navy intercept station was ordered to focus on the messages transmitted by Luftwaffe rather (which were broken on a regular basis) than Kriegsmarine (which had to wait nearly two more years to be broken). He implemented the rational management of the crucial resource—bombe time. His hand can also be recognized in the post-decode stage of intelligence processing of the broken dispatches.

The role of BP came inexorably to an end after the decrypts were handed over to the relevant naval, army or air force staffs and field commanders, whose task was to use the information they received in planning their own operations. The receivers of decrypts were not very hasty in learning this art. Decrypts did not play a role during the French campaign in 1940. A change of Enigma's operational procedure at the beginning of May blinded BP during a crucial period of campaign, and when the decryption resumed, the campaign was already lost. The decrypts played a symbolic role during the Battle of Britain. Luftwaffe planes needed less than an hour to get from airfields in France over targets in England. Even reading their orders in parallel with their nominal receivers left no time for a proper response. Perhaps this is why the architect of Britain's air defense system and head of the Fighter Command, Hugh Dowding, was not on Ultra's list of recipients. The earliest campaign in which Ultra's influence on the operational planning and command is attested was the fighting in Greece in April 1941. One of the earliest SCU/SLU units was operating on the staff of General Henry Maitland Wilson, commanding the British Expeditionary Force. Timely readings of the German intention to outflank the British position at Thermopylae allowed evacuation of troops from the mainland. If defeat was accepted in Greece to avoid disaster, the short campaign on Crete brought both. This time the commander of the British forces on Crete, General Bernard Freyberg, had extremely precise and timely information permitting not only to defend the island, but also inflict a spectacular defeat on the attacking paratroopers. He was a brave soldier, although his principles of war were formed on the fronts of the First World War. His disregard for the role of air force in modern warfare brought his forces defeat in a battle that the information provided by BP allowed to win.

Parallel to the Greek campaign setbacks, the first clashes with German forces in North Africa were developing. As we recall, in their early stages Erwin Rommel had to make use of Luftwaffe networks, which were regularly broken in BP. The operational use of information obtained from decrypts required a shorter time between the interception of the dispatches and the delivery of the information contained in them to the commanders. In May 1941, Colonel Frederick Jacob arrived in Cairo and organized a facility known as the Combined Bureau Middle East (CBME) in the seized King Farouk Museum building in Heliopolis. Its prototype was the GC&CS's Far East Combined Bureau (FECB), which had been in operation since 1935, operating originally in Hong Kong and then moving successively to Singapore, Colombo in Ceylon and Kilindini in Kenya. The purpose of both outposts was to intercept

enemy communications throughout the region and to decipher the intercepted messages locally, using keys broken in BP. General Claude Auchinleck, commanding
the British forces in the Middle East during this period, made it clear in his memoirs
that the operation of the CBME allowed cryptologists to make a real contribution to
the fate of the campaign. However, the most significant element of this contribution
was not the Enigma, but the Italian naval messages encrypted with the C-38 m
machine. Breaking them made it possible to attack convoys delivering supplies to
Africa with precision.

Paradoxically, some Enigma decrypts relating to operations in North Africa may
have had a negative impact on operations planned and executed by British commanders in the region. Since First World War, the Room 40 and his own role as First
Lord of the Admiralty, Winston Churchill was accustomed to having codebreakers
hand over directly to him the most important dispatches in their original form. It
was a practice of dubious value. A single decrypt, even one containing the crucial
information, taken out of context creates a distorted picture of the situation.
Churchill, however, probably insisted on continuing this practice also during the
present conflict, and probably met no resistance to this practice. When the question
of succession to Admiral Hugh Sinclair was decided in November 1939, Stewart
Menzies was not Churchill's favorite. He won the race, but just a few months later
he had to redefine his relations with the new Prime Minister. He had poor cards at
his disposal. In 1940 SIS was unable to follow its glorious traditions. BP decrypts
were one of the few assets he could use to consolidate his position. The easiest way
Menzies could have achieved this was by delivering personally selected decrypts to
the Prime Minister. Churchill, on the other hand, was convinced that his long experience of relations with the secret services entitled him to act as his own intelligence
officer, drawing his own conclusions from original, unprocessed decrypts. This had
questionable impact on the campaign in North Africa. Rommel permanently depre-
ciated the state of his own troops in his reports in the hope of forcing Berlin to
reinforce them. Reading his reports verbatim, Churchill disapproved of the strategy
of his commanders in Africa, who did not flinch from dealing a final blow to a weak
opponent. Under pressure from the Prime Minister, successive military commanders in the Middle East attacked when prudence dictated defense or before they had
accumulated and trained the forces guaranteeing success. Facing defeats, irritated
Churchill was changing commanders, and the cycle kept on repeating. But from the
point of view of the SIS head this strategy of consolidating his position proved
effective.

The importance of Enigma decrypts reached its theoretical apogee in 1942. The
involvement of the US in the war meant the expansion of the Allied intercept stations network. Codebreakers had perfected the theoretical instrumentation of attacks
on the cipher. They also finally had at their disposal the number of bombes to break
the keys of several networks. The Ultra system had been brought to the point of
being able to communicate decrypts quickly and effectively to field headquarters
and commanders. It seemed that the Allies were gaining the upper hand in the cryptological race. Meanwhile, on the fronts, their only task was to persevere until the
effects of US mobilization began to have an impact on the fate of the battles and the

war. For this to happen, millions of soldiers and millions of tons of supplies had to be transported from US training camps and factories to the fronts. It was in this pivotal role and during this crucial period that not only the Enigma decrypts, but also, to some extent, the entire GC&CS as cryptographic service, failed. The introduction of a fourth rotor in the Enigma used by Atlantic U-Boats came as no surprise. It had been foreshadowed by captures aboard U-570 and previously broken messages. Mistakes by German ciphers allowed the new rotor's connections to be recovered before it entered service. Nonetheless, this simple innovation blinded BP for a long ten months of 1942, in which the Atlantic convoys suffered appalling losses.

These losses were largely due to the regular breaking of the convoy code, Naval Cypher No. 3, by Kriegsmarine codebreakers. The first and foremost task of any cryptologic service is to secure its own communications. Any success in attacks on enemy systems is a bonus only. Several examples confirm that the euphoria of breaking an adversary's ciphers causes a reversal of this hierarchy. This has also happened to the GC&CS. Naval Cypher No. 3 was designed as the Royal Navy's backup communications system, whose use was to be limited to emergency situations, by design short-lived. In fact, it was used for almost two years, and the numerous messages exchanged in connection with the organization of Atlantic convoys provided enemy codebreakers with an abundance of material for analysis. GC&CS ignored the numerous indications of possible compromise of the code. Even after final confirmation that the code has been broken by the opponent, it took almost a year to implement a successor. The part of GC&CS responsible for the security of Britain's own codes and ciphers was supervised by Denniston's deputy, Edward Travis. It is hard not to notice that the problems associated with the use of compromised code overlapped with the internal conflict within GC&CS, which culminated in Denniston being sidelined and replaced by Travis. In the view of BP veterans, this was a good change. Travis' energy allowed BP to realize its potential in the second part of the war. More than 72,000 Allied seamen lost their lives in the Battle of the Atlantic, a significant number of them during the fateful year 1942. The knowledge that their deaths were the result of the enemy breaking the convoy code casts a shadow over Travis' memory. On the other hand, the fatal neglect was partly due to the very nature of the organization he led. The success of the codebreakers' attack on the adversary's system is visible and measurable. The security of own systems is generally taken somewhat on faith. If it is ever practically verified, the verification is usually of a negative nature—the service receives proof of breaking its own ciphers. The effort of both parts of the integrated crypto service is infused with a profound asymmetry of risk. If, in addition, one part is spectacularly successful, the importance of the work of the other one inexorably suffers.

The year 1943 finally brought a breakthrough, but not the one that the codebreakers involved in the attacks on the Enigma would have wished for. Breaking the cipher of the four-rotor Enigma did not bring a breakthrough in the struggle in the Atlantic. Earlier successes had made it possible to plan the courses of convoys so that they bypass U-boot patrol lines. By late 1942 and early 1943, the fleet at Dönitz's disposal allowed him to position his ships in several lines that left no significant gaps. Regardless of the of course changes, almost every convoy had to encounter the waiting wolf packs. This is why the greatest convoy battles took place in the first half of 1943. After suffering appalling losses during the first three months of the year, the Allies themselves were surprised by the sudden and unexpected victory in the Atlantic achieved in May. The moment and character of this victory raised the question of its cause. Candidates were plentiful, for during the first half of 1943 the Allies had implemented several innovations in the tactics used against U-Boots. Officers commanding German ships considered an aircraft with a board radar to be their most dangerous adversary. The convoy escort ships were equipped with the Huff-Duff system, which allowed them to pinpoint a ship using her own transmissions. The escorts were equipped with 'Hedgehog', a weapon more precise and effective than depth charges. When the U-Boot was detected, Allied aircraft could now drop an acoustic torpedo homing in on the noise of the ship's propellers. The escort ships were organized into permanent teams and their crews trained in the application of combat tactics based on mathematical principles of operational research. Any attempt to pinpoint a single, dominant cause of the U-Boats' defeat is probably doomed to failure. Enigma was certainly one ingredient in the success, but the very chronology of events indicates that it was unlikely to be single dominant factor.

The year 1943 brought also a change in strategic realities that reduced the importance of Enigma decrypts. The Allies took the strategic initiative on all fronts of the war. Enigma provided information of a tactical nature, its decrypts allowing the weaker side more to avoid defeat than to win victory in battle or campaign. The Allies' seizure of the strategic initiative meant that they would henceforth be the ones to choose the time and place of confrontation, putting the Axis forces in the position of the weaker side. For the party having the initiative, information of a different nature is valuable than for the weaker one. The Allies were henceforth primarily interested in the strategic aspects of the enemy's plans and actions; the balance of its forces in the various theatres of war, the appreciations formulated by adversary's top staffs and commanders, and the relationship between political and military leaders. In a stroke of luck just as the nature of the war changed, Allied codebreakers gained access to a source of information that suited their new needs. They managed to break the ciphers of the teletype network connecting German top-level headquarters. The decrypts of the *Fish* network, as the new source was collectively referred to, definitely reduced the importance of Enigma in the second half of the war. The broken Enigma dispatches could still help win one battle or another, but it was the *Fish* that provided the rationale for shaping the strategy for victory.

How, in the light of the above comments, to assess the actual role of breaking the Enigma ciphers for the fate and outcome of the Second World War? Harry Hinsley's assessment cited earlier probably referred to the entirety of the Ultra operation, which, in addition to the Enigma, included attacks on German teletype and diplomatic ciphers, Italian and Japanese codes and ciphers, and several others. In such a holistic view, the BP veteran's opinion must be accepted. The Enigma alone had prevented Allied defeat in the first part of the war. Its potential to shorten war's duration was not fully realized. The unexpected gift of Polish codebreakers took the

Allied services by surprise. It took them almost a year to undertake systematic decryption. It took another two years to build a workable system for transmitting the decrypts to the field commands and to teach them the proper use of the new weapon. This delay meant that Enigma decrypts began to be properly used just when their importance had diminished as a result of the changing nature of the war. Breaking the Enigma ciphers could indeed have shortened the war by two to three years. It was not the codebreakers' fault that this did not happen.

John Keegan's thesis that Ultra shortened the war by the few months only, needed by Americans to perfect the nuclear bomb, is untenable. It includes the tacit assumption that, without the breaking of Enigma, the picture of the war in 1945 would have been the same as its historical course. For all the controversy and doubt about similar speculation, it is reasonably safe to assume that the breaking of the cipher saved Britain from defeat in the first half of the war. Had it succumbed to the Third Reich during this period, the global conflict in 1945 would probably have looked different. It could have broken up into two independent conflicts; a clash between the dictatorships of Hitler and Stalin and a confrontation between Japan and the US. Given the pragmatism of the dictators, it can be assumed that Hitler and Stalin would have eventually reached some sort of agreement. In such a scenario, the mere readiness of the USA to use the nuclear bomb would not have affected the fate of Europe, and the author of this opinion and some of the readers would not have had the opportunity to formulate or discuss any hypotheses.

So ended the story of the breaking of the Enigma cipher. Years later, the story of its presentation to the world turned out to be no less fascinating. After the end of the hostilities, each of the demobilized codebreakers had to sign again a secrecy pledge. The breaking of the Enigma ciphers seamlessly changed its status from one of the greatest secrets of the Second World War into one of the key secrets of the fledgling Cold War. Adversaries on the other side of the Iron Curtain were using cipher machines similar in design to the Enigma, so the theoretical foundations and practical methods of breaking its cipher remained relevant. Given that nearly ten thousand people worked in BP alone at the center's apogee of activity, the preservation of the secret for almost thirty post-war years qualifies as a miracle. This miracle began to crumble in the second half of the 1960s. After his retirement in 1967 Marian Rejewski wrote down and deposited in the Polish military archive his memories. One of the archive's employees, Władysław Kozaczuk, was just finishing work on a book devoted to the confrontation between the intelligence services of Poland and the Third Reich in the period 1922–1939. When Rejewski's memoirs fell into his hands, he realized he had hit a gold mine. Unfortunately, the publishing cycle in communist Poland allowed him to add literally only one sentence, hinting on the pre-war success of Polish codebreakers. The book in exotic Polish language, published in a country behind the Iron Curtain, did not attract the attention of historians. It only received commentary in a newsletter published by an association of former German intelligence officers. As expected, the commentary denied the possibility of breaking the Enigma cipher. But in the same period, a somewhat controversial British historian, David Irving, was preparing to publish a book on the events leading up to the outbreak of the Second World War. Apparently, he had knowledge of the breaking of Enigma, which must have been a bit of a public secret at that time among British historians. However, he felt that it would be safer to entrust the disclosure of the secret to someone with a more solid position in the academic world. As a result, the first information about the breaking of the cipher, as well as the key role of Polish codebreakers, appeared in the preface to Irving's book, published in 1968, written by Donald Cameron Watt, professor of international history at the London School of Economics. Watt's revelation did not provoke a significant response. It is difficult to judge today whether this was due to the somewhat ambiguous reputation of the book's main author or whether the guardians of the secret felt that it was not yet time to reveal it.

As a result, the secret of breaking the Enigma retained its status until 1973, when Gustave Bertrand, then already a retired general, published his memoirs. He revealed not only the fact of breaking the Enigma, but also the pioneering role of Polish mathematicians in this success. His book, in many ways far from being objective, has caused a stir or even irritation in the UK. The traditional French malice against perfidious Albion present in his book was quite sufficient for this. However, the British secret service apparently also felt that Bertrand had prematurely revealed facts that could still have an impact on the course of the Cold War. Judging by the reaction to Bertrand's revelations, the British services have decided to reach for their traditional weapon, disinformation. In 1974 Frederic Winterbotham published a memoir, "The Ultra Secret", that quickly became a global bestseller. Winterbotham was a significant figure in the Ultra system, but it seems that he never got to know the cryptological kitchen in which the messages he distributed were broken. Probably none of the participants in the Enigma adventure would have been able to recognize their face in the portrait he drew. This was particularly true of the Poles. According to Winterbotham, the role of the Polish cryptologists boiled down to the fact that a Polish worker was supposed to work in a Berlin factory manufacturing Enigmas before the war. Because of his nationality, he was dismissed from his job and expelled to Poland. In Warsaw, he contacted British intelligence service, which transferred him to Paris. There he was given a workshop where, with the help of a French mechanic, he carved an exact model of the machine in wood.

Since the publication of Winterbotham's book, subsequent titles published in the following years have presented a similar version of the story. Polish agents were to ambush a convoy carrying machines, to steal a few Enigma copies, burning the remaining cars thereafter. In yet another version, Polish intelligence was to dispatch its agents to Berlin, who infiltrated the factory producing Enigma and stole the machine's plans. All quoted versions creatively developed Knox's earliest reaction to the Polish disclosure; his claim that the Poles initially had to buy or steal the machine. And it is this common core contained in the early books on the history Enigma that points to deliberate disinformation. All authors unanimously pointed out that the breaking of the Enigma cipher was possible mainly due to the physical compromise of the machine. It was a clear message to the cryptologic adversary; if you are able to ensure the physical security of your machines, your ciphers also remain secure. If this was a conscious act, its message bordered on extreme naivety. However, the disinformation that was let loose began to take on a life of its own and deformed the true story of the breaking of the cipher for many years, including long after the reasons for its dissemination had disappeared.

After nearly a decade of trying to push the genie back into the bottle, there was a twist. It was linked, at least chronologically, to the publication of Gordon Welchman's memoirs in 1982. Disappointed by the criticism from his former colleagues, Welchman died in 1985. We will never know if it was Welchman's sad experience that became the pebble that tipped the scales. However, at the same time, the slow and ongoing process of declassifying source documents relating to the history of the breaking of the Enigma ciphers began. If historians thought that their availability would enable a full reconstruction of events and clarify existing doubts, they were disappointed. The archival documents present pretty well the process of breaking the Enigma ciphers on both sides of the Atlantic during the war. At the same time, they say little or nothing about the most important part of this adventure, namely the original breaking of the cipher. This is quite situation. None of the participants at the meeting where Allied cryptologic cooperation was born were involve in writing of the post-war reports. Dillwyn Knox passed away in 1943, overcome by cancer. Alastair Denniston was sidelined in 1942 and retired immediately after the end of the war. Alan Turing did not attend the 1939 Pyry meeting, but was one of the first people to evaluate and expand the results. He was, incidentally, the only British codebreaker, apart from Knox, to have the opportunity to meet his Polish colleagues directly in January 1940. However, Alan too parted company with BP in 1943, moving to a facility in Hanslope Park. As a result, British codebreakers writing up their reports in 1945 were somewhat helpless before the task of describing the earliest phase of events. Peter Mahon, the last head of Hut 8, admitted that the only information on Polish involvement was taken from Alan Turing, who was reportedly not particularly talkative on the subject. In theory, we should have a source that almost fully describes the extent of the Polish contribution to the breaking of the cipher. During the conference in Pyry, British and French guests received a report drawn up in German, containing full information on the theory and practice of breaking the Enigma cipher. Mahon's report confirms that a copy of the report was still available at BP in August 1945, but he was one of the last people to see it. Despite the efforts of Polish, British and French archivists and historians, to date no copy of the report has been found in the archives of the former Allies.

Given the distance of time, the gaps in the source materials, the disinformation distorting the real story and its long-standing impact, reconstructing a true picture of the Enigma adventure was, is and probably will remain a non-trivial task. Fortunately, fundamental issues were clarified beyond any doubt. These include the distinction between those who had originally broken the cipher, and those who have been breaking it thereafter. Dilly Knox, in his earliest reaction to the success of the Polish codebreakers, judged that 'they must have pinched it or stolen it and then followed the development as anyone could'. We now know that both parts of his opinion are completely wrong. Marian Rejewski, Jerzy Różycki and Henryk Zygalski did not have to climb the walls of the factory manufacturing the machine or blow-up trucks carrying its copies. They achieved their victory over the cipher in a fair fight, purely on the ground of mathematics. After the war, Marian Rejewski wrote in a letter to a friend: "There is a lot of noise now about our breaking the Enigma. And in fact, we did nothing more than put into practice the knowledge passed on to us during the first year (of our studies) by Krygowski and Abramowicz". The authors of the key breakthrough in the struggle with the cipher and the character of their achievement were therefore identified beyond any doubt. However, the second part of Knox's opinion was also misplaced. In fact, it was British and then American codebreakers who were handed over the foundations of the victory over Enigma ciphers. Their role was reduced with keeping up with system developments. The nature and extent of their contribution far exceeded Knox's view that anyone could it.

Basic Bibliography

1. Abrutat David, *Radio War: The Secret Espionage War of the Radio Security Service 1938–1946*, Fonthill, 2019
2. Andrews Geoff, *Agent Moliere. Life of John Cairncross, Fifth Man of Cambridge Spy Circle*, Bloomsbury Academic, 2020
3. Arnold Michael, *Hollow Heroes. Unvarnished Look at the Wartime Careers of Churchill, Montgomery and Mountbatten*, Casemate, 2015
4. Atha Robert I., *Bombe! "I Could Hardly Believe It!"*, Cryptologia, 9:4, 332–336
5. Batey Mavis, *Dilly Knox – A Reminiscence of this Pioneer Enigma Cryptanalyst*, Cryptologia, 32:2, 104–130
6. Bauer Friedrich L., *An Error in the History of Rotor Encryption Devices*, Cryptologia, 23:3, 206–210
7. Beesly Patrick, *Very Special Intelligence. The Story of the Admiralty's Operational Intelligence Centre, 1939–1945*, Sphere, 1978
8. Bloch Gilbert, Deavours C. A., *Enigma Avant Ultra Enigma Before Ultra*, Cryptologia, 12:3, 178–184
9. Bloch Gilbert, Deavours C. A., *Enigma Before Ultra Polish Work and the French Contribution*, Cryptologia, 11:3, 142–155
10. Bloch Gilbert, Deavours C. A., *Enigma Before Ultra The Polish Success and Check (1933–1939)*, Cryptologia, 11:4, 227–234
11. Bloch Gilbert, Erskine Ralph, *Enigma: The Dropping of the Double Encipherment*, Cryptologia, 10:3, 134–141
12. Bouchaudy Jean-François, *Enigma: the spoils of Gustave Bertrand, or "par où tout a commencé"*, Cryptologia, 45–4, 309–341
13. Bruce Robert B., *Petain: Verdun to Vichy*, Potomac Books, 2008
14. Budiansky Stephen, *Codebreaking with IBM Machines in World War II*, Cryptologia, 25:4, 241–255
15. Budiansky Stephen, *The difficult beginnings of US-British codebreaking cooperation*, Intelligence and National Security 2:15, 49–73
16. Bundy, William P., *Some of My Wartime Experiences*, Cryptologia, 11: 2, 65–77
17. Burke Colin, *From the Archives: The last Bombe Run, 1955*, Cryptologia, 32:3, 277–278
18. Carter Frank, *Keith Batey and John Herivel: Two Distinguished Bletchley Park Cryptographers*, Cryptologia, 35:3, 277–281
19. Christensen Chris, *Alan Turing's First Cryptology Textbook and Sinkov's Revision of it*, Cryptologia, 34:27–43

© The Editor(s) (if applicable) and The Author(s), under exclusive license to
Springer Nature Switzerland AG 2025
M. Grajek, *Enigma Myth Deciphered*, History of Information Security,
https://doi.org/10.1007/978-3-031-65475-6

20. Ciechanowski J.S. (ed), *Marian Rejewski 1905–1980. Życie Enigmą pisane*, UM Bydgoszcz, Bydgoszcz, 2005
21. Clark William F., *The Years Between*, Cryptologia, 12:1, 52–58
22. Clarke William F., *Bletchley Park 1941–1945*, Cryptologia, 12:2, 90–97
23. Clarke William F., *Government Code and Cypher School its Foundation and Development with Special Reference to its Naval Side*, Cryptologia, 11: 4, 219–226
24. Clarke William F., *Post War Organization*, Cryptologia, 12:3, 174–177
25. Currier Prescott, *My "Purple" Trip to England in 1941*, Cryptologia, 20:3, 193–201
26. Davies Donald W., *Effectiveness of the Diagonal Board*, Cryptologia, 23: 3, 229–239
27. de Leeuw Karl, *The Dutch Invention of the Rotor Machine*, *1915–1923*, Cryptologia, 27:1, 73–94
28. Deavours C. A., Kruh Louis, *The Turing Bombe: Was it Enough?*, Cryptologia, 14: 4, 331–349
29. Deavours C. A., *The Black Chamber: A Column How the British Broke Enigma*, Cryptologia 4:3, 129–132
30. Deavours C. A., *The Black Chamber: A Column La Methode des Batons*, Cryptologia 4:4, 240–247
31. Deutsch Harold C., *The Influence of ULTRA on World War II*, Parameters. The US Army War College Quarterly, 8:1, 2–15
32. Devours C.A., *Lobsters, Crabs, and the Abwehr Enigma*, Cryptologia, 21:3, 193–199
33. Dobson Alan P., Marsh Steve, *Churchill and the Anglo-American Special Relationship*, Routledge, 2017
34. Donald W. Davies, *The Bombe A Remarkable Logic Machine*, Cryptologia, 23:2, 108–138
35. Dooley John F., *1929–1931: A Transition Period in U.S. Cryptologic History*, Cryptologia, 37:1, 84–98
36. Dooley John F., *The Gambler and the Scholars. Herbert Yardley, William & Elizabeth Friedman, and the Birth of Modern American Cryptology*, 2023, Springer
37. Dubicki Tadeusz, Nałęcz Daria, Stirling Tess (eds), *Polsko-Brytyjska współpraca wywiadowcza podczas II wojny światowej*, Naczelna Dyrekcja Archiwów Państwowych, Warszawa, 2004
38. Ellison Carl M., *A Solution of the Hebern Messages*, Cryptologia, 12:3, 144–158
39. Erskine Ralph, *Captured Kriegsmarine Enigma Documents at Bletchley Park*, Cryptologia, 32: 3, 199–219
40. Erskine Ralph, Freeman Peter, *Brigadier John Tiltman: One of the Britain's Finest Cryptologists*, Cryptologia, 27:4, 289–318
41. Erskine Ralph, *From the Archives GC&CS Mobilizes "Men of the Professor Type"*, Cryptologia, 10:1, 50–59
42. Erskine Ralph, *From the Archives: U-Boat HF WT Signalling*, Cryptologia, 12:2, 98–106,
43. Erskine Ralph, *Kriegsmarine Short Signal Systems – and How Bletchey Park Exploited Them*, Cryptologia, 23:1, 65–92
44. Erskine Ralph, *Kriegsmarine Signal Indicators*, Cryptologia, 20:4, 330–340
45. Erskine Ralph, Marks Philip, *Naval Enigma: Seahorse and Other Kriegsmarine Cipher Blunders*, Cryptologia, 28:3, 211–241
46. Erskine Ralph, *The 1944 Naval BRUSA Agreement and its Aftermath*, Cryptologia, 30: 1, 1–22
47. Erskine Ralph, *The First Naval Enigma Decrypts of World War II*, Cryptologia, 21:1, 42–46
48. Erskine Ralph, *The German Naval Grid in World War II*, Cryptologia, 16:1, 39–51
49. Erskine Ralph, *The Poles Reveal their Secrets: Alastair Denniston's Account of the July 1939 Meeting at Pyry*, Cryptologia, 30:4, 294–305
50. Erskine Ralph, *Ultra and Some U.S. Navy Carrier Operations*, Cryptologia, 19:1, 81–96
51. Erskine Ralph, *Ultra Reveals a Late B-Dienst Success in the Atlantic*, Cryptologia, 34:4, 340–358
52. Erskine Ralph, Weierud Frode, *Naval Enigma: M4 and Its Rotors*, Cryptologia, 11: 4, 235–244
53. Erskine Ralph, *What Dis the Sinkov Mission Receive from Bletchley Park?*, Cryptologia, 24:2, 97–109

54. Faligot Roger, Krop Pascal, *La Piscine: The French Secret Service Since 1944*, Blackwell, 1989
55. Faulkner Marcus, *The Kriegsmarine, Signals Intelligence and the Development of the B-Dienst Before the Second World War*, Intelligence and National Security 25:4, 521–546
56. Foucrier Jean-Charles, *Capitan Honore Louis, Gustave Bertrand's Lost Deputy*, The Enigma Bulletin, No. 13, 2021
57. Foucrier Jean-Charles, Why the French military cryptanalysis failed to break Enigma, Cryptologia, 1–23
58. Friedman William F., *From the Archives Brief History of Signal Intelligence Service*, Cryptologia, 15:3, 263–272
59. Fuensanta Jose Ramon Soler, Lopez-Brea Espiau Francisco Javier, Bonilla Diego Navarro, *A Cryptanalysis Service During the Spanish Civil War*, Cryptologia, 36:263–289
60. Fuensanta José Ramón Soler, López-Brea Espiau Francisco Javier & Weierud Frode, *Spanish Enigma: A History of the Enigma in Spain*, Cryptologia, 34:4, 301–328
61. Fuensanta José Ramón Soler, López-Brea Espiau Francisco Javier & Bonilla Diego Navarro, *Revealing Secrets in Two Wars: The Spanish Codebreakers at PC Bruno and PC Cadix*, Cryptologia, 37:3, 233–249
62. FuensantaJosé Ramón Soler, *Mechanical Cipher Systems in the Spanish Civil War*, Cryptologia, 28:3, 265–276
63. Gaj Krzysztof, *Szyfr Enigmy – metody złamania*, Wydawnictwo Komunikacji i Łączności, Warszawa, 1989
64. Gallehawk John, *Trird Person Singular (Warsaw, 1939)*, Cryptologia, 30:193–198
65. Garliński Józef, *Enigma. Tajemnica drugiej wojny światowej*, Wydawnictwo UMCS, Lublin, 1999
66. Gawłowski Robert, *The First Enigma Codebreaker: The Untold Story of Marian Rejewski who passed the baton to Alan Turing*, Pen and Sword Military, 2023
67. Gilbert James L., Finnegan John P., *U.S. Army Signals Intelligence in World War II. A Documentary History*, Special Publications, 1993
68. Gladwin, Lee A., *Bulldozer: A Cribless Rapid Analytical Machine (RAM) Solution to Enigma and its Variations*, Cryptologia, 31: 4, 305–315
69. GladwinLee A., *Alan Turing's Visit to Dayton*, Cryptologia, 25:1, 11–17
70. Good I.J., *Early Work on Computers at Bletchley*, Cryptologia, 3:2, 65–77
71. Grajek Marek, *An Inventory of Early Inter-Allied Enigma Cooperation*, HistoCrypt 2018, 149:157
72. Grajek Marek, *Enigma History and an Unexpected Treasure Trove*, HistoCrypt 2022, 42–49
73. Grajek Marek, *Mysteries of P.C. Cadix and its evacuation in 1942/43*, HistoCrypt 2023, 63–72
74. Grajek Marek, *Sztafeta Enigmy. Odnaleziony raport polskich kryptologów*, Agencja Bezpieczeństwa Wewnętrznego, Warszawa, 2019
75. Gralewski Leszek, *Złamanie Enigmy. Historia Mariana Rejewskiego*, Wyd. Adam Marszałek, Toruń, 2005
76. Greenberg Joel, *Alastair Denniston: Code-breaking from Room 40 to Berkeley Street and the Birth of GCHQ*, Frontline Books, 2017
77. Greenberg Joel, *The Bletchley Park Codebreakers in Their Own Words*, Greenhill Books, 2022
78. Grey Christopher & Sturdy Andrew, *The 1942 Reorganization of the Government Code and Cypher School*, Cryptologia, 32:4, 311–333
79. Grey Christopher, *From the Archives: Colonel Butler's Satire of Bletchley Park*, Cryptologia, 38:3, 266–275
80. Hamer David H., *The Enigmas — and Other Recovered Artefacts – of U-85*, Cryptologia, 27: 2, 97–110
81. Hamer David M., *Review of Herivelismus and the German Military Enigma by John Herivel*, Cryptologia, 33:1, 95–97
82. Hanyok Robert J., *Eavesdropping on Hell. Historical guide to western communications intelligence and the Holocaust, 1939–1945*, Center for Cryptologic History, National Security Agency, 2005

83. Hastings Max, *Finest Years Churchill as Warlord 1940–45*, HarperPress, 2009
84. Herman John, *Agency Africa: Rygor's Franco-Polish Network and Operation Torch*, Journal of Contemporary History, Volume 22 Issue 4
85. Hore Peter, *Bletchley Park's Secret Source: Churchill's Wrens and the Y Service in World War II*, Greenhill Books, 2021
86. Jackson John, *Solving Enigma's Secrets: The Official History of Bletchley Park's Hut 6*, Book Tower Publishing, 2014
87. José Ramón Soler Fuensanta, *Mensajes Secretos. La Historia de la Criptografía Española desde sus Inicios Hasta los Años 50*, Editorial Tirant lo Blanch, 2016
88. Kahn David, *An Enigma Chronology*, Cryptologia, 17: 3, 237–246
89. Kahn David, *Churchill Pleads for the Intercepts*, Cryptologia, 6:1, 47–49
90. Kahn David, *Codebreaking in World Wars I and II. Major successes and failures, their causes and their effects*, in Andrew Christopher, Dilks David (eds), *The Missing Dimension. Governments and Intelligence Communities in the Twentieth Century*, Springer, 1984
91. Kahn David, *From the Archives, Britain Reveals Its Bombe to America*, Cryptologia, 26:2, 124–128
92. Kahn David, *The Annotated the American Black Chamber*, Cryptologia, 9: 1, 1–37
93. Kahn David, *The Forschungsamt: Nazi Germany's Most Secret Communications Intelligence Agency*, Cryptologia, 2:1, 12–19
94. Kedward Roderic Harry, *Resistance in Vichy France: Study in Ideas and Motivation in the Southern Zone, 1940–42*, Oxford University Press, 1983
95. Keegan John, *Churchill's Generals*, Quill, 1991
96. Kenyon David, Arctic Convoys: Bletchley Park and the War for the Seas, Yale University Press, 2023
97. Kenyon David, Weierud Frode, *Enigma G: The counter Enigma*, Cryptologia, 44–5, 385–420
98. Kitson Simon, *Hunt for Nazi Spies. Fighting Espionage in Vichy France*, The University of Chicago Press, 2008
99. Koot H., *Expert's Opinion on the Enigma Ciphering Machine*, Cryptologia, 26:2, 101–102
100. Kozaczuk Władysław, *Bitwa o tajemnice. Służby wywiadowcze Polski i Niemiec 1918–1939*, Książka i Wiedza, Warszawa, 1967
101. Kozaczuk Władysław, *W kręgu Enigmy*, Książka i Wiedza, Warszawa, 1986
102. Kruh Louis, *British-American Cryptanalytic Cooperation and an Unprecedented Admission by Winston Churchill*, Cryptologia, 13:2, 123–134
103. Kruh Louis, Deavours Cipher, *The Commercial Enigma: Beginnings of Machine Cryptography*, Cryptologia, 26: 1, 1–16
104. Kruh Louis, *Stimson, the Black Chamber, and the "Gentelmen's Mail" Quote*, Cryptologia, 12:2, 65–89
105. Kruh Louis, *Why Was Safford Pessimistic About Breaking the German Enigma Cipher Machine in 1942?*, Cryptologia, 14:3, 253–257
106. Kuhl Alex, *Rejewski's Catalog*, Cryptologia, 31:4, 326–331
107. Kurson Robert, *Shadow Divers: The True Adventure of Two Americans Who Risked Everything to Solve One of the Last Mysteries of World War II*, Random House Publishing Group, 2005
108. Lawrence John, *A Study of Rejewski's Equations*, Cryptologia, 29: 3, 233–247
109. Lawrence John, *Factoring the Plugboard – Was Rejewski's Proposed Solution for Breaking the Enigma Feasible?*, Cryptologia, 29:4, 343–366
110. Liberge Eric, Delalande Arnaud, *The Case of Alan Turing: The Extraordinary and Tragic Story of the Legendary Codebreaker*, Arsenal Pulp Press, 2016
111. Link David, *Resurrecting Bomba Kryptologiczna, Archaeology of Algorithmic Artefacts*, I, Cryptologia, 33:2, 166–182
112. Lintott Brett, *Mediterranean Double-Cross System, 1941–1945*, Routledge, 2019
113. List David, Gallehawk John, *Revelation for Cilli's*, Cryptologia, 38:248–265
114. Lujan Susan M. Lt., *Agnes Meyer Driscoll*, Cryptologia, 15:1, 47–56

115. Macintyre Ben, *Operation Mincemeat. How a Dead Man and a Bizarre Plan Fooled the Nazis and Assured an Allied Victory*, Crown; Reprint edition, 2011
116. Majchrowska Beata, *Więcej niż Enigma. Historia Antoniego Pallutha*, Klinika Języka, Grodzisk Mazowiecki, 2020
117. Marks Philip, *Umkehrwalze D: Enigma's Rewirable Reflector Part I, II, III*, Cryptologia, 25:2, 101–141, 25:3, 177–212, 25:4, 296–310
118. Marks Philip, Weierud Frode, *Recovering the Wiring of Enigma's Umkehrwalze A*, Cryptologia, 24:1, 55–66
119. McKay C. G., *From the Archives Arvid Damm Makes an Offer*, Cryptologia, 18: 3, 243–249
120. McKay Sinclair, *The Lost World of Bletchley Park: An Illustrated History of the Wartime Codebreaking Centre*, Aurum Press, 2013
121. Moorehead Caroline, *Village of Secrets: Defying the Nazis in Vichy France*, Harper Perennial, 2015
122. Napier Stephen, *Churchill. Military Genius or Menace*, History Press Limited, 2020
123. Neiberg Michael S., *When France Fell: The Vichy Crisis and the Fate of the Anglo-American Alliance*, Harvard University Press, 2021
124. Nesbit Roy Conyers, *Ultra Versus U-Boats. Enigma Decrypts in the National Archives*, Pen and Sword Military, 2008
125. Nowik Grzegorz, *Zanim złamano Enigmę. Polski radiowywiad podczas wojny z bolszewicką Rosją 1018–1920*, t. 1, Oficyna Wydawnicza Rytm, Warszawa, 2004
126. Ostwald Olaf, Weierud Frode, *History and Modern Cryptanalysis of Enigma's Pluggable Reflector*, Cryptologia, 40:1, 70–91
127. Parra Mona, *Gustave Bertrand: the Cooperation of British Intelligence with an Officer in Vichy France*, Revue française de civilisation britannique, 2022, Paris-Londres
128. Paterson Michael, *Secret war. Inside story of the code makers and code breakers of World War II*, David and Charles Limited, 2007
129. Paxton Robert O., *Parades and Politics at Vichy: The French Officer Corps Under Marshal Petain*, Princeton University Press, 2016
130. Paxton Robert O., *Vichy France: Old Guard and New Order 1940–1944*, Columbia University Press, 2001
131. Pfeiffer Paul N., *Breaking the German Weather Ciphers in the Mediterranean Detachment G, 849th Signal Intelligence Service*, Cryptologia, 22:4, 354–369
132. Piekalkiewicz Janusz, *Die Deutsche Reichsbahn im Zweiten Weltkrieg*, transpress, 2018
133. Praun Albert, *German Radio Intelligence*, Department of the Army, Office of the Chief of Military History, 1950
134. Przegląd Wojskowo-Historyczny, rok VI (LVII), nr specjalny 5 (210) w stulecie urodzin Mariana Rejewskiego
135. Quirantes Arturo, *Model Z: A numbers-Only Enigma Version*, Cryptologia, 28: 2, 153–156
136. Ratcliff R. A., *How Statistics Led the Germans to Believe Enigma Secure and Why They Were Wrong: Neglecting the Practical Mathematics of Cipher Machines*, Cryptologia, 27:2, 119–131
137. Rejewski Marian, *Jak polscy matematycy rozszyfrowali Enigmę*, Wiadomości Matematyczne, 1980, t.XX, str. 1–28
138. Ridley Norman, *Venlo Sting. MI6's Deadly Fiasco*, Casemate Publishers, 2022
139. Rijmenants Dirk, *Enigma Message Procedures Used by the Heer, Luftwaffe and Kriegsmarine*, Cryptologia, 34:4, 329–339
140. Ruppert Wolfgang A.F., Michor Peter W., *Mathematik in Österreich und die NS-Zeit. 176 Kurzbiographien*, Springer, 2023
141. Safford L.F., Lt, *The Function and Duties of the Cryptography Section, Naval Communications*, Cryptologia, 16:3, 265–281
142. Schmeh Klaus, *Enigma's Contemporary Witness: Gisbert Hasenjaeger*, Cryptologia, 33:343–346

143. Shwedo Bradford J., *XIX Tactical Air Command and Ultra: Patton's Force Enhancers in the 1944 Campaign in France*, Air Univ Pr, 2001

144. Sims John Cary, *The BRUSA Agreement of May 17, 1943*, Cryptologia, 21:1, 30–38

145. Smith Colin, *England's Last War Against France: Fighting Vichy 1940–42*, Orion, 2009

146. Stead Philip John, *Second Bureau. The first full story of Intelligence and Secret Service during the Occupation of France*, Evans Brothers Ltd, 1959

147. Sweets John, *Choices in Vichy France: The French Under Nazi Occupation*, Oxford University Press, 1986

148. Syrett David (ed.), *The Battle of the Atlantic and Signals Intelligence U–Boat Situations and Trends, 1941–1945*, Routledge, 2018

149. Szczuka Elżbieta, *Jerzy Różycki. Jeden z pogromców "Enigmy"*, Oficyna Wydawnicza „Rytm", Warszawa, 2023

150. Thimbleby Harold, *Human factors and missed solutions to Enigma design weaknesses*, Cryptologia, 40:2, 177–202

151. Thompson R.W., *Churchill and the Montgomery Myth*, 2014, M. Evans & Company

152. Turing Alan M., *Visit to National Cash Register Corporation of Dayton, Ohio*, Cryptologia, 25:1, 1–10

153. Turing Dermot, *X, Y & Z: The Real Story of How Enigma Was Broken*, History Press, 2018

154. Ulbricht Heinz, *Enigma-Uhr*, Cryptologia, 23:3, 193–205

155. Van der Meulen Michael, *Werftschluessel A German Navy Hand Cipher Part I and II*, Cryptologia, 19:4, 349–364, 20:1, 37–54

156. Weierud Frode, Zabell Sandy, *German mathematicians and cryptology in WWII*, Cryptologia 44:2, 97–171

157. West Nigel, *Churchill's Spy Files. MI5's Top-Secret Wartime Reports*, The History Press, 2018

158. Wheeler-Bennett John Wheeler (ed.), *Action this Day, Working with Churchill: Memoirs by Lord Normanbrook, John Colville, Sir John Martin, Sir Ian Jacob, Lord Bridges, Sir Leslie Rowan*, 1968, Macmillan

159. Whipple Tom, *The Battle of the Beams: The secret science of radar that turned the tide of the Second World War*, Bantam, 2023

160. Whitehead David, *Cobra and Other Bombes*, Cryptologia, 20:4, 289–307

161. Winton John, *ULTRA at Sea: How Breaking the Nazi Code Affected Allied Naval Strategy During World War II*, HarperCollins Publishers, 1990

162. Woytak Ryszard, *Werble historii*, Związek Powstańców Warszawskich w Bydgoszczy, Bydgoszcz, 1999

163. Wynn Stephen, *Churchill's Flawed Decisions. Errors in Office of the Greatest Briton*, Pen and Sword Military, 2020

164. Zabell Sandy, *Commentary on Alan M. Turing: The Applications of Probability to Cryptography*, Cryptologia, 36:3, 191–214

165. Zamir Meir, *Secret Anglo-French War in Middle East. Intelligence and Decolonization, 1940–1948*, Routledge, 2016

Index

A

Abernethy, Barbara, 352
Abetz, Otto, 134, 135
Abramowicz, Kazimierz Tomasz, 493
Abwehr, 116, 144, 182, 183, 258–260,
    300, 373, 375, 381, 389, 392,
    395, 397–398, 407, 412, 430,
    439, 446, 447
Abwehr Enigma, 259, 260
Adcock, Frank E., 125, 425
Administrative Code, 182
Akin, Spencer, 326
Alberti, Leon Battista, 16
Alexander, Hugh O'Donnel, 138, 178,
    282, 477
ANX, 78, 79, 86, 228
Arnim, Hans-Jürgen von, 437, 438
Asché, *see* Schmidt, Hans-Thilo
Ashcroft, Michael, 273
Atlantis, 288, 409–411
Auchinleck, Claude, 304–308, 310–312,
    315–317, 320, 487
Autoscritcher, 422
AVA, 55, 71, 81, 87, 92, 94, 97, 140,
    142–144, 467

B

Babbage, Charles, 16
Babbage, Dennis, 175, 178
Bacon, Allon, 280, 289, 339
Badoglio, Pietro, 440
Baker-Cresswell, Joe, 279

Balme, David, 279
Banburismus, 229, 231, 232, 234, 236–238,
    270, 272, 273, 338, 458, 465, 468
Barracuda (Kriegsmarine key), 285, 286
Batey, Mavis, 175, 430
Battle of beams, 198, 199
Bedell, Smith Walter, 454
Befehlshaber der Unterseeboote, 27, 231, 264,
    335–337, 357, 358, 412
Benenson, Peter, 478
Beobachtungsdienst (B-Dienst), 182, 353,
    356, 359, 405, 411, 413, 432, 457, 469
Bernstein, Paul, 9, 18
Bertrand, Emilé, 240
Bertrand, Gustave, 62–66, 74, 79, 80,
    103–105, 110, 121, 127, 129–131, 133,
    134, 158–161, 163–165, 185, 187,
    189–192, 239, 240, 242–250, 256,
    371–377, 379–389, 392, 396–398, 400,
    427, 472–475, 491
Betlewski, Czesław, 92, 94, 140
Biffy, *see* Dunderdale, Wilfred Albert
Bigram tables, 90, 220–222, 225, 226, 234,
    236–238, 270, 272–275, 288, 289, 333,
    337–340, 370, 464
Bijleveld, Hendrik, 2, 3
Birch, Frank, 125, 126, 183, 222, 223, 226,
    236, 237, 352
Black Cabinet, 33, 39, 323, 324, 480
Blofield, Berta, 474
Blunt, Anthony, 427
Bochicchio, Phil, 365
Bołdeskuł, Karol, 34, 35, 37

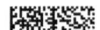